Quantum Mechanics of Charged Particle Beam Optics

Multidisciplinary and Applied Optics

Series Editors:
Vasudevan Lakshminarayanan, University of Waterloo, Ontario, Canada

Understanding Optics with Python
Vasudevan Lakshminarayanan, Hassen Ghalila, Ahmed Ammar, L. Srinivasa Varadharajan
Hassen Ghalila, University Tunis El Manar, Tunisia
Ahmed Ammar, University Tunis El Manar, Tunisia
L. Srinivasa Varadharajan, University of Waterloo, Ontario, Canada

For more information about this series, please visit:
https://www.crcpress.com/Multidisciplinary-and-Applied-Optics/book-series/ CRCMULAPPOPT

Quantum Mechanics of Charged Particle Beam Optics
Understanding Devices from Electron Microscopes to Particle Accelerators

Ramaswamy Jagannathan and Sameen Ahmed Khan

CRC Press
Taylor & Francis Group
Boca Raton London New York

CRC Press is an imprint of the
Taylor & Francis Group, an **informa** business

CRC Press
Taylor & Francis Group
6000 Broken Sound Parkway NW, Suite 300
Boca Raton, FL 33487-2742

© 2019 by Taylor & Francis Group, LLC
CRC Press is an imprint of Taylor & Francis Group, an Informa business

No claim to original U.S. Government works

Printed on acid-free paper

International Standard Book Number-13: 978-1-138-03592-8 (Hardback)

This book contains information obtained from authentic and highly regarded sources. Reasonable efforts have been made to publish reliable data and information, but the author and publisher cannot assume responsibility for the validity of all materials or the consequences of their use. The authors and publishers have attempted to trace the copyright holders of all material reproduced in this publication and apologize to copyright holders if permission to publish in this form has not been obtained. If any copyright material has not been acknowledged please write and let us know so we may rectify in any future reprint.

Except as permitted under U.S. Copyright Law, no part of this book may be reprinted, reproduced, transmitted, or utilized in any form by any electronic, mechanical, or other means, now known or hereafter invented, including photocopying, microfilming, and recording, or in any information storage or retrieval system, without written permission from the publishers.

For permission to photocopy or use material electronically from this work, please access www.copyright.com (http://www.copyright.com/) or contact the Copyright Clearance Center, Inc. (CCC), 222 Rosewood Drive, Danvers, MA 01923, 978-750-8400. CCC is a not-for-profit organization that provides licenses and registration for a variety of users. For organizations that have been granted a photocopy license by the CCC, a separate system of payment has been arranged.

Trademark Notice: Product or corporate names may be trademarks or registered trademarks, and are used only for identification and explanation without intent to infringe.

Visit the Taylor & Francis Web site at
http://www.taylorandfrancis.com

and the CRC Press Web site at
http://www.crcpress.com

Dedication

To the memory of my parents
Thirali Aravamudha Ramaswamy Iyengar & Andal Ramaswamy
—— Ramaswamy Jagannathan

To the memory of my parents
Hamid Ahmed Khan & Nighat Hamid Khan
—— Sameen Ahmed Khan

Contents

Preface xi

Authors xv

Chapter 1   Introduction 1

Chapter 2   An Introductory Review of Classical Mechanics 9

       2.1   Single Particle Dynamics 9
             2.1.1   Lagrangian Formalism 9
                  2.1.1.1   Basic Theory 9
                  2.1.1.2   Example: Motion of a Charged Particle in an Electromagnetic Field 10
             2.1.2   Hamiltonian Formalism 12
                  2.1.2.1   Basic Theory 12
                  2.1.2.2   Example: Motion of a Charged Particle in an Electromagnetic Field 14
             2.1.3   Hamiltonian Formalism in Terms of the Poisson Brackets 16
                  2.1.3.1   Basic Theory 16
                  2.1.3.2   Example: Dynamics of a Charged Particle in a Constant Magnetic Field 18
             2.1.4   Changing the Independent Variable 21
                  2.1.4.1   Basic Theory 21
                  2.1.4.2   Example: Dynamics of a Charged Particle in a Constant Magnetic Field 22
             2.1.5   Canonical Transformations 25
                  2.1.5.1   Basic Theory 25
                  2.1.5.2   Optical Hamiltonian of a Charged Particle Moving Through an Electromagnetic Optical Element with a Straight Axis 29
             2.1.6   Symplecticity of Canonical Transformations 31
                  2.1.6.1   Time-Independent Canonical Transformations 31
                  2.1.6.2   Time-Dependent Canonical Transformations: Hamiltonian Evolution 32
                  2.1.6.3   Canonical Invariants: Poisson Brackets ... 35
       2.2   Dynamics of a System of Particles 36

Chapter 3		An Introductory Review of Quantum Mechanics............41	
	3.1	Introduction............41	
	3.2	General Formalism of Quantum Mechanics............42	
		3.2.1 Single Particle Quantum Mechanics: Foundational Principles 42	
			3.2.1.1 Quantum Kinematics............42
			3.2.1.2 Quantum Dynamics............51
			3.2.1.3 Different Pictures of Quantum Dynamics............64
			3.2.1.4 Ehrenfest's Theorem............68
			3.2.1.5 Spin............70
	3.3	Nonrelativistic Quantum Mechanics 78	
		3.3.1 Nonrelativistic Single Particle Quantum Mechanics 78	
			3.3.1.1 Free Particle............78
			3.3.1.2 Linear Harmonic Oscillator............90
			3.3.1.3 Two-Dimensional Isotropic Harmonic Oscillator............100
			3.3.1.4 Charged Particle in a Constant Magnetic Field............103
			3.3.1.5 Scattering States 107
			3.3.1.6 Approximation Methods, Time-Dependent Systems, and the Interaction Picture 111
			3.3.1.7 Schrödinger–Pauli Equation for the Electron............117
		3.3.2 Quantum Mechanics of a System of Identical Particles............119	
		3.3.3 Pure and Mixed States: Density Operator............126	
	3.4	Relativistic Quantum Mechanics 132	
		3.4.1 Klein–Gordon Equation 132	
			3.4.1.1 Free-Particle Equation and Difficulties in Interpretation 132
			3.4.1.2 Feshbach–Villars Representation 137
			3.4.1.3 Charged Klein–Gordon Particle in a Constant Magnetic Field............139
		3.4.2 Dirac Equation 141	
			3.4.2.1 Free-Particle Equation 141
			3.4.2.2 Zitterbewegung............149
			3.4.2.3 Spin and Helicity of the Dirac Particle............150
			3.4.2.4 Spin Magnetic Moment of the Electron and the Dirac–Pauli Equation............153
			3.4.2.5 Electron in a Constant Magnetic Field ... 154

		3.4.3	Foldy–Wouthuysen Transformation........................ 156
			3.4.3.1 Foldy–Wouthuysen Representation of the Dirac Equation 156
			3.4.3.2 Foldy–Wouthuysen Representation of the Feshbach–Villars form of the Klein–Gordon Equation 168
	3.5	Appendix: The Magnus Formula for the Exponential Solution of a Linear Differential Equation 170

Chapter 4 An Introduction to Classical Charged Particle Beam Optics 173

- 4.1 Introduction: Relativistic Classical Charged Particle Beam Optics 173
- 4.2 Free Propagation 174
- 4.3 Optical Elements with Straight Optic Axis 178
  - 4.3.1 Axially Symmetric Magnetic Lens: Imaging in Electron Microscopy 178
  - 4.3.2 Normal Magnetic Quadrupole 197
  - 4.3.3 Skew Magnetic Quadrupole 202
  - 4.3.4 Axially Symmetric Electrostatic Lens 204
  - 4.3.5 Electrostatic Quadrupole 205
- 4.4 Bending Magnet: An Optical Element with a Curved Optic Axis 206
- 4.5 Nonrelativistic Classical Charged Particle Beam Optics 210

Chapter 5 Quantum Charged Particle Beam Optics: Scalar Theory for Spin-0 and Spinless Particles 213

- 5.1 General Formalism of Quantum Charged Particle Beam Optics 213
- 5.2 Relativistic Quantum Charged Particle Beam Optics Based on the Klein–Gordon Equation 214
  - 5.2.1 General Formalism 214
  - 5.2.2 Free Propagation: Diffraction 230
  - 5.2.3 Axially Symmetric Magnetic Lens: Electron Optical Imaging 232
    - 5.2.3.1 Paraxial Approximation: Point-to-Point Imaging 232
    - 5.2.3.2 Going Beyond the Paraxial Approximation: Aberrations 250
    - 5.2.3.3 Quantum Corrections to the Classical Results 260
  - 5.2.4 Normal Magnetic Quadrupole 262
  - 5.2.5 Skew Magnetic Quadrupole 267
  - 5.2.6 Axially Symmetric Electrostatic Lens 269

		5.2.7	Electrostatic Quadrupole Lens	270
		5.2.8	Bending Magnet	271
	5.3	Effect of Quantum Uncertainties on Aberrations in Electron Microscopy and Nonlinearities in Accelerator Optics		277
	5.4	Nonrelativistic Quantum Charged Particle Beam Optics: Spin-0 and Spinless Particles		281
	5.5	Appendix: Propagator for a System with Time-Dependent Quadratic Hamiltonian		283

Chapter 6 Quantum Charged Particle Beam Optics: Spinor Theory for Spin-$\frac{1}{2}$ Particles ... 287

- 6.1 Relativistic Quantum Charged Particle Beam Optics Based on the Dirac–Pauli Equation ... 287
  - 6.1.1 General Formalism ... 287
    - 6.1.1.1 Free Propagation: Diffraction ... 293
    - 6.1.1.2 Axially Symmetric Magnetic Lens ... 300
    - 6.1.1.3 Bending Magnet ... 306
  - 6.1.2 Beam Optics of the Dirac Particle with Anomalous Magnetic Moment ... 316
    - 6.1.2.1 General Formalism ... 316
    - 6.1.2.2 Lorentz and Stern–Gerlach Forces, and the Thomas–Frenkel–BMT Equation for Spin Dynamics ... 320
    - 6.1.2.3 Phase Space and Spin Transfer Maps for a Normal Magnetic Quadrupole ... 323
    - 6.1.2.4 Phase Space and Spin Transfer Maps for a Skew Magnetic Quadrupole ... 327
- 6.2 Nonrelativistic Quantum Charged Particle Beam Optics: Spin-$\frac{1}{2}$ Particles ... 330

Chapter 7 Concluding Remarks and Outlook on Further Development of Quantum Charged Particle Beam Optics ... 335

Bibliography ... 339

Index ... 349

Preface

It is surprising that only classical mechanics is required in the design and operation of numerous charged particle beam devices, from electron microscopes to particle accelerators, though the microscopic particles of the beam should be obeying quantum mechanics. First, such a success of the classical charged particle beam optics needs an explanation from the point of view of quantum mechanics. Then, it is certain that though the classical charged particle beam optics is extremely successful in the present-day beam devices, future developments in charged particle beam technology would require quantum theory. With this in mind, we present in this book the formalism of quantum charged particle beam optics. Though the formalism of quantum charged particle beam optics presented in this book needs a lot of further development for it to be a complete theory, we hope that it would be the basis for the theory of any future charged particle beam device that would require quantum mechanics for the design and operation of its optical elements.

In a series of papers starting in 1983, Professors N. Mukunda, R. Simon, and E. C. G. Sudarshan had established, using a group theoretical approach, the Fourier optics for the Maxwell Field, which generalizes the paraxial scalar wave optics to paraxial Maxwell wave optics, consistently taking into account polarization. Around 1987, when Professor Sudarshan was the director of The Institute of Mathematical Sciences (IMSc), Chennai, India, he suggested to me, then a junior faculty member of the institute, to study the group theoretical aspects of electron optics based on the Dirac equation, with the aim to extend the traditional nonrelativistic scalar wave theory of electron optics to the Dirac spinor wave theory, taking into account the spin of the electron analogous to the way they had extended the Helmholtz scalar light optics to the Maxwell vector optics, taking into account the polarization of light. Searching the literature on electron optics led to a surprise: electron optics, or in general, charged particle optics, used in beam devices from electron microscopes to particle accelerators, was based essentially on classical mechanics. Of course, in electron microscopy, the image formation was understood on the basis of the nonrelativistic Schrödinger equation, ignoring completely the electron spin and the Dirac equation. In accelerator physics, topics like quantum fluctuations of the classical particle trajectories were treated using quantum mechanical concepts, but the design and operation of the accelerator beam elements were based only on classical mechanics. So, we had to start from scratch to understand the quantum mechanics of electron optics based on the Dirac equation. The first result of this study was a paper by Professors Simon, Sudarshan, and Mukunda, and me, in which we derived the focusing action of the round magnetic lens, the central part of any electron microscope, using the Dirac equation *ab initio*. Professor Sudarshan encouraged me to continue the study further in this direction, and in a subsequent paper, I used the quantum electron beam optics based on the Dirac equation to study other types of electromagnetic lenses. The PhD work of Sameen Ahmed Khan with me, during 1992–1997 at IMSc, led to the formulation of a more general framework of quantum theory of charged particle beam optics using the Heisenberg picture, Ehrenfest's theorem, and quantum transfer maps. Sameen and I are fortunate to have had a collaboration in 1996 with Professors M. Conte and M. Pusterla. This collaboration helped us to extend our theory to accelerator optics. Later work has led to a comprehensive formalism of quantum charged particle beam optics applicable to any charged particle beam device.

We remember Professor Sudarshan with gratitude for initiating our study of quantum charged particle beam optics, and would like to thank Professors Mukunda, Simon, Conte, and Pusterla for fruitful discussions and collaboration in the early stages of our work. I thank the IMSc for support throughout my academic career. We thank the library of IMSc for its excellent resources, and we also thank Dr Paul Pandian, the librarian, and his colleagues, in particular, Dr P. Usha Devi, for timely help in getting the references required during the writing of this book.

I worked on my contribution to the writing of this book during my tenure at the Chennai Mathematical Institute (CMI) as adjunct professor, and I wish to thank CMI for supporting me after retirement from IMSc. I have enjoyed working on the book mostly from the comfort of my home and my daughters' homes and with the kind support of everyone in my family. It gives me great pleasure to thank Malathy (my wife), Subadhra & Sandeep, and Anisha & Ajay (my elder daughter & son-in-law, and granddaughter & grandson), and Sujatha & Srinivas, and Aditi (my younger daughter & son-in-law, and granddaughter). And, I wish to thank my friend Kuppuswamy, friend from my school days, who has been encouraging me constantly throughout my life and, in a way, was responsible for me joining IMSc in 1971 as a PhD student.

Sameen adds: I wish to acknowledge IMSc, where the work on this topic originated, for the support I received as a PhD student (and subsequent visits) and its excellent research environment. During my doctoral study at IMSc, the Internet had just begun to take roots. It was during this time that there was a long-distance collaboration with Professors Conte and Pusterla in Italy, exclusively by e-mail (thanks to the excellent Internet resources at IMSc), leading to the Conte-Jagannathan-Khan-Pusterla paper without a single face-to-face meeting! This collaboration was followed by my postdoctoral work at Istituto Nazionale di Fisica Nucleare (INFN) with Professor Pusterla when some more work was done on accelerator optics. I would also like to acknowledge Professor K. B. Wolf for a fruitful stay as a Consejo Nacional de Ciencia y Tecnologia (CONACyT) postdoctoral fellow at the Universidad Nacional Autnoma de Mxico (UNAM), Cuernavaca, Mxico. I would like to profusely thank my current workplace, Dhofar University, for constant encouragement to pursue research. My contributions to the book would not have been possible without the support of my family. My wife Noama Khan and daughter Hajira Khan always take a keen interest in my writing endeavors. I wish to acknowledge my elder brothers Mohammed Ahmed Khan and Farooq Ahmed Khan, and remember my parents thankfully, for the academic environment in which I grew up. And, I have to thank my sisters, Ayesha Khan and Bushra Khan, who have been very supportive throughout my life. I also wish to thank my friend Azher Majid Siddiqui from my college days for continuous encouragement.

We are grateful to Professor Vasudevan Lakshminarayanan for urging us constantly for a long time to write a book on our work, and the book is here finally!

Ramaswamy Jagannathan,
with
Sameen Ahmed Khan,
November 2018.

Authors

Professor Ramaswamy Jagannathan retired in 2009 as a senior professor of physics from The Institute of Mathematical Sciences (IMSc), Chennai, India. He is currently an adjunct professor of physics at the Chennai Mathematical Institute (CMI), Chennai. He got his PhD (Theor. Phys.) from the University of Madras, Chennai, India, in 1976, working at IMSc. His PhD work on generalized Clifford algebras was done under the guidance of Professor Alladi Ramakrishnan, the founder director of IMSc known popularly as MATSCIENCE at that time. He has authored/coauthored about 80 research papers in various branches of Physical Mathematics, like Generalized Clifford Algebras and Their Physical Applications, Finite-Dimensional Quantum Mechanics, Applications of Classical Groups, Quantum Groups, Nonlinear Dynamics, Deformed Special Functions, and Quantum Theory of Charged Particle Beam Optics with Applications to Electron Microscopy and Accelerator Optics. In particular, his paper with Professors R. Simon, E. C. G. Sudarshan, and N. Mukunda (1989) on the quantum theory of magnetic electron lenses based on the Dirac equation initiated a systematic study of the Quantum Theory of Charged Particle Beam Optics. This theory was subsequently developed vastly by him and his collaborators (in particular, his PhD student Dr Sameen Ahmed Khan).

Dr Sameen Ahmed Khan is an associate professor at the Department of Mathematics and Sciences, College of Arts and Applied Sciences, Dhofar University, Salalah, Sultanate of Oman (http://du.edu.om/). He got his PhD (Theor. Phys.) at the University of Madras, Chennai, India, in 1997. His PhD thesis, done at The Institute of Mathematical Sciences (IMSc), Chennai, under the supervision of Professor Ramaswamy Jagannathan, was on the quantum theory of charged particle beam optics. He did postdoctoral research at INFN, Padova, Italy, and Universidad Nacional Autonoma de Mexico, Cuernavaca, Mexico. He has 16 years of teaching experience in Oman. He has developed a unified treatment of light beam optics and light polarization using quantum methodologies. This formalism describes the beam optics and light polarization from a parent Hamiltonian which is exact and derived from Maxwell's equations. He has authored three books, fifteen book chapters, and about 75 technical publications in journals and proceedings of repute. He has more than 250 publications on science popularization. Dr Sameen is one of the founding members of the Ibn al Haytham LHiSA Light: History, Science, and Applications (LHiSA) International Society set up during the International Year of Light and Light-based Technologies. He is a signatory to six of the reports on the upcoming International Linear Collider.

1 Introduction

This book essentially addresses a question of curiosity: How is *Classical Charged Particle Beam Optics*, entirely based on classical mechanics, so successful in practice in numerous devices from low-energy electron microscopes to high-energy particle accelerators though the microscopic particles of the beam should be obeying quantum mechanics? To get a detailed answer to this question, this book explores *Quantum Charged Particle Beam Optics*.

A charged particle beam is a collection of charged particles of the same type (electrons, protons, same type of ions, etc.), with the same mass, charge, etc., moving almost in the same direction with almost the same momentum. For any beam device, there is a design trajectory, straight or curved, defining its optic axis along which the beam particle is supposed to move with a specific momentum, say $\vec{p}_0 = p_0\vec{s}$, where $\vec{s}$ is the unit vector along the tangent to the design trajectory at any point on it. A charged particle moving along this design trajectory with the specified momentum $\vec{p}_0$, the design momentum, is called the reference particle. If all the particles of a beam propagating through the device along its optic axis have momenta very close to the design momentum, *i.e.*, for any particle of the beam $\vec{p} \approx \vec{p}_0$, then all the particles of the beam would move in almost the same direction. Such a beam is called a paraxial beam. Let $p_0\vec{s}$ be the design momentum of a beam device. If $\vec{p} \nparallel \vec{s}$ is the momentum of a particle of a beam propagating through the device, and $|\vec{p}| = p_0$, then its longitudinal and perpendicular components are $\vec{p}_\parallel = (\vec{s}\cdot\vec{p})\vec{s} = p_\parallel \vec{s}$ and $\vec{p}_\perp = \vec{p} - p_\parallel \vec{s}$, respectively, and $p_\parallel^2 = |\vec{p}|^2 - |\vec{p}_\perp|^2 = p_0^2 - p_\perp^2$. Any particle of a paraxial beam will satisfy the condition $|\vec{p}_\perp| \ll p_0$, and it will have $p_\parallel = \sqrt{p_0^2 - p_\perp^2} \approx p_0 - (p_\perp^2/2p_0)$. At times, for a paraxial beam, we may even take $p_\parallel \approx p_0$. If the paraxial condition is not satisfied strictly, and we can write only $p_\parallel \approx p_0 - (p_\perp^2/2p_0) - (p_\perp^4/8p_0^3)$ then the beam can be called a quasiparaxial beam. If more terms in the Taylor series of $\sqrt{p_0^2 - p_\perp^2}$ are required to get an approximate expression for $p_\parallel$ in terms of p_0 and $|\vec{p}_\perp|$, then the beam has to be called nonparaxial. Any beam violating the paraxial condition is nonparaxial, and hence a quasiparaxial beam may also be classified as nonparaxial.

A charged particle beam, an ordered flow of charged particles in nearly the same direction, is a complex dynamical system. The dynamics of the particles depends on both the external electric and magnetic fields guiding the beam propagation and the internal fields contributed by the beam particles themselves. There will also be collisions between the particles. Even a low-current beam may contain more than 10^{10} particles. So, any exact description of the dynamical behavior of the system is impossible. We can have only approximate theories of the collective behavior of the beam. Such theories have to be built by averaging over the behaviors of the large number of individual charged particles of the beam.

It is clear that the first step in the study of charged particle beam physics is the study of the dynamics of a single charged particle of the beam, ignoring all the interparticle and collective, or statistical, interactions. This single particle dynamics is the basis for the design of any charged particle beam device, *i.e.*, designing the required trajectory of the reference particle. Study of the dynamics of a single charged particle under the influence of electric and magnetic fields, and consequent designs of the beam devices, like electron microscopes, electron beam lithography systems, electron beam welding machines, and particle accelerators, is called charged particle beam optics. The basis of design and operation of any device using charged particle beam technology is classical mechanics, and it has been very successful so far without any exception. This is where one becomes curious to know how and why classical mechanics works so well in this context, though the microscopic particles of the beam, like electrons, protons, and ions, should be obeying quantum mechanics.

Any electron microscope is designed and operated using electron optics based entirely on classical mechanics. Quantum mechanics, or wave mechanics, is being used for understanding the image formation and resolution in electron microscopes since Glaser initiated the work in this direction (see Glaser [62] and references therein; see the encyclopedic three-volume text book of Hawkes and Kasper ([70, 71, 72]) for a comprehensive account of historical aspects and any technical aspect of geometrical electron optics and electron wave optics). In understanding the image formation in electron microscopy, mostly the nonrelativistic Schrödinger equation is used. In high-energy electron microscopy, one starts with the relativistic Klein–Gordon equation and soon approximates it to the nonrelativistic Schrödinger equation, often with a relativistic correction essentially based on replacing the rest mass m of the particle by the so-called relativistic mass $\gamma m = m/\sqrt{1-(v^2/c^2)}$. It is considered that under the conditions obtained in high-energy electron microscopy, use of the Klein–Gordon equation, as approximation of the Dirac equation, is adequate (see, *e.g.*, Ferwerda, Hoenders, and Slump [48, 49], Hawkes and Kasper [72], Groves [68], Lubk [129], and Pozzi [150]).

In accelerator physics, there are several quantum effects, like quantum fluctuations in high-energy electron beams due to synchrotron radiation, which are studied using quantum mechanics (see, *e.g.*, Sokolov and Ternov [173], Ternov [181], Bell and Leinass [9], Hand and Skuja [69], Chao, Mess, Tigner, and Zimmermann [20], Chen [22, 23] and Chen and Reil [24] and references therein). The required quantum effects are treated usually as perturbations to the classical trajectories. Accelerator optics, the basic theory for the design and operation of accelerators, itself is based only on classical mechanics (see, *e.g.*, Berz, Makino, and Wan [12], Conte and MacKay [27], Lee [127], Reiser [158], Rosenzweig [159], Seryi [168], Weidemann [187], Wolski [192], Chao, Mess, Tigner, and Zimmermann [20] and references therein).

To understand the remarkable effectiveness of classical mechanics in charged particle beam optics, we explore in this book the quantum mechanics of charged particle beam optics at the level of single particle dynamics, using the appropriate basic equations, namely, the Klein–Gordon equation for spin-0 particles

Introduction

or high-energy particles without spin, (*i.e.*, when the spin is ignored), the Dirac equation for spin-$\frac{1}{2}$ particles, the nonrelativistic Schrödinger equation for low-energy particles without spin, and the nonrelativistic Schrödinger–Pauli equation for low-energy spin-$\frac{1}{2}$ particles. To this end, we closely follow Jagannathan, Simon, Sudarshan, and Mukunda [79], Jagannathan [80], Khan and Jagannathan [90], Jagannathan and Khan [82], Conte, Jagannathan, Khan, and Pusterla [26], Khan [91], Jagannathan [83, 84, 85], Khan [92, 99], and references therein. This chapter, Introduction, gives a brief overview of what to expect in each of the later chapters.

Chapter 2, An Introductory Review of Classical Mechanics, presents a summary of the basic concepts of classical mechanics required essentially for classical charged particle beam optics. We recall the Lagrangian and the Hamiltonian formalisms of classical mechanics of a single particle. We discuss the Hamiltonian dynamics in terms of the Poisson brackets, changing the independent variable, canonical transformations, and the symplecticity of canonical transformations. The transfer maps for the observables across the system with respect to the coordinate along the optic axis are described in terms of the Lie transfer operator. The dynamics of a charged particle moving in a constant magnetic field serves as an example for the discussions. When time is the independent variable with respect to which we study the dynamical evolution of a given system, the generator of time evolution, the Hamiltonian, corresponds to the energy of the system. When we change the independent variable from time to, say s, the coordinate along the optic axis of the system, the corresponding Hamiltonian generating evolution of the system along the forward s-direction, the optical Hamiltonian, corresponds to $-p_s$, negative of the momentum canonically conjugate to s. We derive the classical optical Hamiltonian of a relativistic charged particle moving through an electromagnetic optical element with a straight optic axis. Though we are concerned only with single particle dynamics, in this chapter, we also give a brief introduction to the formalism of multiparticle dynamics for the sake of completeness. For more details on classical mechanics, see, *e.g.*, Corben and Stehle [28], Goldstein, Poole, and Safko [63], and Sudarshan and Mukunda [174]. Classical charged particle beam optics is based on classical electrodynamics, besides classical mechanics. We have not given any account of classical electrodynamics. Books on electron optics, charged particle optics, accelerator physics, and beam physics, mentioned earlier contain adequate details of classical electrodynamics. For more details on classical electrodynamics, see, *e.g.*, Griffiths [66] and Jackson [78]. For any mathematical details required, see, *e.g.*, Arfken, Weber, and Harris [3] and Byron and Fuller [18]. The Lie transfer operator method of studying the classical dynamical evolution of the observables of the system introduced in this chapter plays a central role in the modern approach to classical charged particle beam optics and has a straightforward generalization in the formalism of quantum charged particle beam optics. Dragt et al. have developed extensive techniques for applying the Lie transfer operator methods to light optics, classical charged particle beam optics, and accelerator optics (see Dragt [34], Dragt and Forest [35], Dragt et al. [36], Forest, Berz, and Irwin [54], Rangarajan, Dragt, and Neri [156], Forest and Hirata [55], Forest [56], Radlička [154], and references therein; see also Berz [11], Mondragón and Wolf [135], Wolf [191], Rangarajan and Sachidanand [157], Lakshminarayanan, Sridhar, and Jagannathan [124], and Wolski [192]).

Chapter 3, An Introductory Review of Quantum Mechanics, recalls the basic concepts of quantum mechanics required essentially for quantum charged particle beam optics. First, the fundamental principles of quantum mechanics like the wave function, probability interpretation, Hermitian operator representations of observables like position, linear momentum, angular momentum, and energy or Hamiltonian, eigenstates of observables, time-dependent Schrödinger equation for the dynamical evolution of a quantum system, measurement of an observable, average (or expectation) values, and the uncertainty principle are recalled. The Schrödinger and Heisenberg pictures of quantum dynamics and the Heisenberg equation of motion for observables are introduced. Ehrenfest's theorem, basic to the formalism of quantum charged particle beam optics, is derived. Time-dependent perturbation theory and the role of interaction picture are explained. The Dyson expression for the time-evolution operator as a time-ordered exponential is given, and its Magnus form as an ordinary exponential, particularly suitable for quantum charged particle beam optics, is presented. As examples of nonrelativistic quantum mechanics, free particle, harmonic oscillator in one and two dimensions, charged particle in a constant magnetic field, Laundau levels, and scattering of a particle by one-dimensional potential well and barrier are treated. Concept of spin as a nonclassical property of a particle is introduced using the angular momentum algebra, and the two-component spinor wave function of a spin-$\frac{1}{2}$ particle is introduced. The nonrelativistic Schrödinger–Pauli equation for the electron, or any spin-$\frac{1}{2}$ particle, is formulated, and the magnetic moment of electron is discussed. Concepts of pure and mixed states are explained, and the formalism of quantum mechanics in terms of density operator is discussed. The scalar relativistic wave equation, the Klein–Gordon equation, and the problems with its interpretation as a single particle theory are analysed. Examples of free particle and a charged particle in a constant magnetic field are treated. The Feshbach–Villars form of the Klein–Gordon equation and the passage from the Klein–Gordon equation to its nonrelativistic limit, the nonrelativistic Schrödinger equation, are discussed. The Dirac equation with four-component spinor wave function is formulated. As examples, free particle and a charged particle in a constant magnetic field are treated. It is shown how a particle obeying the Dirac equation, or a Dirac particle, possesses naturally an intrinsic angular momentum, or spin, with value $\frac{1}{2}\hbar$. The Dirac–Pauli equation incorporating an anomalous magnetic moment is written down. The Foldy–Wouthuysen transformation of the Dirac equation is presented in detail. This transformation leads to a systematic procedure for approximating the Dirac Hamiltonian as a sum of the nonrelativistic Hamiltonian plus higher order relativistic corrections. In the Foldy–Wouthuysen representation, the four-component wave function has large upper pair of components compared to the vanishingly small lower pair of components when the particle has positive energy, and the Hamiltonian becomes approximately a direct sum of two parts such that it is possible to build an effective two-component theory for the positive energy particle. A Foldy–Wouthuysen-like transformation is applied to

Introduction

the Feshbach–Villars form of the Klein–Gordon equation, leading to a systematic procedure for approximating it as the nonrelativistic part plus higher order relativistic corrections. A Foldy–Wouthuysen-like transformation is central to the formalism of quantum charged particle beam optics, in which we study the quantum beam optical Hamiltonian as the sum of a paraxial part plus nonparaxial corrections. For more details on quantum mechanics, see, *e.g.*, Cohen-Tannoudji, Diu, and Loloë [25], Esposito, Marmo, Miele, and Sudarshan [39], Greiner [64], Griffiths and Schroetter [67], Sakurai and Napolitano [166], Shankar [169], Bjorken and Drell [13], Greiner [65], and Parthasarathy [142].

Chapter 4, An Introduction to Classical Charged Particle Beam Optics, gives a brief account of the classical theory of optical behaviors of a few magnetic and electrostatic optical elements which will serve as examples for the formalism of quantum charged particle beam optics. The systems we study are monoenergetic paraxial or quasiparaxial charged particle beams propagating in the forward direction through the optical elements comprising time-independent electromagnetic fields. Starting with the general form of the optical Hamiltonian of a relativistic charged particle moving through an electromagnetic optical element with a straight optic axis, derived in Chapter 2, the relativistic classical charged particle beam optical Hamiltonian, or simply, the classical beam optical Hamiltonian, for any electromagnetic optical element with a straight optic axis is derived. Starting with this classical beam optical Hamiltonian, first we study the propagation of the beam through free space. Then, we study a few optical systems with straight optic axis, namely, axially symmetric magnetic and electrostatic lenses, and magnetic and electrostatic quadrupoles. We derive the well-known results on their optical behaviors by computing the transfer maps for the transverse position coordinates (x, y) and the transverse momentum components (p_x, p_y). These optical systems are studied in the lowest order, the paraxial, approximation. Study of the axially symmetric magnetic lens leads to the understanding of the classical theory of image formation in an electron microscope. Study of the magnetic quadrupole explains its central role in guiding the accelerator beams. Transforming the optical Hamiltonian to a curvilinear coordinate system adapted to the geometry of the design orbit, bending of a charged particle beam by a magnetic dipole element, the simplest optical element with a curved optic axis, is studied in paraxial approximation, and the well-known classical results are obtained. It is found that one can obtain the relativistic results by extending the nonrelativistic results through the replacement of the rest mass m by the so-called relativistic mass γm as is the common practice in classical charged particle beam optics. For more details on classical charged particle beam optics, see, *e.g.*, the references on electron optics, charged particle optics, charged particle beam physics, and accelerator physics, mentioned earlier (Hawkes and Kasper [70, 71], Orloff [141], Groves [68], Lubk [129], Pozzi [150], Berz, Makino, and Wan [12], Conte and MacKay [27], Lee [127], Reiser [158], Rosenzweig [159], Seryi [168], Weidemann [187], Wolski [192], Chao, Mess, Tigner, and Zimmermann [20] and references therein).

Chapter 5, Quantum Charged Particle Beam Optics: Scalar Theory for Spin-0 and Spinless Particles, spells out the general framework of the formalism of quantum charged particle beam optics and develops the scalar theory applicable to spin-0 and spinless particles. Let a monoenergetic paraxial or quasiparaxial charged particle beam be propagating in the forward direction along the optic axis of an optical system comprising a time-independent electromagnetic field. Let s be coordinate along the optic axis of the system and $|\psi(s)\rangle$ represent the quantum state of the beam in the vertical plane at the point s on the optic axis. Since we are interested in studying the evolution of the beam along the optic axis of the system, the quantum charged particle beam optical evolution equation should have the general form:

$$i\hbar \frac{\partial |\psi(s)\rangle}{\partial s} = \widehat{\mathscr{H}_o} |\psi(s)\rangle, \qquad (1.1)$$

where $\widehat{\mathscr{H}_o}$ is the quantum charged particle beam optical Hamiltonian of the system. In quantum charged particle beam optics, Equation (1.1) replaces the Schrödinger equation:

$$i\hbar \frac{\partial |\Psi(t)\rangle}{\partial t} = \widehat{\mathscr{H}} |\Psi(t)\rangle. \qquad (1.2)$$

The quantum systems we are studying are the scattering states corresponding to the beams propagating through optical elements comprising time-independent electromagnetic fields. Once the proper basic quantum mechanical time-evolution equation (nonrelativistic Schrödinger equation, Klein–Gordon equation, Dirac equation, ...) appropriate for the given system is rewritten in the form of (1.1), the quantum charged particle beam optical Schrödinger equation, or in short, the quantum beam optical Schrödinger equation, all the optical behaviors of system can be deduced using the standard rules of quantum mechanics. This chapter presents the scalar theory of quantum charged particle beam optics, which is applicable to spin-0 and spinless (*i.e.*, spin-ignored) particles. It is based on the relativistic Klein–Gordon equation for relativistic particles and the nonrelativistic Schrödinger equation for nonrelativistic particles. In the scalar theory, the wave function has a single component, $\psi(x,y,s) = \langle x,y | \psi(s) \rangle$, where x and y are the coordinates in the vertical plane, or the transverse coordinates, at the point s on the optic axis. The optic axis can be straight or curved. Except for the example of the dipole magnet, or the bending magnet, which has a curved optic axis, the systems we study will have straight optic axis. The coordinate s along a straight optic axis will be taken as z. Starting with the chosen basic time-evolution equation, Klein–Gordon, or the nonrelativistic Schrödinger, equation in the case of the scalar theory, and the Dirac equation in the case of the spinor theory, we shall first write down the beam optical evolution equation for a general electromagnetic optical system as an z-evolution equation. In case the system to be studied has a curved optic axis, we shall transform the z-evolution equation into an s-evolution equation adapted to the geometry of optic axis of the system. Writing the time-independent Klein–Gordon equation in the Feshbach–Villars-like form and using a Foldy–Wouthuysen-like transformation lead to the scalar quantum beam optical Schrödinger equation for a monoenergetic quasiparaxial beam propagating in the forward direction along the optic axis with the quantum beam optical Hamiltonian as the sum of the paraxial part and nonparaxial approximations up to any desired level of accuracy. Using this quantum beam optical Schrödinger equation, the transfer maps for quantum averages of the transverse position coordinates ($\langle x \rangle, \langle y \rangle$) and the transverse momentum components ($\langle \hat{p}_x \rangle, \langle \hat{p}_y \rangle$) across the optical elements are obtained in the paraxial approximation. Propagation through free space explains diffraction. Propagation through an axially symmetric magnetic lens explains image formation in an electron microscope. In the classical limit ($\hbar \longrightarrow 0$), the results coincide exactly with the classical results. There are quantum corrections which are very small compared to the classical results and vanish in the classical limit. This explains the success of classical mechanics in charged particle beam optics. Image aberrations due to the deviation of the beam from paraxial conditions are also studied, leading to the quantum versions of the aberration coefficients which correspond to the well-known classical expressions plus some tiny quantum corrections. Similarly, the quantum theory for other optical elements considered, namely, the axially symmetric electrostatic lens, the magnetic and electrostatic quadrupoles, and the bending magnet, leads to quantum beam optical Hamiltonians corresponding to the exact classical beam optical Hamiltonians plus correction terms, which vanish in the classical limit. The nonrelativistic quantum charged particle beam optics based on the nonrelativistic Schrödinger equation is obtained as the nonrelativistic approximation of the theory based on the Klein–Gordon equation. It is found that the common practice of replacing the rest mass m by γm to get the relativistic results from the corresponding nonrelativistic results is justified in the scalar quantum beam optics. However, as we shall see in the next chapter, this is not true in the spinor theory when the spin of the particle is taken into account.

Chapter 6, Quantum Charged Particle Beam Optics: Spinor Theory for Spin-$\frac{1}{2}$ Particles, presents the spinor theory of quantum charged particle beam optics applicable to spin-$\frac{1}{2}$ particles. For the relativistic spin-$\frac{1}{2}$ particles, the Dirac equation and the Dirac–Pauli equation are the fundamental equations on which the quantum charged particle beam optics should be based. For nonrelativistic spin-$\frac{1}{2}$ particles, the quantum charged particle beam optics can be based on the Schrödinger–Pauli equation. The Dirac equation for a charged particle in an electromagnetic field being linear in $\partial/\partial z$ it is straightforward to write down the z-evolution equation in the time-independent case. Then, using a Foldy–Wouthuysen-like transformation, we build the spinor quantum beam optics for optical elements with straight optic axis. For a monoenergetic quasiparaxial beam propagating in the forward direction along the straight optic axis of an optical element comprising time-independent electromagnetic field, the quantum beam optical spinor, four-component, wave function has the upper pair of components large compared to the vanishingly small lower pair of components. We study the free propagation to understand diffraction. Study of axially symmetric magnetic lens shows that the paraxial part of the spinor quantum beam optical Hamiltonian is same as in the scalar theory, whereas the nonparaxial, aberration parts depend on the spin of the particle. Besides the presence of matrix terms, as is to be expected, the scalar terms in the aberration part of the spinor quantum beam optical Hamiltonian differ from the corresponding terms in the scalar quantum beam optical Hamiltonian. In high-energy electron microscopy, use of the Klein–Gordon equation, as approximation of the Dirac equation, is considered adequate.

However, we find that a scalar approximation of the quantum beam optics based on the Dirac equation by keeping only the scalar terms in the quantum beam optical Hamiltonian and dropping the matrix terms is not the same as the quantum beam optics based on the Klein–Gordon equation. Propagation of a monoenergetic paraxial charged Dirac particle beam through a bending magnet is studied adopting the formalism of general relativity to transform the z-evolution equation into the appropriate s-evolution equation. Quantum beam optics of propagation of a forward-moving monoenergetic paraxial beam of Dirac particles with anomalous magnetic moment through an optical element with straight optic axis is formulated using the accelerator optics framework. Magnetic quadrupoles are studied as examples. This study shows how the spinor quantum beam optics gives a unified treatment of orbital motion, or the Lorentz force, effect of the Stern–Gerlach force, and the Thomas–Frenkel–Bargmann–Michel–Telegdi spin motion. Finally, the two-component spinor theory of quantum beam optics based on the nonrelativistic Schrödinger–Pauli equation is given. We find that while it is possible to approximate the quantum beam optical Dirac Hamiltonian to get the nonrelativistic quantum beam optical Schrödinger–Pauli Hamiltonian by taking the limit $\gamma \longrightarrow 1$, it would be misleading to extend the nonrelativistic results to the corresponding relativistic results when the spin of the particle is also taken into account.

Chapter 7, Concluding Remarks and Outlook on Further Development of Quantum Theory of Charged Particle Beam Optics, lists remarks on certain developments in light optics inspired by the quantum charged particle beam optics, the necessity for the use of quantum charged particle beam optics in future charged particle beam devices and particle accelerators, and the main problems to be addressed in the future development of quantum charged particle beam optics. We hope that the formalism of quantum charged particle beam optics presented here would be the basis for the development of the quantum theory of future charged particle beam devices.

2 An Introductory Review of Classical Mechanics

2.1 SINGLE PARTICLE DYNAMICS
2.1.1 LAGRANGIAN FORMALISM
2.1.1.1 Basic Theory

All charged particle beam devices, from low-energy electron microscopes to high-energy particle accelerators, are designed and operated very successfully on the basis of classical mechanics though the beams propagating through them are beams of microscopic particles, like electrons, protons, and ions, which should be obeying quantum mechanics. To understand in depth, we study the quantum mechanics of charged particle beam optics in detail in this book. In this chapter, we give an introductory review of classical mechanics just at the level necessary for understanding the classical charged particle beam optics. For more details on any topic in classical mechanics and advanced perspectives, see, *e.g.*, Goldstein, Poole, and Safko [63], Corben and Stehle [28], and Sudarshan and Mukunda [174].

Classical mechanics of Newton and its relativistic extension by Einstein follow from a principle of stationary action. Let $\underline{x}(t) = \{x_j(t) \mid j = 1,2,3\}$ be the set of components of the position vector of a particle at time t, in any chosen Cartesian or curvilinear coordinate system, and $\underline{\dot{x}}(t) = \{\dot{x}_j(t) = dx_j(t)/dt \mid j = 1,2,3\}$ be the set of corresponding components of the velocity of the particle at that time. Time t is the independent variable with respect to which we are studying the evolution of the system. An action functional

$$S[\underline{x}(t)] = \int_{t_i}^{t_f} dt L(\underline{x}(t), \underline{\dot{x}}(t), t) \tag{2.1}$$

is associated with any trajectory of the particle, $\underline{x}(t)$, from an initial time t_i to a final time t_f, where L is called the Lagrangian of the particle. There can be several equivalent Lagrangians leading to the same description of the dynamics of the particle. Hamilton's principle of stationary action says that the actual trajectory of the particle from t_i to t_f is along a path for which the action is stationary: *i.e.*, the actual path $\underline{x}(t)$ is such that

$$\delta S = S[\underline{x}(t) + \delta \underline{x}(t)] - S[\underline{x}(t)] = 0, \tag{2.2}$$

to first order in $\delta \underline{x}(t)$, any arbitrary small deviation in the path between the fixed initial and final points, $\underline{x}(t_i)$ and $\underline{x}(t_f)$, respectively. Usually, along the actual path, the action takes the least value, and hence Hamilton's principle is often called the principle of least action. Nature seems to have chosen such variational principles in formulating its basic laws.

Expanding δS, the variation in action, to first order in $\delta \underline{x}$, we have

$$\delta S = \int_{t_i}^{t_f} dt \left\{ \sum_{j=1}^{3} \left[\frac{\partial L}{\partial x_j} \delta x_j + \frac{\partial L}{\partial \dot{x}_j} \delta \dot{x}_j \right] \right\}$$

$$= \int_{t_i}^{t_f} dt \left\{ \sum_{j=1}^{3} \left[\frac{\partial L}{\partial x_j} \delta x_j + \frac{\partial L}{\partial \dot{x}_j} \frac{d}{dt} (\delta x_j) \right] \right\}. \tag{2.3}$$

Using integration by parts, and the boundary conditions $\delta \underline{x}(t_i) = 0$ and $\delta \underline{x}(t_f) = 0$, in the second term, we get

$$\delta S = \int_{t_i}^{t_f} dt \left\{ \sum_{j=1}^{3} \left[\frac{\partial L}{\partial x_j} \delta x_j - \frac{d}{dt} \left(\frac{\partial L}{\partial \dot{x}_j} \right) \delta x_j \right] \right\} + \left[\sum_{j=1}^{3} \frac{\partial L}{\partial \dot{x}_j} \delta x_j \right]_{t_i}^{t_f}$$

$$= \int_{t_i}^{t_f} dt \left\{ \sum_{j=1}^{3} \left[\frac{\partial L}{\partial x_j} - \frac{d}{dt} \left(\frac{\partial L}{\partial \dot{x}_j} \right) \right] \delta x_j \right\}. \tag{2.4}$$

Then, the principle of stationary action (2.2) leads to the set of Euler–Lagrange equations,

$$\frac{d}{dt} \left(\frac{\partial L}{\partial \dot{x}_j} \right) = \frac{\partial L}{\partial x_j}, \qquad j = 1, 2, 3, \tag{2.5}$$

the solutions of which determine the actual paths, given the appropriate initial conditions. The derivation of the Euler–Lagrange equations from the principle of stationary action shows that if we have a different, equivalent Lagrangian, say $\bar{L}$, which also leads to the same Euler–Lagrange equations, then we should have

$$\bar{L} = L + \frac{dF(\underline{x}, \underline{\dot{x}}, t)}{dt}. \tag{2.6}$$

We can write (2.5), Lagrange's equations of motion for the particle, as

$$\frac{d}{dt} \left(\frac{\partial L}{\partial \underline{\dot{x}}} \right) = \frac{\partial L}{\partial \underline{x}}. \tag{2.7}$$

2.1.1.2 Example: Motion of a Charged Particle in an Electromagnetic Field

As an example of the Lagrangian formalism, let us consider the motion of a particle of mass m and electric charge q in an electromagnetic field. Let us denote $\underline{x} = \vec{r} = x\vec{i} + y\vec{j} + z\vec{k}$, the position vector of the particle, and $\underline{\dot{x}} = \vec{\dot{r}} = v_x\vec{i} + v_y\vec{j} + v_z\vec{k} = \vec{v}$, the velocity of the particle in a right-handed Cartesian coordinate system with $(\vec{i}, \vec{j}, \vec{k})$ as the unit vectors along the x-, y-, and z-axes, respectively. Let the electromagnetic

A Review of Classical Mechanics

field, electric field $\vec{E}(\vec{r},t)$, and magnetic field $\vec{B}(\vec{r},t)$ be specified by the scalar potential $\phi(\vec{r},t)$ and the vector potential $\vec{A}(\vec{r},t)$, such that[1]

$$\vec{E}(\vec{r},t) = -\vec{\nabla}\phi(\vec{r},t) - \frac{\partial \vec{A}(\vec{r},t)}{\partial t},$$
$$\vec{B}(\vec{r},t) = \vec{\nabla} \times \vec{A}(\vec{r},t). \tag{2.8}$$

Now, Lagrange's equations of motion become

$$\frac{d}{dt}\left(\frac{\partial L}{\partial \vec{v}}\right) = \frac{\partial L}{\partial \vec{r}} = \vec{\nabla} L. \tag{2.9}$$

The relativistic Lagrangian of the particle is

$$L(\vec{r},\vec{v},t) = -mc^2\sqrt{1-\beta^2} - q(\phi - \vec{v}\cdot\vec{A}), \tag{2.10}$$

where c is the speed of light in vacuum and $\beta = |\vec{v}|/c = v/c$. Then, the x component of (2.9) is

$$\frac{d}{dt}\left(\frac{mv_x}{\sqrt{1-\beta^2}}\right) + q\frac{dA_x}{dt} = q\left(-\frac{\partial \phi}{\partial x} + \frac{\partial}{\partial x}(\vec{v}\cdot\vec{A})\right). \tag{2.11}$$

Similar equations follow for y and z components. These three equations are the components of the following equation:

$$\frac{d(\gamma m\vec{v})}{dt} + q\frac{d\vec{A}}{dt} = q\left(-\vec{\nabla}\phi + \vec{\nabla}(\vec{v}\cdot\vec{A})\right), \tag{2.12}$$

where $\gamma = 1/\sqrt{1-\beta^2}$. Substituting the well-known expressions

$$\frac{d(\cdots)}{dt} = \frac{\partial(\cdots)}{\partial t} + (\vec{v}\cdot\vec{\nabla})(\cdots), \tag{2.13}$$

and

$$\vec{\nabla}(\vec{v}\cdot\vec{A}) = (\vec{v}\cdot\vec{\nabla})\vec{A} + \vec{v}\times(\vec{\nabla}\times\vec{A}), \tag{2.14}$$

we get

$$\frac{d(\gamma m\vec{v})}{dt} = q\left(-\vec{\nabla}\phi - \frac{\partial \vec{A}}{\partial t} + \vec{v}\times(\vec{\nabla}\times\vec{A})\right), \tag{2.15}$$

or

$$\frac{d(\gamma m\vec{v})}{dt} = q(\vec{E} + \vec{v}\times\vec{B}), \tag{2.16}$$

---

[1] Hereafter, in general, we shall denote the electric field $\vec{E}(\vec{r},t)$, magnetic field $\vec{B}(\vec{r},t)$, scalar potential $\phi(\vec{r},t)$, and the vector potential $\vec{A}(\vec{r},t)$, respectively, as $\vec{E}$, $\vec{B}$, ϕ, and $\vec{A}$ suppressing their dependence on $\vec{r}$ and t unless it is required to mention it explicitly.

which is the relativistic equation of motion for the charged particle in an electromagnetic field under the Lorentz force. For details on classical electrodynamics, see, *e.g.*, Griffiths [66], and Jackson [78].

In the nonrelativistic case with $\beta \ll 1$, we have $\gamma \approx 1$, and the equation (2.16) becomes,

$$\frac{d(m\vec{v})}{dt} = q(\vec{E} + \vec{v} \times \vec{B}), \tag{2.17}$$

Newton's equation of motion for the charged particle in an electromagnetic field under the Lorentz force. The Lagrangian of the nonrelativistic particle is[2]

$$L = \frac{1}{2}mv^2 - q(\phi - \vec{v} \cdot \vec{A}), \tag{2.18}$$

obtained by taking $\sqrt{1-\beta^2} \approx 1 - \beta^2/2$ in view of the relation $\beta \ll 1$, in approximating the relativistic Lagrangian (2.10), and dropping the constant rest energy term mc^2.

2.1.2 HAMILTONIAN FORMALISM

2.1.2.1 Basic Theory

Hamiltonian formalism is an equivalent reformulation of the Lagrangian formalism. Let us now define

$$p_j = \frac{\partial L}{\partial \dot{x}_j}, \quad j = 1, 2, 3, \tag{2.19}$$

and

$$H = \sum_{j=1}^{3} \dot{x}_j p_j - L, \tag{2.20}$$

where p_js are called the components of the canonical momentum conjugate to the coordinates x_js and H is called the Hamiltonian of the particle. The pair (x_j, p_j) is known as a pair of canonically conjugate variables. The total differential of H is given by

$$\begin{aligned}
dH &= \sum_{j=1}^{3} \dot{x}_j dp_j + \sum_{j=1}^{3} p_j d\dot{x}_j - \sum_{j=1}^{3} \frac{\partial L}{\partial x_j} dx_j - \sum_{j=1}^{3} \frac{\partial L}{\partial \dot{x}_j} d\dot{x}_j - \frac{\partial L}{\partial t} dt \\
&= \sum_{j=1}^{3} \dot{x}_j dp_j - \sum_{j=1}^{3} \frac{\partial L}{\partial x_j} dx_j - \frac{\partial L}{\partial t} dt + \sum_{j=1}^{3} \left(p_j - \frac{\partial L}{\partial \dot{x}_j} \right) d\dot{x}_j.
\end{aligned} \tag{2.21}$$

In view of the definition (2.19), Lagrange's equations of motion (2.7) can be written as

$$\underline{\dot{p}} = \frac{\partial L}{\partial \underline{x}}, \tag{2.22}$$

---

[2] In general, for any vector quantity, say $\vec{V}$, we shall denote $\vec{V} \cdot \vec{V}$ by $\vec{V}^2$, or V^2, to be understood as $\vec{V} \cdot \vec{V}$ from the context.

A Review of Classical Mechanics

and consequently we get

$$dH = \sum_{j=1}^{3} \dot{x}_j dp_j - \sum_{j=1}^{3} \dot{p}_j dx_j - \frac{\partial L}{\partial t} dt. \qquad (2.23)$$

Note that the definition of p_js in (2.19) allows us to solve for any $\dot{x}_j$ in terms of all x_js, p_js, and t. Thus, the Hamiltonian H defined in (2.20) becomes a function of only x_js, p_js, and t when $\dot{x}_j$s are expressed in terms of x_js, p_js, and t. Hence, writing

$$dH(\underline{x},\underline{p},t) = \sum_{j=1}^{3} \frac{\partial H}{\partial x_j} dx_j + \sum_{j=1}^{3} \frac{\partial H}{\partial p_j} dp_j + \frac{\partial H}{\partial t} dt, \qquad (2.24)$$

and comparing with (2.23), we get

$$\dot{x}_j = \frac{\partial H}{\partial p_j}, \quad \dot{p}_j = -\frac{\partial H}{\partial x_j}, \quad j = 1, 2, 3, \qquad (2.25)$$

and

$$\frac{\partial H}{\partial t} = -\frac{\partial L}{\partial t}. \qquad (2.26)$$

The equations in (2.25) are Hamilton's equations of motion for the particle, which can be written as

$$\underline{\dot{x}} = \frac{\partial H}{\partial \underline{p}}, \quad \underline{\dot{p}} = -\frac{\partial H}{\partial \underline{x}}. \qquad (2.27)$$

The 6-dimensional space formed by the set of coordinates, $\{\underline{x}\}$, and the set of components of the canonically conjugate momentum, $\{\underline{p}\}$, is called the phase space of the particle. Hamilton's equations of motion (2.27), or the canonical equations of motion, are completely equivalent to Lagrange's equations of motion (2.7). However, Hamiltonian formalism has certain advantages over the Lagrangian formalism as we shall see later.

Let us now introduce a 6-dimensional column vector $\underline{\varphi}$ in phase space with components

$$\varphi_j = x_j, \quad j = 1, 2, 3, \quad \varphi_{3+j} = p_j, \quad j = 1, 2, 3. \qquad (2.28)$$

Then, it is obvious that we can write Hamilton's equations of motion (2.27) as

$$\underline{\dot{\varphi}} = J \frac{\partial H}{\partial \underline{\varphi}}, \qquad (2.29)$$

where $\partial H / \partial \underline{\varphi}$ is the 6-dimensional column vector with components

$$\left(\frac{\partial H}{\partial \underline{\varphi}}\right)_j = \frac{\partial H}{\partial x_j}, \quad j = 1, 2, 3, \quad \left(\frac{\partial H}{\partial \underline{\varphi}}\right)_{3+j} = \frac{\partial H}{\partial p_j}, \quad j = 1, 2, 3, \qquad (2.30)$$

and J is the 6×6 antisymmetric matrix given by

$$J = \begin{pmatrix} O & I \\ -I & O \end{pmatrix}, \qquad (2.31)$$

with I and O as 3×3 identity and null matrices, respectively. In (2.29), we have the symplectic form of the canonical equations of motion.

2.1.2.2 Example: Motion of a Charged Particle in an Electromagnetic Field

Let us now use the Hamiltonian formalism to study the motion of a charged particle in an electromagnetic field. As we have seen earlier, the relativistic Lagrangian of the particle is

$$L(\vec{r},\vec{v},t) = -mc^2\sqrt{1-\beta^2} - q(\phi - \vec{v}\cdot\vec{A}). \tag{2.32}$$

The canonical momentum vector conjugate to the position vector is given by

$$\vec{p} = \frac{\partial L}{\partial \vec{v}} = \gamma m\vec{v} + q\vec{A}. \tag{2.33}$$

This shows that

$$\vec{p} - q\vec{A} = \gamma m\vec{v} \tag{2.34}$$

is actually the kinetic, or mechanical, linear momentum of the particle. Hereafter, in general, we shall write

$$\vec{p} - q\vec{A} = \vec{\pi}, \tag{2.35}$$

as a shorthand notation for the kinetic momentum of a particle. Solving for $\vec{v}$ in terms of $\vec{r}$ and $\vec{p}$, we have

$$\vec{v} = \frac{\vec{\pi}}{\sqrt{m^2 + \frac{1}{c^2}\vec{\pi}^2}}. \tag{2.36}$$

Then, the relativistic Hamiltonian is

$$\begin{aligned} H(\vec{r},\vec{p},t) &= \vec{v}\cdot\vec{p} - L \\ &= \vec{v}\cdot\vec{p} + mc^2\sqrt{1-\beta^2} + q(\phi - \vec{v}\cdot\vec{A}) \\ &= \vec{v}\cdot\vec{\pi} + mc^2\sqrt{1-\beta^2} + q\phi \\ &= \sqrt{m^2c^4 + c^2\vec{\pi}^2} + q\phi. \end{aligned} \tag{2.37}$$

In the absence of the electromagnetic field, the particle is free and its Hamiltonian becomes

$$H_F = \sqrt{m^2c^4 + c^2 p^2}. \tag{2.38}$$

Then, note that the Hamiltonian in the presence of the electromagnetic field is obtained from the free particle Hamiltonian by the replacement $H_F \longrightarrow H - q\phi$ and $\vec{p} \longrightarrow \vec{\pi}$, known as the principle of minimal electromagnetic coupling. In the nonrelativistic case, since $\vec{\pi}^2 \ll m^2c^2$, the Hamiltonian becomes

$$H = \frac{\vec{\pi}^2}{2m} + q\phi, \tag{2.39}$$

taking $\sqrt{m^2c^4 + c^2\vec{\pi}^2} \approx mc^2\left[1 + (\vec{\pi}^2/2m^2c^2)\right]$ and dropping the constant rest energy term mc^2. It is clear that the Hamiltonian represents the total energy of the particle.

A Review of Classical Mechanics

Now, Hamilton's equations of motion for the particle are

$$\dot{\vec{r}} = \vec{v} = \frac{\partial H}{\partial \vec{p}},$$

$$\dot{\vec{p}} = -\frac{\partial H}{\partial \vec{r}} = -\vec{\nabla} H. \tag{2.40}$$

It is straightforward to check the first part of this equation:

$$\frac{\partial}{\partial p_x}\left(\sqrt{m^2c^4 + c^2\vec{\pi}^2} + q\phi\right) = \frac{\pi_x}{\sqrt{m^2 + \frac{1}{c^2}\vec{\pi}^2}} = v_x, \tag{2.41}$$

and similar equations follow for y and z components. To understand the second part of the equation, we proceed as follows. For the left-hand side, we have

$$\frac{dp_x}{dt} = \frac{d}{dt}(\gamma m v_x + qA_x)$$

$$= \frac{d(\gamma m v_x)}{dt} + q\left(\frac{\partial A_x}{\partial t} + (\vec{v}\cdot\vec{\nabla})A_x\right), \tag{2.42}$$

and similar equations for y and z components. For the right-hand side, we have

$$-\frac{\partial H}{\partial x} = -\frac{\partial}{\partial x}\left(\sqrt{m^2c^4 + c^2\vec{\pi}^2} + q\phi\right)$$

$$= q\left(\frac{\partial}{\partial x}(\vec{v}\cdot\vec{A}) - \frac{\partial \phi}{\partial x}\right), \tag{2.43}$$

and similar equations for y and z components. Equating the left- and right-hand sides, we get

$$\frac{d(\gamma m \vec{v})}{dt} = q\left[\left(-\vec{\nabla}\phi - \frac{\partial \vec{A}}{\partial t}\right) + \left(\vec{\nabla}(\vec{v}\cdot\vec{A}) - (\vec{v}\cdot\vec{\nabla})\vec{A}\right)\right]$$

$$= q\left(\vec{E} + \vec{v}\times(\vec{\nabla}\times\vec{A})\right)$$

$$= q\left(\vec{E} + \vec{v}\times\vec{B}\right), \tag{2.44}$$

the relativistic equation of motion for the charged particle in an electromagnetic field under the Lorentz force (2.16).

2.1.3 HAMILTONIAN FORMALISM IN TERMS OF THE POISSON BRACKETS

2.1.3.1 Basic Theory

Let $O(\underline{x},\underline{p},t)$ be an observable, or dynamical variable, of a particle with the Hamiltonian $H(\underline{x},\underline{p},t)$. The time evolution of O is given by

$$\frac{dO}{dt} = \sum_{j=1}^{3}\left(\frac{\partial O}{\partial x_j}\dot{x}_j + \frac{\partial O}{\partial p_j}\dot{p}_j\right) + \frac{\partial O}{\partial t}$$

$$= \sum_{j=1}^{3}\left(\frac{\partial O}{\partial x_j}\frac{\partial H}{\partial p_j} - \frac{\partial O}{\partial p_j}\frac{\partial H}{\partial x_j}\right) + \frac{\partial O}{\partial t}, \quad (2.45)$$

because of Hamilton's equations of motion. Defining the Poisson bracket between any two functions $f(\underline{x},\underline{p})$ and $g(\underline{x},\underline{p})$ by

$$\{f,g\} = \sum_{j=1}^{3}\left(\frac{\partial f}{\partial x_j}\frac{\partial g}{\partial p_j} - \frac{\partial f}{\partial p_j}\frac{\partial g}{\partial x_j}\right), \quad (2.46)$$

we can write (2.45) as

$$\frac{dO}{dt} = \{O,H\} + \frac{\partial O}{\partial t}, \quad (2.47)$$

which gives Hamilton's equation of motion for any observable of the particle. Observe that we can write

$$\{f,g\} = \widetilde{\frac{\partial f}{\partial \underline{\varphi}}} J \frac{\partial g}{\partial \underline{\varphi}}, \quad (2.48)$$

where $\widetilde{\partial f/\partial \underline{\varphi}}$ is the row vector obtained by taking the transpose of the column vector $\partial f/\partial \underline{\varphi}$, and

$$\{\varphi_k,\varphi_l\} = J_{kl}, \qquad k,l=1,2,\ldots,6. \quad (2.49)$$

Note that the Poisson bracket is antisymmetric:

$$\{f,g\} = -\{g,f\}. \quad (2.50)$$

Also, note that

$$-\{g,f\} = \{-g,f\}. \quad (2.51)$$

Thus, we can write (2.47) as

$$\frac{dO}{dt} = \{-H,O\} + \frac{\partial O}{\partial t}. \quad (2.52)$$

If an observable O is not explicitly time-dependent, $\partial O/\partial t = 0$, and hence

$$\frac{dO(\underline{x},\underline{p})}{dt} = \{-H,O\}. \quad (2.53)$$

A Review of Classical Mechanics

Thus, the basic Hamilton's equations of motion (2.27) can be written as

$$\dot{\underline{x}} = \{-H, \underline{x}\}, \quad \dot{\underline{p}} = \{-H, \underline{p}\}. \tag{2.54}$$

The Poisson brackets,

$$\{x_j, p_k\} = \delta_{jk}, \quad \{x_j, x_k\} = 0, \quad \{p_j, p_k\} = 0, \quad j,k = 1,2,3, \tag{2.55}$$

are the fundamental Poisson brackets. In calculating the various Poisson brackets, the following basic relations will be often useful:

$$\begin{aligned} &\{f,c\} = 0, \\ &\{af, bg+ch\} = ab\{f,g\} + ac\{f,h\} \\ &\{f, gh\} = \{f,g\}h + g\{f,h\}, \\ &\{f, \{g,h\}\} + \{g, \{h,f\}\} + \{h, \{f,g\}\} = 0, \end{aligned} \tag{2.56}$$

where f, g, and h are functions of $\underline{x}$ and $\underline{p}$, and a, b, and c are constants. The last relation in (2.56) is known as the Jacobi identity. Note that using the antisymmetry of the Poisson bracket (2.50), the Jacobi identity can be rewritten as

$$\{f, \{g,h\}\} - \{g, \{f,h\}\} = \{\{f,g\}, h\}. \tag{2.57}$$

If an observable of the particle $O(\underline{x}, \underline{p})$ is not explicitly time-dependent and has vanishing Poisson bracket with the Hamiltonian, then it will be a constant of motion, *i.e.*, it will be conserved. Thus, for a particle with a Hamiltonian that is not explicitly time-dependent, the energy will be conserved, as one can also verify directly. With $\partial H/\partial t = 0$, it follows from Hamilton's equations that

$$\begin{aligned} \frac{dH(\underline{x}, \underline{p})}{dt} &= \sum_{j=1}^{3} \left(\frac{\partial H}{\partial x_j} \dot{x}_j + \frac{\partial H}{\partial p_j} \dot{p}_j \right) \\ &= \sum_{j=1}^{3} \left(\frac{\partial H}{\partial x_j} \frac{\partial H}{\partial p_j} - \frac{\partial H}{\partial p_j} \frac{\partial H}{\partial x_j} \right) = 0. \end{aligned} \tag{2.58}$$

For a particle with a Hamiltonian not dependent explicitly on time, the equation of motion (2.53) for any time-independent observable $O(\underline{x}, \underline{p})$ can be directly integrated. To this end, define the Lie operator

$$:f: = \sum_{j=1}^{3} \left(\frac{\partial f}{\partial x_j} \frac{\partial}{\partial p_j} - \frac{\partial f}{\partial p_j} \frac{\partial}{\partial x_j} \right). \tag{2.59}$$

Then, one can write

$$\begin{aligned} &:f:^0 g = g, \quad :f: g = \{f,g\}, \quad :f:^2 g = \{f, \{f,g\}\}, \\ &:f:^3 g = \{f, \{f, \{f,g\}\}\}, \quad \ldots. \end{aligned} \tag{2.60}$$

Thus, Hamilton's equations of motion are

$$\dot{\underline{x}} = :-H:\underline{x}, \qquad \dot{\underline{p}} = :-H:\underline{p}. \tag{2.61}$$

Now, for a time-independent observable O, we get on integrating (2.53) or (2.61) directly,

$$O(t) = \left[e^{\{(t-t_i):-H:\}} O \right](t_i)$$
$$= \left[\sum_{n=0}^{\infty} \frac{(t-t_i)^n}{n!} :-H:^n O \right](t_i), \tag{2.62}$$

where t_i and t are the initial and final instants of time, respectively. Here, $[\cdots](t_i)$ means that after the expression $[\cdots]$ has been found, it should be evaluated by substituting the values of the observables in it at the initial time t_i. This equation (2.62) defines the Lie transfer operator generated by the Hamiltonian H, relating the initial and final values of any observable of the particle. For the basic phase space variables, in terms of which any observable of the particle can be written, we get the transfer map, for any $t \geq t_i$, as

$$\underline{\varphi}(t) = \left[e^{\{(t-t_i):-H:\}} \underline{\varphi} \right](t_i). \tag{2.63}$$

We shall see later how this map can be generalized when the Hamiltonian depends on the independent variable. Often, for the sake of notational convenience, we will write the transfer map (2.63) as

$$\underline{\varphi}(t) = e^{\{(t-t_i):-H:\}} \underline{\varphi}(t_i). \tag{2.64}$$

2.1.3.2 Example: Dynamics of a Charged Particle in a Constant Magnetic Field

As an example of how the Poisson bracket formalism works, let us consider the motion of a particle of mass m and charge q in a constant magnetic field. By constant magnetic field, we mean that the magnetic field does not vary in space and time. Without loss of generality, we shall take the magnetic field to be in the z-direction, $\vec{B} = B\vec{k}$, and the corresponding vector potential to be $\vec{A} = \frac{1}{2}(\vec{B} \times \vec{r})$. The relativistic Hamiltonian of the particle is

$$H(\vec{r}, \vec{p}) = \sqrt{m^2 c^4 + c^2 \vec{\pi}^2} \tag{2.65}$$
$$= \left\{ m^2 c^4 + c^2 \left[\left(p_x + \frac{1}{2} qBy \right)^2 + \left(p_y - \frac{1}{2} qBx \right)^2 + p_z^2 \right] \right\}^{1/2},$$

and is time-independent. Thus, the total energy of the particle is a constant of motion. From (2.36), we know that the velocity of the particle is given by

$$\vec{v} = \frac{\vec{\pi}}{\sqrt{m^2 + \frac{1}{c^2} \vec{\pi}^2}}. \tag{2.66}$$

A Review of Classical Mechanics

Noting that

$$\sqrt{m^2 + \frac{1}{c^2}\vec{\pi}^2} = \frac{H}{c^2}, \tag{2.67}$$

we can write, in this case,

$$v_x = \frac{c^2}{H}\left(p_x + \frac{1}{2}qBy\right),$$

$$v_y = \frac{c^2}{H}\left(p_y - \frac{1}{2}qBx\right),$$

$$v_z = \frac{c^2 p_z}{H}, \tag{2.68}$$

where the components of $\vec{v}$ are to be treated as functions of $\vec{r}$ and $\vec{p}$. Note that

$$:-H:\vec{r} = \vec{v}. \tag{2.69}$$

If any $f(\vec{r},\vec{p})$ is a constant of motion, i.e., $:-H:f = 0$, then f^{-1} is also a constant of motion. This follows from the observation

$$:-H:(ff^{-1}) = 0$$
$$= f(:-H:f^{-1}) + (:-H:f)f^{-1}$$
$$= f(:-H:f^{-1}), \tag{2.70}$$

where the relations in (2.56) have been used. Then, we find

$$:-H:v_z =:-H:\frac{c^2 p_z}{H} = \frac{c^2}{H}:-H:p_z = 0,$$

$$:-H:v_x =:-H:\left[\frac{c^2}{H}\left(p_x + \frac{1}{2}qBy\right)\right]$$
$$= \frac{c^2}{H}\left[:-H:\left(p_x + \frac{1}{2}qBy\right)\right] = \left(\frac{c^2 qB}{H}\right)v_y,$$

$$:-H:v_y =:-H:\left[\frac{c^2}{H}\left(p_y - \frac{1}{2}qBx\right)\right]$$
$$= \frac{c^2}{H}\left[:-H:\left(p_y - \frac{1}{2}qBx\right)\right] = -\left(\frac{c^2 qB}{H}\right)v_x. \tag{2.71}$$

It follows that

$$:-H:(v_x + iv_y) = -i\left(\frac{c^2 qB}{H}\right)(v_x + iv_y), \tag{2.72}$$

and

$$:-H:^n(v_x + iv_y) = \left(-i\frac{c^2 qB}{H}\right)^n(v_x + iv_y), \quad n = 0, 1, 2, 3, \ldots. \tag{2.73}$$

Let us take $t_i = 0$ as the time when the particle enters the magnetic field, and t as the time of observation. From (2.62), (2.69), (2.71), and (2.73), it follows that

$$[x(t)+iy(t)] = \left[e^{t:-H:}(x+iy)\right](0)$$
$$= [x(0)+iy(0)] + \left[\sum_{n=1}^{\infty}(-i\omega_c)^{n-1}\frac{t^n}{n!}\right][v_x+iv_y](0)$$
$$= [x(0)+iy(0)] + \frac{i}{\omega_c}\left[e^{-i\omega_c t}-1\right][v_x(0)+iv_y(0)].$$
$$z(t) = \left[e^{t:-H:}z\right](0) = z(0)+v_z(0)t, \quad (2.74)$$

where $\omega_c = c^2 qB/H(0)$. Let us now take, without loss of generality, that the particle enters the magnetic field at $t=0$ in the yz-plane: $v_x(0) = 0, v_y(0) = v_\perp, v_z(0) = v_\parallel$. Then, the position of the particle at any later time t is given by

$$x(t) = (x(0)+\rho_L) - \rho_L \cos(\omega_c t),$$
$$y(t) = y(0) + \rho_L \sin(\omega_c t),$$
$$z(t) = z(0) + v_\parallel t, \quad (2.75)$$

where $\rho_L = v_\perp/\omega_c$. From this, we get

$$v_x(t) = -v_\perp \sin(\omega_c t), \quad v_y(t) = v_\perp \cos(\omega_c t), \quad v_z(t) = v_\parallel. \quad (2.76)$$

Since the particle enters the field at $t=0$ as a free particle with velocity $\vec{v}(0) = (0, v_\perp, v_\parallel)$, we know that its total energy, or Hamiltonian, at the initial time to be

$$H(0) = \sqrt{m^2 c^4 + c^2 p(0)^2} = \frac{mc^2}{\sqrt{1-\beta(0)^2}} = \gamma(0)mc^2. \quad (2.77)$$

where $\beta(0) = v(0)/c$. From (2.76), it is clear that

$$v_\perp(t)^2 = v_x(t)^2 + v_y(t)^2 = v_x(0)^2 + v_y(0)^2 = v_\perp(0)^2,$$
$$v_\parallel(t) = v_z(t) = v_z(0) = v_\parallel(0), \quad (2.78)$$

and hence $v(t)^2 = v_\parallel(t)^2 + v_\perp(t)^2$ has at any time t the same value as at the initial time, namely $v(0)^2$. So, γ is a constant of motion for the particle and $c^2/H(0) = 1/\gamma m$. Thus, we have

$$\omega_c = \frac{qB}{\gamma m} \quad (2.79)$$

known as the cyclotron frequency or the frequency of gyration. Equation (2.75) shows that the particle moves in a helical trajectory of radius $\rho_L = v_\perp/\omega_c$, the radius of gyration called the Larmor radius, and pitch $\wp = 2\pi v_\parallel/\omega_c$. The central axis of the helix cuts the xy-plane at the point $(x_0, y_0) = (x(0)+\rho_L, y(0))$, the guiding center, where $(x(0), y(0), z(0))$ is the point at which the particle enters the magnetic field at the initial time.

A Review of Classical Mechanics

2.1.4 CHANGING THE INDEPENDENT VARIABLE

2.1.4.1 Basic Theory

Let us rederive Hamilton's equations of motion by formulating the principle of stationary action in phase space. We can write (2.2) in phase space as

$$\delta S = \delta \left\{ \int_{t_i}^{t_f} dt \left[\sum_{j=1}^{3} p_j \dot{x}_j - H(\underline{x}, \underline{p}, t) \right] \right\} = 0, \quad (2.80)$$

where the Lagrangian has been substituted in terms of the Hamiltonian. Expanding (2.80), we get

$$\begin{aligned}
\delta S &= \int_{t_i}^{t_f} dt \left[\sum_{j=1}^{3} \left(p_j \delta \dot{x}_j + \dot{x}_j \delta p_j - \frac{\partial H}{\partial x_j} \delta x_j - \frac{\partial H}{\partial p_j} \delta p_j \right) \right] \\
&= \int_{t_i}^{t_f} dt \left[\sum_{j=1}^{3} \left(-\dot{p}_j \delta x_j + \dot{x}_j \delta p_j - \frac{\partial H}{\partial x_j} \delta x_j - \frac{\partial H}{\partial p_j} \delta p_j \right) \right] \\
&\quad + \left[\sum_{j=1}^{3} p_j \delta x_j \right]_{t_i}^{t_f} \\
&= \int_{t_i}^{t_f} dt \left\{ \sum_{j=1}^{3} \left[-\left(\dot{p}_j + \frac{\partial H}{\partial x_j} \right) \delta x_j + \left(\dot{x}_j - \frac{\partial H}{\partial p_j} \right) \delta p_j \right] \right\} \\
&= 0, \quad (2.81)
\end{aligned}$$

where in the second step, integration by parts has been used and the boundary conditions at t_i and t_f make the last term vanish. Now, for the action to be stationary for arbitrary variations in the phase space trajectory $(\delta \underline{x}, \delta \underline{p})$, we must have

$$\dot{p}_j = -\frac{\partial H}{\partial x_j}, \qquad \dot{x}_j = \frac{\partial H}{\partial p_j}, \qquad j = 1, 2, 3, \quad (2.82)$$

which are Hamilton's equations of motion.

Let us now write the principle of stationary action in phase space (2.80) as

$$\delta S = \delta \left\{ \int_{t_i}^{t_f} \left[\sum_{j=1}^{3} p_j dx_j + H(\underline{x}, \underline{p}, t) d(-t) \right] \right\} = 0. \quad (2.83)$$

It is clear from this that we can regard $(-t, H)$ as a canonically conjugate pair, and all the canonically conjugate pairs $\{(-t, H), (x_j, p_j) \mid j = 1, 2, 3\}$ have equal status. This shows that instead of time, we can choose any other coordinate as the independent variable, and the corresponding Hamilton's equations will give the evolution of the state of the particle in terms of that variable. If we choose, say, x_k as the independent variable, then we can write the variational principle (2.80) as

$$\delta S = \delta \left\{ \int_{x_k(t_i)}^{x_k(t)} dx_k \left[\sum_{j=1, \neq k}^{3} \left(p_j \frac{dx_j}{dx_k} + E \frac{d(-t)}{dx_k} - \mathcal{H} \right) \right] \right\} = 0, \quad (2.84)$$

with the new Hamiltonian

$$\mathscr{H} = -p_k(\underline{x}, p_1, p_2, \ldots, p_{k-1}, p_{k+1}, \ldots, p_n, E, t), \quad (2.85)$$

where p_k has been obtained by solving the equation $H(\underline{x},\underline{p},t) = E$, and the old Hamiltonian H has been denoted by E, representing the total energy of the particle. Equation (2.84) has the same form as (2.80), except for the difference in the independent variable, and hence this will lead to Hamilton's equations of motion

$$\frac{dx_j}{dx_k} = \frac{\partial \mathscr{H}}{\partial p_j} = -\frac{\partial p_k}{\partial p_j}, \quad \frac{dp_j}{dx_k} = -\frac{\partial \mathscr{H}}{\partial x_j} = \frac{\partial p_k}{\partial x_j}, \quad j(\neq k) = 1,2,3,$$

$$-\frac{dt}{dx_k} = \frac{\partial \mathscr{H}}{\partial E} = -\frac{\partial p_k}{\partial E}, \quad \frac{dE}{dx_k} = \frac{\partial \mathscr{H}}{\partial t} = -\frac{\partial p_k}{\partial t}. \quad (2.86)$$

Introducing the notations $x_0 = -t$ and $p_0 = E$, we can define a Poisson bracket

$$\{f,g\}_{x_k} = \sum_{j=0, \neq k}^{3} \left(\frac{\partial f}{\partial x_j} \frac{\partial g}{\partial p_j} - \frac{\partial f}{\partial p_j} \frac{\partial g}{\partial x_j} \right), \quad (2.87)$$

and write the equations of motion (2.86) as

$$\frac{dx_j}{dx_k} = \{-\mathscr{H}, x_j\}_{x_k} = \{p_k, x_j\}_{x_k},$$

$$\frac{dp_j}{dx_k} = \{-\mathscr{H}, p_j\}_{x_k} = \{p_k, p_j\}_{x_k},$$

$$j(\neq k) = 0, 1, 2, 3. \quad (2.88)$$

It is straightforward to write down the equations analogous to (2.52) and (2.64) in this case.

2.1.4.2 Example: Dynamics of a Charged Particle in a Constant Magnetic Field

As an example of changing the independent variable, we shall consider the case of a charged particle moving in a constant magnetic field. We shall write $\underline{x} = \vec{r} = (\vec{r}_\perp, z)$ and $\underline{p} = \vec{p} = (\vec{p}_\perp, p_z)$. We can write the principle of stationary action as

$$\delta S = \delta \left[\int_{t_i}^{t_f} dt \, (p_x \dot{x} + p_y \dot{y} + p_z \dot{z} - H(\vec{r}, \vec{p}, t)) \right] = 0. \quad (2.89)$$

If we now choose z as the independent variable, instead of t, we can write (2.89) as follows: with $z_i = z(t_i)$, $z_f = z(t_f)$,

$$\delta S = \delta \left\{ \int_{z_i}^{z_f} dz \left[p_x \frac{dx}{dz} + p_y \frac{dy}{dz} - E \frac{dt}{dz} - \mathscr{H} \right] \right\} = 0, \quad (2.90)$$

where

$$\mathscr{H} = -p_z(\vec{r}_\perp, \vec{p}_\perp, E, t, z), \quad (2.91)$$

A Review of Classical Mechanics

with p_z obtained by solving the equation $H(\vec{r},\vec{p},t) = E$ and E denoting the total energy of the particle. The new Hamiltonian is $\mathcal{H}$, and the corresponding Hamilton's equations of motion are

$$\frac{dx}{dz} = \frac{\partial \mathcal{H}}{\partial p_x} = -\frac{\partial p_z}{\partial p_x}, \qquad \frac{dy}{dz} = \frac{\partial \mathcal{H}}{\partial p_y} = -\frac{\partial p_z}{\partial p_y},$$
$$\frac{dp_x}{dz} = -\frac{\partial \mathcal{H}}{\partial x} = \frac{\partial p_z}{\partial x}, \qquad \frac{dp_y}{dz} = -\frac{\partial \mathcal{H}}{\partial y} = \frac{\partial p_z}{\partial y},$$
$$-\frac{dt}{dz} = \frac{\partial \mathcal{H}}{\partial E} = -\frac{\partial p_z}{\partial E}, \qquad \frac{dE}{dz} = \frac{\partial \mathcal{H}}{\partial t} = -\frac{\partial p_z}{\partial t}. \qquad (2.92)$$

Note that the evolution of a system from a time t_i to a time t_f through Hamilton's equations of motion with a Hamiltonian H corresponds to $t_f > t_i$, *i.e.*, in the forward time direction. Similarly, the z-evolution of a system with $-p_z$ as the Hamiltonian would correspond to the propagation of the system in the forward z, or $+z$, direction.

For a charged particle moving in a constant magnetic field $\vec{B} = B\vec{k}$, the Hamiltonian is time-independent and is given by

$$H(\vec{r},\vec{p}) = \left\{ m^2 c^4 + c^2 \left[\left(p_x + \frac{1}{2} qBy \right)^2 + \left(p_y - \frac{1}{2} qBx \right)^2 + p_z^2 \right] \right\}^{1/2}, \qquad (2.93)$$

representing the total energy E of the particle. Solving for p_z, we can write the Hamiltonian for z-evolution as

$$\mathcal{H} = -p_z$$
$$= -\frac{1}{c} \left\{ E^2 - m^2 c^4 - c^2 \left[\left(p_x + \frac{1}{2} qBy \right)^2 + \left(p_y - \frac{1}{2} qBx \right)^2 \right] \right\}^{1/2}. \qquad (2.94)$$

The corresponding canonical equations of motion (2.92) for z-evolution become

$$\frac{dx}{dz} = -\frac{\partial p_z}{\partial p_x} = \frac{1}{p_z} \left(p_x + \frac{1}{2} qBy \right) = \frac{v_x}{v_z},$$
$$\frac{dy}{dz} = -\frac{\partial p_z}{\partial p_y} = \frac{1}{p_z} \left(p_y - \frac{1}{2} qBx \right) = \frac{v_y}{v_z},$$
$$\frac{dp_x}{dz} = \frac{\partial p_z}{\partial x} = \frac{qB}{2p_z} \left(p_y - \frac{1}{2} qBx \right) = \frac{qBv_y}{2v_z},$$
$$\frac{dp_y}{dz} = \frac{\partial p_z}{\partial y} = -\frac{qB}{2p_z} \left(p_x + \frac{1}{2} qBy \right) = -\frac{qBv_x}{2v_z},$$
$$-\frac{dt}{dz} = -\frac{\partial p_z}{\partial E} = -\frac{E}{c^2 p_z} = -\frac{1}{v_z},$$
$$\frac{dE}{dz} = -\frac{\partial p_z}{\partial t} = 0. \qquad (2.95)$$

These equations can be easily verified with the help of the earlier discussions on the dynamics of a charged particle in an electromagnetic field and a constant magnetic field (see (2.42), (2.44, (2.68)). The last equation expresses the constancy of the total energy of the particle along the trajectory.

For any observable of the charged particle moving in the constant magnetic field, $O(\vec{r}_\perp, \vec{p}_\perp, t, E, z)$, we can write

$$\frac{dO}{dz} = \{-\mathscr{H}, O\}_z + \frac{\partial O}{\partial z} = \{p_z, O\}_z + \frac{\partial O}{\partial z}, \qquad (2.96)$$

where

$$\{f, g\}_z = \left[\left(\frac{\partial f}{\partial x}\frac{\partial g}{\partial p_x} - \frac{\partial f}{\partial p_x}\frac{\partial g}{\partial x}\right) + \left(\frac{\partial f}{\partial y}\frac{\partial g}{\partial p_y} - \frac{\partial f}{\partial p_y}\frac{\partial g}{\partial y}\right) \right.$$
$$\left. - \left(\frac{\partial f}{\partial t}\frac{\partial g}{\partial E} - \frac{\partial f}{\partial E}\frac{\partial g}{\partial t}\right)\right]. \qquad (2.97)$$

For an observable O not explicitly dependent on z,

$$\frac{dO}{dz} = \{-\mathscr{H}, O\}_z = \{p_z, O\}_z. \qquad (2.98)$$

Introducing the notation

$$\{f, g\}_z =: f :_z g, \qquad (2.99)$$

we would have, on integrating (2.98),

$$O(z) = \left[e^{(z-z_i):-\mathscr{H}:_z}O\right](z_i)$$
$$= \left[e^{(z-z_i):p_z:_z}O\right](z_i)$$
$$= \left[\sum_{n=0}^{\infty} \frac{(z-z_i)^n}{n!} : p_z :_z^n O\right](z_i), \qquad (2.100)$$

where z_i is the initial point of the trajectory, and z is any point of observation on the trajectory.

Applying the relation (2.100) to $O = x + iy$, we have, with $z_i = 0$, and with the help of (2.95),

$$[x(z) + iy(z)] = [e^{z:p_z:_z}(x+iy)](0)$$
$$= [x(0) + iy(0)]$$
$$+ \left[\sum_{n=1}^{\infty} (-i\omega_c)^{n-1} \frac{z^n}{n! v_z^n}\right][v_x + iv_y](0)$$
$$= [x(0) + iy(0)]$$
$$+ \frac{i}{\omega_c}\left[e^{(-i\omega_c z/v_z)} - 1\right][v_x(0) + iv_y(0)]. \qquad (2.101)$$

Similarly, we get
$$E(z) = [e^{z:p_z:z}E](0) = E(0), \tag{2.102}$$
showing that the total energy of the particle is a constant of motion, and
$$t(z) = [e^{z:p_z:z}t](0) = t(0) + z\left(\frac{E}{c^2 p_z}\right)(0) = \frac{z}{v_z(0)}, \tag{2.103}$$

where $t(z)$ refers to the time when the particle reaches the position at z if it starts at $z = 0$ at $t = 0$. From (2.101) and (2.103), we have

$$x(z) = (x(0) + \rho_L) - \rho_L \cos\left(\frac{\omega_c z}{v_\parallel}\right),$$

$$y(z) = y(0) + \rho_L \sin\left(\frac{\omega_c z}{v_\parallel}\right),$$

$$t(z) = \frac{z}{v_\parallel}, \tag{2.104}$$

where we have taken $v_x(0) = 0, v_y(0) = v_\perp, v_z(0) = v_\parallel$ as earlier. The trajectory is seen to be a helix of radius $\rho_L = v_\perp/\omega_c$ and pitch $\wp = 2\pi v_\parallel/\omega_c$ such that $x(z+\wp) = x(z), y(z+\wp) = y(z)$.

2.1.5 CANONICAL TRANSFORMATIONS

2.1.5.1 Basic Theory

Once we have a Hamiltonian of a particle with a set of phase space variables and an independent variable, it is sometimes desirable to change the set of phase space variables to suit a particular description of the evolution of the state of the particle preserving the form of Hamilton's equations of motion. Such a transformation of the phase space variables is called a canonical transformation. We shall denote the independent variable as t generally; it can be z or any other independent variable of the particle.

For a particle with the Hamiltonian $H(\underline{x}, \underline{p}, t)$, and the equations of motion

$$\underline{\dot{x}} = \frac{\partial H}{\partial \underline{p}}, \qquad \underline{\dot{p}} = -\frac{\partial H}{\partial \underline{x}}, \tag{2.105}$$

let us consider a change of phase space coordinates given by

$$x_j \longrightarrow X_j(\underline{x}, \underline{p}, t), \qquad p_j \longrightarrow P_j(\underline{x}, \underline{p}, t), \qquad j = 1, 2, 3. \tag{2.106}$$

If there exists a corresponding Hamiltonian, say $K(\underline{X}, \underline{P}, t)$, such that the resulting Hamilton's equations of motion are in the canonical form, i.e.,

$$\underline{\dot{X}} = \frac{\partial K}{\partial \underline{P}}, \qquad \underline{\dot{P}} = -\frac{\partial K}{\partial \underline{X}}, \tag{2.107}$$

then the transformation (2.106) is said to be a canonical transformation. As a simple example, let us consider the following scale transformation:

$$x_j \longrightarrow X_j = \frac{x_j}{\mu_x}, \qquad p_j \longrightarrow P_j = \frac{p_j}{\mu_p}, \qquad \text{for all } j = 1,2,3. \qquad (2.108)$$

If we take the corresponding Hamiltonian to be given by

$$K(\underline{X},\underline{P},t) = \frac{1}{\mu_x \mu_p} H(\mu_x \underline{X}, \mu_p \underline{P}, t), \qquad (2.109)$$

then the equations in (2.107) are obviously satisfied in view of (2.105). This implies that we must have

$$\delta \left\{ \int_{t_i}^{t_f} dt \left[\sum_{j=1}^{3} P_j \dot{X}_j - K(\underline{X},\underline{P},t) \right] \right\} = 0, \qquad (2.110)$$

where $\underline{\delta X}(t_i) = 0$, $\underline{\delta P}(t_i) = 0$, $\underline{\delta X}(t_f) = 0$, $\underline{\delta P}(t_f) = 0$, since $\underline{\delta x}(t_i) = 0$, $\underline{\delta p}(t_i) = 0$, $\underline{\delta x}(t_f) = 0$, $\underline{\delta p}(t_f) = 0$. This is seen to happen because

$$\left[\sum_{j=1}^{3} P_j \dot{X}_j - K(\underline{X},\underline{P},t) \right] = \frac{1}{\mu_x \mu_p} \left[\sum_{j=1}^{3} p_j \dot{x}_j - H(\underline{x},\underline{p},t) \right]. \qquad (2.111)$$

Now, let us consider a nonsimple example. Let a function of the old coordinates and new coordinates, and possibly time, $F_1(\underline{x},\underline{X},t)$, be such that

$$p_j = \frac{\partial F_1}{\partial x_j}, \qquad P_j = -\frac{\partial F_1}{\partial X_j}, \qquad j = 1,2,3. \qquad (2.112)$$

Define

$$K(\underline{X},\underline{P},t) = H(\underline{x},\underline{p},t) + \frac{\partial F_1}{\partial t}, \qquad (2.113)$$

where $\underline{x}$ and $\underline{p}$ have been substituted in terms of $\underline{X}$ and $\underline{P}$ on the right-hand side. Note that from the first part of the equation (2.112), it is possible to solve for all X_js in terms of all x_js and p_js, and from the second part of the equation (2.112), it is possible to solve for all P_js. Now, observe that

A Review of Classical Mechanics

$$\sum_{j=1}^{3} P_j \dot{X}_j - K + \frac{dF_1}{dt} = \sum_{j=1}^{3} P_j \dot{X}_j - \left(H + \frac{\partial F_1}{\partial t} \right)$$

$$+ \left(\sum_{j=1}^{3} \frac{\partial F_1}{\partial x_j} \dot{x}_j + \sum_{j=1}^{3} \frac{\partial F_1}{\partial X_j} \dot{X}_j + \frac{\partial F_1}{\partial t} \right)$$

$$= \sum_{j=1}^{3} \frac{\partial F_1}{\partial x_j} \dot{x}_j - H$$

$$- \frac{\partial F_1}{\partial t} + \left[\frac{\partial F_1}{\partial t} + \sum_{j=1}^{3} \left(P_j \dot{X}_j + \frac{\partial F_1}{\partial X_j} \dot{X}_j \right) \right]$$

$$= \sum_{j=1}^{3} p_j \dot{x}_j - H. \qquad (2.114)$$

Thus, we have

$$\delta \left\{ \int_{t_i}^{t_f} dt \left[\sum_{j=1}^{3} p_j \dot{x}_j - H \right] \right\} = \delta \left\{ \int_{t_i}^{t_f} dt \left[\sum_{j=1}^{3} P_j \dot{X}_j - K + \frac{dF_1}{dt} \right] \right\}$$

$$= \delta \left\{ \int_{t_i}^{t_f} dt \left[\sum_{j=1}^{3} P_j \dot{X}_j - K \right] \right\} + \delta F_1 \Big|_{t_i}^{t}$$

$$= \delta \left\{ \int_{t_i}^{t_f} dt \left[\sum_{j=1}^{3} P_j \dot{X}_j - K \right] \right\} = 0, \qquad (2.115)$$

leading to the canonical equations of motion (2.107) in terms of the new phase space coordinates $\underline{X}$ and $\underline{P}$. Here, $\delta F_1 \big|_{t_i}^{t_f}$ vanishes because $\underline{\delta x}(t_i) = 0$, and $\underline{\delta X}(t_i) = 0$, since we also have $\underline{\delta p}(t_i) = 0$. The function $F_1(\underline{x}, \underline{X}, t)$ is called a generating function of a canonical transformation.

Let us see another example of a generating function of a canonical transformation. Let $F_2(\underline{x}, \underline{P}, t)$ be such that

$$p_j = \frac{\partial F_2}{\partial x_j}, \qquad X_j = \frac{\partial F_2}{\partial P_j}, \qquad j = 1, 2, 3, \qquad (2.116)$$

and define

$$K(\underline{X}, \underline{P}, t) = H(\underline{x}, \underline{p}, t) + \frac{\partial F_2}{\partial t}. \qquad (2.117)$$

Now, observe

$$\sum_{j=1}^{3} P_j \dot{X}_j - K + \frac{dF}{dt} = \sum_{j=1}^{3} p_j \dot{x}_j - H, \qquad (2.118)$$

where

$$F = F_2(\underline{x}, \underline{P}, t) - \sum_{j=1}^{3} X_j P_j. \qquad (2.119)$$

Thus,

$$\delta\left\{\int_{t_i}^{t_f} dt \left[\sum_{j=1}^{3} p_j \dot{x}_j - H\right]\right\} = \delta\left\{\int_{t_i}^{t_f} dt \left[\sum_{j=1}^{3} P_j \dot{X}_j - K + \frac{dF}{dt}\right]\right\}$$

$$= \delta\left\{\int_{t_i}^{t_f} dt \left[\sum_{j=1}^{3} P_j \dot{X}_j - K\right]\right\} + \delta F\Big|_{t_i}^{t_f}$$

$$= \delta\left\{\int_{t_i}^{t_f} dt \left[\sum_{j=1}^{3} P_j \dot{X}_j - K\right]\right\} = 0, \quad (2.120)$$

leading to the canonical equations of motion (2.107). In this case, F_2 is the generating function. Note that if we take the generating function to be

$$F_2(\underline{x}, \underline{P}, t) = \sum_{j=1}^{3} x_j P_j \quad (2.121)$$

then it leads to a canonical transformation

$$p_j = \frac{\partial F_2}{\partial x_j} = P_j, \qquad X_j = \frac{\partial F_2}{\partial P_j} = x_j, \qquad j = 1, 2, 3, \quad (2.122)$$

i.e., $\underline{X} = \underline{x}$ and $\underline{P} = \underline{p}$, which is the identity transformation.

There are other choices for the generating functions. One can choose

$$F = F_3(\underline{p}, \underline{X}, t) + \sum_{j=1}^{3} x_j p_j, \quad (2.123)$$

with

$$x_j = -\frac{\partial F_3}{\partial p_j}, \qquad P_j = -\frac{\partial F_3}{\partial X_j}, \qquad j = 1, 2, 3, \quad (2.124)$$

and

$$K = H + \frac{\partial F_3}{\partial t}. \quad (2.125)$$

Another choice is

$$F = F_4(\underline{p}, \underline{P}, t) + \sum_{j=1}^{3} x_j p_j - \sum_{j=1}^{3} X_j P_j, \quad (2.126)$$

with

$$x_j = -\frac{\partial F_4}{\partial p_j}, \qquad X_j = -\frac{\partial F_4}{\partial P_j}, \qquad j = 1, 2, 3 \quad (2.127)$$

and

$$K = H + \frac{\partial F_4}{\partial t}. \quad (2.128)$$

The aforementioned four types do not exhaust all the possibilities. One may also choose a mixture of the four types. For example, for a particle moving in a plane, one can choose $F = F'(x_1, X_2, P_1, p_2, t) - X_1 P_1 + x_2 p_2$. Then, to have

$$P_1 \dot{X}_1 + P_2 \dot{X}_2 - K + \frac{dF}{dt} = p_1 \dot{x}_1 + p_2 \dot{x}_2 - H, \tag{2.129}$$

we must have

$$p_1 = \frac{\partial F'}{\partial x_1}, \quad x_2 = -\frac{\partial F'}{\partial p_2},$$
$$X_1 = \frac{\partial F'}{\partial P_1}, \quad P_2 = -\frac{\partial F'}{\partial X_2},$$
$$K = H + \frac{\partial F'}{\partial t}. \tag{2.130}$$

2.1.5.2 Optical Hamiltonian of a Charged Particle Moving Through an Electromagnetic Optical Element with a Straight Axis

As we have seen earlier, the Hamiltonian of a charged particle moving in an electromagnetic field is given by

$$H = \sqrt{m^2 c^4 + c^2 \vec{\pi}^2} + q\phi. \tag{2.131}$$

Let us consider an optical element of a charged particle beam optical system comprising a time-independent electromagnetic field and having its optic axis along the z-direction. If we now change the independent variable from t to z and replace H by E, as discussed earlier, the z-evolution Hamiltonian for a particle of a charged particle beam propagating through the optical element in the forward z-direction will be

$$\mathcal{H} = -p_z = -\frac{1}{c}\sqrt{(E - q\phi)^2 - m^2 c^4 - c^2 \vec{\pi}_\perp^2} - qA_z. \tag{2.132}$$

This $\mathcal{H}$ is the basic Hamiltonian, the classical charged particle beam optical Hamiltonian, for studying the optical behavior of any electromagnetic optical element with a straight optic axis along the z-direction. Often one makes further transformations for convenience. If we make a scale transformation as

$$\vec{p} = \frac{\vec{p}}{p_0}, \quad \vec{a} = \frac{q\vec{A}}{p_0}, \tag{2.133}$$

where p_0 is the magnitude of the design momentum, the corresponding z-evolution Hamiltonian, or the beam optical Hamiltonian, will become

$$\mathcal{H} = -\sqrt{\frac{(E - q\phi)^2 - m^2 c^4}{c^2 p_0^2} - \left[(\bar{p}_x - a_x)^2 + (\bar{p}_y - a_y)^2\right]} - a_z. \tag{2.134}$$

Note that $E = p_t$ and $-t$ are canonically conjugate variables. A further change of phase space variables is made usually in accelerator beam dynamics:

$$x \longrightarrow x, \qquad y \longrightarrow y, \qquad -t \longrightarrow \zeta = z - v_0 t$$
$$\bar{p}_x \longrightarrow \bar{p}_x, \qquad \bar{p}_y \longrightarrow \bar{p}_y, \qquad \bar{p}_t \longrightarrow p_\zeta = \frac{E - E_0}{v_0 p_0}, \qquad (2.135)$$

where E_0 is the energy of the reference particle with momentum p_0 and velocity v_0. This transformation is effected through the generating function

$$F_2(x, y, -t, \bar{p}_x, \bar{p}_y, p_\zeta, z) = x\bar{p}_x + y\bar{p}_y + (z - v_0 t)\left(p_\zeta + \frac{E_0}{v_0 p_0}\right), \qquad (2.136)$$

such that

$$\frac{\partial F_2}{\partial x} = \bar{p}_x, \qquad \frac{\partial F_2}{\partial y} = \bar{p}_y, \qquad \frac{\partial F_2}{\partial (-t)} = \bar{p}_t = \frac{E}{p_0},$$
$$\frac{\partial F_2}{\partial \bar{p}_x} = x, \qquad \frac{\partial F_2}{\partial \bar{p}_y} = y, \qquad \frac{\partial F_2}{\partial p_\zeta} = \zeta. \qquad (2.137)$$

The resulting z-evolution Hamiltonian becomes

$$\widetilde{\mathscr{H}}(x, y, \zeta, \bar{p}_x, \bar{p}_y, p_\zeta, z) = \mathscr{H} + \frac{\partial F_2}{\partial z}$$
$$= p_\zeta - a_z - \sqrt{\frac{(E - q\phi)^2 - m^2 c^4}{c^2 p_0^2} - \left[(\bar{p}_x - a_x)^2 + (\bar{p}_y - a_y)^2\right]}, \qquad (2.138)$$

after dropping the constant additional term $E_0/v_0 p_0$. The longitudinal position coordinate $\zeta = z - v_0 t$ is a measure of delay in the arrival time of the tracked particle relative to the reference particle at the position z of the optics axis, namely, the z-axis. The longitudinal momentum coordinate p_ζ is a measure of deviation in the energy of the tracked particle relative to the reference particle, i.e., $v_0 p_0 p_\zeta = \Delta E = E - E_0$.

In what follows, we shall consider only the propagation of monoenergetic charged particle beams through optical elements comprising constant electromagnetic fields. In such systems, the energy will be conserved, and the beam optics is concerned only with the transfer maps across the optical elements for the 4-dimensional phase space comprising the transverse position coordinates and momentum components. For studying the beam dynamics of systems in which the energy of the particle changes during the propagation of the beam, like in an accelerating cavity, we will have to consider the transfer maps for the 6-dimensional phase space, including the longitudinal position coordinate ζ and the longitudinal momentum coordinate p_ζ.

A Review of Classical Mechanics

2.1.6 SYMPLECTICITY OF CANONICAL TRANSFORMATIONS

2.1.6.1 Time-Independent Canonical Transformations

Let us now analyse the conditions which a phase space coordinate transformation has to satisfy in order to be canonical. First, we shall consider a transformation

$$(\underline{x},\underline{p}) \longrightarrow (\underline{X}(\underline{x},\underline{p}),\underline{P}(\underline{x},\underline{p})) \quad \text{or,} \quad \underline{\varphi} \longrightarrow \underline{\Phi}(\underline{\varphi}) \qquad (2.139)$$

which does not involve time, or the independent variable, which we will generally denote by t. The transformation is independent of the particle Hamiltonian. If the dynamics of a particle is described in terms of the phase space variables $\underline{\varphi}$ and the Hamiltonian $H(\underline{\varphi},t)$, then its canonical equations of motion are given by

$$\underline{\dot{\varphi}} = J \frac{\partial H}{\partial \underline{\varphi}}. \qquad (2.140)$$

Let the Hamiltonian of the particle become $K(\underline{\Phi},t)$ under the given transformation (2.139). For this transformation to be canonical, we should have

$$\underline{\dot{\Phi}} = J \frac{\partial K}{\partial \underline{\Phi}}. \qquad (2.141)$$

From the transformation relations (2.139), and the equations of motion (2.140) for $\underline{\varphi}$, we can write

$$\dot{\Phi}_j = \sum_{k=1}^{6} \frac{\partial \Phi_j}{\partial \varphi_k} \dot{\varphi}_k = \sum_{k=1}^{6} \sum_{l=1}^{6} \frac{\partial \Phi_j}{\partial \varphi_k} J_{kl} \frac{\partial H}{\partial \varphi_l}, \qquad (2.142)$$

or

$$\underline{\dot{\Phi}} = M J \frac{\partial H}{\partial \underline{\varphi}}, \qquad (2.143)$$

where the elements of the 6×6 matrix M, the Jacobian matrix of the transformation, are given by

$$M_{jk} = \frac{\partial \Phi_j}{\partial \varphi_k}, \qquad j,k = 1,2,\ldots,6. \qquad (2.144)$$

Observing that

$$\frac{\partial H}{\partial \varphi_j} = \sum_{k=1}^{6} \frac{\partial H}{\partial \Phi_k} \frac{\partial \Phi_k}{\partial \varphi_j} = \sum_{k=1}^{6} M_{kj} \frac{\partial H}{\partial \Phi_k}, \qquad (2.145)$$

or

$$\frac{\partial H}{\partial \underline{\varphi}} = \widetilde{M} \frac{\partial H}{\partial \underline{\Phi}}, \qquad (2.146)$$

we get

$$\underline{\dot{\Phi}} = M J \widetilde{M} \frac{\partial H}{\partial \underline{\Phi}}. \qquad (2.147)$$

From the discussion on canonical transformations in terms of generating functions, we know already that for any time-independent transformation, the new Hamiltonian is the same as the old Hamiltonian expressed in terms of the new variables, using the inverse of the transformation (2.139), *i.e.* $K(\underline{\Phi}) = H\left(\underline{\varphi}(\underline{\Phi})\right)$. Thus, the equation (2.147) becomes

$$\underline{\dot{\Phi}} = MJ\widetilde{M}\frac{\partial K}{\partial \underline{\Phi}}. \qquad (2.148)$$

Comparison of this equation with (2.141) shows that for any time-independent transformation of the phase space coordinates to be canonical, the corresponding Jacobian matrix M should satisfy the condition

$$MJ\widetilde{M} = J, \qquad (2.149)$$

known as the symplectic condition. Any matrix M satisfying the condition (2.149) is called a symplectic matrix. Note that in view of the relations

$$\left(MJ\widetilde{M}\right)^{-1} = \widetilde{M}^{-1}J^{-1}M^{-1} = J^{-1}, \quad J^{-1} = -J, \qquad (2.150)$$

the symplectic condition (2.149) can also be written as

$$\widetilde{M}JM = J. \qquad (2.151)$$

As we shall see later, this is the condition also for time-dependent transformations.

We have taken $K(\underline{\Phi}) = H\left(\underline{\varphi}(\underline{\Phi})\right)$ in the above discussion. In case the transformation involves a scale transformation, we should take $K(\underline{\Phi}) = \lambda H\left(\underline{\varphi}(\underline{\Phi})\right)$ where λ is a constant (see (2.109), where $\lambda = 1/\mu_x\mu_p$). Then, the condition for canonical transformation (2.149) becomes, obviously, $MJ\widetilde{M} = \lambda J$, or $\widetilde{M}JM = \lambda J$. If a canonical transformation corresponds to $\lambda \neq 1$, it is called an extended canonical transformation.

2.1.6.2 Time-Dependent Canonical Transformations: Hamiltonian Evolution

Let us now consider a phase space coordinate transformation

$$\underline{\varphi} = (\underline{x},\underline{p}) \longrightarrow \underline{\Phi}\left(\underline{\varphi},t\right) = \left(\underline{X}(\underline{x},\underline{p},t),\underline{P}(\underline{x},\underline{p},t)\right) \qquad (2.152)$$

which depends continuously on t, representing time or the chosen independent variable. In the transformation, $\underline{\Phi}(t) \longrightarrow \underline{\Phi}(t+\delta t)$ during an infinitesimal time interval δt the change in $\underline{\Phi}$, namely $\delta\underline{\Phi}$, should result from a transformation close to identity. So, we shall take the generating function of this transformation to be

$$F_2 = \sum_{j=1}^{3} x_j P_j + \delta t G(\underline{x},\underline{P},t), \qquad (2.153)$$

A Review of Classical Mechanics

differing only infinitesimally from the identity transformation. Then, the relation between the old and the new coordinates is given by

$$X_j = x_j(t + \delta t) = \frac{\partial F_2}{\partial P_j} = x_j + \delta t \frac{\partial G}{\partial P_j}, \qquad j = 1,2,3,$$

$$p_j(t) = \frac{\partial F_2}{\partial x_j} = P_j + \delta t \frac{\partial G}{\partial x_j} = p_j(t + \delta t) + \delta t \frac{\partial G}{\partial x_j},$$

$$j = 1,2,3. \qquad (2.154)$$

In other words,

$$\delta x_j = x_j(t + \delta t) - x_j = \delta t \frac{\partial G}{\partial P_j}, \qquad j = 1,2,3,$$

$$\delta p_j = p_j(t + \delta t) - p_j = -\delta t \frac{\partial G}{\partial x_j}, \qquad j = 1,2,3. \qquad (2.155)$$

Since P_j differs from p_j only infinitesimally, we can take $G(\underline{x}, \underline{P}, t) = G(\underline{x}, \underline{p}, t)$, and $\partial G / \partial P_j = \partial G / \partial p_j$ for all $j = 1, 2, 3$. The function $G(\underline{x}, \underline{p}, t)$ is called the generating function of the infinitesimal canonical transformation. Then, we can write (2.155) as

$$\delta \underline{\varphi} = \delta t J \frac{\partial G}{\partial \underline{\varphi}}. \qquad (2.156)$$

It is clear that

$$\underline{\Phi} = \underline{\varphi}(t + \delta t) = \underline{\varphi} + \delta \underline{\varphi} = \underline{\varphi} + \delta t J \frac{\partial G}{\partial \underline{\varphi}}. \qquad (2.157)$$

Thus, the Jacobian matrix of the transformation is

$$M = \frac{\partial \underline{\Phi}}{\partial \underline{\varphi}} = I + \frac{\partial \delta \underline{\varphi}}{\partial \underline{\varphi}} = I + \delta t J \frac{\partial^2 G}{\partial \underline{\varphi} \partial \underline{\varphi}}, \qquad (2.158)$$

where I is the 6×6 identity matrix. Since J is an antisymmetric matrix and $\partial^2 G / \partial \underline{\varphi} \partial \underline{\varphi}$, with elements $\partial^2 G / \partial \varphi_j \partial \varphi_k$, is a symmetric matrix, we have

$$\widetilde{M} = I - \delta t \frac{\partial^2 G}{\partial \underline{\varphi} \partial \underline{\varphi}} J. \qquad (2.159)$$

As a result we have, upto first order in δt,

$$MJ\widetilde{M} = \left(I + \delta t J \frac{\partial^2 G}{\partial \underline{\varphi} \partial \underline{\varphi}} \right) J \left(I - \delta t \frac{\partial^2 G}{\partial \underline{\varphi} \partial \underline{\varphi}} J \right)$$

$$\approx J + \delta t J \frac{\partial^2 G}{\partial \underline{\varphi} \partial \underline{\varphi}} J - \delta t J \frac{\partial^2 G}{\partial \underline{\varphi} \partial \underline{\varphi}} J$$

$$= J, \qquad (2.160)$$

which is the symplectic condition for the transformation to be canonical. Thus, a canonical transformation during an infinitesimal time interval δt satisfies the symplectic condition, i.e., $M J \widetilde{M} = J$. Let M_1 and M_2 be two symplectic matrices, i.e., $M_1 J \widetilde{M_1} = J$ and $M_2 J \widetilde{M_2} = J$. Note that $M_2 M_1 J \widetilde{M_2 M_1} = M_2 M_1 J \widetilde{M_1} \widetilde{M_2} = J$, and similarly $M_1 M_2 J \widetilde{M_1 M_2} = J$, i.e., any product of symplectic matrices is also a symplectic matrix. From this, it follows that since a succession of infinitesimal transformations leads to a finite transformation, with the Jacobian matrix of the finite transformation being equal to the product of the symplectic Jacobian matrices of the individual transformations, any time-dependent canonical transformation $\underline{\varphi}(t_i) \longrightarrow \underline{\Phi}\left(\underline{\varphi}, t_i + t\right)$ will satisfy the symplectic condition.

Let us now take the Hamiltonian of the particle $H(\underline{x}, \underline{p}, t)$ to be the generating function of an infinitesimal canonical transformation depending on time continuously. Then, for a time interval δt, we have, as seen from (2.156),

$$\delta \underline{\varphi} = \delta t J \frac{\partial H}{\partial \underline{\varphi}}, \qquad (2.161)$$

which implies

$$\frac{\delta \underline{\varphi}}{\delta t} = J \frac{\partial H}{\partial \underline{\varphi}}. \qquad (2.162)$$

On taking the limit $\delta t \longrightarrow 0$, these equations become just Hamilton's equations of motion:

$$\underline{\dot{\varphi}} = J \frac{\partial H}{\partial \underline{\varphi}}. \qquad (2.163)$$

This shows that the Hamiltonian evolution of the state of a particle, i.e., time evolution of the trajectory of the particle following Hamilton's equations of motion, is a continuous canonical transformation generated by the Hamiltonian. Thus, if the phase space variables of a particle have values $(\underline{x}(t_i), \underline{p}(t_i))$ at time t_i, and values $(\underline{x}(t_i + t), \underline{p}(t_i + t))$ at time $t_i + t$, then the time evolution in phase space, $(\underline{x}(t_i), \underline{p}(t_i)) \longrightarrow (\underline{x}(t_i + t), \underline{p}(t_i + t))$, is a canonical transformation obeying the symplectic condition. This makes the transfer map (2.64), a symplectic map.

While deriving the optical Hamiltonian of a charged particle moving through an electromagnetic optical element with a straight optic axis, we have made two transformations of the corresponding phase space. Let us check how these transformations satisfy the conditions for being canonical. The first is a scale transformation:

$$\underline{\varphi} = (x, y, z, p_x, p_y, p_z) \longrightarrow \underline{\Phi} = \left(x, y, z, \frac{p_x}{p_0}, \frac{p_y}{p_0}, \frac{p_z}{p_0}\right). \qquad (2.164)$$

The Jacobian matrix of this transformation has the elements

$$M_{jk} = \begin{cases} \delta_{jk} & j = 1, 2, 3 \ k = 1, 2, \ldots, 6, \\ \frac{\delta_{jk}}{p_0} & j = 4, 5, 6 \ k = 1, 2, \ldots, 6. \end{cases} \qquad (2.165)$$

A Review of Classical Mechanics

satisfying the condition $MJ\widetilde{M} = \lambda J$ with $\lambda = 1/p_0$, i.e., it is an extended canonical transformation. The second is the transformation:

$$\underline{\varphi} = \left(x, y, -t, \frac{p_x}{p_0}, \frac{p_y}{p_0}, \frac{E}{p_0}\right) \longrightarrow \underline{\Phi} = \left(x, y, z - v_0 t, \frac{p_x}{p_0}, \frac{p_y}{p_0}, \frac{E - E_0}{v_0 p_0}\right). \quad (2.166)$$

The corresponding Jacobian matrix has the elements

$$M_{jk} = \begin{cases} \delta_{jk} & j = 1, 2, 4, 5 \ \ k = 1, 2, \ldots, 6, \\ v_0 \delta_{jk} & j = 3 \ \ k = 1, 2, \ldots, 6, \\ \frac{\delta_{jk}}{v_0} & j = 6 \ \ k = 1, 2, \ldots, 6, \end{cases} \quad (2.167)$$

and satisfies the symplectic condition for a canonical transformation, namely, $MJ\widetilde{M} = J$.

2.1.6.3 Canonical Invariants: Poisson Brackets

As we have seen earlier, the Poisson bracket of two functions of the phase space variables of a particle, say f and g, representing two observables of the particle, is given by

$$\{f, g\}_{\underline{\varphi}} = \widetilde{\frac{\partial f}{\partial \underline{\varphi}}} J \frac{\partial g}{\partial \underline{\varphi}}, \quad (2.168)$$

where the subscript indicates that the Poisson bracket is evaluated with respect to the set of phase space variables $\underline{\varphi}$. If we make a canonical transformation $\underline{\varphi} \longrightarrow \underline{\Phi}(\underline{\varphi})$, then the Poisson bracket of f and g with respect to $\underline{\Phi}$ is given by

$$\{f, g\}_{\underline{\Phi}} = \widetilde{\frac{\partial f}{\partial \underline{\Phi}}} J \frac{\partial g}{\partial \underline{\Phi}}. \quad (2.169)$$

If M is the Jacobian matrix of the transformation, i.e., $M_{jk} = \partial \Phi_j / \partial \varphi_k$, then

$$\widetilde{\frac{\partial f}{\partial \underline{\varphi}}} = \widetilde{\frac{\partial f}{\partial \underline{\Phi}}} M, \quad \frac{\partial f}{\partial \underline{\varphi}} = \widetilde{M} \frac{\partial f}{\partial \underline{\Phi}}, \quad (2.170)$$

and M obeys the symplectic condition $MJ\widetilde{M} = J$ since the transformation is canonical. In view of these relations, Equation (2.168) becomes

$$\{f, g\}_{\underline{\varphi}} = \widetilde{\frac{\partial f}{\partial \underline{\Phi}}} M J \widetilde{M} \frac{\partial g}{\partial \underline{\Phi}} = \widetilde{\frac{\partial f}{\partial \underline{\Phi}}} J \frac{\partial g}{\partial \underline{\Phi}} = \{f, g\}_{\underline{\Phi}}, \quad (2.171)$$

showing that the Poisson bracket of any two functions of the phase space variables is an invariant under canonical transformations, i.e., it has the same value irrespective of with respect to which set of phase space variables it is calculated if the sets of variables are related to each other by canonical transformations.

For example, let us calculate the fundamental Poisson brackets between the new phase space variables $\underline{\Phi}$ with respect to the old phase space variables $\underline{\varphi}$. We have

$$\{\Phi_k, \Phi_l\}_{\underline{\varphi}} = \widetilde{\frac{\partial \Phi_k}{\partial \underline{\varphi}}} J \frac{\partial \Phi_l}{\partial \underline{\varphi}} = \sum_{r=1}^{6} \sum_{s=1}^{6} M_{kr} J_{rs} M_{ls} = \left(M J \widetilde{M}\right)_{kl}$$
$$= J_{kl} = \{\Phi_k, \Phi_l\}_{\underline{\Phi}}, \qquad k, l = 1, 2, \ldots, 6, \qquad (2.172)$$

showing the invariance of the fundamental Poisson brackets under canonical transformations. This fact can also be used to test whether a given transformation of the phase space coordinates is canonical. For example, for the canonical transformation (2.166), it is easy to verify that

$$\{\Phi_k, \Phi_l\}_{\underline{\varphi}} = \{\Phi_k, \Phi_l\}_{\underline{\Phi}} = J_{kl}, \qquad k, l = 1, 2, \ldots, 6. \qquad (2.173)$$

2.2 DYNAMICS OF A SYSTEM OF PARTICLES

Though we shall be dealing only with single particle dynamics, for the sake of completeness, let us see how to extend the Lagrangian and the Hamiltonian formalisms of single particle dynamics to a system of particles like a charged particle beam, which is a system of identical particles. In classical mechanics, identical particles can be distinguished, labeled, and tracked individually.

Let us consider a system of N particles. The Lagrangian of the system will be a function of $3N$ configuration space coordinates $\underline{x} = \left\{x_j^{(i)} \mid i = 1, 2, \ldots, N;\ j = 1, 2, 3\right\}$ and their time derivatives $\underline{\dot{x}} = \left\{\dot{x}_j^{(i)}\right\}$ where the subscript j corresponds to the position coordinates of a particle and the superscript (i) corresponds to the labeling of the particles. The trajectories of all the particles of the system will follow the same principle of stationary action where the action has the same definition

$$S[\underline{x}(t)] = \int_{t_i}^{t_f} dt L(\underline{x}(t), \underline{\dot{x}}(t), t), \qquad (2.174)$$

as in the single particle case, except that now $\underline{x}$ and $\underline{\dot{x}}$ have $3N$ components. Consequently, the Euler–Lagrange equations of motion for the system become

$$\frac{d}{dt}\left(\frac{\partial L}{\partial \dot{x}_j^{(i)}}\right) = \frac{\partial L}{\partial x_j^{(i)}}, \qquad i = 1, 2, \ldots, N;\ j = 1, 2, 3, \qquad (2.175)$$

or

$$\frac{d}{dt}\left(\frac{\partial L}{\partial \underline{\dot{x}}}\right) = \frac{\partial L}{\partial \underline{x}}. \qquad (2.176)$$

Defining the canonically conjugate momentum coordinates by

$$p_j^{(i)} = \frac{\partial L}{\partial \dot{x}_j^{(i)}}, \qquad i = 1, 2, \ldots, N;\ j = 1, 2, 3, \qquad (2.177)$$

the Hamiltonian of the system is given by

$$H(\underline{x},\underline{p}) = \sum_{i=1}^{N}\sum_{j=1}^{3} \dot{x}_j^{(i)} p_j^{(i)} - L(\underline{x},\underline{\dot{x}}), \qquad (2.178)$$

where $\underline{p} = \{p_j^{(i)} \mid i = 1,2,\ldots,N;\ j = 1,2,3\}$. As a consequence of Hamilton's principle of stationary action in phase space, the canonical equations of motion for the system become

$$\dot{x}_j^{(i)} = \frac{\partial H}{\partial p_j^{(i)}}, \qquad \dot{p}_j^{(i)} = -\frac{\partial H}{\partial x_j^{(i)}}, \qquad i = 1,2,\ldots,N;\ j = 1,2,3, \qquad (2.179)$$

or

$$\underline{\dot{x}} = \frac{\partial H}{\partial \underline{p}}, \qquad \underline{\dot{p}} = -\frac{\partial H}{\partial \underline{x}}. \qquad (2.180)$$

Now, the phase space of the system is $6N$-dimensional. Let us represent a point in the phase space by

$$\underline{\varphi} = (\underline{x},\underline{p}), \qquad (2.181)$$

where

$$\underline{x} = \{x_1, x_2, \ldots, x_{3N}\}, \qquad x_{3(i-1)+j} = x_j^{(i)},$$
$$i = 1, 2, \ldots, N;\ j = 1, 2, 3,$$
$$\underline{p} = \{p_1, p_1, \ldots, p_{3N}\}, \qquad p_{3(i-1)+j} = p_j^{(i)},$$
$$i = 1, 2, \ldots, N;\ j = 1, 2, 3. \qquad (2.182)$$

Then,

$$\varphi_j = x_j, \quad j = 1, 2, \ldots, 3N, \qquad \varphi_{3N+j} = p_j, \quad j = 1, 2, \ldots, 3N. \qquad (2.183)$$

Hamilton's equations of motion can be written as

$$\underline{\dot{\varphi}} = J \frac{\partial H}{\partial \underline{\varphi}}, \qquad (2.184)$$

where J is the antisymmetric matrix

$$J = \begin{pmatrix} O & I \\ -I & O \end{pmatrix}, \qquad (2.185)$$

with O as the $3N \times 3N$ null matrix and I as the $3N \times 3N$ identity matrix.

A coordinate transformation in the phase space,

$$\underline{\varphi} = (\underline{x},\underline{p}) \longrightarrow \underline{\Phi}(\underline{\varphi},t) = (\underline{X}(\underline{x},\underline{p},t), \underline{P}(\underline{x},\underline{p},t)), \qquad (2.186)$$

will be a canonical transformation, *i.e.*, will preserve the form of Hamilton's equations of motion (2.184), if the symplectic condition is satisfied:

$$MJ\widetilde{M} = J, \tag{2.187}$$

where M is the $6N$-dimensional Jacobian matrix of the transformation with elements

$$M_{kl} = \frac{\partial \Phi_k}{\partial \varphi_l}, \quad k,l = 1,2,\ldots,6N. \tag{2.188}$$

For any two functions $f(\underline{x},\underline{p})$ and $g(\underline{x},\underline{p})$, the Poisson bracket is defined by

$$\{f,g\}_{\underline{\varphi}} = \widetilde{\frac{\partial f}{\partial \varphi}} J \frac{\partial g}{\partial \varphi} = \sum_{j=1}^{3N} \left\{ \frac{\partial f}{\partial x_j} \frac{\partial g}{\partial p_j} - \frac{\partial f}{\partial p_j} \frac{\partial g}{\partial x_j} \right\}$$

$$= \sum_{i=1}^{N} \sum_{j=1}^{3} \left\{ \frac{\partial f}{\partial x_j^{(i)}} \frac{\partial g}{\partial p_j^{(i)}} - \frac{\partial f}{\partial p_j^{(i)}} \frac{\partial g}{\partial x_j^{(i)}} \right\}. \tag{2.189}$$

The Poisson bracket is a canonical invariant, *i.e.*, its value is the same whether it is calculated with respect to a set of phase space variables $\underline{\varphi}$ or with respect to a set of phase space variables $\underline{\Phi}$ related to $\underline{\varphi}$ by a canonical transformation. Time evolution of the system under Hamilton's equations of motion is a continuous canonical transformation. The transfer map,

$$\underline{\varphi}(t) = \left[e^{(t-t_i):-H:} \underline{\varphi} \right](t_i)$$

$$= \left[\sum_{n=0}^{\infty} \frac{(t-t_i)^n}{n!} :-H:^n \underline{\varphi} \right](t_i), \tag{2.190}$$

with $:-H:f = \{-H,f\}$, is a symplectic map.

Besides the Poisson brackets, there are several other canonical invariants. Let us look at one more such canonical invariant, namely, the magnitude of a volume element in the phase space of the system of particles. When we make a canonical transformation, $\underline{\varphi} = (\underline{x},\underline{p}) \longrightarrow \underline{\Phi}(\underline{\varphi},t) = (\underline{X}(\underline{x},\underline{p},t),\underline{P}(\underline{x},\underline{p},t))$ the volume element

$$(d\varphi) = dx_1 dx_2 \ldots dx_{3N} dp_1 dp_2 \ldots dp_{3N} \tag{2.191}$$

transforms into

$$(d\Phi) = dX_1 dX_2 \ldots dX_{3N} dP_1 dP_2 \ldots dP_{3N}. \tag{2.192}$$

As is well known from the rule for change of variables in multivariate calculus, these volume elements are related:

$$(d\Phi) = |\det(M)|(d\varphi), \tag{2.193}$$

where M is the Jacobian matrix of the transformation, $\det(M)$ is the determinant of M, and $|\det(M)|$ is the absolute value of $\det(M)$. From the relation (2.187), we get

$$\det\left(MJ\widetilde{M}\right) = \det(M)\det(J)\det\left(\widetilde{M}\right) = \det(M)^2 \det(J) = \det(J), \tag{2.194}$$

and hence we find that $|\det(M)| = 1$. Thus, $(d\Phi) = (d\varphi)$, or

$$dx_1 dx_2 \ldots dx_{3N} dp_1 dp_2 \ldots dp_{3N} = dX_1 dX_2 \ldots dX_{3N} dP_1 dP_2 \ldots dP_{3N}. \quad (2.195)$$

This proves the canonical invariance of the volume element in phase space for an N-particle system evolving in time according to Hamilton's equations. It follows that the volume of any arbitrary region in phase space $\int \cdots \int (d\varphi)$ occupied by an N-particle system evolving in time according to Hamilton's equations is a canonical invariant.

3 An Introductory Review of Quantum Mechanics

3.1 INTRODUCTION

All physical phenomena are quantum mechanical at the fundamental level. Classical mechanics we observe in the macroscopic world is an approximation. That is why it raises a curiosity when we find classical mechanics to be very successful at a microscopic level in the design and operation of beam devices like an electron microscope or a particle accelerator. In this chapter, we give an introductory review of quantum mechanics just at the level necessary for presenting the formalism of quantum charged particle beam optics. For more details on any topic in quantum mechanics at introductory and advanced levels, see, *e.g.*, Cohen-Tannoudji, Diu, and Loloë [25], Esposito, Marmo, Miele, and Sudarshan [39], Greiner [64], Griffiths and Schroetter [67], Sakurai and Napolitano [166], Shankar [169], Bjorken and Drell [13], Greiner [65], and Parthasarathy [142].

In classical mechanics, we specify the state of a particle by its position and velocity vectors in configuration space (Lagrangian description), or by its position and canonical momentum vectors in phase space (Hamiltonian description). This classical description, based on experience in the macroscopic domain, breaks down in the microscopic—molecular, atomic, nuclear, and subnuclear—domains. Experimental realization of the failure of classical mechanics in the atomic domain led to the discovery of quantum mechanics, which shows that it is not possible to know, or determine precisely, both the position and momentum of a particle like a single electron or even a single atom in a molecule. Of course, quantum mechanics approaches classical mechanics, as should be, in the macroscopic domain. With the impossibility of knowing precisely both the position $\vec{r}$ and canonical momentum $\vec{p}$ of a particle, at any time t, it has been only possible to specify the state of the particle by a function, $\Psi(\vec{r},t)$, which is in general a complex function of $\vec{r}$ and t. We shall be using the right-handed Cartesian coordinate system unless stated otherwise. This function $\Psi(\vec{r},t)$ is called the wave function of the particle. It does not represent any time-dependent physical wave in space but represents the amplitude of a probability wave. It has been found that $\Psi(\vec{r},t)$ is the probability amplitude for the particle to have its position at $\vec{r}$, at time t, if its position is determined, *i.e.*, $|\Psi(\vec{r},t)|^2 = \Psi^*(\vec{r},t)\Psi(\vec{r},t)$ gives the probability of finding the particle at $\vec{r}$ if its position is determined at time t. Since the particle has to be found somewhere in the entire space the wave function $\Psi(\vec{r},t)$ of a particle has to satisfy, at any time t, the normalization condition

$$\int_{-\infty}^{\infty}\int_{-\infty}^{\infty}\int_{-\infty}^{\infty} dx\,dy\,dz\, |\Psi(\vec{r},t)|^2 = 1. \tag{3.1}$$

Since $|\Psi(\vec{r},t)|^2 dxdydz$ can be interpreted as the probability of finding the particle within an infinitesimal volume element $dxdydz$ around the point at $\vec{r}$, the expression $|\Psi(\vec{r},t)|^2$ is called the position probability density. If there are N particles with the same wave function $\Psi(\vec{r},t)$, then we can say that the number of particles found at time t in an infinitesimal volume $dxdydz$ around the point $\vec{r}$ will be $N|\Psi(\vec{r},t)|^2 dxdydz$. Thus, when dealing with a beam of particles, we can identify $|\psi(\vec{r},t)|^2$ with the intensity of the beam at the position $\vec{r}$ at time t. Hereafter, we shall write

$$\int_{-\infty}^{\infty}\int_{-\infty}^{\infty}\int_{-\infty}^{\infty} dxdydz \ \{ \quad \} = \int d^3r \ \{ \quad \}, \qquad (3.2)$$

with $d^3r = dxdydz$, without specifying the limits whenever the integral is over the entire space. We shall now see how the quantum mechanics of a particle is described in terms of the wave function $\Psi(\vec{r},t)$. Quantum theory is essentially based on a set of postulates, supported by experimental verifications of their consequences, without any indication of any violation so far. First, we shall summarize the principles of quantum kinematics and then consider the quantum dynamics.

3.2 GENERAL FORMALISM OF QUANTUM MECHANICS

3.2.1 SINGLE PARTICLE QUANTUM MECHANICS: FOUNDATIONAL PRINCIPLES

3.2.1.1 Quantum Kinematics

▶ Any observable of a particle, taking real values on measurement, is to be represented by a Hermitian operator, say $\widehat{O}$. Time t is a parameter as in classical physics.

Notes:
The Hermitian conjugate, or adjoint, of an operator $\widehat{O}$, denoted by $\widehat{O}^\dagger$, is defined by the relation

$$\int d^3r \, f^*(\vec{r})\widehat{O}g(\vec{r}) = \int d^3r \left(\widehat{O}^\dagger f(\vec{r})\right)^* g(\vec{r}), \qquad (3.3)$$

for any arbitrary $f(\vec{r})$ and $g(\vec{r})$. Here $*$ denotes the complex conjugate. It is understood that the integral is over the entire space, as in (3.2), if the limits are not specified explicitly. Taking complex conjugate of both sides of this equation (3.3), we have

$$\int d^3r \left(\widehat{O}g(\vec{r})\right)^* f(\vec{r}) = \int d^3r \, g(\vec{r})^* \widehat{O}^\dagger f(\vec{r})$$
$$= \int d^3r \left(\left(\widehat{O}^\dagger\right)^\dagger g(\vec{r})\right)^* f(\vec{r}), \qquad (3.4)$$

implying that $\left(\widehat{O}^\dagger\right)^\dagger = \widehat{O}$. A Hermitian operator $\widehat{H}$ is defined by the condition

$$\widehat{H}^\dagger = \widehat{H}, \qquad (3.5)$$

or
$$\int d^3r \, f^*(\vec{r})\widehat{H}g(\vec{r}) = \left\{\int d^3r \, g^*(\vec{r})\widehat{H}f(\vec{r})\right\}^* \tag{3.6}$$

for any arbitrary $f(\vec{r})$ and $g(\vec{r})$. This implies that for a Hermitian operator $\widehat{H}$, $\int d^3r \, f^*(\vec{r})\widehat{H}f(\vec{r})$ is real. Such Hermitian operators have only real eigenvalues, *i.e.*, the equation

$$\widehat{H}\varphi_n(\vec{r}) = \lambda_n \varphi_n(\vec{r}) \tag{3.7}$$

admits only real eigenvalues λ_n. Further, the eigenfunctions $\varphi_n(\vec{r})$ and $\varphi_{n'}(\vec{r})$ corresponding to two distinct eigenvalues λ_n and $\lambda_{n'}$, respectively, are orthogonal in the sense that

$$\int d^3r \, \varphi_n^*(\vec{r})\varphi_{n'}(\vec{r}) = 0. \tag{3.8}$$

A nondegenerate eigenvalue λ_n will have only one eigenfunction $\varphi_n(\vec{r})$. A degenerate eigenvalue has more than one eigenfunctions, and all those degenerate eigenfunctions can be chosen to be orthogonal to each other. Thus, all the eigenfunctions of a Hermitian operator can be chosen to be orthonormal:

$$\int d^3r \, \varphi_n^*(\vec{r})\varphi_{n'}(\vec{r}) = \delta_{nn'}. \tag{3.9}$$

Also, the set of all orthonormal eigenfunctions of a Hermitian operator form a complete set, such that

$$\sum_n \varphi_n(\vec{r})\varphi_n^*(\vec{r}') = \delta(\vec{r}-\vec{r}'), \tag{3.10}$$

where $\delta(\vec{r}-\vec{r}')$ is the three-dimensional Dirac delta function given by the definition

$$\psi(\vec{r}) = \int d^3r' \, \delta(\vec{r}-\vec{r}') \, \psi(\vec{r}'), \tag{3.11}$$

for any $\psi(\vec{r})$. As a result, any function $\psi(\vec{r})$ can be expanded in terms of the complete set of orthonormal eigenfunctions of a Hermitian operator, $\{\varphi_n(\vec{r})\}$, as

$$\psi(\vec{r}) = \sum_n a_n \varphi_n(\vec{r}), \tag{3.12}$$

with

$$a_n = \int d^3r \, \varphi_n(\vec{r})^* \psi(\vec{r}). \tag{3.13}$$

This is seen to follow from (3.10) and (3.11). If $\psi(\vec{r})$ is a zero function, then all a_ns are zero, showing the linear independence of the basis functions $\{\varphi_n(\vec{r})\}$. For more details on Hermitian operators, see *e.g.*, Arfken, Weber, and Harris [3], and Byron and Fuller [18].

The set of all real eigenvalues $\{\lambda_n\}$ of the Hermitian operator $\widehat{O}$ corresponding to an observable O is called the spectrum of the observable O. The spectrum of an observable can be completely discrete or continuous or mixed. If the wave function

$\Psi(\vec{r},t)$ of a particle is expanded in terms of $\{\varphi_n(\vec{r})\}$, the complete set of orthonormal eigenfunctions of an observable, we have

$$\Psi(\vec{r},t) = \sum_n a_n(t)\varphi_n(\vec{r}), \tag{3.14}$$

with

$$a_n(t) = \int d^3r\, \varphi_n(\vec{r})^*\Psi(\vec{r},t). \tag{3.15}$$

Now, the normalization condition (3.1) becomes, in view of (3.9),

$$\int d^3r\, \Psi^*(\vec{r},t)\Psi(\vec{r},t) = \sum_n \sum_{n'} a_n^*(t)a_{n'}(t)\int d^3r\, \varphi_n^*(\vec{r})\varphi_{n'}(\vec{r})$$

$$= \sum_n |a_n(t)|^2 = \sum_n \left|\int d^3r\, \varphi_n^*(\vec{r})\Psi(\vec{r},t)\right|^2 = 1. \tag{3.16}$$

▶ The operators corresponding to the Cartesian position coordinates of a particle (x,y,z) are just multiplication by (x,y,z), respectively, i.e.,

$$\widehat{x} = x\times, \qquad \widehat{y} = y\times, \qquad \widehat{z} = z\times, \tag{3.17}$$

or

$$\widehat{x}\psi(\vec{r}) = x\psi(\vec{r}), \qquad \widehat{y}\psi(\vec{r}) = y\psi(\vec{r}), \qquad \widehat{z}\psi(\vec{r}) = z\psi(\vec{r}), \tag{3.18}$$

for any arbitrary $\psi(\vec{r})$. The operators corresponding to the Cartesian components of linear momentum of the particle, canonically conjugate to the respective Cartesian position coordinates, are

$$\widehat{p}_x = -i\hbar\frac{\partial}{\partial x}, \qquad \widehat{p}_y = -i\hbar\frac{\partial}{\partial y}, \qquad \widehat{p}_z = -i\hbar\frac{\partial}{\partial z}, \tag{3.19}$$

or

$$\widehat{p}_x\psi(\vec{r}) = -i\hbar\frac{\partial\psi(\vec{r})}{\partial x}, \qquad \widehat{p}_y\psi(\vec{r}) = -i\hbar\frac{\partial\psi(\vec{r})}{\partial y}, \qquad \widehat{p}_z\psi(\vec{r}) = -i\hbar\frac{\partial\psi(\vec{r})}{\partial z}, \tag{3.20}$$

where $\hbar = h/2\pi$, and $h = 6.62607004 \times 10^{-34}$ J·s is Planck's constant. We can write $\vec{\widehat{r}}\psi(\vec{r}) = \vec{r}\psi(\vec{r})$ and $\vec{\widehat{p}}\psi(\vec{r}) = -i\hbar\vec{\nabla}\psi(\vec{r})$.

Notes:
Hermiticity of position operators is obvious. For $\widehat{x}$, we have

$$\int dx\, f^*(x)\widehat{x}g(x) = \int dx\, f^*(x)xg(x)$$

$$= \left(\int dx\, g^*(x)xf(x)\right)^* = \left(\int dx\, g^*(x)\widehat{x}f(x)\right)^*, \tag{3.21}$$

A Review of Quantum Mechanics

and we get similar results for $\hat{y}$ and $\hat{z}$. Hermiticity of $\hat{p}_x$ is seen as follows:

$$\int dx\, f^*(x)\hat{p}_x g(x) = \int dx\, f^*(x)\left(-i\hbar\frac{\partial g(x)}{\partial x}\right)$$

$$= \int dx\left(-i\hbar\frac{\partial}{\partial x}\right)(f^*(x)g(x)) + \int dx\left(i\hbar\frac{\partial f^*(x)}{\partial x}\right)g(x)$$

$$= f^*(x)g(x)\big|_{-\infty}^{\infty} + \int dx\left(i\hbar\frac{\partial f^*(x)}{\partial x}\right)g(x)$$

$$= \left[\int dx\, g^*(x)\left(-i\hbar\frac{\partial f(x)}{\partial x}\right)\right]^*$$

$$= \left(\int dx\, g^*(x)\hat{p}_x f(x)\right)^*, \tag{3.22}$$

where we have assumed that $f(x) \longrightarrow 0$ and $g(x) \longrightarrow 0$ when $x \longrightarrow \pm\infty$ as required of functions normalizable in the sense of (3.1). Similar results for $\hat{p}_y$ and $\hat{p}_z$ prove their Hermiticity. Representing $\hat{\vec{r}}$ by $\vec{r}$ and $\hat{\vec{p}}$ by $-i\hbar\vec{\nabla}$ is known as the position representation. It is seen that the operators $\hat{x}$ and $\hat{p}_x$ do not commute with each other, *i.e.*, $\hat{x}\hat{p}_x\psi(\vec{r}) \neq \hat{p}_x\hat{x}\psi(\vec{r})$ for any arbitrary $\psi(\vec{r})$. Similarly, $\hat{y}$ and $\hat{z}$ do not commute with $\hat{p}_y$ and $\hat{p}_z$, respectively. The precise commutation relation between $\hat{x}$ and $\hat{p}_x$ is seen to be

$$(\hat{x}\hat{p}_x - \hat{p}_x\hat{x})\psi(\vec{r}) = \hat{x}\hat{p}_x\psi(\vec{r}) - \hat{p}_x\hat{x}\psi(\vec{r})$$

$$= -i\hbar\left[x\frac{\partial\psi(\vec{r})}{\partial x} - \frac{\partial}{\partial x}(x\psi(\vec{r}))\right] = i\hbar\psi(\vec{r}). \tag{3.23}$$

Since this relation is valid for any arbitrary $\psi(\vec{r})$, we write

$$\hat{x}\hat{p}_x - \hat{p}_x\hat{x} = [\hat{x},\hat{p}_x] = i\hbar, \tag{3.24}$$

which is known as the Heisenberg canonical commutation relation. The expression

$$\left[\hat{O}_1,\hat{O}_2\right] = \hat{O}_1\hat{O}_2 - \hat{O}_2\hat{O}_1 \tag{3.25}$$

is known as the commutator of $\hat{O}_1$ and $\hat{O}_2$. Similarly, we have

$$[\hat{y},\hat{p}_y] = i\hbar, \qquad [\hat{z},\hat{p}_z] = i\hbar. \tag{3.26}$$

The commutators $[\hat{x},\hat{p}_y]$, $[\hat{y},\hat{p}_x]$, etc., vanish. If we write $(\hat{x},\hat{y},\hat{z}) = (\hat{x}_1,\hat{x}_2,\hat{x}_3)$ and $(\hat{p}_x,\hat{p}_y,\hat{p}_z) = (\hat{p}_1,\hat{p}_2,\hat{p}_3)$, then we have

$$[\hat{x}_i,\hat{x}_j] = 0, \quad [\hat{p}_i,\hat{p}_j] = 0, \quad [\hat{x}_i,\hat{p}_j] = i\hbar\delta_{ij}, \qquad i,j = 1,2,3. \tag{3.27}$$

To evaluate $\left[\hat{O}_1,\hat{O}_2\right]$, one has to find $\left(\hat{O}_1\hat{O}_2 - \hat{O}_2\hat{O}_1\right)\psi(\vec{r})$ for an arbitrary $\psi(\vec{r})$. If we replace the minus $(-)$ sign in the commutator bracket by the plus $(+)$ sign, we get the anticommutator bracket:

$$\left\{\hat{O}_1,\hat{O}_2\right\} = \hat{O}_1\hat{O}_2 + \hat{O}_2\hat{O}_1. \tag{3.28}$$

If $\hat{O}_1$ and $\hat{O}_2$ are two operators $\left(\hat{O}_1\hat{O}_2\right)^\dagger = \hat{O}_2^\dagger\hat{O}_1^\dagger$. So, if two Hermitian operators $\hat{O}_1$ and $\hat{O}_2$ commute with eath other, the product operator $\hat{O}_1\hat{O}_2$ is a Hermitian operator: $\left(\hat{O}_1\hat{O}_2\right)^\dagger = \hat{O}_2^\dagger\hat{O}_1^\dagger = \hat{O}_2\hat{O}_1 = \hat{O}_1\hat{O}_2$. If the two Hermitian operators $\hat{O}_1$ and $\hat{O}_2$ do not commute with each other, the product operator $\hat{O}_1\hat{O}_2$ is not Hermitian operator. This has to be taken into account while forming the quantum mechanical operators corresponding to classical observables. If an operator, say $\hat{O}$, is not Hermitian, it can be written as the sum of a Hermitian and an anti-Hermitian operator:

$$\hat{O} = \hat{O}_H + \hat{O}_A, \quad \hat{O}_H = \frac{1}{2}\left(\hat{O}+\hat{O}^\dagger\right), \quad \hat{O}_A = \frac{1}{2}\left(\hat{O}-\hat{O}^\dagger\right),$$
$$\hat{O}_H^\dagger = \hat{O}_H, \quad \hat{O}_A^\dagger = -\hat{O}_A. \tag{3.29}$$

We shall call $\hat{O}_H$ and $\hat{O}_A$ as the Hermitian and anti-Hermitian parts of $\hat{O}$, respectively. For example, corresponding to the classical observable xp_x, the quantum operator cannot be $\hat{x}\hat{p}_x$, since $\hat{x}$ and $\hat{p}_x$ do not commute and hence $\hat{x}\hat{p}_x$ is not Hermitian. We can take the corresponding quantum operator to be the symmetrical combination $\frac{1}{2}\left(\hat{x}\hat{p}_x + (\hat{x}\hat{p}_x)^\dagger\right) = \frac{1}{2}(\hat{x}\hat{p}_x + \hat{p}_x\hat{x}) = \frac{1}{2}\{\hat{x},\hat{p}_x\}$. Note that the antisymmetrical combination, $\frac{1}{2}\left(\hat{x}\hat{p}_x - (\hat{x}\hat{p}_x)^\dagger\right) = \frac{1}{2}(\hat{x}\hat{p}_x - \hat{p}_x\hat{x}) = \frac{1}{2}[\hat{x},\hat{p}_x] = \frac{1}{2}i\hbar$ is anti-Hermitian. In general, the 'Classical $\longrightarrow$ Quantum' correspondence rule used for getting the Hermitian quantum operator corresponding to any classical observable $O(\vec{r},\vec{p},t)$ is: Replace $\vec{r}$ and $\vec{p}$ by $\hat{\vec{r}}$ and $\hat{\vec{p}}$, respectively, and make the resulting expression Hermitian, if it is not already Hermitian. To make an operator Hermitian, one may use the result that $\frac{1}{2}\left(\hat{O}+\hat{O}^\dagger\right)$ is Hermitian, or some ordering rule like the Weyl ordering rule:

$$\text{classical } x^m p_x^n \longrightarrow \text{quantum } \frac{1}{2^m}\sum_{j=0}^{m}\binom{m}{j}\hat{x}^{m-j}\hat{p}_x^n\hat{x}^j. \tag{3.30}$$

The algebra of commutators is exactly similar to the algebra of the Poisson brackets (2.56). We have

$$[\hat{A},\hat{B}] = -[\hat{B},\hat{A}],$$
$$[\hat{A},\hat{B}\hat{C}] = [\hat{A},\hat{B}]\hat{C} + \hat{B}[\hat{A},\hat{C}],$$
$$[a\hat{A},b\hat{B}+c\hat{C}] = ab[\hat{A},\hat{B}] + ac[\hat{A},\hat{C}],$$
$$[\hat{A},[\hat{B},\hat{C}]] + [\hat{B},[\hat{C},\hat{A}]] + [\hat{C},[\hat{A},\hat{B}]] = 0, \tag{3.31}$$

where a, b, and c are constants, and the last relation is the Jacobi identity for commutators. Note that, using the antisymmetry of the commutator, we can rewrite the Jacobi identity as

$$[\hat{A},[\hat{B},\hat{C}]] - [\hat{B},[\hat{A},\hat{C}]] = [[\hat{A},\hat{B}],\hat{C}]. \tag{3.32}$$

It may be noted that while the commutator of any two Hermitian operators is anti-Hermitian, the anticommutator of any two Hermitian operators is Hermitian.

A Review of Quantum Mechanics

Besides the position representation, we can have other equivalent representations too. For example, we have the momentum representation in which, with $\widetilde{\psi}(\vec{p})$ as any momentum space function, we have

$$\widehat{p}_x \widetilde{\psi}(\vec{p}) = p_x \widetilde{\psi}(\vec{p}), \quad \widehat{p}_y \widetilde{\psi}(\vec{p}) = p_y \widetilde{\psi}(\vec{p}), \quad \widehat{p}_z \widetilde{\psi}(\vec{p}) = p_z \widetilde{\psi}(\vec{p}), \tag{3.33}$$

and

$$\widehat{x}\widetilde{\psi}(\vec{p}) = i\hbar \frac{\partial \widetilde{\psi}(\vec{p})}{\partial p_x}, \quad \widehat{y}\widetilde{\psi}(\vec{p}) = i\hbar \frac{\partial \widetilde{\psi}(\vec{p})}{\partial p_y}, \quad \widehat{z}\widetilde{\psi}(\vec{p}) = i\hbar \frac{\partial \widetilde{\psi}(\vec{p})}{\partial p_z}. \tag{3.34}$$

Thus, in momentum representation

$$\vec{\widehat{r}} = i\hbar \vec{\nabla}_{\vec{p}}, \quad \vec{\widehat{p}} = \vec{p} \times, \tag{3.35}$$

acting on functions in $\vec{p}$-space. Note that the Heisenberg commutation relations are valid in the same form (3.27) in this representation also. One can work in any representation of the operators which preserves the Heisenberg canonical commutation relations. In the momentum representation, the wave function would be a function of momentum of the particle, say $\widetilde{\Psi}(\vec{p},t)$, such that $\left|\widetilde{\Psi}(\vec{p},t)\right|^2$ gives the probability of finding the particle to have the momentum $\vec{p}$, at time t, if its momentum is measured.

It would be useful to note down the following commutation relations:

$$[\widehat{p}_x, f(\vec{r})] = -i\hbar \frac{\partial f(\vec{r})}{\partial x}, \quad [\widehat{p}_y, f(\vec{r})] = -i\hbar \frac{\partial f(\vec{r})}{\partial y}, \quad [\widehat{p}_z, f(\vec{r})] = -i\hbar \frac{\partial f(\vec{r})}{\partial z},$$

$$\left[\widehat{x}, f\left(\vec{\widehat{p}}\right)\right] = i\hbar \frac{\partial \left(f\left(\vec{\widehat{p}}\right)\right)}{\partial \widehat{p}_x}, \quad \left[\widehat{y}, f\left(\vec{\widehat{p}}\right)\right] = i\hbar \frac{\partial \left(f\left(\vec{\widehat{p}}\right)\right)}{\partial \widehat{p}_y}, \quad \left[\widehat{z}, f\left(\vec{\widehat{p}}\right)\right] = i\hbar \frac{\partial \left(f\left(\vec{\widehat{p}}\right)\right)}{\partial \widehat{p}_z}. \tag{3.36}$$

The first set of relations follow by finding $(\widehat{p}_x f(\vec{r}) - f(\vec{r}) \widehat{p}_x) \psi(\vec{r})$, etc., for an arbitrary $\psi(\vec{r})$. The second set of relations follow from the observation that in the momentum representation $f\left(\vec{\widehat{p}}\right) = f(\vec{p})$ and $\widehat{x} = i\hbar(\partial/\partial p_x)$, etc. Thus, the second set of relations are got by calculating them first in the momentum representation and then converting the results to the position representation. Another way to obtain the second set of relations is to use, repeatedly, the relation $[\widehat{x}, \widehat{p}_x^n] = i\hbar \widehat{p}_x^{n-1} + \widehat{p}_x [\widehat{x}, \widehat{p}_x^{n-1}]$, got using the second of the relations in (3.31), and note that the final result can be written as $[\widehat{x}, \widehat{p}_x^n] = i\hbar (\partial \widehat{p}_x^n/\partial \widehat{p}_x)$.[1]

▶ If a particle has the wave function $\Psi(\vec{r},t)$, at time t, and an observable O of the particle is measured, then O will be found to have only one of the eigenvalues, $\{\lambda_n\}$, of the corresponding Hermitian operator $\widehat{O}$. Let us assume that the eigenvalues $\{\lambda_n\}$ are all distinct and nondegenerate, *i.e.*, to each eigenvalue λ_n, there is only one

---

[1] Since $\widehat{x}$, $\widehat{y}$, and $\widehat{z}$ are simply multiplications by x, y, and z, respectively, we shall hereafter write them simply as x, y, and z, and write $\vec{\widehat{r}}$ as $\vec{r}$, unless it is necessary to use the operator ($\widehat{}$) notation.

eigenfunction $\varphi_n(\vec{r})$ such that $\hat{O}\varphi_n(\vec{r}) = \lambda_n \varphi_n(\vec{r})$. The probability that O is found to have a value λ_n is given by

$$P(O=\lambda_n) = \left| \int d^3r \, \varphi_n^*(\vec{r}) \Psi(\vec{r},t) \right|^2. \tag{3.37}$$

If the eigenvalue λ_n is d_n-fold degenerate, there would be d_n eigenfunctions, say $\{\varphi_{nj}(\vec{r}) | j=1,2,\ldots,d_n\}$, such that $\hat{O}\varphi_{nj}(\vec{r}) = \lambda_n \varphi_{nj}(\vec{r})$, for all $j=1,2,\ldots,d_n$. In that case, the probability that O is found to have a value λ_n is given by

$$P(O=\lambda_n) = \sum_{j=1}^{d_n} \left| \int d^3r \, \varphi_{nj}^*(\vec{r}) \Psi(\vec{r},t) \right|^2. \tag{3.38}$$

Note that

$$\sum_n P(O=\lambda_n) = 1, \tag{3.39}$$

as should be, in view of the normalization of the wave function (3.16).

Notes:

For the position operator $\hat{x}$, the eigenvalue equation $\hat{x}\xi(x',x) = x'\xi(x',x)$ has the solution $\xi(x',x) = \delta(x-x')$, i.e., $x\delta(x-x') = x'\delta(x-x')$, $-\infty < x' < \infty$, where $\delta(x-x')$ is the one-dimensional Dirac delta function defined by

$$\int_{-\infty}^{\infty} dx' \, \delta(x-x')\psi(x') = \psi(x). \tag{3.40}$$

Taking $\psi(x') = 1$, we see that

$$\int_{-\infty}^{\infty} dx' \, \delta(x-x') = 1. \tag{3.41}$$

Effectively,

$$\delta(x-x') = \begin{cases} \longrightarrow \infty & \text{at } x=x', \\ 0 & \text{for } x \neq x'. \end{cases} \tag{3.42}$$

The Dirac delta function is a generalization of the Kronecker delta symbol $\delta_{nn'}$ and the defining relation (3.11), or (3.40), is analogous to $\sum_{n'} \delta_{nn'} \psi_{n'} = \psi_n$. Note that $\delta^*(x-x') = \delta(x-x')$. Also, $\delta(x-x') = \delta(x'-x)$, and hence we can write (3.40) also as

$$\int_{-\infty}^{\infty} dx' \, \delta(x'-x)\psi(x') = \psi(x). \tag{3.43}$$

The Dirac delta function can arise in several ways. For example,

$$\delta(x-x') = \frac{1}{2\pi} \int_{-\infty}^{\infty} dk \, e^{ik(x-x')},$$

$$\delta(x-x') = \lim_{\sigma \to 0} \frac{1}{\sqrt{2\pi}\sigma} e^{-\frac{(x-x')^2}{2\sigma^2}}. \tag{3.44}$$

A Review of Quantum Mechanics

Further, the orthonormality relation for these position eigenfunctions takes the form

$$\int_{-\infty}^{\infty} dx\, \delta^*(x-x')\delta(x-x'') = \delta(x'-x''). \tag{3.45}$$

The eigenfunctions of $\hat{y}$ and $\hat{z}$ will be similarly given by $\{\delta(y-y')\,|\,-\infty < y' < \infty\}$ and $\{\delta(z-z')\,|\,-\infty < z' < \infty\}$, respectively, with similar orthonormality properties. With

$$\delta(\vec{r}-\vec{r}') = \delta(x-x')\delta(y-y')\delta(z-z'), \tag{3.46}$$

we have the orthonormality relation

$$\int d^3r\, \delta^*(\vec{r}-\vec{r}')\delta(\vec{r}-\vec{r}'') = \delta(\vec{r}'-\vec{r}''). \tag{3.47}$$

Note that the position eigenfunction $\delta(\vec{r}-\vec{r}')$ cannot be normalized in the sense of (3.1) since the position eigenvalues $\{\vec{r}'\}$ span the entire continuous space and $\int d^3r\, |\delta(\vec{r}-\vec{r}')|^2 \longrightarrow \infty$ for any $\vec{r}'$. This is because if a particle has the wave function $\delta(\vec{r}-\vec{r}')$, at any time t, it is localized at the position $\vec{r}'$ at that instant, and the probability of finding it at that position becomes ∞ (!), and the probability of finding it elsewhere is zero. This happens whenever we have continuous eigenvalues for any observable, and the normalization of corresponding eigenfunctions has to be done in the sense of (3.47). It is to be observed that the three position coordinates of any particle are still assumed to take all continuous values from $-\infty$ to ∞, like in classical physics, without any quantization. May be the space itself is quantized at a deeper level (see, e.g., Singh and Carroll [172]).

Let $\hat{O}_1$ and $\hat{O}_2$ be two Hermitian operators with a complete set of common orthonormal eigenfunctions $\varphi_{\lambda_n^{(1)},\lambda_m^{(2)}}(\vec{r})$ such that $\hat{O}_1 \varphi_{\lambda_n^{(1)},\lambda_m^{(2)}}(\vec{r}) = \lambda_n^{(1)} \varphi_{\lambda_n^{(1)},\lambda_m^{(2)}}(\vec{r})$ and $\hat{O}_2 \varphi_{\lambda_n^{(1)},\lambda_m^{(2)}}(\vec{r}) = \lambda_m^{(2)} \varphi_{\lambda_n^{(1)},\lambda_m^{(2)}}(\vec{r})$. Then, $\hat{O}_1\hat{O}_2 \varphi_{\lambda_n^{(1)},\lambda_m^{(2)}}(\vec{r}) = \lambda_n^{(1)}\lambda_m^{(2)} \varphi_{\lambda_n^{(1)},\lambda_m^{(2)}}(\vec{r})$ and $\hat{O}_2\hat{O}_1 \varphi_{\lambda_n^{(1)},\lambda_m^{(2)}}(\vec{r}) = \lambda_m^{(2)}\lambda_n^{(1)} \varphi_{\lambda_n^{(1)},\lambda_m^{(2)}}(\vec{r})$ for any $\lambda_n^{(1)}$ and $\lambda_m^{(2)}$. Or, for any $\varphi_{\lambda_n^{(1)},\lambda_m^{(2)}}(\vec{r})$, $\hat{O}_1\hat{O}_2 \varphi_{\lambda_n^{(1)},\lambda_m^{(2)}}(\vec{r}) = \hat{O}_2\hat{O}_1 \varphi_{\lambda_n^{(1)},\lambda_m^{(2)}}(\vec{r})$. Consequently, for any arbitrary $\psi(\vec{r}) = \sum_{n,m} C_{\lambda_n^{(1)},\lambda_m^{(2)}} \varphi_{\lambda_n^{(1)},\lambda_m^{(2)}}(\vec{r})$, we have $\hat{O}_1\hat{O}_2 \psi(\vec{r}) = \hat{O}_2\hat{O}_1 \psi(\vec{r})$. In other words, $\hat{O}_1$ and $\hat{O}_2$ must commute with each other. So, only commuting Hermitian operators can have a complete set of simultaneous orthonormal eigenfunctions.

The three components of the position operator $\vec{\hat{r}}$ commute with each other and have $\{\delta(\vec{r}-\vec{r}')\}$ as their simultaneous eigenfunctions. We can write the three eigenvalue equations

$$\hat{x}\delta(\vec{r}-\vec{r}') = x'\delta(\vec{r}-\vec{r}'), \quad \hat{y}\delta(\vec{r}-\vec{r}') = y'\delta(\vec{r}-\vec{r}'), \quad \hat{z}\delta(\vec{r}-\vec{r}') = z'\delta(\vec{r}-\vec{r}'), \tag{3.48}$$

as

$$\vec{\hat{r}}\delta(\vec{r}-\vec{r}') = \vec{r}'\delta(\vec{r}-\vec{r}'). \tag{3.49}$$

Taking $\widehat{O} = \vec{r}$ and $\lambda_n = \vec{r}'$ in (3.37), we see that the probability of finding the particle at the position $\vec{r}'$ is given by

$$P(\vec{r} = \vec{r}') = \left| \int d^3r \, \delta^*(\vec{r} - \vec{r}') \Psi(\vec{r}, t) \right|^2 = |\Psi(\vec{r}', t)|^2, \qquad (3.50)$$

as stated in the beginning.

The three components of the momentum operator $\widehat{p}$ commute with each other. Their common eigenfunction is given by

$$\phi_{\vec{p}}(\vec{r}) = \frac{1}{(2\pi\hbar)^{3/2}} e^{\frac{i}{\hbar} \vec{p} \cdot \vec{r}}, \qquad (3.51)$$

such that

$$\widehat{p}_x \phi_{\vec{p}}(\vec{r}) = p_x \phi_{\vec{p}}(\vec{r}), \qquad \widehat{p}_y \phi_{\vec{p}}(\vec{r}) = p_y \phi_{\vec{p}}(\vec{r}), \qquad \widehat{p}_z \phi_{\vec{p}}(\vec{r}) = p_z \phi_{\vec{p}}(\vec{r}), \qquad (3.52)$$

where the eigenvalues of the momentum components, (p_x, p_y, p_z), take all real values from $-\infty$ to ∞. We can write

$$\vec{\widehat{p}} \phi_{\vec{p}}(\vec{r}) = \vec{p} \phi_{\vec{p}}(\vec{r}). \qquad (3.53)$$

These momentum eigenfunctions $\{\phi_{\vec{p}}(\vec{r})\}$ corresponding to a continuous spectrum of momentum eigenvalues, like the position eigenfunctions, cannot be normalized in the sense of (3.1) and are to be normalized only in the sense of (3.47). They are orthonormal in the sense

$$\int d^3r \, \phi_{\vec{p}}^*(\vec{r}) \phi_{\vec{p}'}(\vec{r}) = \delta(\vec{p} - \vec{p}'). \qquad (3.54)$$

They form a complete set such that any $\psi(\vec{r})$ can be expanded in terms of $\{\phi_{\vec{p}}(\vec{r})\}$ as

$$\psi(\vec{r}) = \int d^3p \, \widetilde{\psi}(\vec{p}) \phi_{\vec{p}}(\vec{r}), \qquad (3.55)$$

where $d^3p = dp_x dp_y dp_z$, the integral is over the entire momentum space, and

$$\widetilde{\psi}(\vec{p}) = \int d^3r \, \phi_{\vec{p}}^*(\vec{r}) \psi(\vec{r}) = \frac{1}{(2\pi\hbar)^{3/2}} \int d^3r \, e^{-\frac{i}{\hbar} \vec{p} \cdot \vec{r}} \psi(\vec{r}). \qquad (3.56)$$

This shows that for a particle with the wave function $\Psi(\vec{r}, t)$, which can be expanded in terms of the momentum eigenfunctions as

$$\Psi(\vec{r}, t) = \int d^3p \, \widetilde{\Psi}(\vec{p}, t) \phi_{\vec{p}}(\vec{r}), \qquad (3.57)$$

the probability of finding it with momentum $\vec{p}'$ is given by

$$P(\vec{p} = \vec{p}') = \left| \widetilde{\Psi}(\vec{p}', t) \right|^2 = \left| \int d^3r \, \phi_{\vec{p}'}^*(\vec{r}) \Psi(\vec{r}, t) \right|^2$$

$$= \left| \frac{1}{(2\pi\hbar)^{3/2}} \int d^3r \, e^{-\frac{i}{\hbar} \vec{p}' \cdot \vec{r}} \Psi(\vec{r}, t) \right|^2. \qquad (3.58)$$

Note that $\widetilde{\Psi}(\vec{p}, t)$ is the Fourier transform of the wave function $\Psi(\vec{r}, t)$ and represents the state in the momentum representation.

A Review of Quantum Mechanics

3.2.1.2 Quantum Dynamics

▶ The wave function $\Psi(\vec{r},t)$ evolves in time according to the Schrödinger equation

$$i\hbar \frac{\partial}{\partial t}\Psi(\vec{r},t) = \widehat{H}(\vec{r},\vec{p},t)\Psi(\vec{r},t), \tag{3.59}$$

where $\widehat{H}(\vec{r},\vec{p},t)$ is the Hamiltonian operator corresponding to the energy of the system, when the system is left undisturbed by any action on the system like a measurement of an observable of the system. When an observable O of the system is measured, the wave function collapses, immediately after the measurement, to an eigenfunction of the observable corresponding to the observed eigenvalue. From that moment onwards, the wave function evolves deterministically according to the Schrödinger equation (3.59) until a further measurement.

Notes:

The quantum Hamiltonian operator $\widehat{H}(\vec{r},\vec{p}.t)$ is derived from the classical Hamiltonian function $H(\vec{r},\vec{p},t)$ by the Classical $\longrightarrow$ Quantum transition, or correspondence, rule mentioned earlier, for getting the Hermitian quantum operator corresponding to any classical observable. This procedure works well for nonrelativistic quantum mechanics. But, in the case of relativistic quantum mechanics, the quantum Hamiltonian is to be obtained by a different approach as we shall see later. Also, the quantum operators are to be expressed first in Cartesian coordinates and then transformed to curvilinear coordinates if needed.

If a particle has the wave function $\Psi(\vec{r},t)$ and its position is determined to be $\vec{r}=\vec{r}'$, at time t, then immediately after the observation, its wave function collapses to $\delta(\vec{r}-\vec{r}')$. If at that instant the momentum of the particle is measured, the probability for it to be $\vec{p}'$ is given by

$$P(\vec{p}=\vec{p}') = \left| \frac{1}{(2\pi\hbar)^{3/2}} \int d^3 r\, e^{-\frac{i}{\hbar}\vec{p}'\cdot\vec{r}} \delta(\vec{r}-\vec{r}') \right|^2$$

$$= \left| \frac{1}{(2\pi\hbar)^{3/2}} e^{-\frac{i}{\hbar}\vec{p}'\cdot\vec{r}'} \right|^2 = \frac{1}{(2\pi\hbar)^3}, \tag{3.60}$$

showing that the probability of finding it to be any value $\vec{p}'$ is the same (ignoring the value of probability given by this equation since the position eigenfunction is not normalized in the sense of (3.1)). In other words, if we try to locate the particle precisely at a particular position, its momentum will be spread over its infinite spectrum of values with equal probability. Similarly, if a particle is found to have a definite momentum, say $\vec{p}'$, after a momentum measurement, it will have its wave function collapsed to $\frac{1}{(2\pi\hbar)^{3/2}} e^{\frac{i}{\hbar}\vec{p}'\cdot\vec{r}}$ immediately after the measurement. If at that instant the position of the particle is measured, the probability of finding it to be $\vec{r}'$ is given by

$$P(\vec{r}=\vec{r}') = \left| \frac{1}{(2\pi\hbar)^{3/2}} \int d^3r \, \delta(\vec{r}-\vec{r}') e^{\frac{i}{\hbar}\vec{p}'\cdot\vec{r}} \right|^2$$

$$= \left| \frac{1}{(2\pi\hbar)^{3/2}} e^{\frac{i}{\hbar}\vec{p}'\cdot\vec{r}'} \right|^2 = \frac{1}{(2\pi\hbar)^3}, \quad (3.61)$$

showing that the probability of finding it at any position $\vec{r}'$ is the same (again, ignoring the value of probability given by this equation since the momentum eigenfunction is not normalized in the sense of (3.1)). This means that a particle with a well-defined value of momentum is likely to be found at any position with equal probability. In general, it is impossible to say that a particle has well-defined values for position and momentum both, or for any two observables associated with noncommuting Hermitian operators that cannot have a complete set of simultaneous eigenfunctions. When the value of one of them is determined, the state of the particle collapses to the eigenfunction corresponding to the eigenvalue obtained for that observable. Since this eigenfunction is not an eigenfunction of the other observable for any of its eigenvalues, it cannot take a definite value on a measurement of the other observable. This is the essence of Heisenberg's uncertainty principle.

To proceed further, we have to recall some basic concepts of finite-dimensional vector spaces. An N-dimensional vector

$$|v\rangle = \sum_{j=1}^{N} v_j |j\rangle, \quad (3.62)$$

where $\{|j\rangle | j = 1, 2, \ldots, N\}$ are N linearly independent vectors forming a basis, can be represented as

$$|\underline{v}\rangle = \begin{pmatrix} v_1 \\ v_2 \\ \vdots \\ v_N \end{pmatrix}. \quad (3.63)$$

Following Dirac, $|v\rangle$ or its representative column vector $|\underline{v}\rangle$ is called a ket vector, and its adjoint, or the transpose conjugate, row vector

$$|\underline{v}\rangle^{\dagger} = (v_1^* \ v_2^* \ \cdots \ v_N^*), \quad (3.64)$$

is called a bra vector and represents

$$\langle v| = \sum_{j=1}^{N} v_j^* \langle j|. \quad (3.65)$$

The inner product of two vectors, say $|u\rangle$ and $|v\rangle$, is given by

$$\langle u|v\rangle = \sum_{j=1}^{N} u_j^* v_j, \quad (3.66)$$

and the norm of any vector is

$$\| |v\rangle \|^2 = \langle v|v\rangle = \sum_{j=1}^{N} |v_j|^2. \tag{3.67}$$

It follows that

$$\langle v|v\rangle = 0 \implies |v\rangle = 0, \tag{3.68}$$

where $|v\rangle = 0$ means $v_j = 0$, for $j = 1, 2, \ldots, N$. Note that

$$\langle v|u\rangle = \langle u|v\rangle^*. \tag{3.69}$$

The basis vectors $\{|j\rangle | j = 1, 2, \ldots, N\}$ are orthonormal:

$$\langle j|k\rangle = \delta_{jk}, \qquad j, k = 1, 2, \ldots, N. \tag{3.70}$$

Such a basis is called a unitary basis. The basis bra vectors $\{\langle j|\}$ are said to be dual to the basis ket vectors $\{|j\rangle\}$ in the sense of (3.70). From this, it follows that for any vector $|v\rangle$,

$$v_j = \langle j|v\rangle, \qquad j = 1, 2, \ldots, N, \tag{3.71}$$

and hence

$$|v\rangle = \sum_{j=1}^{N} |j\rangle\langle j|v\rangle = \left(\sum_{j=1}^{N} |j\rangle\langle j|\right) |v\rangle. \tag{3.72}$$

Thus, the completeness of the basis vectors $\{|j\rangle | j = 1, 2, \ldots, N\}$ is expressed as

$$\sum_{j=1}^{N} |j\rangle\langle j| = I, \tag{3.73}$$

where I is the identity operator, or $N \times N$ identity matrix in this case. This relation (3.73) is known as the resolution of identity. Note that $|j\rangle\langle j|$ is the projection operator that projects out of $|v\rangle$ its j-th component $v_j |j\rangle$.

We can change the unitary basis from the given set of complete orthonormal vectors $\{|j\rangle | j = 1, 2, \ldots, N\}$ to another set of complete orthonormal vectors, say, $\{|\varphi_j\rangle | j = 1, 2, \ldots, N\}$. We can write

$$|\varphi_j\rangle = \sum_{k=1}^{N} \varphi_{kj} |k\rangle = \sum_{k=1}^{N} |k\rangle \langle k| \varphi_j\rangle, \quad j = 1, 2, \ldots,$$

$$\langle \varphi_j| = \sum_{k=1}^{N} \varphi_{kj}^* \langle k| = \sum_{k=1}^{N} \langle \varphi_j|k\rangle \langle k|, \quad j = 1, 2, \ldots,$$

$$\langle \varphi_m | \varphi_n\rangle = \sum_{j=1}^{N} \langle \varphi_m | j\rangle\langle j| \varphi_n\rangle = \sum_{j=1}^{N} \varphi_{jm}^* \varphi_{jn} = \delta_{mn},$$

$$m, n = 1, 2, \ldots, N. \tag{3.74}$$

The completeness of the new basis vectors means, we have

$$I = \sum_{j=1}^{N} |\varphi_j\rangle\langle\varphi_j|. \tag{3.75}$$

To see this, let us observe

$$\sum_{j=1}^{N} \varphi_{jm}^* \varphi_{jn} = \delta_{mn} \implies \sum_{j=1}^{N} \varphi_{jn}^* \varphi_{jm} = \delta_{mn}$$

$$\implies \sum_{j=1}^{N} \varphi_{nj} \varphi_{mj}^* = \delta_{mn} \implies \sum_{j=1}^{N} \langle n | \varphi_j \rangle \langle \varphi_j | m \rangle = \delta_{mn}$$

$$\implies \sum_{j=1}^{N} |\varphi_j\rangle\langle\varphi_j| = I. \tag{3.76}$$

Then, for any vector $|v\rangle$, we have

$$|v\rangle = \sum_{j=1}^{N} |\varphi_j\rangle\langle\varphi_j|v\rangle, \qquad \langle v| = \sum_{j=1}^{N} \langle v|\varphi_j\rangle\langle\varphi_j|, \tag{3.77}$$

and the inner product of $|u\rangle$ and $|v\rangle$ becomes

$$\langle u|v\rangle = \sum_{j=1}^{N} \langle u|\varphi_j\rangle\langle\varphi_j|v\rangle = \langle u| \left(\sum_{j=1}^{N} |\varphi_j\rangle\langle\varphi_j| \right) |v\rangle. \tag{3.78}$$

Note that if we form an $N \times N$ matrix U with the N orthonormal basis vectors as its columns, i.e.,

$$U = \left(|\underline{\varphi_1}\rangle, |\underline{\varphi_2}\rangle, \ldots, |\underline{\varphi_N}\rangle \right),$$

$$= \begin{pmatrix} \varphi_{11} & \varphi_{12} & \cdot & \cdot & \cdot & \varphi_{1N} \\ \varphi_{21} & \varphi_{22} & \cdot & \cdot & \cdot & \varphi_{2N} \\ \cdot & \cdot & \cdot & \cdot & \cdot & \cdot \\ \cdot & \cdot & \cdot & \cdot & \cdot & \cdot \\ \varphi_{N1} & \varphi_{N2} & \cdot & \cdot & \cdot & \varphi_{NN} \end{pmatrix}, \tag{3.79}$$

then, it is a unitary matrix:

$$\sum_{j=1}^{N} \varphi_{jm}^* \varphi_{jn} = \delta_{mn} \implies U^\dagger U = I,$$

$$\sum_{j=1}^{N} \varphi_{nj} \varphi_{mj}^* = \delta_{mn} \implies UU^\dagger = I. \tag{3.80}$$

If M is an $N \times N$ matrix with elements $\{M_{jk}|j,k = 1,2,\ldots,N\}$, we can write

$$M = \sum_{j=1}^{N} \sum_{k=1}^{N} M_{jk} |j\rangle\langle k|, \tag{3.81}$$

A Review of Quantum Mechanics

and
$$M_{jk} = \langle j|M|k\rangle, \qquad j,k = 1,2,\ldots,N. \tag{3.82}$$

For any vector $|v\rangle$, we have

$$M|v\rangle = \sum_{j=1}^{N}\sum_{k=1}^{N} |j\rangle M_{jk}\langle k|v\rangle = \sum_{j=1}^{N}\left(\sum_{k=1}^{N} M_{jk}v_k\right)|j\rangle, \tag{3.83}$$

as expected. If the basis is changed from the old basis to any other unitary basis, $\{|\varphi_j\rangle | j = 1,2,\ldots,N\}$, then for any vector $|v\rangle$ represented in the new basis as

$$|v\rangle = \sum_{k=1}^{N} |\varphi_k\rangle\langle\varphi_k|v\rangle \tag{3.84}$$

the action of M becomes

$$M|v\rangle = \sum_{j=1}^{N}\sum_{k=1}^{N} |\varphi_j\rangle\langle\varphi_j|M|\varphi_k\rangle\langle\varphi_k|v\rangle. \tag{3.85}$$

Thus, the matrix elements of the linear operator M in the new basis become

$$\langle\varphi_j|M|\varphi_k\rangle = \sum_{l=1}^{N}\sum_{m=1}^{N} \langle\varphi_j|l\rangle\langle l|M|m\rangle\langle m|\varphi_k\rangle, \qquad j,k = 1,2,\ldots,N., \tag{3.86}$$

where $\{\langle l|M|m\rangle = M_{lm} | l,m = 1,2,\ldots,N\}$ are the matrix elements of M in the old basis. If we denote the matrix in the old basis as M, and in the new basis as M_φ, then it follows from (3.86) that

$$M_\varphi = U^\dagger M U. \tag{3.87}$$

For a Hermitian matrix H, one should have in any basis $H^\dagger = H$, or

$$H_{jk}^\dagger = H_{kj}^* = H_{jk}. \tag{3.88}$$

If $H \longrightarrow H_\varphi$ under the change to the new unitary basis, then its hermiticity is seen to be preserved:

$$H_\varphi^\dagger = \left(U^\dagger H U\right)^\dagger = U^\dagger H U = H_\varphi. \tag{3.89}$$

An arbitrary vector $|v\rangle$ can be written as

$$|v\rangle = \sum_{j=1}^{N}\sum_{k=1}^{N} |\varphi_j\rangle\langle\varphi_j|k\rangle\langle k|v\rangle. \tag{3.90}$$

This equation becomes, in terms of the components of $|v\rangle$ in the old and new bases,

$$\langle\varphi_j|v\rangle = \sum_{k=1}^{N} \langle\varphi_j|k\rangle\langle k|v\rangle = \sum_{k=1}^{N} \varphi_{kj}^*\langle k|v\rangle, \qquad j = 1,2,\ldots,N., \tag{3.91}$$

or
$$\left|v_{\underline{\varphi}}\right\rangle = U^\dagger |\underline{v}\rangle, \tag{3.92}$$

where U is the unitary matrix defined in (3.79). The inverse of this relation

$$|\underline{v}\rangle = U \left|v_{\underline{\varphi}}\right\rangle, \tag{3.93}$$

gives the components of $|v\rangle$ in the old basis in terms of its components in the new basis. The two equations (3.92) and (3.93) imply that if all the vectors of the vector space are transformed by premultiplication by a unitary matrix $\mathbb{U}$ (in the above case $\mathbb{U} = U^\dagger$), the resulting vectors represent the same vectors in a new basis consisting of the columns of the unitary matrix $\mathbb{U}^\dagger$ (in the above case $\mathbb{U}^\dagger = U$). Correspondingly, any linear transformation matrix M acting on the vector space gets transformed into $\mathbb{U}M\mathbb{U}^\dagger$ in the new representation. In fact, one can generalize this result to nonunitary transformations. Let us make a linear transformation of an arbitrary vector $|\underline{v}\rangle$ by a nonsingular matrix, say T, as

$$\left|\underline{v}'\right\rangle = T|\underline{v}\rangle, \qquad \det(T) \neq 0, \tag{3.94}$$

or

$$v'_j = \sum_{k=1}^N T_{jk} v_j, \qquad j = 1, 2, \ldots, N. \tag{3.95}$$

The inverse relation,

$$|\underline{v}\rangle = T^{-1} \left|\underline{v}'\right\rangle, \tag{3.96}$$

or

$$v_j = \sum_{k=1}^N \left(T^{-1}\right)_{jk} v'_k, \qquad j = 1, 2, \ldots, N, \tag{3.97}$$

implies that the components of the new vector $|\underline{v}'\rangle$ are the components of the original vector $|\underline{v}\rangle$ in a new representation in which the columns of T^{-1} are the basis vectors. Since T^{-1} is not unitary, its columns will not be orthonormal, and hence the linear transformation effected by T changes the original basis to a nonorthogonal basis. If M is a linear transformation acting on the vector space, it will become TMT^{-1} in the new representation, since

$$\left|(Mv)'\right\rangle = T(M|v\rangle) = TMT^{-1} \left|v'\right\rangle. \tag{3.98}$$

For a nonorthonormal basis, with a set of basis vectors $\{|\xi_j\rangle\}$, it is possible to construct a dual, or reciprocal, basis with the set of dual basis vectors $\{\langle \tilde{\xi}_j |\}$, such that

$$\left\langle \tilde{\xi}_j \mid \xi_k \right\rangle = \sum_\ell \tilde{\xi}^*_{j\ell} \xi_{\ell k} = \delta_{jk},$$

$$\sum_j |\xi_j\rangle \left\langle \tilde{\xi}_j \right| = I. \tag{3.99}$$

A Review of Quantum Mechanics

Hence, any vector $|v\rangle$ can the expanded in the nonorthonormal basis

$$|v\rangle = \sum_j v_j |\xi_j\rangle, \qquad (3.100)$$

in which

$$v_j = \langle \tilde{\xi}_j | v \rangle. \qquad (3.101)$$

Let us now extend the above discussion to the space of functions of $\vec{r}$, which can be regarded as an infinite-dimensional vector space, *i.e.*, $N \longrightarrow \infty$ which can be discrete or continuous. In the finite-dimensional vector space, the canonical basis vector $|j\rangle$ has its components as $\langle k|j\rangle = \delta_{jk}$. In the function space we can take the canonical basis vectors as $\{|\vec{r}\rangle\}$ with $\langle \vec{r}'|\vec{r}\rangle = \delta(\vec{r}' - \vec{r})$. Then, identifying $f(\vec{r}')$ with $\langle \vec{r}'|f\rangle$, the relation

$$f(\vec{r}') = \int d^3r\, \delta(\vec{r}' - \vec{r}) f(\vec{r}) \qquad (3.102)$$

translates to the representation

$$|f\rangle = \int d^3r\, |\vec{r}\rangle \langle \vec{r}|f\rangle, \qquad (3.103)$$

with the completeness relation reading

$$\int d^3r\, |\vec{r}\rangle \langle \vec{r}| = I. \qquad (3.104)$$

Corresponding to the ket $|f\rangle$, we can write the dual bra vector as

$$\langle f| = \int d^3r \langle f|\vec{r}\rangle \langle \vec{r}| \qquad (3.105)$$

where $\langle f|\vec{r}\rangle = \langle \vec{r}|f\rangle^*$. Now, the inner product of two functions $f(\vec{r})$ and $g(\vec{r})$ can be written as

$$\int d^3r\, f^*(\vec{r}) g(\vec{r}) = \langle f|g\rangle. \qquad (3.106)$$

Note that $\langle f|g\rangle = \langle g|f\rangle^*$. The norm of a function $f(\vec{r})$, $\|f\|$, is given by $\|f\|^2 = \langle f|f\rangle$, and $\langle f|f\rangle = 0$ implies that $f(\vec{r}) = 0$. The normalization condition for the wave function of a particle (3.1) now reads

$$\langle \Psi(t)|\Psi(t)\rangle = 1, \qquad (3.107)$$

where

$$\Psi(\vec{r},t) = \langle \vec{r}|\Psi(t)\rangle, \qquad \Psi^*(\vec{r},t) = \langle \Psi(t)|\vec{r}\rangle, \qquad (3.108)$$

and $|\psi(t)\rangle$ is called the state vector of the particle. Sometimes, we will be using the terms state vector and wave function interchangeably.

If we now change the basis to another complete set of orthonormal functions, $\{\varphi_j(\vec{r}) = \langle \vec{r}|\varphi_j\rangle\}$, which can be the eigenfunctions of a Hermitian operator representing some observable of a particle, we can write any arbitrary $|\psi\rangle$, with $\langle \vec{r}|\psi\rangle = \psi(\vec{r})$, as

$$|\psi\rangle = \sum_j |\varphi_j\rangle \langle \varphi_j|\psi\rangle, \qquad (3.109)$$

with the completeness relation, or the resolution of identity, for the new basis, say φ-basis, reading

$$\sum_j |\varphi_j\rangle\langle\varphi_j| = I. \tag{3.110}$$

In the position representation, the resolution of identity reads

$$\sum_j \langle \vec{r}| \varphi_j\rangle\langle\varphi_j|\vec{r}'\rangle = \sum_j \varphi_j(\vec{r})\varphi_j(\vec{r}')^* = \langle\vec{r}|\vec{r}'\rangle = \delta(\vec{r}-\vec{r}'). \tag{3.111}$$

Here, $\sum_j$ runs over all the infinite values of j, discrete, continuous, or mixed; it will become an integral over ranges where j is continuous. A basis with a complete set of orthonormal functions is called a unitary basis for the same reason as in the finite-dimensional case. If $\widehat{O}$ is a linear operator, we shall write

$$\langle\vec{r}|\widehat{O}f\rangle = \langle\vec{r}|\widehat{O}|f\rangle, \tag{3.112}$$

or, independent of representation,

$$|\widehat{O}f\rangle = \widehat{O}|f\rangle. \tag{3.113}$$

In the φ-basis, we can write this as

$$\langle\varphi_j|\widehat{O}f\rangle = \sum_k \langle\varphi_j|\widehat{O}|\varphi_k\rangle\langle\varphi_k|f\rangle, \tag{3.114}$$

where $\left\{\langle\varphi_j|\widehat{O}|\varphi_k\rangle\right\}$ are the matrix elements of $\widehat{O}$ in the φ-basis. For any Hermitian operator $\widehat{H}$, we should have

$$\langle\varphi_j|\widehat{H}|\varphi_k\rangle = \langle\varphi_k|\widehat{H}|\varphi_j\rangle^*, \tag{3.115}$$

for all j and k, such that for any arbitrary $|f\rangle = \sum_j |\varphi_j\rangle\langle\varphi_j|f\rangle$ and $|g\rangle = \sum_k |\varphi_k\rangle\langle\varphi_k|g\rangle$ we will have

$$\langle f|\widehat{H}|g\rangle = \langle g|\widehat{H}|f\rangle^*, \tag{3.116}$$

as demanded in (3.6). This implies that for any Hermitian operator $\widehat{H}$, $\langle f|\widehat{H}|f\rangle$ is real.

When we change the basis from one complete orthonormal set $\left\{\varphi_j(\vec{r})\right\}$ to another complete orthonormal set $\left\{\varphi'_j(\vec{r})\right\}$, we go from the representation

$$|f\rangle = \sum_j |\varphi_j\rangle\langle\varphi_j|f\rangle, \tag{3.117}$$

to the representation

$$|f\rangle = \sum_j |\varphi'_j\rangle\langle\varphi'_j|f\rangle. \tag{3.118}$$

The components of $|f\rangle$ in the two representations are related by

$$\langle\varphi'_j|f\rangle = \sum_k \langle\varphi'_j|\varphi_k\rangle\langle\varphi_k|f\rangle. \tag{3.119}$$

Observe that the connection coefficients between the two representations, *i.e.*, $\{\langle\varphi'_j|\varphi_k\rangle\}$ satisfy the relation

$$\sum_k \langle\varphi'_j|\varphi_k\rangle\langle\varphi_k|\varphi'_m\rangle = \delta_{jm}, \qquad (3.120)$$

since both the bases are complete orthonormal bases. Writing $\langle\varphi'_j|\varphi_k\rangle = U_{jk}$, the jk-th element of a matrix U, the above relation becomes

$$\sum_k U_{jk}U^*_{mk} = \left(UU^\dagger\right)_{jm} = \delta_{jm}, \qquad (3.121)$$

showing that U is a unitary matrix. In other words, the change of basis effected is a unitary transformation. The matrix elements of a Hermitian operator $\widehat{O}$, corresponding to an observable O, given in the two bases by $\{\langle\varphi_j|\widehat{O}|\varphi_k\rangle\}$ and $\{\langle\varphi'_j|\widehat{O}|\varphi'_k\rangle\}$ are related by

$$\langle\varphi'_j|\widehat{O}|\varphi'_k\rangle = \sum_l \sum_m \langle\varphi'_j|\varphi_l\rangle\langle\varphi_l|\widehat{O}|\varphi_m\rangle\langle\varphi_m|\varphi'_k\rangle. \qquad (3.122)$$

This shows that if we call the matrices representing $\widehat{O}$ in φ-basis and φ'-basis as $[O]$ and $[O']$, respectively, then the two matrices are related by a unitary transformation

$$[O'] = U[O]U^\dagger. \qquad (3.123)$$

It is seen that under such unitary transformations the commutator relations between the operators remain invariant, *i.e.*, if $[[O_1],[O_2]] = [O_3]$ in φ-basis, then $[[O'_1],[O'_2]] = [O'_3]$ in φ'-basis. Thus on change of basis, the commutation relations like $[\widehat{x},\widehat{p}_x] = i\hbar$ remain invariant. In other words, classical canonical transformations, under which the Poisson bracket relations are invariant, become in quantum mechanics unitary transformations under which the corresponding commutator bracket relations remain invariant.

In the position representation, or the canonical basis, we can write the matrix elements of any operator $\widehat{O}(\vec{r},\vec{p},t)$ as

$$\langle\vec{r}|\widehat{O}(\vec{r},\vec{p},t)|\vec{r}'\rangle = \widehat{O}(\vec{r},\vec{p},t)\delta(\vec{r}-\vec{r}'). \qquad (3.124)$$

In the φ-basis, we get

$$\langle\varphi_j|\widehat{O}(\vec{r},\vec{p},t)|\varphi_k\rangle = \int\int d^3r\, d^3r'\, \langle\varphi_j|\vec{r}\rangle\langle\vec{r}|\widehat{O}(\vec{r},\vec{p},t)|\vec{r}'\rangle\langle\vec{r}'|\varphi_k\rangle$$

$$= \int\int d^3r\, d^3r'\, \varphi^*_j(\vec{r})\widehat{O}(\vec{r},\vec{p},t)\delta(\vec{r}-\vec{r}')\varphi_k(\vec{r}')$$

$$= \int d^3r\, \varphi^*_j(\vec{r})\widehat{O}(\vec{r},\vec{p},t)\varphi_k(\vec{r}). \qquad (3.125)$$

Let us make a unitary transformation of the function space such that any vector $|\psi\rangle$ in it gets transformed as

$$|\psi'\rangle = \widehat{U}|\psi\rangle. \qquad (3.126)$$

In a representation with $\{|\phi_j\rangle\}$ as the complete set of orthonormal basis kets, we have

$$|\psi\rangle = \sum_j |\phi_j\rangle\langle\phi_j|\psi\rangle, \qquad |\psi'\rangle = \sum_j |\phi_j\rangle\langle\phi_j|\psi'\rangle. \qquad (3.127)$$

Inverting the relation (3.126), we get

$$|\psi\rangle = \widehat{\mathbb{U}}^\dagger |\psi'\rangle, \qquad (3.128)$$

which can be written as

$$|\psi\rangle = \sum_j \widehat{\mathbb{U}}^\dagger |\phi_j\rangle\langle\phi_j|\psi'\rangle, \qquad (3.129)$$

showing that $|\psi'\rangle$ in the old representation represents the vector $|\psi\rangle$ in a new representation for which the complete set of orthonormal basis kets is given by $\{\widehat{\mathbb{U}}^\dagger |\phi_j\rangle\}$. If a linear operator is represented by $\widehat{O}$ in the old representation, with $\{|\phi_j\rangle\}$ as the basis kets, in the new representation with $\{\widehat{\mathbb{U}}^\dagger |\phi_j\rangle\}$ as the basis kets, it will be represented by $\widehat{\mathbb{U}}\widehat{O}\widehat{\mathbb{U}}^\dagger$. This is seen as follows:

$$\left|\left(\widehat{O}\psi\right)'\right\rangle = \widehat{\mathbb{U}}\left(\widehat{O}|\psi\rangle\right) = \widehat{\mathbb{U}}\widehat{O}\widehat{\mathbb{U}}^\dagger |\psi'\rangle = \widehat{O}'|\psi'\rangle. \qquad (3.130)$$

Analogous to what we found in the finite-dimensional case, we can also effect a linear transformation of the function space using any invertible operator. Let such a transformation in the function space lead to the transformation of any vector in it as

$$|\psi'\rangle = \widehat{T}|\psi\rangle, \qquad (3.131)$$

where $\widehat{T}$ has an inverse $\widehat{T}^{-1}$ such that $\widehat{T}^{-1}\widehat{T} = I$. Then, we can write

$$|\psi\rangle = \widehat{T}^{-1}|\psi'\rangle, \qquad (3.132)$$

or, in any orthonormal basis $\{|\phi_j\rangle\}$,

$$|\psi\rangle = \sum_j \left(\widehat{T}^{-1}|\phi_j\rangle\right)\langle\phi_j|\psi'\rangle, \qquad (3.133)$$

showing that $|\psi'\rangle$ represents $|\psi\rangle$ in a new representation with a nonorthonormal basis given by $\{\widehat{T}^{-1}|\phi_j\rangle\}$. In this case, the relation between the expressions for any linear operator in the old representation ($\widehat{O}$), and the new representation ($\widehat{O}'$) becomes

$$\widehat{O}' = \widehat{T}\widehat{O}\widehat{T}^{-1}. \qquad (3.134)$$

This is seen as follows:

$$\left|\left(\widehat{O}\psi\right)'\right\rangle = \widehat{T}\left(\widehat{O}|\psi\rangle\right) = \widehat{T}\widehat{O}\widehat{T}^{-1}|\psi'\rangle = \widehat{O}'|\psi'\rangle. \qquad (3.135)$$

A Review of Quantum Mechanics

As in the finite-dimensional case, it is possible to construct a dual basis for a nonorthonormal basis. If a set of functions $\{\xi_j(x)\}$ form a nonorthonormal basis, then a set of dual basis functions $\{\tilde{\xi}_j(x)\}$ can be defined such that

$$\langle \tilde{\xi}_j | \xi_k \rangle = \int dx\, \tilde{\xi}_j(x)^* \xi_k(x) = \delta_{jk}, \qquad \sum_j \xi_j(x) \tilde{\xi}_j(x')^* = \delta(x-x'). \tag{3.136}$$

Then, any function $\phi(x)$ can be expanded as

$$\phi(x) = \sum_j f_j \xi_j(x), \qquad f_j = \int dx\, \tilde{\xi}_j(x)^* \phi(x). \tag{3.137}$$

In connection with unitary and nonunitary transformations, it is useful to recall here an operator identity. For any pair of operators $\widehat{A}$ and $\widehat{B}$,

$$e^{\widehat{A}} \widehat{B} e^{-\widehat{A}} = \widehat{B} + \left[\widehat{A}, \widehat{B}\right] + \frac{1}{2!}\left[\widehat{A}, \left[\widehat{A}, \widehat{B}\right]\right] + \frac{1}{3!}\left[\widehat{A}, \left[\widehat{A}, \left[\widehat{A}, \widehat{B}\right]\right]\right] + \cdots. \tag{3.138}$$

This can be proved as follows. Let

$$\widehat{f}(\lambda) = e^{\lambda \widehat{A}} \widehat{B} e^{-\lambda \widehat{A}} = \widehat{f}(0) + \lambda \left.\frac{d\widehat{f}}{d\lambda}\right|_{\lambda=0} + \frac{\lambda^2}{2!}\left.\frac{d^2 \widehat{f}}{d\lambda^2}\right|_{\lambda=0} + \frac{\lambda^3}{3!}\left.\frac{d^3 \widehat{f}}{d\lambda^3}\right|_{\lambda=0} + \cdots, \tag{3.139}$$

where the infinite series represents the Taylor series expansion of $\widehat{f}(\lambda)$. Observe that

$$\left.\frac{d\widehat{f}}{d\lambda}\right|_{\lambda=0} = \left(A e^{\lambda \widehat{A}} \widehat{B} e^{-\lambda \widehat{A}} - e^{\lambda \widehat{A}} \widehat{B} e^{-\lambda \widehat{A}} A\right)\bigg|_{\lambda=0}$$

$$= \left[\widehat{A}, \widehat{f}(\lambda)\right]\bigg|_{\lambda=0} = \left[\widehat{A}, \widehat{B}\right],$$

$$\left.\frac{d^2 \widehat{f}}{d\lambda^2}\right|_{\lambda=0} = \left[\widehat{A}, \left[\widehat{A}, \widehat{B}\right]\right], \text{ and so on.} \tag{3.140}$$

Substituting these results in the Taylor series for $\widehat{f}(\lambda)$ and setting $\lambda = 1$, we get the identity (3.138). Introducing the notation

$$:\widehat{A}:\widehat{B} = \left[\widehat{A},,\widehat{B}\right], \tag{3.141}$$

analogous to the case of the Poisson bracket, we can write the identity (3.138) as

$$e^{\widehat{A}} \widehat{B} e^{-\widehat{A}} = \left(I + :\widehat{A}: + \frac{1}{2!}:\widehat{A}:^2 + \frac{1}{3!}:\widehat{A}:^3 + \cdots\right) \widehat{B} = e^{:\widehat{A}:} \widehat{B}. \tag{3.142}$$

Also note that for any constant a

$$e^{:a\widehat{A}:} = e^{a:\widehat{A}:}. \tag{3.143}$$

Using the Dirac ⟨bra|c|ket⟩ notation, we can write the Schrödinger quantum dynamical time-evolution equation (3.59) in a representation-independent way[2] as

$$i\hbar \frac{\partial |\Psi(t)\rangle}{\partial t} = \widehat{H}|\Psi(t)\rangle. \tag{3.144}$$

Taking the adjoint of the above equation on both sides, we get the time-evolution equation for ⟨Ψ(t)| as

$$i\hbar \frac{\partial \langle \Psi(t)|}{\partial t} = -\langle \Psi(t)|\widehat{H}. \tag{3.145}$$

If we choose to work in a φ-representation, with an orthonormal basis, then multiplying both sides of (3.144) from left by $I = \sum_j |\varphi_j\rangle\langle\varphi_j|$ and equating the coefficients of the linearly independent basis vectors $|\varphi_j\rangle$ on both sides, we get

$$i\hbar \frac{\partial}{\partial t}\langle \varphi_j|\Psi(t)\rangle = \langle \varphi_j|\widehat{H}|\Psi(t)\rangle = \sum_k \langle \varphi_j|\widehat{H}|\varphi_k\rangle\langle \varphi_k|\Psi(t)\rangle. \tag{3.146}$$

Or, with $\Psi_j(t) = \langle \varphi_j|\Psi(t)\rangle$ and $H_{jk} = \langle \varphi_j|\widehat{H}|\varphi_k\rangle$, we have

$$i\hbar \frac{\partial}{\partial t} \begin{pmatrix} \vdots \\ \Psi_{j-1}(t) \\ \Psi_j(t) \\ \Psi_{j+1}(t) \\ \vdots \end{pmatrix}$$

$$= \begin{pmatrix} \cdot & \cdot & \cdot & \cdot & \cdot & \cdot & \cdot & \cdot \\ \cdot & \cdot & \cdot & \cdot & \cdot & \cdot & \cdot & \cdot \\ \cdot & \cdot & \cdot & H_{j-1j-1} & H_{j-1j} & H_{j-1j+1} & \cdot & \cdot \\ \cdot & \cdot & \cdot & H_{jj-1} & H_{jj} & H_{jj+1} & \cdot & \cdot \\ \cdot & \cdot & \cdot & H_{j+1j-1} & H_{j+1j} & H_{j+1j+1} & \cdot & \cdot \\ \cdot & \cdot & \cdot & \cdot & \cdot & \cdot & \cdot & \cdot \\ \cdot & \cdot & \cdot & \cdot & \cdot & \cdot & \cdot & \cdot \end{pmatrix} \begin{pmatrix} \vdots \\ \Psi_{j-1}(t) \\ \Psi_j(t) \\ \Psi_{j+1}(t) \\ \vdots \end{pmatrix}. \tag{3.147}$$

---

[2]Hereafter we will write, in general, $\widehat{H}(\vec{r},\vec{p},t)$ and other observables, such as $\widehat{O}(\vec{r},\vec{p},t)$, simply as $\widehat{H}$ and $\widehat{O}$, respectively, without mentioning their dependence on $\vec{r},\vec{p}$ and t whenever such dependence is clear from the context. Similarly, we shall write $\Psi(\vec{r},t)$ and $\psi(\vec{r})$ simply as Ψ and ψ, respectively, unless for the sake of better clarity, we have to indicate their dependence on space, and time, explicitly.

A Review of Quantum Mechanics

Note that in the position representation, with

$$\langle \varphi_j | \Psi(t) \rangle \longrightarrow \langle \vec{r} | \Psi(t) \rangle = \Psi(\vec{r},t)$$

$$H_{jk} = \langle \varphi_j | \widehat{H} | \varphi_k \rangle \longrightarrow \langle \vec{r} | \widehat{H} | \vec{r}' \rangle = \widehat{H} \delta(\vec{r}-\vec{r}')$$

$$\sum_k \langle \varphi_j | \widehat{H} | \varphi_k \rangle \langle \varphi_k | \Psi(t) \rangle \longrightarrow \int d^3 r' \, \langle \vec{r} | \widehat{H}(\vec{r},\vec{p},t) | \vec{r}' \rangle \langle \vec{r}' | \Psi(t) \rangle$$

$$= \int d^3 r' \, \widehat{H}(\vec{r},\vec{p},t) \delta(\vec{r}-\vec{r}') \Psi(\vec{r}',t)$$

$$= \widehat{H}(\vec{r},\vec{p},t) \Psi(\vec{r},t), \qquad (3.148)$$

the equation (3.147) becomes the differential equation of Schrödinger (3.59).

For a particle in a state $|\Psi(t)\rangle$, the average (or expectation, or mean) value for any observable O, at time t, can be defined as

$$\langle O \rangle_{\Psi(t)} = \langle \Psi(t) | \widehat{O} | \Psi(t) \rangle. \qquad (3.149)$$

Sometimes, we shall write $\langle O \rangle_\Psi$ also as $\langle \widehat{O} \rangle_\Psi$. The definition (3.149) can be understood as follows. Let the Hermitian quantum operator $\widehat{O}$ corresponding to the observable O have real eigenvalues $\{\lambda_n\}$, where λ_n is d_n-fold degenerate. Let the corresponding eigenvectors be $\{|\varphi_{nj}\rangle | j=1,2,\ldots,d_n\}$ such that $\widehat{O}|\varphi_{nj}\rangle = \lambda_n |\varphi_{nj}\rangle$ for all $j=1,2,\ldots,d_n$. These eigenvectors would form a complete orthonormal set, such that

$$\langle \varphi_{nj} | \varphi_{mk} \rangle = \delta_{nm} \delta_{jk}, \qquad j=1,2,\ldots,d_n, \; k=1,2,\ldots,d_m, \qquad (3.150)$$

with the resolution of identity

$$\sum_n \sum_{j=1}^{d_n} |\varphi_{nj}\rangle \langle \varphi_{nj}| = I. \qquad (3.151)$$

From this we get the spectral decomposition of $\widehat{O}$ as

$$\widehat{O} = \widehat{O} I = \widehat{O} \sum_n \sum_{j=1}^{d_n} |\varphi_{nj}\rangle \langle \varphi_{nj}| = \sum_n \sum_{j=1}^{d_n} \lambda_n |\varphi_{nj}\rangle \langle \varphi_{nj}|. \qquad (3.152)$$

Substituting this expression for $\widehat{O}$ in the right-hand side of (3.149), we get

$$\langle O \rangle_{\Psi(t)} = \langle \Psi(t) | \sum_n \sum_{j=1}^{d_n} \lambda_n |\varphi_{nj}\rangle \langle \varphi_{nj} | \Psi(t) \rangle$$

$$= \sum_n \lambda_n \left\{ \sum_{j=1}^{d_n} |\langle \varphi_{nj} | \Psi(t) \rangle|^2 \right\} = \sum_n \lambda_n P(O=\lambda_n), \qquad (3.153)$$

as claimed.

3.2.1.3 Different Pictures of Quantum Dynamics

We can formally integrate the Schrödinger equation (3.144) as

$$|\Psi(t)\rangle = \widehat{U}(t,t_i)|\Psi(t_i)\rangle, \qquad t \geq t_i, \tag{3.154}$$

where t_i is the initial time at which the system starts evolving with the initial state vector given as $|\Psi(t_i)\rangle$, and $\widehat{U}(t,t_i)$ is the time-evolution operator depending on the Hamiltonian $\widehat{H}$ of the system. The Hamiltonian may or may not depend explicitly on time. The adjoint of (3.154) is seen to be

$$\langle\Psi(t)| = \langle\Psi(t_i)|\widehat{U}^\dagger(t,t_i). \tag{3.155}$$

Obviously, we should have

$$\widehat{U}(t_i,t_i) = I, \tag{3.156}$$

as the initial condition for $\widehat{U}(t_i,t_i)$. From the normalization condition for the state vector, namely $\langle\Psi(t)|\Psi(t)\rangle = 1$ at any time t, or the conservation of probability, we get

$$\langle\Psi(t)|\Psi(t)\rangle = \langle\Psi(t_i)|\widehat{U}^\dagger(t,t_i)\widehat{U}(t,t_i)|\Psi(t_i)\rangle = \langle\Psi(t_i)|\Psi(t_i)\rangle = 1. \tag{3.157}$$

This shows that $\widehat{U}(t,t_i)$ must be a unitary operator such that

$$\widehat{U}^\dagger(t,t_i)\widehat{U}(t,t_i) = I. \tag{3.158}$$

Then, note that the relations

$$\widehat{U}^\dagger(t,t_i)|\Psi(t)\rangle = |\Psi(t_i)\rangle,$$
$$\widehat{U}(t,t_i)\widehat{U}^\dagger(t,t_i)|\Psi(t)\rangle = \widehat{U}(t,t_i)|\Psi(t_i)\rangle = |\Psi(t)\rangle, \tag{3.159}$$

for any $|\Psi(t)\rangle$, imply that

$$\widehat{U}(t,t_i)\widehat{U}^\dagger(t,t_i) = I. \tag{3.160}$$

Substituting the solution (3.154) for $|\Psi(t)\rangle$ in the Schrödinger equation (3.144), we have

$$i\hbar\frac{\partial}{\partial t}\left(\widehat{U}(t,t_i)|\Psi(t_i)\rangle\right) = \left(i\hbar\frac{\partial}{\partial t}\widehat{U}(t,t_i)\right)|\Psi(t_i)\rangle$$
$$= \widehat{H}\widehat{U}(t,t_i)|\Psi(t_i)\rangle. \tag{3.161}$$

This implies that $\widehat{U}(t,t_i)$ must satisfy the differential equation

$$i\hbar\frac{\partial}{\partial t}\widehat{U}(t,t_i) = \widehat{H}\widehat{U}(t,t_i), \tag{3.162}$$

with the initial condition

$$\widehat{U}(t_i,t_i) = I. \tag{3.163}$$

A Review of Quantum Mechanics

Taking the adjoint of both sides of (3.162) implies that

$$i\hbar \frac{\partial \widehat{U}^{\dagger}(t,t_i)}{\partial t} = -\widehat{U}^{\dagger}(t,t_i)\widehat{H}. \tag{3.164}$$

Multiplying both sides of this equation from left by $\langle \Psi(t_i)|$, we get the Schrödinger equation for $\langle \Psi(t)|$ as given in (3.145). Another important property of the time-evolution operator is to be noted. Let $t > t' > t_i$. From

$$|\Psi(t)\rangle = \widehat{U}(t,t')|\Psi(t')\rangle, \quad |\Psi(t')\rangle = \widehat{U}(t',t_i)|\Psi(t_i)\rangle,$$
$$|\Psi(t)\rangle = \widehat{U}(t,t_i)|\Psi(t_i)\rangle, \tag{3.165}$$

it is clear that

$$\widehat{U}(t,t_i) = \widehat{U}(t,t')\widehat{U}(t',t_i). \tag{3.166}$$

In general, for $t > t_n > t_{n-1} > \cdots > t_2 > t_1 > t_i$, we have

$$\widehat{U}(t,t_i) = \widehat{U}(t,t_n)\widehat{U}(t_n,t_{n-1})\ldots\widehat{U}(t_2,t_1)\widehat{U}(t_1,t_i). \tag{3.167}$$

This property of the time-evolution operator is referred to as semigroup property.

For a system with a time-independent Hamiltonian $\widehat{H}$, we can integrate (3.162) directly to get

$$\widehat{U}(t,t_i) = e^{-\frac{i}{\hbar}(t-t_i)\widehat{H}}, \tag{3.168}$$

which obviously satisfies the initial condition (3.163). The Hamiltonian can be written, using the spectral decomposition, as

$$\widehat{H} = \sum_n E_n |n\rangle\langle n|, \tag{3.169}$$

in terms of its eigenvalues $\{E_n\}$ and the corresponding eigenkets $\{|n\rangle\}$ and bras $\{\langle n|\}$. Note that the projection operators $\{P_n = |n\rangle\langle n|\}$ satisfy the relation

$$P_n P_{n'} = \delta_{nn'} P_n. \tag{3.170}$$

Using the expression (3.169) for $\widehat{H}$, and the relation (3.170), we get

$$\widehat{U}(t,t_i) = e^{-\frac{i}{\hbar}(t-t_i)\sum_n E_n |n\rangle\langle n|} = \sum_n e^{-\frac{i}{\hbar}(t-t_i)E_n} |n\rangle\langle n|, \tag{3.171}$$

an expression for the time-evolution operator, $\widehat{U}(t,t_i)$, in terms of the eigenvalues and eigenvectors of the time-independent Hamiltonian. We shall see later how $\widehat{U}(t,t_i)$ can be calculated when the Hamiltonian depends explicitly on time.

In position representation, we can write (3.154) as

$$\langle \vec{r}|\Psi(t)\rangle = \int d^3 r_i \langle \vec{r}|\widehat{U}(t,t_i)|\vec{r}_i\rangle \langle \vec{r}_i|\Psi(t_i)\rangle, \tag{3.172}$$

or

$$\Psi(\vec{r},t) = \int d^3 r_i K(\vec{r},t;\vec{r}_i,t_i)\Psi(\vec{r}_i,t_i), \tag{3.173}$$

where the kernel of propagation

$$K(\vec{r},t;\vec{r}_i,t_i) = \langle \vec{r}|\widehat{U}(t,t_i)|\vec{r}_i\rangle \tag{3.174}$$

is called the propagator. From (3.173), it is clear that

$$K(\vec{r},t;\vec{r}_i,t_i) = \widehat{U}(t,t_i)\delta(\vec{r}-\vec{r}_i), \qquad t \geq t_i, \tag{3.175}$$

which means that $K(\vec{r},t;\vec{r}_i,t_i)$ would be the wave function of a particle with the Hamiltonian $\widehat{H}$ at time t if at an earlier time t_i it was localized at $\vec{r}_i$ with the wave function $\delta(\vec{r}-\vec{r}_i)$. This means that the propagator $K(\vec{r},t;\vec{r}_i,t_i)$ can be interpreted as the probability amplitude for finding the particle at $\vec{r}$ at time t if it was found at $\vec{r}_i$ at time t_i. In position representation, the equation (3.162) becomes

$$\left(i\hbar\frac{\partial}{\partial t} - \widehat{H}\right) K(\vec{r},t;\vec{r}_i,t_i) = 0. \tag{3.176}$$

as can be seen by taking $\left\langle \vec{r}\left|\widehat{H}\right|\vec{r}_i\right\rangle = \widehat{H}\left(\vec{r},\vec{p},t\right)\delta(\vec{r}-\vec{r}_i)$. In other words, $K(\vec{r},t;\vec{r}_i,t_i)$ satisfies the Schrödinger equation. The initial condition $\widehat{U}(t_i,t_i) = I$ implies that

$$K(\vec{r},t_i;\vec{r}_i,t_i) = \delta(\vec{r}-\vec{r}_i). \tag{3.177}$$

When the Hamiltonian is time-independent, we have explicitly, using (3.171),

$$\begin{aligned}K(\vec{r},t;\vec{r}_i,t_i) &= \sum_n e^{-\frac{i}{\hbar}(t-t_i)E_n}\langle\vec{r}|n\rangle\langle n|\vec{r}_i\rangle \\ &= \sum_n e^{-\frac{i}{\hbar}(t-t_i)E_n}\psi_n(\vec{r})\psi_n^*(\vec{r}_i),\end{aligned} \tag{3.178}$$

where $\psi_n(\vec{r})$ is the n-th eigenfunction of $\widehat{H}$ corresponding to the eigenvalue E_n. The validity of equations (3.176) and (3.177) can be directly checked in this case.

One can absorb the condition $t \geq t_i$ in the definition of the propagator and write

$$G(\vec{r},t;\vec{r}_i,t_i) = \theta(t-t_i)K(\vec{r},t;\vec{r}_i,t_i), \tag{3.179}$$

where $\theta(t-t_i)$ is the Heaviside step function defined by

$$\theta(t-t_i) = \begin{cases} 0 & \text{for } t < t_i, \\ 1 & \text{for } t \geq t_i. \end{cases} \tag{3.180}$$

The step function satisfies the differential equation

$$\frac{d\theta(t-t_i)}{dt} = \delta(t-t_i). \tag{3.181}$$

Then,

$$\begin{aligned}\left(i\hbar\frac{\partial}{\partial t} - \widehat{H}\right) G(\vec{r},t;\vec{r}_i,t_i) &= i\hbar\delta(t-t_i)K(\vec{r},t;\vec{r}_i,t_i) \\ &\quad + \theta(t-t_i)\left(i\hbar\frac{\partial}{\partial t} - \widehat{H}\right)K(\vec{r},t;\vec{r}_i,t_i) \\ &= i\hbar\delta(t-t_i)\delta(\vec{r}-\vec{r}_i),\end{aligned} \tag{3.182}$$

A Review of Quantum Mechanics

where, in the last step, we have used equations (3.176) and (3.181), and the fact that $\delta(t-t_i) K(\vec{r},t;\vec{r}_i,t_i)$ vanishes for $t \neq t_i$ and becomes $\delta(t-t_i)\delta(\vec{r}-\vec{r}_i)$ as $t \longrightarrow t_i$. The function $G(\vec{r},t;\vec{r}_i,t_i)$ is known as Green's function for the operator $[(i\hbar\partial/\partial t) - \hat{H}]$.

Substituting the formal solution (3.154) of the Schrödinger equation in the expression (3.149) for the average of an observable, we get

$$\langle O \rangle_{\Psi(t)} = \langle \Psi(t_i) | \hat{U}^\dagger(t,t_i) \hat{O} \hat{U}(t,t_i) | \Psi(t_i) \rangle. \tag{3.183}$$

This expression (3.183) for the average of O can be reinterpreted as the average of the time-dependent operator

$$\hat{O}_{\text{H}}(t) = \hat{U}^\dagger(t,t_i) \hat{O} \hat{U}(t,t_i), \tag{3.184}$$

in the fixed state of the system at the initial time t_i, namely $|\Psi(t_i)\rangle$. Note that

$$i\hbar \frac{d\hat{O}_{\text{H}}(t)}{dt} = i\hbar \left\{ \left(\frac{\partial}{\partial t} \hat{U}^\dagger(t,t_i) \right) \hat{O} \hat{U}(t,t_i) + \hat{U}^\dagger(t,t_i) \left(\frac{\partial \hat{O}}{\partial t} \right) \hat{U}(t,t_i) \right.$$

$$\left. + \hat{U}^\dagger(t,t_i) \hat{O} \left(\frac{\partial}{\partial t} \hat{U}(t,t_i) \right) \right\}$$

$$= \left\{ -\hat{U}^\dagger(t,t_i) \hat{H} \hat{O} \hat{U}(t,t_i) + \hat{U}^\dagger(t,t_i) \hat{O} \hat{H} \hat{U}(t,t_i) \right\}$$

$$+ i\hbar \hat{U}^\dagger(t,t_i) \frac{\partial \hat{O}}{\partial t} \hat{U}(t,t_i) \tag{3.185}$$

It is easy to see that we can rewrite this equation as

$$\frac{d\hat{O}_{\text{H}}(t)}{dt} = \frac{i}{\hbar} \left(\left[\hat{H},\hat{O}\right] \right)_{\text{H}} + \left(\frac{\partial \hat{O}}{\partial t} \right)_{\text{H}}, \tag{3.186}$$

or

$$\frac{d\hat{O}_{\text{H}}(t)}{dt} = \frac{i}{\hbar} \left[\hat{H}_{\text{H}},\hat{O}_{\text{H}}\right] + \left(\frac{\partial \hat{O}}{\partial t} \right)_{\text{H}}, \tag{3.187}$$

since $(O_1 O_2)_{\text{H}} = (O_1)_{\text{H}} (O_2)_{\text{H}}$. Note that a time-dependent $\hat{H}$ need not commute with $\hat{U}(t,t_i)$ and hence $\hat{H}_{\text{H}} \neq \hat{H}$, unless $\hat{H}$ is time-independent. For any Hermitian operator $\hat{O}$, the operator $\hat{O}_{\text{H}}$ defined by (3.184) is Hermitian $\left(\left(\hat{U}^\dagger \hat{O} \hat{U}\right)^\dagger = \hat{U}^\dagger \hat{O} \hat{U} \right)$ and is called the Heisenberg operator corresponding to the observable O. The equation (3.187) is the Heisenberg equation of motion for the quantum observables. This is the Heisenberg picture of quantum dynamics. From this picture, it follows that an observable without an explicit time dependence will be a conserved quantity if its operator commutes with the Hamiltonian. In that case, $\hat{O}_{\text{H}}(t) = \hat{U}^\dagger(t,t_i) \hat{O} \hat{U}(t,t_i) = \hat{O}$ independent of time.

In the Schrödinger picture, the state vector $|\Psi(t)\rangle$ evolves in time according to the Schrödinger equation (3.144), and the quantum observables do not change with time unless they have an explicit time dependence. In the Heisenberg picture the state vector is fixed at the initial time, say t_i, as $|\Psi(t_i)\rangle$, and the observables evolve in time according to the Heisenberg equation of motion (3.187). Rewriting the Heisenberg equation of motion (3.187) as

$$\frac{d\widehat{O}_{\mathrm{H}}(t)}{dt} = \frac{1}{i\hbar}\left[-\widehat{H}_{\mathrm{H}},\widehat{O}_{\mathrm{H}}\right] + \left(\frac{\partial \widehat{O}}{\partial t}\right)_{\mathrm{H}}, \qquad (3.188)$$

we see its striking resemblance to the classical Hamilton's equation of motion in the Poisson bracket formalism (2.52),

$$\frac{dO}{dt} = \{-H, O\} + \frac{\partial O}{\partial t}. \qquad (3.189)$$

Dirac's rule of correspondence between classical and quantum mechanics is

$$\text{quantum } \frac{1}{i\hbar}\left[\widehat{O}_1, \widehat{O}_2\right] \longrightarrow \text{classical } \{O_1, O_2\}. \qquad (3.190)$$

It is seen that the passage from the Heisenberg equation of motion for the observables (3.188) to the classical Hamilton's equation of motion for the observables in the Poisson bracket formalism (3.189) is in accordance with Dirac's rule of correspondence between classical and quantum mechanics (3.190). Writing the Heisenberg equation of motion (3.188) for observables in matrix form, with the matrix elements of the operators calculated in some orthonormal basis, takes us to Heisenberg's matrix mechanics discovered earlier to Schrödinger's wave mechanics based on differential equations. Anyway, it was soon realized that the two forms of quantum mechanics were the same. Heisenberg's matrix mechanics was the closest to classical mechanics in form because of the correspondence (3.190).

Besides the Schrödinger and the Heisenberg pictures, there is an important third picture, namely, the Dirac picture or the interaction picture. In this picture, intermediate between the Schrödinger and the Heisenberg pictures, both the state and the observables change with time. We shall consider this picture later.

3.2.1.4 Ehrenfest's Theorem

Let us now find the equation of motion for the quantum average of an observable. To this end, let us start with the Heisenberg equation of motion written as

$$\frac{d\widehat{O}_{\mathrm{H}}(t)}{dt} = \frac{i}{\hbar}\left([\widehat{H},\widehat{O}]\right)_{\mathrm{H}} + \left(\frac{\partial \widehat{O}}{\partial t}\right)_{\mathrm{H}}. \qquad (3.191)$$

A Review of Quantum Mechanics

Taking averages on both sides of this equation in the state $|\Psi(t_i)\rangle$, we get

$$\left\langle \Psi(t_i) \left| \frac{d}{dt}\hat{O}_H \right| \Psi(t_i) \right\rangle = \frac{d}{dt}\left\langle \Psi(t_i) \left| \hat{O}_H \right| \Psi(t_i) \right\rangle$$

$$= \frac{i}{\hbar}\left\langle \Psi(t_i) \left| \left([\hat{H},\hat{O}]\right)_H \right| \Psi(t_i) \right\rangle$$

$$+ \left\langle \Psi(t_i) \left| \left(\frac{\partial \hat{O}}{\partial t}\right)_H \right| \Psi(t_i) \right\rangle, \quad (3.192)$$

where d/dt has been pulled out of the term in the left-hand side since $\langle \Psi(t_i)|$ and $|\Psi(t_i)\rangle$ are time-independent vectors. Since

$$\left\langle \Psi(t_i) |\hat{O}_H| \Psi(t_i) \right\rangle = \left\langle \Psi(t_i) |\hat{U}^\dagger(t,t_i)\hat{O}\hat{U}(t,t_i)| \Psi(t_i) \right\rangle$$

$$= \langle \Psi(t)|\hat{O}|\Psi(t)\rangle = \langle O \rangle_{\Psi(t)}, \quad (3.193)$$

the average of the observable O in the Schrödinger picture, we can write (3.192) as

$$\frac{d}{dt}\langle O \rangle_{\Psi(t)} = \frac{i}{\hbar}\langle [\hat{H},\hat{O}]\rangle_{\Psi(t)} + \left\langle \frac{\partial O}{\partial t} \right\rangle_{\Psi(t)}. \quad (3.194)$$

This is the equation of motion for the average of a quantum observable, representing the general form of the Ehrenfest theorem.

To see what the equation (3.194) implies, let us take $O = x$ and

$$\hat{H} = \frac{\hat{p}_x^2}{2m} + V(x), \quad (3.195)$$

the quantum Hamiltonian of a particle of mass m moving nonrelativistically in one dimension (x-direction) in the field of a potential $V(x)$. Then, from (3.194), we find

$$\frac{d\langle x \rangle_{\Psi(t)}}{dt} = \frac{i}{\hbar}\left\langle \left[\frac{\hat{p}_x^2}{2m} + V(x), x\right]\right\rangle_{\Psi(t)}$$

$$= \frac{i}{\hbar}\left\langle \left[\frac{\hat{p}_x^2}{2m}, x\right]\right\rangle_{\Psi(t)} = \frac{1}{m}\langle p_x \rangle_{\Psi(t)}, \quad (3.196)$$

where we have used the fact that x is time-independent, and the commutation relations

$$\left[\frac{\hat{p}_x^2}{2m}, x\right] = -i\hbar\frac{\hat{p}_x}{m}, \quad [V(x), x] = 0, \quad (3.197)$$

derived with the help of (3.31). For the same system, if we take $O = p_x$, we find

$$\frac{d\langle p_x \rangle_{\Psi(t)}}{dt} = \frac{i}{\hbar}\left\langle \left[\frac{\hat{p}_x^2}{2m} + V(x), \hat{p}_x\right]\right\rangle_{\Psi(t)}$$

$$= \frac{i}{\hbar}\langle [V(x),\hat{p}_x]\rangle_{\Psi(t)} = \left\langle -\frac{\partial V(x)}{\partial x}\right\rangle_{\Psi(t)}$$

$$= \langle F(x)\rangle_{\Psi(t)}, \quad (3.198)$$

where $F(x)$ is the force experienced by the particle. Equations (3.196) and (3.198) are the classical Newton's equations if we replace the quantum averages $\langle x \rangle_{\Psi(t)}$, $\langle p_x \rangle_{\Psi(t)}$, and $\langle F(x) \rangle_{\Psi(t)}$, for position of the particle, momentum of the particle, and the force acting on the particle, respectively, by the corresponding classical observables. Thus, the Ehrenfest theorem suggests that since classical mechanics is only an approximation of quantum mechanics, classical observables are the averages of the corresponding quantum operators.

3.2.1.5 Spin

In classical mechanics, the orbital angular momentum of a particle is given by

$$\vec{L} = \vec{r} \times \vec{p}. \tag{3.199}$$

Explicitly writing, the components of $\vec{L}$ are

$$L_x = yp_z - zp_y, \qquad L_y = zp_x - xp_z, \qquad L_z = xp_y - yp_x, \tag{3.200}$$

with the Poisson bracket relations,

$$\{L_x, L_y\} = L_z, \qquad \{L_y, L_z\} = L_x, \qquad \{L_z, L_x\} = L_y. \tag{3.201}$$

In quantum mechanics, the components of the orbital angular momentum are represented by Hermitian operators

$$\begin{aligned}\widehat{L}_x &= y\widehat{p}_z - z\widehat{p}_y = -i\hbar \left(y\frac{\partial}{\partial z} - z\frac{\partial}{\partial y} \right), \\ \widehat{L}_y &= z\widehat{p}_x - x\widehat{p}_z = -i\hbar \left(z\frac{\partial}{\partial x} - x\frac{\partial}{\partial z} \right), \\ \widehat{L}_z &= x\widehat{p}_y - y\widehat{p}_x = -i\hbar \left(x\frac{\partial}{\partial y} - y\frac{\partial}{\partial x} \right),\end{aligned} \tag{3.202}$$

as obtained by the classical $\longrightarrow$ quantum transition rule. We can write

$$\vec{\widehat{L}} = \vec{r} \times \vec{\widehat{p}} = -i\hbar \vec{r} \times \vec{\nabla}. \tag{3.203}$$

The commutation relations between the components of the angular momentum operator become

$$\left[\widehat{L}_x, \widehat{L}_y \right] = i\hbar \widehat{L}_z, \qquad \left[\widehat{L}_y, \widehat{L}_z \right] = i\hbar \widehat{L}_x, \qquad \left[\widehat{L}_z, \widehat{L}_x \right] = i\hbar \widehat{L}_y. \tag{3.204}$$

A Review of Quantum Mechanics

Let us collect here, for later use, the commutation relations between the components of the angular momentum operator and the components of the position and momentum operators:

$$\left[\hat{L}_x, x\right] = 0, \quad \left[\hat{L}_x, y\right] = i\hbar z, \quad \left[\hat{L}_x, z\right] = -i\hbar y,$$
$$\left[\hat{L}_y, x\right] = -i\hbar z, \quad \left[\hat{L}_y, y\right] = 0, \quad \left[\hat{L}_y, z\right] = i\hbar x,$$
$$\left[\hat{L}_z, x\right] = i\hbar y, \quad \left[\hat{L}_z, y\right] = -i\hbar x, \quad \left[\hat{L}_z, z\right] = 0,$$
$$\left[\hat{L}_x, \hat{p}_x\right] = 0, \quad \left[\hat{L}_x, \hat{p}_y\right] = i\hbar \hat{p}_z, \quad \left[\hat{L}_x, \hat{p}_z\right] = -i\hbar \hat{p}_y,$$
$$\left[\hat{L}_y, \hat{p}_x\right] = -i\hbar \hat{p}_z, \quad \left[\hat{L}_y, \hat{p}_y\right] = 0, \quad \left[\hat{L}_y, \hat{p}_z\right] = i\hbar \hat{p}_x,$$
$$\left[\hat{L}_z, \hat{p}_x\right] = i\hbar \hat{p}_y, \quad \left[\hat{L}_z, \hat{p}_y\right] = -i\hbar \hat{p}_x, \quad \left[\hat{L}_z, \hat{p}_z\right] = 0,$$
$$\left[\vec{\hat{L}}, r^2\right] = 0, \quad \left[\vec{\hat{L}}, \hat{p}^2\right] = 0. \tag{3.205}$$

In deriving these relations, we have to use the commutator algebra (3.31).

Since $\hat{L}_x$, $\hat{L}_y$, and $\hat{L}_z$ do not commute with each other, they are incompatible observables, *i.e.*, it is not possible to measure more than one of them simultaneously and they cannot have simultaneous eigenstates. There exists an operator that commutes with all of them: the square of the total angular momentum

$$\hat{L}^2 = \hat{L}_x^2 + \hat{L}_y^2 + \hat{L}_z^2. \tag{3.206}$$

Using the commutation relations (3.204) and the commutator algebra (3.31), it is straightforward to verify that

$$\left[\hat{L}^2, \hat{L}_x\right] = 0, \quad \left[\hat{L}^2, \hat{L}_y\right] = 0, \quad \left[\hat{L}^2, \hat{L}_z\right] = 0. \tag{3.207}$$

Thus, it is possible to have simultaneous eigenstates for $\hat{L}^2$ and any one of the angular momentum components $(\hat{L}_x, \hat{L}_y, \hat{L}_z)$. The traditional choice is to use the simultaneous eigenstates of $\hat{L}^2$ and $\hat{L}_z$. Let us denote a simultaneous eigenket of $\hat{L}^2$ and $\hat{L}_z$ by $|\lambda m_\lambda\rangle$ which satisfies the eigenvalue equations

$$\hat{L}^2|\lambda m_\lambda\rangle = \lambda \hbar^2 |\lambda m_\lambda\rangle, \quad \hat{L}_z|\lambda m_\lambda\rangle = m_\lambda \hbar |\lambda m_\lambda\rangle. \tag{3.208}$$

We have taken the eigenvalues of $\hat{L}_z$ and $\hat{L}^2$ in units of $\hbar$ and $\hbar^2$, respectively, since the angular momentum has the same dimensions as $\hbar$, as seen from (3.202). Now, define

$$\hat{L}_\pm = \hat{L}_x \pm i\hat{L}_y. \tag{3.209}$$

Note that $\left(\hat{L}_\pm\right)^\dagger = \hat{L}_\mp$. Observe that

$$\left[\hat{L}_z, \hat{L}_\pm\right] = \pm \hbar \hat{L}_\pm, \quad \left[\hat{L}^2, \hat{L}_\pm\right] = 0. \tag{3.210}$$

As a consequence of these relations, we get

$$\widehat{L}^2\left(\widehat{L}_\pm|\lambda m_\lambda\rangle\right) = \widehat{L}_\pm \widehat{L}^2|\lambda m_\lambda\rangle = \lambda\hbar^2\left(\widehat{L}_\pm|\lambda m_\lambda\rangle\right),$$
$$\widehat{L}_z\left(\widehat{L}_\pm|\lambda m_\lambda\rangle\right) = \widehat{L}_\pm\left(\left(\widehat{L}_z \pm \hbar\right)|\lambda m_\lambda\rangle\right)$$
$$= (m_\lambda \pm 1)\hbar\left(\widehat{L}_\pm|\lambda m_\lambda\rangle\right). \qquad (3.211)$$

This shows that $\widehat{L}_+|\lambda m_\lambda\rangle$ and $\widehat{L}_-|\lambda m_\lambda\rangle$ are simultaneous eigenkets of $\widehat{L}^2$ and $\widehat{L}_z$ corresponding to the same eigenvalue $\lambda\hbar^2$ for $\widehat{L}^2$ and eigenvalues $(m_\lambda + 1)\hbar$ and $(m_\lambda - 1)\hbar$, respectively, for $\widehat{L}_z$. For this reason $\widehat{L}_+$ and $\widehat{L}_-$ are, respectively, called the raising and lowering operators. $\widehat{L}_z$ being a component of $\vec{\widehat{L}}$, its value cannot become greater than the total value of angular momentum. Hence, for some maximum value of m_λ, say ℓ, we should have

$$\widehat{L}^2|\lambda\ell\rangle = \lambda\hbar^2|\lambda\ell\rangle, \qquad \widehat{L}_z|\lambda\ell\rangle = \ell\hbar|\lambda\ell\rangle, \qquad \widehat{L}_+|\lambda\ell\rangle = 0. \qquad (3.212)$$

Now, from the identity

$$\widehat{L}^2 = \widehat{L}_-\widehat{L}_+ + \widehat{L}_z\left(\widehat{L}_z + \hbar\right), \qquad (3.213)$$

it follows that

$$\widehat{L}^2|\lambda\ell\rangle = \ell(\ell+1)\hbar^2|\lambda\ell\rangle. \qquad (3.214)$$

Hence,

$$\lambda = \ell(\ell+1). \qquad (3.215)$$

When the lowering operator $\widehat{L}_-$ acts repeatedly on any eigenket $|\lambda m_\lambda\rangle$, the eigenvalue $m_\lambda\hbar$ is reduced by $\hbar$ at each step. This process also cannot go on and should stop for some minimum value of m_λ, say $\underline{\ell}$, such that

$$\widehat{L}^2|\lambda\underline{\ell}\rangle = \lambda\hbar^2|\lambda\underline{\ell}\rangle, \qquad \widehat{L}_z|\lambda\underline{\ell}\rangle = \underline{\ell}\hbar|\lambda\underline{\ell}\rangle, \qquad \widehat{L}_-|\lambda\underline{\ell}\rangle = 0. \qquad (3.216)$$

Now, from the identity

$$\widehat{L}^2 = \widehat{L}_+\widehat{L}_- + \widehat{L}_z\left(\widehat{L}_z - \hbar\right), \qquad (3.217)$$

it follows that

$$\widehat{L}^2|\lambda\underline{\ell}\rangle = \underline{\ell}(\underline{\ell}-1)\hbar^2|\lambda\underline{\ell}\rangle. \qquad (3.218)$$

Hence,

$$\lambda = \underline{\ell}(\underline{\ell}-1). \qquad (3.219)$$

For this result to be consistent with (3.215), it must be that either $\underline{\ell} = \ell+1$ or $\underline{\ell} = -\ell$. The first choice is obviously absurd, and so we have $\underline{\ell} = -\ell$. Then, we get $\ell(\ell+1)\hbar^2$ as the eigenvalue of $\widehat{L}^2$ in terms of the maximum value of m_λ, namely, ℓ. The minimum value of m_λ is $-\ell$. Thus, m_λ takes all values from $-\ell$ to ℓ in integer steps.

A Review of Quantum Mechanics

If the number of integer steps from $-\ell$ to ℓ is N, then $2\ell = N$. Thus, relabeling the eigenket $|\lambda m_\lambda\rangle$ as $|\ell m_\ell\rangle$, we have

$$\hat{L}^2|\ell m_\ell\rangle = \ell(\ell+1)\hbar^2|\ell m\rangle, \qquad \hat{L}_z|\ell m_\ell\rangle = m_\ell \hbar |\ell m_\ell\rangle,$$
$$\ell = 0, 1/2, 1, 3/2, \ldots,$$
$$m_\ell = -\ell, -(\ell-1), \ldots, (\ell-1), \ell. \tag{3.220}$$

Being eigenkets of Hermitian operators, $\{|\ell m_\ell\rangle \mid \ell = 0, 1/2, 1, \ldots, m_\ell = -\ell, \ldots, \ell\}$ are orthonormal:

$$\langle \ell' m_{\ell'} | \ell m_\ell \rangle = \delta_{\ell\ell'} \delta_{m_\ell m_{\ell'}}. \tag{3.221}$$

Thus, we have solved completely the eigenvalue problem of quantum angular momentum algebraically.

From the identity (3.213), we get

$$\langle \ell m_\ell | \hat{L}_- \hat{L}_+ | \ell m_\ell \rangle = \langle \ell m_\ell | \hat{L}^2 - \hat{L}_z \left(\hat{L}_z + \hbar \right) | \ell m_\ell \rangle$$
$$= \hbar^2 (\ell(\ell+1) - m_\ell(m_\ell+1)). \tag{3.222}$$

This shows that we can take

$$\hat{L}_+ |\ell m_\ell\rangle = \hbar \sqrt{\ell(\ell+1) - m_\ell(m_\ell+1)} |\ell(m_\ell+1)\rangle,$$
$$\hat{L}_- |\ell m_\ell\rangle = \hbar \sqrt{\ell(\ell+1) - m_\ell(m_\ell-1)} |\ell(m_\ell-1)\rangle. \tag{3.223}$$

From this, we can write down the matrix elements of $\hat{L}^2$ and $\hat{L}_\pm$ between the various eigenkets as

$$\langle \ell' m_{\ell'} | \hat{L}^2 | \ell m_\ell \rangle = \ell(\ell+1)\hbar^2 \delta_{\ell\ell'} \delta_{m_\ell m_{\ell'}},$$
$$\langle \ell' m_{\ell'} | \hat{L}_+ | \ell m_\ell \rangle = \hbar \sqrt{\ell(\ell+1) - m_\ell(m_\ell+1)} \delta_{\ell\ell'} \delta_{m_{\ell'}(m_\ell+1)},$$
$$\langle \ell' m_{\ell'} | \hat{L}_- | \ell m_\ell \rangle = \hbar \sqrt{\ell(\ell+1) - m_\ell(m_\ell-1)} \delta_{\ell\ell'} \delta_{m_{\ell'}(m_\ell-1)},$$
$$\langle \ell' m_{\ell'} | \hat{L}_z | \ell m_\ell \rangle = m\hbar \delta_{\ell\ell'} \delta_{m_\ell m_{\ell'}}. \tag{3.224}$$

Note that, for any ℓ, $\hat{L}^2$ and $\hat{L}_z$ matrices are Hermitian and $\hat{L}_+$ and $\hat{L}_-$ matrices are such that $\left(\hat{L}_+\right)^\dagger = \hat{L}_-$. It is clear from these matrix elements that the action of $\hat{L}_\pm$ and $\hat{L}_z$ on any linear combination of the $2\ell+1$ degenerate eigenkets of $\hat{L}^2$ corresponding to the eigenvalue $\ell(\ell+1)\hbar^2$, namely $\{|\ell m\rangle\}$, takes it to another linear combination of the same $2\ell+1$ eigenkets. Thus, the $(2\ell+1)$-dimensional vector space spanned by the kets $\{|\ell m_\ell\rangle\}$ is invariant under the action of angular momentum operators. Hence, the vector space spanned by $\{|\ell m_\ell\rangle\}$ carries a $(2\ell+1)$-dimensional representation of the angular momentum operators $\left(\hat{L}_\pm, \hat{L}_z\right)$, or $\left\{\hat{L}_x = \left(\hat{L}_+ - \hat{L}_-\right)/2, \hat{L}_y = i\left(\hat{L}_- - \hat{L}_+\right)/2, \hat{L}_z\right\}$. Let us spell out these angular momentum matrices $\left\{L_x^{(\ell)}, L_y^{(\ell)}, L_z^{(\ell)}\right\}$ and $\left(L^{(\ell)}\right)^2$ explicitly for $\ell = 0, 1/2, 1.$:

$\ell = 0$: In this case, there is only one basis vector $|00\rangle$. In the corresponding one-dimensional representation,

$$L_x^{(0)} = 0, \qquad L_y^{(0)} = 0, \qquad L_z^{(0)} = 0, \qquad (3.225)$$

and

$$\left(L^{(0)}\right)^2 = 0. \qquad (3.226)$$

Note that, though this representation satisfies the algebra (3.204) trivially, it is not a faithful representation.

$\ell = \frac{1}{2}$: In this case, there are two basis vectors $|\frac{1}{2}\ \frac{1}{2}\rangle$ and $|\frac{1}{2}\ -\frac{1}{2}\rangle$, which can be represented by $\begin{pmatrix} 1 \\ 0 \end{pmatrix}$ and $\begin{pmatrix} 0 \\ 1 \end{pmatrix}$, respectively. The corresponding 2-dimensional representation can be worked out from (3.224) to be

$$L_x^{(\frac{1}{2})} = \frac{\hbar}{2}\begin{pmatrix} 0 & 1 \\ 1 & 0 \end{pmatrix}, \quad L_y^{(\frac{1}{2})} = \frac{\hbar}{2}\begin{pmatrix} 0 & -i \\ i & 0 \end{pmatrix}, \quad L_z^{(\frac{1}{2})} = \frac{\hbar}{2}\begin{pmatrix} 1 & 0 \\ 0 & -1 \end{pmatrix}, \qquad (3.227)$$

and

$$\left(L^{(\frac{1}{2})}\right)^2 = \frac{3}{4}\hbar^2 \begin{pmatrix} 1 & 0 \\ 0 & 1 \end{pmatrix}. \qquad (3.228)$$

Note that all the above 2×2 matrices are Hermitian. This representation satisfies the algebra (3.204) exactly and is a faithful representation.

$\ell = 1$: In this case, there are three basis vectors, namely $|1\ 1\rangle$, $|1\ 0\rangle$, and $|1\ -1\rangle$, which can be represented by $\begin{pmatrix} 1 \\ 0 \\ 0 \end{pmatrix}$, $\begin{pmatrix} 0 \\ 1 \\ 0 \end{pmatrix}$, and $\begin{pmatrix} 0 \\ 0 \\ 1 \end{pmatrix}$, respectively. Now the corresponding 3-dimensional representation becomes

$$L_x^{(1)} = \frac{\hbar}{\sqrt{2}}\begin{pmatrix} 0 & 1 & 0 \\ 1 & 0 & 1 \\ 0 & 1 & 0 \end{pmatrix}, \quad L_y^{(1)} = \frac{i\hbar}{\sqrt{2}}\begin{pmatrix} 0 & -1 & 0 \\ 1 & 0 & -1 \\ 0 & 1 & 0 \end{pmatrix},$$

$$L_z^{(1)} = \hbar \begin{pmatrix} 1 & 0 & 0 \\ 0 & 0 & 0 \\ 0 & 0 & -1 \end{pmatrix}, \quad \left(L^{(1)}\right)^2 = 2\hbar^2 \begin{pmatrix} 1 & 0 & 0 \\ 0 & 1 & 0 \\ 0 & 0 & 1 \end{pmatrix}. \qquad (3.229)$$

Note that all the above 3×3 matrices are Hermitian. This representation is a faithful representation of the algebra (3.204).

Let us now look at the angular momentum eigenvalue equations (3.220) as partial differential equations in position representation. To this end, let us transform the angular momentum operators in Cartesian coordinates (3.202) to spherical polar coordinates defined by

$$x = r\sin\theta\cos\phi, \quad y = r\sin\theta\sin\phi, \quad z = r\cos\theta. \qquad (3.230)$$

where $0 \leq \theta \leq \pi$ and $0 \leq \phi \leq 2\pi$. The result is

$$\widehat{L}_x = -i\hbar \left(-\sin\phi \frac{\partial}{\partial \theta} - \cos\phi \cot\theta \frac{\partial}{\partial \phi} \right),$$

$$\widehat{L}_x = -i\hbar \left(\cos\phi \frac{\partial}{\partial \theta} - \sin\phi \cot\theta \frac{\partial}{\partial \phi} \right),$$

$$\widehat{L}_z = -i\hbar \frac{\partial}{\partial \phi}, \qquad (3.231)$$

and

$$\widehat{L}^2 = -\hbar^2 \left[\frac{1}{\sin\theta} \frac{\partial}{\partial \theta} \left(\sin\theta \frac{\partial}{\partial \theta} \right) + \frac{1}{\sin^2\theta} \frac{\partial^2}{\partial \phi^2} \right]. \qquad (3.232)$$

Note that the angular momentum operators in spherical polar coordinates depend only on θ and ϕ and not on r. The eigenvalue equations for $\widehat{L}^2$ and $\widehat{L}_z$ can now be written as

$$-\hbar^2 \left[\frac{1}{\sin\theta} \frac{\partial}{\partial \theta} \left(\sin\theta \frac{\partial}{\partial \theta} \right) + \frac{1}{\sin^2\theta} \frac{\partial^2}{\partial \phi^2} \right] \langle \theta\phi | \ell m_\ell \rangle$$
$$= \ell(\ell+1)\hbar^2 \langle \theta\phi | \ell m_\ell \rangle,$$

$$-i\hbar \frac{\partial}{\partial \phi} \langle \theta\phi | \ell m_\ell \rangle = m_\ell \hbar \langle \theta\phi | \ell m_\ell \rangle. \qquad (3.233)$$

The solution for $\langle \theta\phi | \ell m_\ell \rangle$ is given by

$$\langle \theta\phi | \ell m_\ell \rangle = Y_\ell^{m_\ell}(\theta, \phi)$$
$$= \varepsilon \sqrt{\frac{(2\ell+1)(\ell-|m_\ell|)!}{4\pi(\ell+|m_\ell|)!}} P_\ell^{m_\ell}(\cos\theta) e^{im_\ell \phi} \qquad (3.234)$$

where

$$P_\ell^{m_\ell}(x) = (1-x^2)^{|m_\ell|/2} \left(\frac{d}{dx} \right)^{|m_\ell|} P_\ell(x),$$

$$P_\ell(x) = \frac{1}{2^\ell \ell!} \left(\frac{d}{dx} \right)^\ell (x^2-1)^\ell, \qquad (3.235)$$

ε is equal to $(-1)^{m_\ell}$ for $m_\ell \geq 0$, and 1 for $m_\ell \leq 0$. $Y_\ell^{m_\ell}(\theta, \phi)$ is called a spherical harmonic, $P_\ell^{m_\ell}(x)$ is the associated Legendre function, and $P_\ell(x)$ is the Legendre polynomial (for the details on these special functions and polynomials, see, *e.g.*, Arfken, Weber, and Harris [3], Byron and Fuller [18], and the book of Lakshminarayanan and Varadharajan [126], which is an excellent source for all important special functions from a practical point of view with computational codes in Python). Now, it is to be seen that when ϕ increases by 2π we return to the same point in space so that we have to require

$$\langle \theta, \phi + 2\pi | \ell m_\ell \rangle = \langle \theta\phi | \ell m_\ell \rangle. \qquad (3.236)$$

This implies that we must have

$$e^{im_\ell(\phi+2\pi)} = e^{im_\ell\phi} \qquad (3.237)$$

In other words, $e^{2\pi i m_\ell} = 1$, *i.e.*,

$$m_\ell = 0, \pm 1, \pm 2, \ldots. \qquad (3.238)$$

This means that ℓ, called the orbital angular momentum quantum number, can take only nonnegative integer values. Also, only for nonnegative integer values of ℓ, the definition of $P_\ell(x)$ in the solution (3.234) makes sense. The orbital angular momentum of the particle is given by $\sqrt{\ell(\ell+1)}\hbar$, the square root of the eigenvalue of $\widehat{L}^2$.

If the orbital angular momentum quantum number can take only nonnegative integer values, is there any physical significance for the matrix representations of $\vec{L}$ for half integer values of ℓ? The answer is that there is an intrinsic $\vec{L}$-like dynamical variable associated with particles, called the spin, which can take any one of the allowed values of $\ell = 0, \frac{1}{2}, 1, \frac{3}{2}, \ldots$. While the orbital angular momentum quantum number of a particle ℓ can take all nonnegative integer values, the spin quantum number of a particle takes only a fixed value. Let us relabel ℓ as s to refer to the spin quantum number and relabel the corresponding matrices (L_x, L_y, L_z) as (S_x, S_y, S_z). Then, for any particle the spin quantum number s has a fixed value 0, or $\frac{1}{2}$, or 1, or $\frac{3}{2}$, or $\ldots$.

For a particle with spin $s = 0$ there is only one spin state, and so we can consider it simply as a spectator variable. In other words, we can continue to use the single-component state vector $|\Psi(t)\rangle$ in single particle dynamical problems, ignoring the presence of spin. However, in dealing with the physics of multiple identical spin-0 particles, one has to consider them as bosons obeying the Bose–Einstein statistics. Any number of identical bosons can occupy a single quantum state. All integer spin particles, *i.e.*, particles with $s = 0, 1, 2, \ldots$, are bosons.

For a spin-$\frac{1}{2}$ particle spin is an observable vector quantity $\vec{S}$ with three components (S_x, S_y, S_z). Since spin has no classical analog, we have to only postulate the corresponding Hermitian operators. Since the z-component of the spin of electron is found, experimentally, to have two values, namely $\pm\frac{1}{2}$, and to get coupled to the orbital angular momentum in atomic systems, the components of $\vec{S}$ are taken to obey the same algebra as the components of $\vec{L}$. Thus, for a spin-$\frac{1}{2}$ particle, we take

$$S_x = \frac{\hbar}{2}\begin{pmatrix} 0 & 1 \\ 1 & 0 \end{pmatrix}, \quad S_y = \frac{\hbar}{2}\begin{pmatrix} 0 & -i \\ i & 0 \end{pmatrix}, \quad S_z = \frac{\hbar}{2}\begin{pmatrix} 1 & 0 \\ 0 & -1 \end{pmatrix}, \qquad (3.239)$$

following the two-dimensional matrix representation of the angular momentum algebra corresponding to $\ell = 1/2$ as given in (3.227). In terms of the Pauli matrices,

$$\sigma_x = \begin{pmatrix} 0 & 1 \\ 1 & 0 \end{pmatrix}, \quad \sigma_y = \begin{pmatrix} 0 & -i \\ i & 0 \end{pmatrix}, \quad \sigma_z = \begin{pmatrix} 1 & 0 \\ 0 & -1 \end{pmatrix}, \qquad (3.240)$$

we write

$$S_x = \frac{\hbar}{2}\sigma_x, \quad S_y = \frac{\hbar}{2}\sigma_y, \quad S_z = \frac{\hbar}{2}\sigma_z. \qquad (3.241)$$

Hence for a spin-$\frac{1}{2}$ particle, there are two spin states: $|\frac{1}{2}\,\frac{1}{2}\rangle$ and $|\frac{1}{2}\,-\frac{1}{2}\rangle$. Since s has a fixed value $\frac{1}{2}$, we can drop it in the specification of states and write simply $|\frac{1}{2}\,\frac{1}{2}\rangle = |\frac{1}{2}\rangle$ and $|\frac{1}{2}\,-\frac{1}{2}\rangle = |-\frac{1}{2}\rangle$. Physically, when a magnetic field is switched on in the z-direction, a charged spin-$\frac{1}{2}$ particle aligns its spin either in the z-direction, *i.e.*, parallel to the field, or in the $-z$-direction, *i.e.*, antiparallel to the field. When its spin is aligned parallel to the field, its spin state is $|\frac{1}{2}\rangle$ and $S_z = \frac{\hbar}{2}$, and when it is aligned antiparallel to the field, its spin state is $|-\frac{1}{2}\rangle$ and $S_z = -\frac{\hbar}{2}$. So the states $|\frac{1}{2}\rangle$ and $|-\frac{1}{2}\rangle$ are also referred to as up and down states and written as $|\uparrow\rangle$ and $|\downarrow\rangle$, respectively. We can also write $|\uparrow\rangle = \begin{pmatrix} 1 \\ 0 \end{pmatrix}$ and $|\downarrow\rangle = \begin{pmatrix} 0 \\ 1 \end{pmatrix}$. Then, we have

$$S_z|\uparrow\rangle = \frac{\hbar}{2}|\uparrow\rangle, \quad S_z|\downarrow\rangle = -\frac{\hbar}{2}|\downarrow\rangle,$$
$$S^2|\uparrow\rangle = \frac{3}{4}\hbar^2|\uparrow\rangle, \quad S^2|\downarrow\rangle = \frac{3}{4}\hbar^2|\downarrow\rangle. \tag{3.242}$$

The state vector of a spin-$\frac{1}{2}$ particle can be represented as

$$|\Psi(t)\rangle = |\Psi_1(t)\rangle|\uparrow\rangle + |\Psi_2(t)\rangle|\downarrow\rangle = \begin{pmatrix} |\Psi_1(t)\rangle \\ |\Psi_2(t)\rangle \end{pmatrix}. \tag{3.243}$$

Note that the quantum operator of any classical observable O commutes with all the components of spin, $\vec{S}$, and so any one component of $\vec{S}$ can be measured precisely along with O. If position and spin of the particle are measured simultaneously in the above state $|\Psi(t)\rangle$, the probability for the result {position $= \vec{r}$, spin $=\uparrow$} is $|\langle\vec{r}|\Psi_1(t)\rangle|^2$ and the probability for the result {position $= \vec{r}$, spin $=\downarrow$} is $|\langle\vec{r}|\Psi_2(t)\rangle|^2$. Hence, the normalization condition now becomes

$$\int d^3r \left(|\langle\vec{r}|\Psi_1(t)\rangle|^2 + |\langle\vec{r}|\Psi_2(t)\rangle|^2\right)$$
$$= \int d^3r \left(\langle\Psi_1(t)|\vec{r}\rangle\langle\vec{r}|\Psi_1(t)\rangle + \langle\Psi_2(t)|\vec{r}\rangle\langle\vec{r}|\Psi_2(t)\rangle\right)$$
$$= \langle\Psi_1(t)|\Psi_1(t)\rangle + \langle\Psi_2(t)|\Psi_2(t)\rangle$$
$$= (\langle\Psi_1(t)|\langle\Psi_2(t)|) \begin{pmatrix} |\Psi_1(t)\rangle \\ |\Psi_2(t)\rangle \end{pmatrix}$$
$$= \langle\Psi(t)|\Psi(t)\rangle = 1. \tag{3.244}$$

Note that $|\langle\vec{r}|\Psi_1(t)\rangle|^2 + |\langle\vec{r}|\Psi_2(t)\rangle|^2$ is the probability for finding the particle at $\vec{r}$ with its spin up or down. The probability for finding the particle with its spin up, independent of its position, is $\langle\Psi_1(t)|\Psi_1(t)\rangle$ and the probability for finding the particle with its spin down, independent of its position, is $\langle\Psi_2(t)|\Psi_2(t)\rangle$.

Particles with half odd integer spin, *i.e.*, $s = \frac{1}{2}, \frac{3}{2}, \ldots$, are fermions that obey the Fermi–Dirac statistics. They obey the Pauli exclusion principle according to which two identical fermions cannot be in the same quantum state.

3.3 NONRELATIVISTIC QUANTUM MECHANICS

All physical systems are basically quantum mechanical, and the classical behavior of macroscopic bodies is an approximation. Similarly, basic physics is relativistic, and nonrelativistic phenomena are approximations. The classical Hamiltonian of a particle of mass m and charge q moving in an electromagnetic field and the field of a potential V is given by

$$H = \sqrt{m^2 c^4 + c^2 \vec{\pi}^2} + q\phi + V, \qquad (3.245)$$

where ϕ and $\vec{A}$ are the scalar and vector potentials of the electromagnetic field, respectively, and the potential V gives rise to the force $-\vec{\nabla} V(\vec{r}, t)$ acting on the particle. The Hamiltonian H represents the total energy of the particle. When the motion of the particle is nonrelativistic ($|\vec{v}| \ll c$), such that its kinetic energy $\left(\sqrt{m^2 c^4 + c^2 \vec{\pi}^2} - mc^2 \right)$ is very small compared to its rest energy mc^2, we can approximate the classical Hamiltonian as

$$H \approx mc^2 + \frac{\vec{\pi}^2}{2m} + q\phi + V. \qquad (3.246)$$

Since the rest energy term mc^2 is a constant, we can set it as the zero of the energy scale and take $H - mc^2$ as the nonrelativistic Hamiltonian of the particle. Hence, we take the classical nonrelativistic Hamiltonian of the particle to be

$$H = \frac{\vec{\pi}^2}{2m} + q\phi + V. \qquad (3.247)$$

To study the quantum mechanics of the system, we have to get the quantum Hamiltonian operator $\widehat{H}$ from the classical Hamiltonian $H(\vec{r}, \vec{p}, t)$ by replacing in it $\vec{p}$ by $\widehat{\vec{p}}$ and making the resulting expression Hermitian by suitable symmetrization procedure. Thus, we get the quantum Hamiltonian of a particle of mass m and charge q moving nonrelativistically in the field of a potential V and an electromagnetic field with ϕ and $\vec{A}$ as the scalar and vector potentials to be

$$\widehat{H} = \frac{\widehat{\vec{\pi}}^2}{2m} + q\phi + V. \qquad (3.248)$$

Note that in expanding $\widehat{\vec{\pi}}^2 = (\widehat{\vec{p}} - q\vec{A}) \cdot (\widehat{\vec{p}} - q\vec{A})$, one has to be careful, since the components of $\vec{r}$ and $\widehat{\vec{p}}$ do not commute. We shall now look at how nonrelativistic quantum mechanics works.

3.3.1 NONRELATIVISTIC SINGLE PARTICLE QUANTUM MECHANICS

3.3.1.1 Free Particle

Single free particle is the simplest physical system. For a particle moving in free space, with $\vec{A} = 0$, $\phi = 0$, and $V = 0$, the Hamiltonian (3.248) becomes

$$\widehat{H} = \frac{\widehat{\vec{p}}^2}{2m}. \qquad (3.249)$$

A Review of Quantum Mechanics

The dynamics is given by the Schrödinger equation

$$i\hbar \frac{\partial |\Psi(t)\rangle}{\partial t} = \widehat{H} |\Psi(t)\rangle. \tag{3.250}$$

Since the Hamiltonian is time-independent, we can integrate this equation immediately to get

$$|\Psi(t)\rangle = \widehat{U}(t,t_i) |\Psi(t_i)\rangle = e^{-\frac{i}{\hbar}(t-t_i)\widehat{H}} |\Psi(t_i)\rangle. \tag{3.251}$$

Now, we have to choose a representation. Let us choose the energy representation in which the eigenvectors of the Hamiltonian form the complete orthonormal basis. To this end, let us look at the eigenvalue equation for the Hamiltonian

$$\widehat{H} |\psi\rangle = \frac{\widehat{p}^2}{2m} |\psi\rangle = E |\psi\rangle. \tag{3.252}$$

We have already solved the eigenvalue equation for the momentum operator. We know that

$$\vec{\widehat{p}} |\phi_{\vec{p}}\rangle = \vec{p} |\phi_{\vec{p}}\rangle, \tag{3.253}$$

with

$$\langle \phi_{\vec{p}} | \phi_{\vec{p}'} \rangle = \delta(\vec{p} - \vec{p}'). \tag{3.254}$$

It is obvious that

$$\frac{\widehat{p}^2}{2m} |\phi_{\vec{p}}\rangle = \frac{p^2}{2m} |\phi_{\vec{p}}\rangle, \quad \text{with } p^2 = |\vec{p}|^2. \tag{3.255}$$

Hence, we find that the free-particle Hamiltonian $\widehat{H}$ has a continuous eigenvalue spectrum $\{E(p) = p^2/2m \mid 0 \le p \le \infty\}$.

In position representation

$$\langle \vec{r} | \phi_{\vec{p}} \rangle = \phi_{\vec{p}}(\vec{r}) = \frac{1}{(2\pi\hbar)^{3/2}} e^{\frac{i}{\hbar} \vec{p} \cdot \vec{r}}. \tag{3.256}$$

The free-particle wave function, corresponding to a particle of momentum $\vec{p}$ and energy $p^2/2m$, is

$$\Psi(\vec{r},t) = \frac{1}{(2\pi\hbar)^{3/2}} e^{\frac{i}{\hbar}\left(\vec{p}\cdot\vec{r} - \frac{p^2}{2m}t\right)}. \tag{3.257}$$

In one dimension, a particle of momentum p moving along the $+x$-direction has the wave function

$$\Psi(x,t) = \frac{1}{(2\pi\hbar)^{1/2}} e^{\frac{i}{\hbar}\left(px - \frac{p^2}{2m}t\right)}, \tag{3.258}$$

and the wave function

$$\Psi(x,t) = \frac{1}{(2\pi\hbar)^{1/2}} e^{-\frac{i}{\hbar}\left(px + \frac{p^2}{2m}t\right)} \tag{3.259}$$

represents a particle of momentum p moving along the $-x$-direction.

Note that each eigenvalue $E(p)$ is infinitely degenerate with the set of eigenvectors $\{|\phi_{\vec{p}}\rangle \,|\, |\vec{p}| = p\}$ belonging to the same eigenvalue $E(p)$. This means that a ket vector defined by

$$|\psi_p\rangle = \int_{|\vec{p}|=p} d^3p\, C(\vec{p}) |\phi_{\vec{p}}\rangle, \tag{3.260}$$

with arbitrary coefficients $\{C(\vec{p})\}$, subject to normalization of $|\psi_p\rangle$, will be an eigenvector of $\widehat{H}$ corresponding to the eigenvalue $E(p) = p^2/2m$. If we specify the eigenvalues of both $\widehat{H}$ and $\vec{\hat{p}}$, then the degeneracy is lifted, and we get a unique eigenvector $|\phi_{\vec{p}}\rangle$.

Now, let us take $t_i = 0$ in (3.251) and consider

$$|\Psi(0)\rangle = \int d^3p\, C(\vec{p},0) |\phi_{\vec{p}}\rangle, \tag{3.261}$$

with the coefficients $\{C(\vec{p},0) = \langle \phi_{\vec{p}} | \Psi(0) \rangle\}$ satisfying the relation

$$\int d^3p\, |C(\vec{p},0)|^2 = 1, \tag{3.262}$$

as required by the normalization condition

$$\begin{aligned}\langle \Psi(0)|\Psi(0)\rangle &= \int\int d^3p'd^3p\, C^*(\vec{p}',0)C(\vec{p},0)\langle \phi_{\vec{p}'}|\phi_{\vec{p}}\rangle \\ &= \int\int d^3p'd^3p\, C^*(\vec{p}',0)C(\vec{p},0)\delta(\vec{p}-\vec{p}') \\ &= \int d^3p\, |C(\vec{p},0)|^2 = 1. \end{aligned} \tag{3.263}$$

Then, we have

$$|\Psi(t)\rangle = \int d^3p\, e^{-\frac{ip^2t}{2m\hbar}} C(\vec{p},0) |\phi_{\vec{p}}\rangle = \int d^3p\, C(\vec{p},t) |\phi_{\vec{p}}\rangle. \tag{3.264}$$

If we represent the state vectors $|\Psi(t)\rangle$ and $|\Psi(0)\rangle$ by the column vectors of the coefficients, then the above equation becomes

$$\begin{pmatrix} \vdots \\ C(\vec{p},t) \\ \vdots \end{pmatrix} = \begin{pmatrix} \vdots & \vdots & \vdots & \vdots & \vdots \\ \cdot & \cdot & e^{-ip^2t/2m\hbar} & \cdot & \cdot \\ \vdots & \vdots & \vdots & \vdots & \vdots \end{pmatrix} \begin{pmatrix} \vdots \\ C(\vec{p},0) \\ \vdots \end{pmatrix}, \tag{3.265}$$

showing that the matrix representing the unitary time-evolution operator $\widehat{U}(t,0) = e^{-it\widehat{H}/\hbar}$ is diagonal in the energy representation. In other words, the Hamiltonian is diagonal in the energy representation.

In the position representation, corresponding to (3.261), we have

$$\langle \vec{r}|\Psi(0)\rangle = \int d^3p\, C(\vec{p},0) \langle \vec{r}|\phi_{\vec{p}}\rangle, \tag{3.266}$$

or

$$\Psi(\vec{r},0) = \int d^3p\, C(\vec{p},0)\phi_{\vec{p}}(\vec{r})$$
$$= \frac{1}{(2\pi\hbar)^{3/2}} \int d^3p\, C(\vec{p},0) e^{\frac{i}{\hbar}\vec{p}\cdot\vec{r}}. \tag{3.267}$$

For the particle with this initial wave function $\Psi(\vec{r},0)$ at time 0, the wavefunction at a later time t will be, following (3.264),

$$\Psi(\vec{r},t) = \frac{1}{(2\pi\hbar)^{3/2}} \int d^3p\, C(\vec{p},0) e^{\frac{i}{\hbar}(\vec{p}\cdot\vec{r}-E(p)t)}, \tag{3.268}$$

where $E(p) = p^2/2m$. This represents a wave packet solution of the time-dependent Schrödinger equation (3.250) in the position representation. This wave packet is not an eigenstate of the Hamiltonian, since it is a linear superposition of several eigenstates with different energy eigenvalues. It has an average energy

$$\langle E \rangle_{\Psi(t)} = \langle \Psi(t)|\hat{H}|\Psi(t)\rangle$$
$$= \frac{1}{2m} \int d^3r\, \Psi^*(\vec{r},t)\hat{p}^2\Psi(\vec{r},t)$$
$$= \frac{1}{2m} \int d^3p\, |C(\vec{p},0)|^2 p^2 = \frac{1}{2m}\langle p^2\rangle_{\Psi(0)}. \tag{3.269}$$

Note that the average energy is a constant of motion, in accordance with (3.194), since the quantum operator for energy, the time-independent Hamiltonian, $\hat{H}$, commutes with itself. Similarly, the averages of the momentum components will also be constants of motion for this state, since the momentum operator commutes with the Hamiltonian, i.e., $\langle \vec{p}\rangle_{\Psi(t)} = \langle \vec{p}\rangle_{\Psi(0)}$. For the average position of the particle, represented by the above wave packet, the equation of motion (3.194) shows that

$$\frac{d\langle \vec{r}\rangle_{\Psi(t)}}{dt} = \frac{1}{m}\langle \vec{p}\rangle_{\Psi(t)}, \tag{3.270}$$

in accordance with the classical relation between the velocity and momentum, $d\vec{r}/dt = \vec{p}/m$, (á la Ehrenfest's theorem).

In the Heisenberg picture, we can write the time-evolution, or transfer, map of $\left(\vec{\hat{r}},\vec{\hat{p}}\right)$ as

$$\begin{pmatrix} \vec{\hat{r}}(t) \\ \vec{\hat{p}}(t) \end{pmatrix} = \begin{pmatrix} e^{\frac{it\hat{p}^2}{2m\hbar}} \vec{\hat{r}} e^{\frac{-it\hat{p}^2}{2m\hbar}} \\ e^{\frac{it\hat{p}^2}{2m\hbar}} \vec{\hat{p}} e^{\frac{-it\hat{p}^2}{2m\hbar}} \end{pmatrix}$$
$$= \begin{pmatrix} \vec{\hat{r}}(0) + \frac{t}{m}\vec{\hat{p}}(0) \\ \vec{\hat{p}}(0) \end{pmatrix} = \begin{pmatrix} 1 & \frac{t}{m} \\ 0 & 1 \end{pmatrix} \begin{pmatrix} \vec{\hat{r}}(0) \\ \vec{\hat{p}}(0) \end{pmatrix}. \tag{3.271}$$

Note that $e^{it\hat{p}^2/2m\hbar}\vec{\hat{r}}e^{-it\hat{p}^2/2m\hbar}$ and $e^{it\hat{p}^2/2m\hbar}\vec{\hat{p}}e^{-it\hat{p}^2/2m\hbar}$ are of the form $e^{\hat{A}}\hat{B}e^{-\hat{A}}$, which can be calculated using the identity (3.142)

$$e^{\hat{A}}\hat{B}e^{-\hat{A}} = e^{:\hat{A}:}\hat{B} = \left(I + :\hat{A}: + \frac{1}{2!}:\hat{A}:^2 + \frac{1}{3!}:\hat{A}:^3 + \cdots\right)\hat{B}, \quad (3.272)$$

by substituting $it\hat{p}^2/2m\hbar$ for $\hat{A}$, and $\hat{r}$ and $\hat{p}$ for $\hat{B}$. If the series terminates because of the vanishing of a commutator at any step, we get an exact expression as in the present case.

Let us now have a closer look at the Heisenberg uncertainty principle in the case of the conjugate pair of dynamical variables x and p_x. For any observable O, the uncertainty in a state $|\Psi\rangle$ is defined by

$$(\Delta O)_\Psi = \sqrt{\left\langle \left(\hat{O} - \langle O\rangle_\Psi\right)^2 \right\rangle_\Psi}, \quad (3.273)$$

which is a measure of the spread, or dispersion, of its values about the average, or the mean value, and $(\Delta O)^2_\Psi$ is known as variance, second moment about the mean, or the second-order central moment. Note that

$$\left\langle \left(\hat{O} - \langle O\rangle_\Psi\right)\right\rangle_\Psi = 0. \quad (3.274)$$

Since

$$\begin{aligned}\left\langle \left(\hat{O} - \langle O\rangle_\Psi\right)^2 \right\rangle_\Psi &= \left\langle \hat{O}^2 + \langle O\rangle^2_\Psi - 2\hat{O}\langle O\rangle_\Psi\right\rangle_\Psi \\ &= \left\langle \hat{O}^2 + \langle O\rangle^2_\Psi - 2\langle O\rangle^2_\Psi\right\rangle_\Psi \\ &= \langle O^2\rangle_\Psi - \langle O\rangle^2_\Psi, \quad (3.275)\end{aligned}$$

we can also write

$$(\Delta O)_\Psi = \sqrt{\langle O^2\rangle_\Psi - \langle O\rangle^2_\Psi}. \quad (3.276)$$

When $|\Psi\rangle$ is an eigenstate of $\hat{O}$, $(\Delta O)_\Psi = 0$.

Let O_1 and O_2 be two incompatible observables (*i.e.*, $\hat{O}_1\hat{O}_2 \neq \hat{O}_2\hat{O}_1$) with uncertainties $(\Delta O_1)_\Psi$ and $(\Delta O_2)_\Psi$ in a state $|\Psi\rangle$. We shall drop the subscript Ψ to be understood from the context. Let

$$\widehat{\delta O_1} = \hat{O}_1 - \langle O_1\rangle, \quad \widehat{\delta O_2} = \hat{O}_2 - \langle O_2\rangle, \quad (3.277)$$

which are Hermitian operators. Now, we can write

$$(\Delta O_1)^2 = \left(\langle\Psi|\widehat{\delta O_1}\right)\left(\widehat{\delta O_1}|\Psi\rangle\right), \quad (\Delta O_2)^2 = \left(\langle\Psi|\widehat{\delta O_2}\right)\left(\widehat{\delta O_2}|\Psi\rangle\right). \quad (3.278)$$

A Review of Quantum Mechanics

According to the Cauchy–Schwarz inequality, or the Schwarz inequality (see, *e.g.*, Arfken, Weber, and Harris [3], and Byron and Fuller [18]), for any two functions f and g,

$$\langle f|f\rangle \langle g|g\rangle \geq |\langle f|g\rangle|^2. \tag{3.279}$$

Taking $|f\rangle = \widehat{\delta O_1}|\Psi\rangle$ and $|g\rangle = \widehat{\delta O_2}|\Psi\rangle$, we get

$$(\Delta O_1)^2 (\Delta O_2)^2 \geq |\langle \Psi|\widehat{\delta O_1}\widehat{\delta O_2}|\Psi\rangle|^2. \tag{3.280}$$

Note that $\langle \Psi|\widehat{\delta O_1}\widehat{\delta O_2}|\Psi\rangle$ is a complex number and recall that $\langle f|g\rangle^* = \langle g|f\rangle$. For any complex number $z = x+iy$, $|z|^2 = x^2+y^2 \geq y^2 = [(z-z^*)/2i]^2$. Thus, we have

$$(\Delta O_1)^2 (\Delta O_2)^2 \geq \left(\frac{1}{2i} \left[\left\langle \Psi \left| \widehat{\delta O_1}\widehat{\delta O_2} \right| \Psi \right\rangle - \left\langle \Psi \left| \widehat{\delta O_2}\widehat{\delta O_1} \right| \Psi \right\rangle \right] \right)^2$$

$$= \left(\frac{1}{2i} \left[\left\langle \Psi \left| \left[\widehat{\delta O_1}, \widehat{\delta O_2} \right] \right| \Psi \right\rangle \right] \right)^2. \tag{3.281}$$

Since

$$\left[\widehat{\delta O_1}, \widehat{\delta O_2} \right] = \left[\hat{O}_1 - \langle O_1\rangle, \hat{O}_2 - \langle O_2\rangle \right] = \left[\hat{O}_1, \hat{O}_2 \right], \tag{3.282}$$

we get

$$(\Delta O_1)^2 (\Delta O_2)^2 \geq \left(\left\langle \frac{1}{2i} \left[\hat{O}_1, \hat{O}_2 \right] \right\rangle \right)^2, \tag{3.283}$$

or

$$(\Delta O_1)(\Delta O_2) \geq \left| \left\langle \frac{1}{2i} \left[\hat{O}_1, \hat{O}_2 \right] \right\rangle \right|. \tag{3.284}$$

Note that for any two Hermitian operators $\hat{O}_1$ and $\hat{O}_2$, $\left(\left[\hat{O}_1, \hat{O}_2 \right]/2i \right)$ is a Hermitian operator.

If we now take $O_1 = x$ and $O_2 = p_x$ in (3.284), we get the Heisenberg uncertainty relation

$$(\Delta x)(\Delta p_x) \geq \frac{\hbar}{2}. \tag{3.285}$$

The free-particle eigenstate with specific momentum and energy eigenvalues, $\vec{p}$ and $E(\vec{p}) = |\vec{p}|^2/2m$, given in position representation by,

$$\Psi(\vec{r},t) = \frac{1}{(2\pi\hbar)^{3/2}} e^{\frac{i}{\hbar}(\vec{p}\cdot\vec{r}-E(\vec{p})t)}, \tag{3.286}$$

has $\Delta p_x = \Delta p_y = \Delta p_z = 0$ and $\Delta x = \Delta y = \Delta z \longrightarrow \infty$. Note that this free-particle wave function represents a plane wave

$$\Psi(\vec{r},t) = \frac{1}{(2\pi\hbar)^{3/2}} e^{i(\vec{k}\cdot\vec{r}-\omega(\vec{k})t)}, \tag{3.287}$$

with

$$\hbar\vec{k} = \vec{p}, \qquad \hbar\omega(\vec{k}) = E(\vec{p}) = \frac{\hbar^2 k^2}{2m}. \tag{3.288}$$

This takes us to the beginning of wave mechanics, or the early quantum mechanics, when de Broglie associated this plane wave with a free particle. The wavelength of this wave $\lambda = 2\pi/|\vec{k}| = 2\pi\hbar/|\vec{p}| = h/p$ is called the de Broglie wavelength of the particle, and the frequency of the wave is $\nu = \omega/2\pi$ in accordance with the relation $E = h\nu$ postulated by Planck for a harmonic oscillator in the context of blackbody radiation, and adopted by Einstein and Bohr in their theories of light and atomic spectra, respectively. History of quantum mechanics starts with Planck's relation $E = h\nu$. In case the motion of the particle is not nonrelativistic, the associated de Broglie wave will have the same plane waveform (3.287) with $\hbar\omega(\vec{k}) = \sqrt{m^2c^4 + c^2p^2}$ and will obey a relativistic generalization of the nonrelativistic Schrödinger equation known as the Klein–Gordon equation, which we shall study later. Association of a physical wave with a particle was abandoned later when Born's interpretation of $|\Psi(\vec{r},t)|^2$ as the position probability density was established.

As an example of the Heisenberg uncertainty principle, let us consider the dynamics of a free-particle wave packet in one dimension (x-direction). Let the particle be, at $t = 0$, in the state

$$|\Psi(0)\rangle = \int_{-\infty}^{\infty} dp_x\, C(p_x,0)|\phi_{p_x}\rangle, \quad (3.289)$$

where

$$C(p_x,0) = \left(\frac{2}{\pi}\right)^{1/4} \sqrt{\frac{(\Delta x)_0}{\hbar}} e^{-\frac{(\Delta x)_0^2 (p_x - p_0)^2}{\hbar^2}}, \quad (3.290)$$

a Gaussian function such that

$$\int_{-\infty}^{\infty} dp_x\, |C(p_x,0)|^2 = 1. \quad (3.291)$$

To verify (3.291), use the well-known result on the Fourier transform of the Gaussian function, (see *e.g.*, Arfken, Weber, and Harris [3], and Byron and Fuller [18])

$$\int_{-\infty}^{\infty} d\chi\, e^{-a\chi^2 + i\chi\eta} = \sqrt{\frac{\pi}{a}} e^{-\frac{\eta^2}{4a}}, \quad (3.292)$$

with the assumption that a has a positive real part and its square root $\sqrt{a}$ is also chosen to have a positive real part. In position representation, the wave function of the particle is, with $\bar{p} = p_x - p_0$,

$$\begin{aligned}
\Psi(x,0) &= \frac{1}{\sqrt{2\pi\hbar}} \left(\frac{2}{\pi}\right)^{1/4} \sqrt{\frac{(\Delta x)_0}{\hbar}} \int_{-\infty}^{\infty} dp_x\, e^{-\frac{(\Delta x)_0^2 (p_x - p_0)^2}{\hbar^2} + \frac{i}{\hbar} p_x x} \\
&= \sqrt{\frac{(\Delta x)_0}{2\pi\hbar^2}} \left(\frac{2}{\pi}\right)^{1/4} e^{\frac{i}{\hbar} p_0 x} \int_{-\infty}^{\infty} d\bar{p}\, e^{-\frac{(\Delta x)_0^2 \bar{p}^2}{\hbar^2} + \frac{i}{\hbar}\bar{p}x} \\
&= \frac{1}{(\sqrt{2\pi}(\Delta x)_0)^{1/2}} e^{-\frac{x^2}{4(\Delta x)_0^2} + \frac{i}{\hbar} p_0 x}, \quad (3.293)
\end{aligned}$$

where we have used again (3.292). Using (3.268), the wave function of the wave packet at $t > 0$ is found to be given by

$$\Psi(x,t) = \left(\frac{2}{\pi}\right)^{1/4} \sqrt{\frac{(\Delta x)_0}{\hbar}} \frac{1}{\sqrt{2\pi\hbar}} \int_{-\infty}^{\infty} dp_x \, e^{-\frac{(\Delta x)_0^2(p_x-p_0)^2}{\hbar^2}} e^{\frac{i}{\hbar}\left(p_x x - \frac{p_x^2}{2m}t\right)}$$

$$= \left(\frac{2}{\pi}\right)^{1/4} \sqrt{\frac{(\Delta x)_0}{\hbar}} \frac{1}{\sqrt{2\pi\hbar}} e^{\frac{i}{\hbar}\left(p_0 x - \frac{p_0^2}{2m}t\right)}$$

$$\times \int_{-\infty}^{\infty} d\bar{p} \, e^{-\frac{(\Delta x)_0^2 \bar{p}^2}{\hbar^2}} e^{\frac{i}{\hbar}\left[-\frac{\bar{p}^2}{2m}t + \bar{p}\left(x - \frac{p_0}{m}t\right)\right]}$$

$$= \left(\frac{2}{\pi}\right)^{1/4} \sqrt{\frac{(\Delta x)_0}{2\pi\hbar^2}} e^{\frac{i}{\hbar}\left(p_0 x - \frac{p_0^2}{2m}t\right)}$$

$$\times \int_{-\infty}^{\infty} d\bar{p} \, e^{-\frac{\bar{p}^2(\Delta x)_0^2}{\hbar^2}\left(1+\frac{i\hbar t}{2m(\Delta x)_0^2}\right)} e^{i\frac{\bar{p}}{\hbar}\left(x - \frac{p_0}{m}t\right)}$$

$$= \frac{1}{\left(\sqrt{2\pi}(\Delta x)_0(1+i\delta)\right)^{1/2}} e^{-\frac{\bar{x}^2}{4(\Delta x)_0^2(1+i\delta)} + \frac{i}{\hbar}\left(p_0 x - \frac{p_0^2}{2m}t\right)}, \qquad (3.294)$$

with

$$\delta = \frac{\hbar t}{2m(\Delta x)_0^2}, \qquad \bar{x} = x - \frac{p_0}{m}t. \qquad (3.295)$$

The last step can be verified again using (3.292). Note that as $t \longrightarrow 0$, $\Psi(x,t) \longrightarrow \Psi(x,0)$.

For the initial wave packet $\Psi(x,0)$, the average value of x is given by

$$\langle x \rangle(0) = \int_{-\infty}^{\infty} dx \, \Psi(x,0)^* x \Psi(x,0) = \int_{-\infty}^{\infty} dx \, x |\Psi(x,0)|^2$$

$$= \frac{1}{\sqrt{2\pi}(\Delta x)_0} \int_{-\infty}^{\infty} dx \, x e^{-\frac{x^2}{2(\Delta x)_0^2}} = 0, \qquad (3.296)$$

where the integral vanishes since the integrand is an odd function of x. This result shows that initially, at $t = 0$, the wave packet is centered at $x = 0$. The average of x^2 for $\Psi(x,0)$ is given by

$$\langle x^2 \rangle(0) = \int_{-\infty}^{\infty} dx \, \Psi(x,0)^* x^2 \Psi(x,0) = \int_{-\infty}^{\infty} dx \, x^2 |\Psi(x,0)|^2$$

$$= \frac{1}{\sqrt{2\pi}(\Delta x)_0} \int_{-\infty}^{\infty} dx \, x^2 e^{-\frac{x^2}{2(\Delta x)_0^2}} = (\Delta x)_0^2, \qquad (3.297)$$

where we have used the integral

$$\int_{-\infty}^{\infty} d\chi \, \chi^2 e^{-a\chi^2} = \sqrt{\frac{\pi}{4a^3}}, \qquad (3.298)$$

which follows by differentiating the integral in (3.292), with $\eta = 0$, with respect to a. Thus, the uncertainty in x for the wave packet $\Psi(x,0)$ is

$$(\Delta x)(0) = \sqrt{\langle x^2\rangle(0) - \langle x\rangle(0)^2} = (\Delta x)_0. \tag{3.299}$$

The average value of p_x for $\Psi(x,0)$ is

$$\begin{aligned}\langle p_x\rangle(0) &= \int_{-\infty}^{\infty} dx\, \Psi^*(x,0)\widehat{p}_x\Psi(x,0) \\ &= \int_{-\infty}^{\infty} dx\, \Psi^*(x,0)\left(-i\hbar\frac{\partial}{\partial x}\Psi(x,0)\right) \\ &= \frac{1}{\sqrt{2\pi}(\Delta x)_0}\int_{-\infty}^{\infty} dx\, e^{-\frac{x^2}{4(\Delta x)_0^2}-\frac{i}{\hbar}p_0 x} \\ &\quad \times \left[-i\hbar\frac{\partial}{\partial x}\left(e^{-\frac{x^2}{4(\Delta x)_0^2}+\frac{i}{\hbar}p_0 x}\right)\right] \\ &= \frac{1}{\sqrt{2\pi}(\Delta x)_0}\int_{-\infty}^{\infty} dx\, \left(\frac{i\hbar x}{2(\Delta x)_0^2}+p_0\right) e^{-\frac{x^2}{2(\Delta x)_0^2}} \\ &= p_0. \end{aligned} \tag{3.300}$$

Calculating the average of p_x^2 for $\Psi(x,0)$, we get

$$\begin{aligned}\langle p_x^2\rangle(0) &= \int_{-\infty}^{\infty} dx\, \Psi^*(x,0)\widehat{p}_x^2\Psi(x,0) \\ &= \int_{-\infty}^{\infty} dx\, \Psi^*(x,0)\left(-\hbar^2\frac{\partial^2}{\partial x^2}\Psi(x,0)\right) \\ &= \frac{1}{\sqrt{2\pi}(\Delta x)_0}\int_{-\infty}^{\infty} dx\, e^{-\frac{x^2}{4(\Delta x)_0^2}-\frac{i}{\hbar}p_0 x} \\ &\quad \times \left[-\hbar^2\frac{\partial^2}{\partial x^2}\left(e^{-\frac{x^2}{4(\Delta x)_0^2}+\frac{i}{\hbar}p_0 x}\right)\right] \\ &= \frac{1}{\sqrt{2\pi}(\Delta x)_0}\int_{-\infty}^{\infty} dx\, \left[\frac{\hbar^2}{2(\Delta x)_0^2}-\frac{\hbar^2 x^2}{4(\Delta x)_0^4}+\frac{i\hbar p_0 x}{(\Delta x)_0^2}+p_0^2\right] \\ &\quad \times e^{-\frac{x^2}{2(\Delta x)_0^2}} \\ &= p_0^2 + \frac{\hbar^2}{4(\Delta x)_0^2}. \end{aligned} \tag{3.301}$$

Thus the uncertainty in p_x for the wave packet $\Psi(x,0)$ is

$$(\Delta p_x)(0) = \sqrt{\langle p_x^2\rangle(0) - \langle p_x\rangle(0)^2} = \frac{\hbar}{2(\Delta x)_0}. \tag{3.302}$$

A Review of Quantum Mechanics

This shows that for the initial wave packet $\Psi(x,0)$, the uncertainty product is

$$(\Delta x \Delta p_x)(0) = (\Delta x)_0 \frac{\hbar}{2(\Delta x)_0} = \frac{\hbar}{2}, \qquad (3.303)$$

i.e., $\Psi(x,0)$ is a minimum uncertainty wave packet.

Calculating the average value of x for the wave packet $\Psi(x,t)$, we find

$$\begin{aligned}
\langle x \rangle(t) &= \int_{-\infty}^{\infty} dx\, \Psi(x,t)^* x \Psi(x,t) = \int_{-\infty}^{\infty} dx\, x|\Psi(x,t)|^2 \\
&= \frac{1}{\sqrt{2\pi}(\Delta x)_t} \int_{-\infty}^{\infty} d\bar{x} \left(\bar{x} + \frac{p_0}{m}t\right) e^{-\frac{\bar{x}^2}{2(\Delta x)_t^2}} \\
&= \frac{p_0}{m} t,
\end{aligned} \qquad (3.304)$$

where

$$(\Delta x)_t = (\Delta x)_0 \sqrt{(1+\delta^2)}. \qquad (3.305)$$

The result (3.304) shows that the center of the wave packet is moving with the velocity p_0/m as if it corresponds to the position of a classical free particle of mass m moving with momentum p_0. This is Ehrenfest's theorem. The average of x^2 for $\Psi(x,t)$ is

$$\begin{aligned}
\langle x^2 \rangle(t) &= \int_{-\infty}^{\infty} dx\, \Psi(x,t)^* x^2 \Psi(x,t) = \int_{-\infty}^{\infty} dx\, x^2 |\Psi(x,t)|^2 \\
&= \frac{1}{\sqrt{2\pi}(\Delta x)_t} \int_{-\infty}^{\infty} d\bar{x} \left(\bar{x} + \frac{p_0}{m}t\right)^2 e^{-\frac{\bar{x}^2}{2(\Delta x)_t^2}} \\
&= \frac{1}{\sqrt{2\pi}(\Delta x)_t} \int_{-\infty}^{\infty} d\bar{x} \left[\bar{x}^2 + \left(\frac{p_0}{m}t\right)^2 + \frac{2 p_0 \bar{x}}{m}t\right] e^{-\frac{\bar{x}^2}{2(\Delta x)_t^2}} \\
&= (\Delta x)_t^2 + \left(\frac{p_0}{m}t\right)^2.
\end{aligned} \qquad (3.306)$$

Thus, the uncertainty in x for $\Psi(x,t)$ is

$$\begin{aligned}
(\Delta x)(t) &= \sqrt{\langle x^2 \rangle(t) - \langle x \rangle(t)^2} = (\Delta x)_t \\
&= (\Delta x)(0) \left[1 + \left(\frac{\hbar t}{2m(\Delta x)(0)^2}\right)^2\right]^{1/2},
\end{aligned} \qquad (3.307)$$

showing that the wave packet is spreading or dissipating. This spreading of the free-particle wave packet implies that we cannot think of the de Broglie waves as real physical waves as we don't observe any such dissipation of a free particle.

The average of p_x for $\Psi(x,t)$ is given by

$$\langle p_x \rangle (t) = \int_{-\infty}^{\infty} dx\, \Psi^*(x,t) \hat{p}_x \Psi(x,t)$$

$$= \int_{-\infty}^{\infty} dx\, \Psi^*(x,t) \left(-i\hbar \frac{\partial}{\partial x} \Psi(x,t) \right)$$

$$= \frac{1}{\sqrt{2\pi}(\Delta x)_t} \int_{-\infty}^{\infty} dx\, e^{-\frac{\tilde{x}^2}{4(\Delta x)_0^2 (1-i\delta)} - \frac{i}{\hbar}\left(p_0 x - \frac{p_0^2}{2m}t\right)}$$

$$\times \left\{ -i\hbar \frac{\partial}{\partial x} \left[e^{-\frac{\tilde{x}^2}{4(\Delta x)_0^2 (1+i\delta)} + \frac{i}{\hbar}\left(p_0 x - \frac{p_0^2}{2m}t\right)} \right] \right\}$$

$$= \frac{1}{\sqrt{2\pi}(\Delta x)_t} \int_{-\infty}^{\infty} d\tilde{x} \left[p_0 + \frac{i\hbar \tilde{x}}{2(\Delta x)_0^2 (1+i\delta)} \right] e^{-\frac{\tilde{x}^2}{2(\Delta x)_t^2}}$$

$$= p_0, \qquad (3.308)$$

as should be, since the momentum is conserved for a free particle. This is also in accordance with Ehrenfest's theorem. Similarly, we can find the average of p_x^2 for $\Psi(x,t)$:

$$\langle p_x^2 \rangle (t) = \int_{-\infty}^{\infty} dx\, \Psi^*(x,t) \hat{p}_x^2 \Psi(x,t)$$

$$= \int_{-\infty}^{\infty} dx\, \Psi^*(x,t) \left(-\hbar^2 \frac{\partial^2}{\partial x^2} \Psi(x,t) \right)$$

$$= \frac{1}{\sqrt{2\pi}(\Delta x)_t} \int_{-\infty}^{\infty} dx\, e^{-\frac{\tilde{x}^2}{4(\Delta x)_0^2 (1-i\delta)} - \frac{i}{\hbar}\left(p_0 x - \frac{p_0^2}{2m}t\right)}$$

$$\times \left\{ -\hbar^2 \frac{\partial^2}{\partial x^2} \left[e^{-\frac{\tilde{x}^2}{4(\Delta x)_0^2 (1+i\delta)} + \frac{i}{\hbar}\left(p_0 x - \frac{p_0^2}{2m}t\right)} \right] \right\}$$

$$= \frac{1}{\sqrt{2\pi}(\Delta x)_t} \int_{-\infty}^{\infty} d\tilde{x} \left[p_0^2 + \frac{1}{(\Delta x)_0^2 (1+i\delta)} \left(\frac{\hbar^2}{2} + i\hbar p_0 \tilde{x} \right. \right.$$

$$\left. \left. - \frac{\hbar^2 \tilde{x}^2}{4(\Delta x)_0^2 (1+i\delta)} \right) \right] e^{-\frac{\tilde{x}^2}{2(\Delta x)_t^2}}$$

$$= p_0^2 + \frac{\hbar^2}{4(\Delta x)_0^2}, \qquad (3.309)$$

same as for the initial wave packet $\Psi(x,0)$. This is again due to conservation of momentum for the free particle. Thus, for $\Psi(x,t)$ the uncertainty in p_x is

$$(\Delta p_x)(t) = \sqrt{\langle p_x^2 \rangle (t) - \langle p_x \rangle (t)^2} = \frac{\hbar}{2(\Delta x)_0}, \qquad (3.310)$$

same as for the initial wave packet $\Psi(x,0)$. The uncertainty product for $\Psi(x,t)$ is

$$(\Delta x \Delta p_x)(t) = (\Delta x)(0)\left[1+\left(\frac{\hbar t}{2m(\Delta x)(0)^2}\right)^2\right]^{1/2} \frac{\hbar}{2(\Delta x)(0)}$$

$$= \frac{\hbar}{2}\left[1+\left(\frac{\hbar t}{2m(\Delta x)(0)^2}\right)^2\right]^{1/2} > \frac{\hbar}{2}, \quad \text{for } t>0, \qquad (3.311)$$

showing that as the minimum uncertainty free-particle wave packet evolves in time, it does not remain a minimum uncertainty wave packet. The uncertainty principle implies that we cannot assign definite values for both the position and the momentum of particle at any time. Thus, the concept of trajectory loses its meaning since we can only specify the probability of finding the particle at any position at any time.

Let us now look at the propagator for a free particle. From (3.178), we have

$$K_{\text{free}}(\vec{r},t;\vec{r}_i,t_i) = \sum_n e^{-\frac{i}{\hbar}(t-t_i)E_n} \langle \vec{r}|n\rangle \langle n|\vec{r}_i\rangle$$

$$= \int d^3p \, e^{-\frac{ip^2(t-t_i)}{2m\hbar}} \phi_{\vec{p}}(\vec{r})\phi_{\vec{p}}^*(\vec{r}_i)$$

$$= \frac{1}{(2\pi\hbar)^3} \int d^3p \, e^{\frac{i}{\hbar}\left[\vec{p}\cdot(\vec{r}-\vec{r}_i)-\frac{p^2(t-t_i)}{2m}\right]}. \qquad (3.312)$$

The integral in the last step can be performed exactly using the integral in (3.292). The result is

$$K_{\text{free}}(\vec{r},t;\vec{r}_i,t_i) = \left[\frac{m}{2\pi i\hbar(t-t_i)}\right]^{3/2} e^{\frac{im|\vec{r}-\vec{r}_i|^2}{2\hbar(t-t_i)}}. \qquad (3.313)$$

As an example, let us find $\Psi(x,t)$ when $\Psi(x,0)$ is the free-particle minimum uncertainty wave packet (3.293):

$$\Psi(x,t) = \int_{-\infty}^{\infty} dx' \, K_{\text{free}}(x,t;x',0)\,\Psi(x',0)$$

$$= \left(\frac{m}{(2\pi)^{3/2}i\hbar(\Delta x)_0}\right)^{1/2} \int_{-\infty}^{\infty} dx' \, e^{\frac{im(x-x')^2}{2\hbar}} e^{-\frac{x'^2}{4(\Delta x)_0^2}+\frac{i}{\hbar}p_0 x'}$$

$$= \left(\frac{m}{(2\pi)^{3/2}i\hbar(\Delta x)_0}\right)^{1/2} e^{\frac{im}{2\hbar}x^2} \int_{-\infty}^{\infty} dx' \, e^{-\left(\frac{1}{4(\Delta x)_0^2}-\frac{im}{2\hbar t}\right)x'^2-\frac{im}{\hbar t}\left(x-\frac{p_0 t}{m}\right)x'}$$

$$= \frac{1}{\left(\sqrt{2\pi}(\Delta x)_0\left(1+\frac{i\hbar t}{2m(\Delta x)_0^2}\right)\right)^{1/2}} e^{-\frac{\left(x-\frac{p_0}{m}t\right)^2}{4(\Delta x)_0^2\left(1+\frac{i\hbar t}{2m(\Delta x)_0^2}\right)}+\frac{i}{\hbar}\left(p_0 x - \frac{p_0^2}{2m}t\right)}, \qquad (3.314)$$

where, in the last step, we have to use the integral (3.292) and some algebraic manipulations. Thus, using the propagator, we have arrived at the same result for $\Psi(x,t)$ as in (3.294) starting with $\Psi(x,0)$.

3.3.1.2 Linear Harmonic Oscillator

Before considering the quantum mechanics of a nonrelativistic linear harmonic oscillator, let us recall its classical mechanics. Consider a particle of mass m moving along the x-axis under the influence of a restoring force $F(x) = -\kappa x$, where κ is the spring constant. The classical Hamiltonian of the particle is

$$H = \frac{p^2}{2m} + \frac{1}{2}\kappa x^2, \qquad (3.315)$$

where $p^2/2m$ is the kinetic energy and $V(x) = \kappa x^2/2$ is the potential energy of the particle at the position x. Since the particle is moving only along the x-direction, we have denoted p_x simply by p. Hamilton's equations of motion are

$$\begin{aligned}
\frac{dx}{dt} &=: -H : x = \left\{ -\left(\frac{p^2}{2m} + \frac{1}{2}\kappa x^2\right), x \right\} \\
&= \left\{ -\frac{p^2}{2m}, x \right\} = \frac{p}{m}, \\
\frac{dp}{dt} &=: -H : p = \left\{ -\left(\frac{p^2}{2m} + \frac{1}{2}\kappa x^2\right), p \right\} \\
&= \left\{ -\frac{1}{2}\kappa x^2, p \right\} = -\kappa x. \qquad (3.316)
\end{aligned}$$

Defining $\kappa = m\omega^2$, we can rewrite these equations as

$$\frac{d}{dt}\begin{pmatrix} x \\ \frac{p}{m\omega} \end{pmatrix} = \omega \begin{pmatrix} 0 & 1 \\ -1 & 0 \end{pmatrix} \begin{pmatrix} x \\ \frac{p}{m\omega} \end{pmatrix}. \qquad (3.317)$$

This equation can be readily integrated to give the classical phase space transfer map

$$\begin{aligned}
\begin{pmatrix} x(t) \\ \frac{p(t)}{m\omega} \end{pmatrix} &= e^{\omega t \begin{pmatrix} 0 & 1 \\ -1 & 0 \end{pmatrix}} \begin{pmatrix} x(0) \\ \frac{p(0)}{m\omega} \end{pmatrix} \\
&= \begin{pmatrix} \cos\omega t & \sin\omega t \\ -\sin\omega t & \cos\omega t \end{pmatrix} \begin{pmatrix} x(0) \\ \frac{p(0)}{m\omega} \end{pmatrix}, \qquad (3.318)
\end{aligned}$$

where $(x(0), p(0))$ and $(x(t), p(t))$ are the values of x and p at times 0 and t, respectively.

Since the Hamiltonian is time-independent, the energy of the oscillator is a constant of motion, say E. When the particle is at the extreme position of oscillation $x = A$, the amplitude of motion, its energy is completely potential energy, and the

A Review of Quantum Mechanics

kinetic energy is zero since $p = 0$. Thus, $E = m\omega^2 A^2/2$. When the particle is at any other position x, between 0 and A, its energy will be distributed between the kinetic and potential energies such that

$$\frac{p^2}{2m} + \frac{1}{2}m\omega^2 x^2 = \frac{1}{2}m\omega^2 A^2. \tag{3.319}$$

Dividing both sides of this equation by $m\omega^2/2$, we get

$$\left(\frac{p}{m\omega}\right)^2 + x^2 = A^2, \tag{3.320}$$

a relation to be satisfied at any position of the particle. Thus, at any position, we can write $x = A\cos\varphi$, and $p/m\omega = \pm A\sin\varphi$ depending on whether the particle is moving in the $+x$-direction or $-x$-direction. Let us now choose the initial conditions at $t = 0$ as $x(0) = A\cos\varphi$, with $0 \leq \varphi \leq \pi/2$, and $p(0)/m\omega = A\sin\varphi$, without loss of generality. Then, we get

$$x(t) = A\cos(\omega t - \varphi), \qquad p(t) = -m\omega A \sin(\omega t - \varphi), \tag{3.321}$$

as the general solution for the position and momentum of the oscillating particle. Note that $p(t) = m\dot{x}(t)$. The constant φ is the initial phase of the oscillator, fixed by the initial conditions at $t = 0$, and the circular frequency ω gives the frequency with which the phase changes, *i.e.*, $\omega T = 2\pi$, where T is the period of the oscillator such that $x(t+T) = x(t)$, $p(t+T) = p(t)$. The energy of the oscillator is, at any time,

$$H(t) = \frac{p(t)^2}{2m} + \frac{1}{2}m\omega^2 x(t)^2 = \frac{1}{2}m\omega^2 A^2, \tag{3.322}$$

constant, equal to the potential energy of the particle at its extreme positions, or the turning points, where the kinetic energy vanishes. Let us observe that we can write $H(t)$ in a factorized form as

$$H(t) = \left(\frac{m\omega x(t) + ip(t)}{\sqrt{2m}}\right)\left(\frac{m\omega x(t) - ip(t)}{\sqrt{2m}}\right) = a(t)a^*(t). \tag{3.323}$$

Substituting the solutions for $x(t)$ and $p(t)$, we have

$$a(t) = \sqrt{\frac{m}{2}}\,\omega A e^{-i(\omega t - \varphi)} = a(0)e^{-i\omega t},$$

$$a^*(t) = \sqrt{\frac{m}{2}}\,\omega A e^{i(\omega t - \varphi)} = a^*(0)e^{i\omega t}, \tag{3.324}$$

such that the relation (3.323) is consistent with (3.322). Since the amplitude of oscillation A can take continuous values, we see that the energy of the classical harmonic oscillator can vary continuously. Any oscillatory motion is approximately a linear harmonic motion as long as the amplitude is small and that is what makes the linear harmonic oscillator a basic paradigm for understanding a wide variety of physical phenomena.

Let us now look at the quantum mechanics of the linear harmonic oscillator. The quantum Hamiltonian of the nonrelativistic linear harmonic oscillator of mass m and circular frequency ω along the x-axis is given by

$$\widehat{H} = \frac{\widehat{p}^2}{2m} + \frac{1}{2}m\omega^2 x^2. \tag{3.325}$$

Let us work in the position representation. Then, the Schrödinger equation for $\Psi(x,t)$ reads

$$i\hbar \frac{\partial \Psi}{\partial t} = \left(-\frac{\hbar^2}{2m}\frac{\partial^2}{\partial x^2} + \frac{1}{2}m\omega^2 x^2\right)\Psi. \tag{3.326}$$

This partial differential equation can be solved by the method of separation of variables. To this end, let $\Psi(x,t) = T(t)\psi(x)$. This turns (3.326) into

$$\frac{i\hbar}{T}\frac{dT}{dt} = \frac{1}{\psi}\left(-\frac{\hbar^2}{2m}\frac{d^2}{dx^2} + \frac{1}{2}m\omega^2 x^2\right)\psi. \tag{3.327}$$

Since the left-hand side of this equation is a function of only t and the right-hand side is a function of only x, both sides should be equal to some constant, say E. Thus, we get

$$i\hbar \frac{dT}{dt} = ET, \tag{3.328}$$

$$\left(-\frac{\hbar^2}{2m}\frac{d^2}{dx^2} + \frac{1}{2}m\omega^2 x^2\right)\psi = E\psi. \tag{3.329}$$

The first equation is solved immediately: $T(t) = e^{-iEt/\hbar}$. The second equation is the eigenvalue equation for the Hamiltonian of the oscillator. Hence, we identify the constant E with the energy eigenvalue. The equation (3.329) is called the time-independent Schrödinger equation for the harmonic oscillator. In general, the eigenvalue equation for the Hamiltonian is known as the time-independent Schrödinger equation of the system. Let $\{E_n\}$ be the set of all eigenvalues and let $\{\psi_n(x)\}$ be the respective eigenfunctions allowed by (3.329). In this case, we will find that the eigenvalues are discrete, nondegenerate, and can be labeled as $n = 0, 1, 2, \ldots,$. Since the Hamiltonian operator is Hermitian, these eigenvalues will be real, and the corresponding eigenfunctions will form a complete orthonormal set. Note that the Hamiltonian of any nonrelativistic system, of the type $\widehat{p}^2/2m + V(\vec{r},t)$, will be Hermitian. Thus, the particular solutions of (3.326) are given by

$$\Psi_n(x,t) = e^{-\frac{i}{\hbar}E_n t}\psi_n(x), \qquad n = 0, 1, 2, \ldots. \tag{3.330}$$

These wave functions correspond to stationary states in which the position probability density, $|\Psi_n(x,t)|^2$, is independent of time since the time dependence cancels out. Similarly, the mean value of any time-independent observable $\langle O \rangle_{\Psi_n(t)}$ is also seen to

A Review of Quantum Mechanics

be time independent. The general solution of the time-dependent Schrödinger equation (3.326), $\Psi(x,t)$, is any linear combination of the particular solutions, subject to the only condition of normalization $\langle \Psi(t)|\Psi(t)\rangle = 1$. Thus, the general solution is

$$\Psi(x,t) = \sum_n C_n \Psi_n(x,t), \qquad \text{with} \quad \sum_n |C_n|^2 = 1. \tag{3.331}$$

Let us now look at the solutions $\psi_n(x)$ of the time-independent Schrödinger equation (3.329), the eigenvalue equation for the time-independent Hamiltonian of the system. To this end, we have to solve the differential equation

$$\left(-\frac{\hbar^2}{2m}\frac{d^2}{dx^2} + \frac{1}{2}m\omega^2 x^2\right)\psi(x) = E\psi(x). \tag{3.332}$$

Now, introducing the dimensionless variable $\xi = \sqrt{m\omega/\hbar}\, x$, this equation becomes

$$\left(-\frac{d^2}{d\xi^2} + \xi^2\right)\psi(\xi) = K\psi(\xi), \tag{3.333}$$

where $K = 2E/\hbar\omega$ is the energy in units of $\hbar\omega/2$. Normalizable solutions for $\psi(x)$ exist only when $K = 2n+1$, with $n = 0,1,2,\ldots,$. Thus, the quantized energy spectrum of the oscillator consists of eigenvalues

$$E_n = \left(n + \frac{1}{2}\right)\hbar\omega, \qquad n = 0,1,2,\ldots. \tag{3.334}$$

All these energy eigenvalues are nondegenerate. Note that except for the addition of the extra energy $h\nu/2$ to each energy level, and the zero-point energy $E_0 = h\nu/2$, the formula for the energy spectrum of the harmonic oscillator (3.334) coincides exactly with the postulate of Planck that the energy of oscillators in a black body is quantized as $E = nh\nu$, with $n = 1,2,3,\ldots$. The origin of quantum mechanics in 1900 lies in this postulate of Planck. When the oscillator makes a transition, or jumps down, from the $(n+1)$-th energy level to the n-th energy level, it emits a quantum of energy $h\nu$ (photon) and by absorbing a quantum of energy $h\nu$ it can move, or jump up, from a lower energy level to the next higher energy level.

The normalized eigenfunctions, corresponding to the energy eigenvalues E_n, orthogonal to each other, are

$$\psi_n(x) = \left(\frac{m\omega}{\pi\hbar}\right)^{1/4} \frac{1}{\sqrt{2^n n!}} H_n(\xi) e^{-\frac{1}{2}\xi^2}, \qquad n = 0,1,2,\ldots, \tag{3.335}$$

where $H_n(\xi)$ are called the Hermite polynomials given by the Rodrigues formula (for details, see, *e.g.*, Arfken, Weber, and Harris [3], and Byron and Fuller [18], and Lakshminarayanan and Varadharajan [126]):

$$H_n(\xi) = (-1)^n e^{\xi^2}\left(\frac{d}{d\xi}\right)^n e^{-\xi^2}. \tag{3.336}$$

The first few Hermite polynomials are

$$H_0(\xi) = 1, \quad H_1(\xi) = 2\xi, \quad H_2(\xi) = 4\xi^2 - 2, \quad H_3(\xi) = 8\xi^3 - 12\xi. \quad (3.337)$$

The time-dependent energy eigenstates, the stationary states, are given by

$$\Psi_n(x,t) = e^{-\frac{i}{\hbar}E_n t}\psi_n(x) = e^{-i\left(n+\frac{1}{2}\right)\omega t}\psi_n(x). \quad (3.338)$$

The general solution (3.331) for the time-dependent Schrödinger equation (3.326) can now be written explicitly as

$$\begin{aligned}\Psi(x,t) &= \sum_n C_n e^{-i\left(n+\frac{1}{2}\right)\omega t}\psi_n(x) \\ &= e^{-\frac{i}{\hbar}t\widehat{H}}\sum_n C_n \psi_n(x) = e^{-\frac{i}{\hbar}t\widehat{H}}\Psi(x,0),\end{aligned} \quad (3.339)$$

where $\sum_n |C_n|^2 = 1$. The lowest eigenstate, the ground state ($n = 0$), corresponding to energy $E_0 = \hbar\omega/2$, has the wave function

$$\Psi_0(x,t) = \left(\frac{m\omega}{\pi\hbar}\right)^{1/4} e^{-\frac{1}{2}\left(\frac{m\omega}{\hbar}x^2 + i\omega t\right)}. \quad (3.340)$$

The nonclassical nature of the quantum oscillator, besides the quantization of energy, is evident now. If we think classically, for the oscillator in its ground state with energy $E_0 = \hbar\omega/2$, the amplitude of oscillation would be $A = \sqrt{2E_0/m\omega^2} = \sqrt{\hbar/m\omega}$. Classically, for the oscillating particle, $x = \pm A$ are the turning points beyond which the particle cannot go. But, for the quantum oscillator in the ground state, we have

$$|\Psi_0(x,t)|^2 = \left(\frac{m\omega}{\pi\hbar}\right)^{1/2} e^{-\frac{m\omega}{\hbar}x^2}, \quad (3.341)$$

a Gaussian function, showing that there is a definite probability for the particle to be found beyond the classical turning points, in principle up to $x = \pm\infty$. Higher ($n > 0$), excited, states of the oscillator also have this property. Further, it is seen that the wave function of n-th excited level, with energy $E_n = \left(n + \frac{1}{2}\right)\hbar\omega$, vanishes at n points, called nodes, where the probability of finding the particle is zero! We have found that all the wave functions of the linear harmonic oscillator corresponding to discrete energy eigenvalues are normalizable in the sense $\int dx\, |\Psi(x,t)|^2 = 1$ and $\Psi(x,t) \longrightarrow 0$ as $x \longrightarrow \pm\infty$. Such states are known as bound states. Note that free-particle eigenstates are not bound states.

There is an alternative way of finding the energy spectrum and the corresponding eigenstates of the harmonic oscillator, the algebraic method, similar to the case of angular momentum we have already seen. To this end, let us define the quantum operators corresponding to the complex classical dynamical variables a and a^* in (3.323) as follows:

$$\widehat{a} = \frac{m\omega x + i\widehat{p}}{\sqrt{2m\hbar\omega}}, \quad \widehat{a}^\dagger = \frac{m\omega x - i\widehat{p}}{\sqrt{2m\hbar\omega}}. \quad (3.342)$$

A Review of Quantum Mechanics

Since $\hat{a}$ and $\hat{a}^\dagger$ correspond to the classical complex a and a^*, they are not Hermitian operators. Note that $\hat{a}$ and $\hat{a}^\dagger$ are dimensionless, $\hat{a}^\dagger \hat{a}$ is Hermitian, and in terms of them the Hamiltonian $\hat{H}$ becomes

$$\hat{H} = \left(\hat{a}^\dagger \hat{a} + \frac{1}{2}\right)\hbar\omega, \qquad (3.343)$$

as can be verified directly. Further, we have the following commutation relations:

$$[\hat{a}, \hat{a}^\dagger] = 1, \qquad (3.344)$$

and

$$[\hat{a}^\dagger \hat{a}, \hat{a}^\dagger] = \hat{a}^\dagger, \qquad [\hat{a}^\dagger \hat{a}, \hat{a}] = -\hat{a}. \qquad (3.345)$$

An operator for which the expectation value in any state is nonnegative is called a positive operator. Now, $\hat{a}^\dagger \hat{a}$ is seen to be a positive operator because

$$\langle \varphi | \hat{a}^\dagger \hat{a} | \varphi \rangle = \langle \hat{a}\varphi | \hat{a}\varphi \rangle \geq 0, \qquad (3.346)$$

and

$$\langle \varphi | \hat{a}^\dagger \hat{a} | \varphi \rangle = 0 \quad \text{implies} \quad \hat{a}|\varphi\rangle = 0. \qquad (3.347)$$

Let $\hat{P}$ be a positive operator and $|\varphi_\lambda\rangle$ be its normalized eigenvector corresponding to an eigenvalue λ. Then, we have $\langle \varphi_\lambda | \hat{P} | \varphi_\lambda \rangle = \lambda \geq 0$. Thus, all the eigenvalues of any positive operator are ≥ 0. Now, from (3.343), it follows that the eigenvalues of $\hat{H}$ should be $\geq \hbar\omega/2$ since the lowest eigenvalue of the positive operator $\hat{a}^\dagger \hat{a}$ should be ≥ 0. From (3.343) and (3.345), it follows that

$$\hat{H}\hat{a}^\dagger = \hat{a}^\dagger(\hat{H} + \hbar\omega), \qquad \hat{H}\hat{a} = \hat{a}(\hat{H} - \hbar\omega). \qquad (3.348)$$

These relations imply that if $|n\rangle$ is an eigenstate of $\hat{H}$ with the eigenvalue E_n, i.e., $\hat{H}|n\rangle = E_n|n\rangle$, then

$$\hat{H}\left(\hat{a}^\dagger|n\rangle\right) = (E_n + \hbar\omega)\left(\hat{a}^\dagger|n\rangle\right), \qquad \hat{H}\left(\hat{a}|n\rangle\right) = (E_n - \hbar\omega)\left(\hat{a}|n\rangle\right). \qquad (3.349)$$

For this reason $\hat{a}^\dagger$ and $\hat{a}$, called the raising and lowering operators, respectively, are known as the ladder operators. Since the eigenvalue of $\hat{H}$ cannot become negative, the action of the lowering operator must stop at some stage. Let $|0\rangle$ be the ground state corresponding to this lowest eigenvalue E_0, such that

$$\hat{a}|0\rangle = 0. \qquad (3.350)$$

If $\hat{a}|0\rangle$ were nonzero, then it has to be an eigenstate with the eigenvalue $E_0 - \hbar\omega$ contradicting the assumption that E_0 is the lowest eigenvalue. Then, it is obvious that $E_0 = \hbar\omega/2$ since

$$\hat{H}|0\rangle = \hbar\omega\left(\hat{a}^\dagger \hat{a} + \frac{1}{2}\right)|0\rangle = \frac{\hbar}{2}\omega|0\rangle. \qquad (3.351)$$

Now, applying the raising operator repeatedly on the ground state, we get the complete spectrum and the eigenstates of $\widehat{H}$. Thus, we have

$$\widehat{H}|n\rangle = \left(n + \frac{1}{2}\right)\hbar\omega|n\rangle, \qquad |n\rangle = N_n\left(\widehat{a}^\dagger\right)^n|0\rangle, \qquad n = 0, 1, 2, \ldots, \qquad (3.352)$$

where N_n is the normalization constant that can be found by assuming that the ground state is normalized, *i.e.*, $\langle 0|0\rangle = 1$. The normalization constant N_n can be found as follows. First, let us observe that

$$\widehat{a}^\dagger \widehat{a}|n\rangle = \left(\frac{\widehat{H}}{\hbar\omega} - \frac{1}{2}\right)|n\rangle = n|n\rangle, \qquad n = 0, 1, 2, \ldots. \qquad (3.353)$$

Thus, the spectrum of $\widehat{a}^\dagger \widehat{a}$ consists of all integers ≥ 0. Hence, $\widehat{a}^\dagger \widehat{a}$, denoted by $\widehat{N}$, is called the number operator. Let $|n\rangle$ be the normalized eigenvector of $\widehat{H}$, or $\widehat{a}^\dagger \widehat{a}$. The normalized eigenvector $|n+1\rangle$ can be written as

$$|n+1\rangle = N_{n+1}\left(\widehat{a}^\dagger\right)^{n+1}|0\rangle = N_{n+1}\widehat{a}^\dagger\left(\widehat{a}^\dagger\right)^n|0\rangle = \frac{N_{n+1}}{N_n}\widehat{a}^\dagger|n\rangle. \qquad (3.354)$$

Now, with $\langle n|\widehat{a}^\dagger \widehat{a}|n\rangle = n$, the normalization condition for $|n+1\rangle$ becomes

$$\begin{aligned}\langle n+1|n+1\rangle &= \left(\frac{N_{n+1}}{N_n}\right)^2 \langle n|\widehat{a}\widehat{a}^\dagger|n\rangle \\ &= \left(\frac{N_{n+1}}{N_n}\right)^2 \langle n|\left(1 + \widehat{a}^\dagger \widehat{a}\right)|n\rangle \\ &= \left(\frac{N_{n+1}}{N_n}\right)^2 (n+1) = 1, \end{aligned} \qquad (3.355)$$

where we have used the relation $\widehat{a}\widehat{a}^\dagger = 1 + \widehat{a}^\dagger \widehat{a}$, following from the commutation relation (3.344). This implies that $N_{n+1} = N_n/\sqrt{n+1}$. By definition $N_0 = 1$. Hence, we get

$$N_n = \frac{1}{\sqrt{n!}}, \qquad n = 0, 1, 2, \ldots. \qquad (3.356)$$

Thus, the normalized eigenstates of $\widehat{H}$ can be written as

$$|n\rangle = \frac{1}{\sqrt{n!}}\left(\widehat{a}^\dagger\right)^n|0\rangle, \quad \text{with } \widehat{a}|0\rangle = 0. \qquad (3.357)$$

Since $\widehat{H}$ is Hermitian, its eigenvectors $\{|n\rangle\}$ are orthogonal. In the orthonormal basis $\{|n\rangle\}$ we find, from (3.357),

$$\widehat{a}^\dagger |n\rangle = \frac{1}{\sqrt{n!}} \left(\widehat{a}^\dagger\right)^{n+1} |0\rangle$$
$$= \sqrt{n+1}\left[\frac{1}{\sqrt{(n+1)!}}\left(\widehat{a}^\dagger\right)^{n+1}|0\rangle\right]$$
$$= \sqrt{n+1}\,|n+1\rangle. \tag{3.358}$$

Similarly,

$$\widehat{a}|n\rangle = \widehat{a}\left[\frac{1}{\sqrt{n!}}\left(\widehat{a}^\dagger\right)^n|0\rangle\right]$$
$$= \frac{\widehat{a}\widehat{a}^\dagger}{\sqrt{n}}\left[\frac{1}{\sqrt{(n-1)!}}\left(\widehat{a}^\dagger\right)^{n-1}|0\rangle\right]$$
$$= \frac{(1+\widehat{a}^\dagger\widehat{a})}{\sqrt{n}}|n-1\rangle = \sqrt{n}\,|n-1\rangle, \tag{3.359}$$

where we have used the relation $\widehat{a}^\dagger\widehat{a}|n-1\rangle = (n-1)|n-1\rangle$. The eigenfunctions of $\widehat{H}$, namely $\psi_n(x) = \langle x|n\rangle$, can be obtained from the above formalism as follows. In the position representation, the defining equation for the ground state, $\widehat{a}|0\rangle = 0$, becomes the simple equation

$$\left(m\omega x + \hbar\frac{d}{dx}\right)\psi_0(x) = 0, \tag{3.360}$$

with the solution

$$\psi_0(x) = \mathcal{N}_0 e^{-\frac{m\omega}{2\hbar}x^2}, \tag{3.361}$$

where $\mathcal{N}_0$ is the normalization constant such that

$$\int dx\,|\psi_0(x)|^2 = \mathcal{N}_0^2 \int dx\, e^{-\frac{m\omega}{\hbar}x^2} = \mathcal{N}_0^2\sqrt{\frac{\pi\hbar}{m\omega}} = 1, \tag{3.362}$$

using the result on the Gaussian integral we have already seen (3.292). Thus, $\mathcal{N}_0 = (m\omega/\pi\hbar)^{1/4}$ and the ground state eigenfunction of $\widehat{H}$ is

$$\psi_0(x) = \left(\frac{m\omega}{\pi\hbar}\right)^{1/4} e^{-\frac{m\omega}{2\hbar}x^2}. \tag{3.363}$$

The excited state eigenfunctions can be obtained from (3.357), which becomes in position representation

$$\psi_n(x) = \frac{1}{\sqrt{n!}}\left(\sqrt{\frac{m\omega}{2\hbar}}x - \sqrt{\frac{\hbar}{2m\omega}}\frac{d}{dx}\right)^n \psi_0(x). \tag{3.364}$$

Using the dimensionless variable $\xi = (\sqrt{m\omega/\hbar})x$, we have

$$\psi_n(x) = \left(\frac{m\omega}{\pi\hbar}\right)^{1/4} \frac{1}{\sqrt{2^n n!}} \left(\xi - \frac{d}{d\xi}\right)^n e^{-\frac{1}{2}\xi^2}. \tag{3.365}$$

Using the identity

$$\left(\xi - \frac{d}{d\xi}\right)^n e^{-\frac{1}{2}\xi^2} = (-1)^n e^{\frac{1}{2}\xi^2} \left(\frac{d}{d\xi}\right)^n e^{-\xi^2}, \tag{3.366}$$

easily proved by induction, we can rewrite

$$\psi_n(x) = \left(\frac{m\omega}{\pi\hbar}\right)^{1/4} \frac{1}{\sqrt{2^n n!}} e^{-\frac{1}{2}\xi^2} H_n(\xi), \tag{3.367}$$

where

$$H_n(\xi) = (-1)^n e^{\xi^2} \left(\frac{d}{d\xi}\right)^n e^{-\xi^2}. \tag{3.368}$$

Thus, we have reconstructed the earlier result (3.335).

The existence of a nonzero ground state energy, so-called zero-point energy, can be understood intuitively in terms of the uncertainty principle. As seen in (3.363), the ground state wave function is a Gaussian with a finite width. It is not localized means that the position of the oscillating particle is fluctuating around the equilibrium even in the ground state unlike the classical oscillator which will be at rest in the equilibrium position in the ground state. As a consequence of this uncertainty in position, say Δx, there will be a finite uncertainty in the momentum of the particle, $\sim \hbar/\Delta x$, as dictated by the Heisenberg uncertainty principle. This intrinsic ground state motion implies the existence of a minimum energy.

Let us make a precise calculation of the uncertainty product $\Delta x \Delta p$ for the harmonic oscillator eigenstates. From the definitions of $\hat{a}$ and $\hat{a}^\dagger$ in (3.342), we have

$$x = \sqrt{\frac{\hbar}{2m\omega}} \left(\hat{a}^\dagger + \hat{a}\right), \qquad \hat{p} = i\sqrt{\frac{m\hbar\omega}{2}} \left(\hat{a}^\dagger - \hat{a}\right). \tag{3.369}$$

Calculating the averages of x and p in the n-th eigenstate, we get

$$\langle x \rangle_n = \langle n|x|n \rangle = \sqrt{\frac{\hbar}{2m\omega}} \langle n|(\hat{a}^\dagger + \hat{a})|n \rangle = 0,$$

$$\langle p \rangle_n = \langle n|\hat{p}|n \rangle = i\sqrt{\frac{m\hbar\omega}{2}} \langle n|(\hat{a}^\dagger - \hat{a})|n \rangle = 0, \tag{3.370}$$

where we have used the relations (3.358) and (3.359) and the orthogonality of the eigenstates $\{|n\rangle\}$. Averages of x^2 and p^2 can be calculated as follows:

$$\langle x^2 \rangle_n = \langle n|x^2|n \rangle$$

$$= \frac{\hbar}{2m\omega} \left\langle n \left|(\hat{a}^\dagger + \hat{a})^2\right| n \right\rangle$$

$$= \frac{\hbar}{2m\omega} \left\langle n \left| \left[(\hat{a}^\dagger)^2 + \hat{a}^2 + \hat{a}^\dagger \hat{a} + \hat{a} \hat{a}^\dagger \right] \right| n \right\rangle$$

$$= \frac{\hbar}{2m\omega} \left\langle n \left| (\hat{a}^\dagger \hat{a} + \hat{a} \hat{a}^\dagger) \right| n \right\rangle$$

$$= \frac{\hbar}{m\omega} \left(n + \frac{1}{2} \right), \tag{3.371}$$

$$\langle p^2 \rangle_n = \langle n | \hat{p}^2 | n \rangle$$

$$= \frac{m\hbar\omega}{2} \left\langle n \left| \left[-(\hat{a}^\dagger - \hat{a})^2 \right] \right| n \right\rangle$$

$$= \frac{m\hbar\omega}{2} \left\langle n \left| \left[-(\hat{a}^\dagger)^2 - \hat{a}^2 + \hat{a}^\dagger \hat{a} + \hat{a} \hat{a}^\dagger \right] \right| n \right\rangle$$

$$= \frac{m\hbar\omega}{2} \left\langle n \left| (\hat{a}^\dagger \hat{a} + \hat{a} \hat{a}^\dagger) \right| n \right\rangle$$

$$= m\hbar\omega \left(n + \frac{1}{2} \right), \tag{3.372}$$

where we have used the relations (3.358) and (3.359) and the orthonormality of the eigenstates $\{|n\rangle\}$. From (3.276), we find that the uncertainties in x and p in the n-th eigenstate are given by

$$(\Delta x)_n = \sqrt{\langle x^2 \rangle_n - \langle x \rangle_n^2} = \sqrt{\langle x^2 \rangle_n} = \sqrt{\frac{\hbar}{m\omega} \left(n + \frac{1}{2} \right)}$$

$$(\Delta p)_n = \sqrt{\langle p^2 \rangle_n - \langle p \rangle_n^2} = \sqrt{\langle p^2 \rangle_n} = \sqrt{m\hbar\omega \left(n + \frac{1}{2} \right)}. \tag{3.373}$$

Now, for the n-th eigenstate $|n\rangle$, the uncertainty product $\Delta x \Delta p$ becomes

$$(\Delta x)_n (\Delta p)_n = \left(n + \frac{1}{2} \right) \hbar. \tag{3.374}$$

For the ground state $|0\rangle$

$$(\Delta x)_0 (\Delta p)_0 = \frac{\hbar}{2}. \tag{3.375}$$

Thus, the ground state is a minimum uncertainty state. It should be noted that, unlike in the case of a minimum uncertainty free-particle wave packet, the ground state of the harmonic oscillator is a minimum uncertainty state at any time since it is a stationary state and hence Δx and Δp do not depend on time.

The propagator for the linear harmonic oscillator (lho), for $t \geq t_i$, is given by

$$K_{\text{lho}}(x,t;x',t_i) = \sum_{n=0}^{\infty} e^{-\frac{i}{\hbar} E_n (t - t_i)} \psi_n(x) \psi_n^*(x')$$

$$= \sqrt{\frac{m\omega}{\pi\hbar}} \sum_{n=0}^{\infty} \frac{1}{2^n n!} e^{-i(n+\frac{1}{2})\omega(t-t_i)} H_n(\xi) H_n(\xi') e^{-\frac{1}{2}(\xi^2 + \xi'^2)}, \tag{3.376}$$

where $\xi = \sqrt{m\omega/\hbar}\, x$ and $\xi' = \sqrt{m\omega/\hbar}\, x'$. Now, it is possible to sum the series in the above expression exactly, using the Mehler formula (for a derivation of this formula, see, *e.g.*, Lakshminarayanan and Varadharajan [126]),

$$\sum_{n=0}^{\infty} \frac{1}{2^n n!} H_n(z) H_n(z') \zeta^n e^{-\frac{1}{2}\left(z^2+z'^2\right)} = \frac{1}{\sqrt{1-\zeta^2}} e^{\frac{\left(1+\zeta^2\right)\left(z^2+z'^2\right)-4\zeta z z'}{2\left(1-\zeta^2\right)}}, \qquad (3.377)$$

taking $z = \xi$, $z' = \xi'$, and $\zeta = e^{-i\omega(t-t_i)}$. Thus, we get

$$K_{\text{lho}}(x,t;x',t_i) = \left(\frac{m\omega}{2\pi i\hbar \sin\omega(t-t_i)}\right)^{1/2} e^{\frac{im\omega}{2\hbar \sin\omega(t-t_i)}\left[\left(x^2+x'^2\right)\cos\omega(t-t_i)-2xx'\right]}. \qquad (3.378)$$

When $\omega \longrightarrow 0$, the linear harmonic oscillator becomes a free particle in one dimension. Note that when $\omega \longrightarrow 0$, $K_{\text{lho}}(x,t;x',t_i) \longrightarrow K_{\text{free}}(x,t;x',t_i)$ as should be.

3.3.1.3 Two-Dimensional Isotropic Harmonic Oscillator

A classical two-dimensional isotropic harmonic oscillator in the xy-plane has the Hamiltonian

$$H(x,y,p_x,p_y) = \frac{1}{2m}\left(p_x^2 + p_y^2\right) + \frac{1}{2}m\omega^2\left(x^2+y^2\right). \qquad (3.379)$$

Hamilton's equations are given by

$$\frac{d}{dt}\begin{pmatrix} x \\ \frac{p_x}{m\omega} \\ y \\ \frac{p_y}{m\omega} \end{pmatrix} = \omega \begin{pmatrix} 0 & 1 & 0 & 0 \\ -1 & 0 & 0 & 0 \\ 0 & 0 & 0 & 1 \\ 0 & 0 & -1 & 0 \end{pmatrix} \begin{pmatrix} x \\ \frac{p_x}{m\omega} \\ y \\ \frac{p_y}{m\omega} \end{pmatrix}, \qquad (3.380)$$

showing the independence of motions in the x- and y-directions. From the treatment of the linear harmonic oscillator, we can write down the solution of (3.380) as

$$\begin{aligned} x(t) &= A_x \cos(\omega t - \varphi_x), & p_x(t) &= -A_x m\omega \sin(\omega t - \varphi_x), \\ y(t) &= A_y \cos(\omega t - \varphi_y), & p_y(t) &= -A_y m\omega \sin(\omega t - \varphi_y), \end{aligned} \qquad (3.381)$$

where A_x and A_y are the amplitudes of motion in the x- and y-directions, respectively, and φ_x and φ_y are the corresponding initial phases. If there is no relative phase between the x and y motions initially, *i.e.*, $\varphi_x = \varphi_y$, then at all times $y(t) = (A_y/A_x)x(t)$ and so the motion of the particle will be a straight line. If the initial phase difference $\varphi_x - \varphi_y$ is $\pm\pi/2$, then at all times $(x(t)/A_x)^2 + (y(t)/A_y)^2 = 1$, and hence the trajectory of the particle will be an ellipse with its axes parallel to the x and y axes; if $A_x = A_y$, the motion will be circular. In general, the particle will have an elliptical orbit with the orientation of its axes depending on the initial phase difference $\varphi_x - \varphi_y$.

A Review of Quantum Mechanics

The two-dimensional isotropic quantum harmonic oscillator in the xy-plane has the Hamiltonian

$$\widehat{H} = \frac{1}{2m}\left(\widehat{p}_x^2 + \widehat{p}_y^2\right) + \frac{1}{2}m\omega^2\left(x^2 + y^2\right)$$
$$= -\frac{\hbar^2}{2m}\left(\frac{\partial^2}{\partial x^2} + \frac{\partial^2}{\partial y^2}\right) + \frac{1}{2}m\omega^2\left(x^2 + y^2\right). \tag{3.382}$$

We can write this as

$$\widehat{H} = \widehat{H}_x + \widehat{H}_y, \tag{3.383}$$

where

$$\widehat{H}_x = \frac{\widehat{p}_x^2}{2m} + \frac{1}{2}m\omega^2 x^2, \qquad \widehat{H}_y = \frac{\widehat{p}_y^2}{2m} + \frac{1}{2}m\omega^2 y^2. \tag{3.384}$$

Note that $\widehat{H}_x$ and $\widehat{H}_y$, Hamiltonians of linear oscillators along x and y axes, respectively, commute with each other. Hence, the complete set of orthonormal solutions of the eigenvalue equation for $\widehat{H}$,

$$\widehat{H}\psi(x,y) = E\psi(x,y), \tag{3.385}$$

the time-independent Schrödinger equation of the system, are given by

$$\psi_{n_x n_y}(x,y) = \psi_{n_x}(x)\psi_{n_y}(y), \qquad E_{n_x n_y} = (n_x + n_y + 1)\hbar\omega, \tag{3.386}$$

where

$$\widehat{H}_x\psi_{n_x}(x) = \left(n_x + \frac{1}{2}\right)\hbar\omega, \qquad \widehat{H}_y\psi_{n_y}(y) = \left(n_y + \frac{1}{2}\right)\hbar\omega. \tag{3.387}$$

Explicitly,

$$\psi_{n_x n_y}(x,y) = \left(\frac{m\omega}{\pi\hbar 2^{n_x+n_y}n_x!n_y!}\right)^{1/2} e^{-\frac{m\omega}{2\hbar}(x^2+y^2)}$$
$$\times H_{n_x}\left(\sqrt{\frac{m\omega}{\hbar}}\,x\right) H_{n_y}\left(\sqrt{\frac{m\omega}{\hbar}}\,y\right). \tag{3.388}$$

The energy eigenvalue $E_{n_x n_y}$ is seen to be $(n_x + n_y + 1)$-fold degenerate since $n_x + n_y = n$ can be partitioned into an ordered pair of nonnegative integers in $(n+1)$ ways.

One can define the lowering and raising operators

$$\widehat{a}_x = \frac{m\omega x + i\widehat{p}_x}{\sqrt{2m\hbar\omega}}, \qquad \widehat{a}_x^\dagger = \frac{m\omega x - i\widehat{p}_x}{\sqrt{2m\hbar\omega}}$$
$$\widehat{a}_y = \frac{m\omega y + i\widehat{p}_y}{\sqrt{2m\hbar\omega}}, \qquad \widehat{a}_y^\dagger = \frac{m\omega y - i\widehat{p}_y}{\sqrt{2m\hbar\omega}} \tag{3.389}$$

which obey the commutation relations

$$\left[\hat{a}_j, \hat{a}_k^\dagger\right] = \delta_{jk}, \quad \left[\hat{a}_j, \hat{a}_k\right] = 0, \quad \left[\hat{a}_j^\dagger, \hat{a}_k^\dagger\right] = 0, \quad j,k = x,y. \tag{3.390}$$

Then, the Hamiltonian can be written as

$$\widehat{H} = \left(\hat{a}_x^\dagger \hat{a}_x + \hat{a}_y^\dagger \hat{a}_y + 1\right)\hbar\omega, \tag{3.391}$$

so that the energy eigenvalues are given by $(n_x + n_y + 1)\hbar\omega$. The corresponding eigenstates can be written as

$$|n_x, n_y\rangle = \frac{1}{\sqrt{n_x! n_y!}} \left(\hat{a}_x^\dagger\right)^{n_x} \left(\hat{a}_y^\dagger\right)^{n_y} |0,0\rangle, \tag{3.392}$$

where the ground state is defined by

$$\hat{a}_x|0,0\rangle = 0, \quad \hat{a}_y|0,0\rangle = 0. \tag{3.393}$$

The Hamiltonian $\widehat{H}$ is seen to have rotational symmetry, *i.e.*, if the coordinates x and y are replaced by $x\cos\phi + y\sin\phi$ and $-x\sin\phi + y\cos\phi$, respectively, it remains invariant. This suggests using plane polar coordinates for solving the differential equation to get the eigenvalues and eigenfunctions of $\widehat{H}$. Changing from (x,y) to the plane polar coordinates (r,ϕ), defined by $x = r\cos\phi$ and $y = r\sin\phi$, we have

$$\widehat{H} = -\frac{\hbar^2}{2m}\left[\frac{1}{r}\frac{\partial}{\partial r}\left(r\frac{\partial}{\partial r}\right) + \frac{1}{r^2}\frac{\partial^2}{\partial\phi^2}\right] + \frac{1}{2}m\omega^2 r^2 \tag{3.394}$$

and the time-independent Schrödinger equation (3.385) becomes

$$\left\{-\frac{\hbar^2}{2m}\left[\frac{1}{r}\frac{\partial}{\partial r}\left(r\frac{\partial}{\partial r}\right) + \frac{1}{r^2}\frac{\partial^2}{\partial\phi^2}\right] + \frac{1}{2}m\omega^2 r^2\right\}\psi(r,\phi) = E\psi(r,\phi). \tag{3.395}$$

The complete set of orthonormal eigenfunctions of this equation are given by

$$\psi_{n_r,m_\phi}(r,\phi) \sim r^{|m_\phi|} e^{-\frac{m\omega r^2}{2\hbar}} L_{n_r}^{|m_\phi|}\left(\frac{m\omega r^2}{\hbar}\right) e^{im_\phi \phi},$$

$$n_r = 0, 1, 2, \ldots, \quad m_\phi = 0, \pm 1, \pm 2, \ldots, \tag{3.396}$$

apart from normalization factors, where

$$L_n^\alpha(x) = \frac{x^{-\alpha} e^x}{n!} \frac{d^n}{dx^n}\left(e^{-x} x^{n+\alpha}\right), \tag{3.397}$$

is called an associated Laguerre polynomial (see, *e.g.*, Arfken, Weber, and Harris [3], Byron and Fuller [18], and Lakshminarayanan and Varadharajan [126]). The respective energy eigenvalues are

$$E_{n_r,m_\phi} = \left(2n_r + |m_\phi| + 1\right)\hbar\omega, \tag{3.398}$$

A Review of Quantum Mechanics

where n_r is the radial quantum number and m_ϕ is the two-dimensional angular momentum quantum number, with $m_\phi \hbar$ being the eigenvalue of $\widehat{L}_z = x\widehat{p}_y - y\widehat{p}_x = -i\hbar \partial/\partial \phi$ corresponding to the eigenfunction $e^{im_\phi \phi}$. We can write $E_{n_r, m_\phi} = (n+1)\hbar\omega$ with $n = (2n_r + |m_\phi|)$. It is seen that for a fixed value of n, there are $n+1$ choices for the pair (n_r, m_ϕ): e.g., $n=0$ corresponds to $(n_r = 0, m_\phi = 0)$, $n=1$ corresponds to $(n_r = 0, m_\phi = \pm 1)$, $n=2$ corresponds to $(n_r = 0, m_\phi = \pm 2)$ and $(n_r = 1, m_\phi = 0)$, $n=3$ corresponds to $(n_r = 0, m_\phi = \pm 3)$ and $(n_r = 1, m_\phi = \pm 1)$, $n=4$ corresponds to $(n_r = 0, m_\phi = \pm 4)$, $(n_r = 1, m_\phi = \pm 2)$, and $(n_r = 2, m_\phi = 0)$, etc. Thus, the energy eigenvalues of the two-dimensional isotropic oscillator are given by $E_n = (n+1)\hbar\omega$, with $n = 0, 1, 2, \ldots$, and the eigenvalue E_n is $(n+1)$-fold degenerate, as we already found earlier in the treatment using the Cartesian coordinates. Note that the position probability distribution, $\left|\psi_{n_r, m_\phi}(r, \phi)\right|^2$, has rotational symmetry in the xy-plane with only r-dependence and no ϕ-dependence for any state $|n_r m_\phi\rangle$. This is to be compared with the classical mechanics where the particle has, in general, elliptical orbits.

3.3.1.4 Charged Particle in a Constant Magnetic Field

We have found that, according to classical mechanics, a charged particle moving in a constant magnetic field has, in general, a helical trajectory with the direction of the magnetic field as its axis and its conserved energy can take any nonzero value. Such a helical trajectory is a superposition of free-particle motion along the axis of the magnetic field and a circular motion in the plane perpendicular to the axis. Now, we shall see how a charged particle moving in a constant magnetic field behaves when it obeys quantum mechanics.

For a particle of mass m and charge q moving in an electromagnetic field with ϕ and $\vec{A}$ as the scalar and vector potentials, the nonrelativistic Schrödinger equation is

$$i\hbar \frac{\partial \Psi}{\partial t} = \left(\frac{1}{2m} \vec{\pi}^2 + q\phi \right) \Psi \tag{3.399}$$

related to the free-particle equation through the principle of minimal electromagnetic coupling, i.e., $i\hbar \partial/\partial t \longrightarrow (i\hbar \partial/\partial t) - q\phi$, $\vec{p} \longrightarrow \vec{\pi} = \vec{p} - q\vec{A}$. For a charged particle moving in a constant magnetic field $\vec{B}$, there is no scalar potential, and the time-independent vector potential can be taken in the symmetric gauge as $\vec{A}(\vec{r}) = \frac{1}{2}\vec{B} \times \vec{r}$, which is such that $\vec{\nabla} \cdot \vec{A} = 0$. The corresponding time-independent Schrödinger equation is

$$\widehat{H}\psi(\vec{r}) = \frac{1}{2m} \left(\vec{p} - q\vec{A} \right)^2 \psi(\vec{r})$$
$$= \frac{1}{2m} \left[\vec{p}^2 + q^2 A^2 - q\left(\vec{A} \cdot \vec{p} + \vec{p} \cdot \vec{A} \right) \right] \psi(\vec{r})$$
$$= \left(\frac{\vec{p}^2}{2m} + \frac{q^2}{2m} A^2 - \frac{q}{m} \vec{A} \cdot \vec{p} \right) \psi(\vec{r})$$

$$= \left(\frac{\hat{p}^2}{2m} + \frac{q^2}{8m}\left[B^2 r^2 - (\vec{B}\cdot\vec{r})^2 \right] - \frac{q}{2m}\vec{B}\cdot\vec{L} \right)\psi(\vec{r})$$
$$= E\psi(\vec{r}), \tag{3.400}$$

where $\hat{\vec{L}} = \vec{r}\times\vec{p}$ is the angular momentum operator, and we have used the identity $(\vec{A}\times\vec{B})^2 = A^2 B^2 - (\vec{A}\cdot\vec{B})^2$, and the relation $\left(\vec{A}\cdot\vec{\nabla} + \vec{\nabla}\cdot\vec{A}\right)\psi(\vec{r}) = 2\vec{A}\cdot\vec{\nabla}\psi(\vec{r})$ since $\vec{\nabla}\cdot\vec{A} = 0$. This shows that the Hamiltonian of the particle is

$$\hat{H} = \left(\frac{\hat{p}^2}{2m} + \frac{q^2}{8m}\left[B^2 r^2 - (\vec{B}\cdot\vec{r})^2 \right] - \frac{q}{2m}\vec{B}\cdot\vec{L} \right), \tag{3.401}$$

in which the third term represents the potential energy due to the interaction of the orbital magnetic dipole moment, or simply called the orbital magnetic moment, of the charged particle,

$$\hat{\vec{\mu}} = \frac{q}{2m}\vec{L}, \tag{3.402}$$

with the magnetic field. Note that $\hat{\vec{\mu}}$ is the quantum operator corresponding to the classical magnetic moment of a charged particle $\vec{\mu} = (q/2m)\vec{L}$, where $\vec{L}$ is the angular momentum, and recall that a magnetic moment $\vec{\mu}$ placed in a magnetic field $\vec{B}$ has the interaction energy $-\vec{\mu}\cdot\vec{B}$.

Let us now take the magnetic field to be in the z-direction: $\vec{B} = B\vec{k}$. The corresponding vector potential in the symmetric gauge is given by $\vec{A} = (-By/2, Bx/2, 0)$. Then, the Hamiltonian is

$$\hat{H} = \frac{1}{2m}\left[\left(\hat{p}_x + \frac{1}{2}qBy\right)^2 + \left(\hat{p}_y - \frac{1}{2}qBx\right)^2 + \hat{p}_z^2 \right]$$
$$= \frac{1}{2m}\hat{p}_z^2 + \frac{1}{2m}\left(\hat{p}_x^2 + \hat{p}_y^2\right) + \frac{1}{2}m\left(\frac{qB}{2m}\right)^2 (x^2+y^2) - \frac{qB}{2m}\hat{L}_z, \tag{3.403}$$

which, apart from the last magnetic interaction term, represents a free particle moving along the z-direction, the direction of the magnetic field, and a two-dimensional isotropic harmonic oscillator in the perpendicular xy-plane with a frequency $qB/2m$. Let us write

$$\hat{H} = \hat{H}_z + \hat{H}_{xy}, \tag{3.404}$$

with

$$\hat{H}_z = \frac{1}{2m}\hat{p}_z^2,$$
$$\hat{H}_{xy} = \frac{1}{2m}\left(\hat{p}_x^2 + \hat{p}_y^2\right) + \frac{1}{2}m\omega_L^2(x^2+y^2) - \omega_L(x\hat{p}_y - y\hat{p}_x), \tag{3.405}$$

where $\omega_L = qB/2m$, called the Larmor frequency, is half of the nonrelativistic cyclotron frequency ω_c. Since $\hat{H}_z$ and $\hat{H}_{xy}$ commute with each other, the eigenfunctions of $\hat{H}$ can be chosen to be simultaneous eigenfunctions of $\hat{H}_z$ and

$\widehat{H}_{xy}$, and the eigenvalues of $\widehat{H}$ can be expressed as the sum of the eigenvalues of $\widehat{H}_z$ and $\widehat{H}_{xy}$. The eigenvalues of $\widehat{H}_z$ form a continuous spectrum given by $\{p_z^2/2m|-\infty < p_z < \infty\}$ corresponding to the free-particle eigenfunctions in the z-direction $\{e^{ip_z z/\hbar}/\sqrt{2\pi\hbar}|-\infty < p_z < \infty\}$. The spectrum of $\widehat{H}_{xy}$ can be found by applying the algebraic method used earlier for the linear harmonic oscillator and the two-dimensional isotropic oscillator. To this end, let us define

$$\widehat{a}_x = \frac{m\omega_L x + i\widehat{p}_x}{\sqrt{2m\hbar\omega_L}}, \qquad \widehat{a}_x^\dagger = \frac{m\omega_L x - i\widehat{p}_x}{\sqrt{2m\hbar\omega_L}}$$
$$\widehat{a}_y = \frac{m\omega_L y + i\widehat{p}_y}{\sqrt{2m\hbar\omega_L}}, \qquad \widehat{a}_y^\dagger = \frac{m\omega_L y - i\widehat{p}_y}{\sqrt{2m\hbar\omega_L}} \qquad (3.406)$$

which obey the commutation relations

$$\left[\widehat{a}_j, \widehat{a}_k^\dagger\right] = \delta_{jk}, \qquad [\widehat{a}_j, \widehat{a}_k] = 0, \qquad \left[\widehat{a}_j^\dagger, \widehat{a}_k^\dagger\right] = 0, \qquad j,k = x,y. \qquad (3.407)$$

In terms of these operators, we have

$$\widehat{H}_{xy} = \hbar\omega_L \left[\widehat{a}_x^\dagger \widehat{a}_x + \widehat{a}_y^\dagger \widehat{a}_y + 1 + i\left(\widehat{a}_x^\dagger \widehat{a}_y - \widehat{a}_y^\dagger \widehat{a}_x\right)\right]. \qquad (3.408)$$

Let us introduce the operators

$$\widehat{A} = \frac{1}{\sqrt{2}}(\widehat{a}_x + i\widehat{a}_y), \qquad \widehat{A}^\dagger = \frac{1}{\sqrt{2}}(\widehat{a}_x^\dagger - i\widehat{a}_y^\dagger), \qquad (3.409)$$

which obey the commutation relation

$$\left[\widehat{A}, \widehat{A}^\dagger\right] = 1. \qquad (3.410)$$

Now, it is found that

$$\widehat{H}_{xy} = 2\hbar\omega_L \left(\widehat{A}^\dagger \widehat{A} + \frac{1}{2}\right) = \hbar\omega_c \left(\widehat{A}^\dagger \widehat{A} + \frac{1}{2}\right), \qquad (3.411)$$

with ω_c as the nonrelativistic cyclotron frequency. Then, the eigenvalues of $\widehat{H}_{xy}$ are seen to be $\{(n+(1/2))\hbar\omega_c \mid n = 0,1,2,\ldots\}$. Thus, the eigenvalues of $\widehat{H}$ are

$$E_{n,p_z} = \left(n + \frac{1}{2}\right)\hbar\omega_c + \frac{p_z^2}{2m}, \qquad n = 0,1,2,\ldots, \quad -\infty < p_z < \infty. \qquad (3.412)$$

The story of the spectrum of $\widehat{H}$ is not over. Let us solve the equation

$$\widehat{A}\psi(x,y) \sim [m\omega_L(x+iy) + (i\widehat{p}_x - \widehat{p}_y)]\psi(x,y)$$
$$= \left[\frac{m\omega_c}{2\hbar}(x+iy) + \left(\frac{\partial}{\partial x} + i\frac{\partial}{\partial y}\right)\right]\psi(x,y) = 0. \qquad (3.413)$$

If we transform this equation to plane polar coordinates (r,ϕ), such that $x = r\cos\phi$ and $y = r\sin\phi$, we get

$$e^{i\phi}\left[\frac{m\omega_c}{2\hbar}r + \left(\frac{\partial}{\partial r} + \frac{i}{r}\frac{\partial}{\partial \phi}\right)\right]\psi(r,\phi) = 0. \tag{3.414}$$

Writing $\psi(r,\phi) = R(r)\Phi(\phi)$ leads to separate equations for $R(r)$ and $\Phi(\phi)$. Solving these equations, we find that

$$\psi_{0,m_\phi}(r,\phi) = \left(re^{i\phi}\right)^{m_\phi} e^{-\frac{m\omega_c}{4\hbar}r^2}, \qquad m_\phi = 0,1,2,\ldots. \tag{3.415}$$

The value of m_ϕ is restricted to nonnegative integers since solution is to be single valued for any ϕ and $\phi + 2\pi$ and should be regular at the origin $r = 0$. Note that $m_\phi \hbar$ is the eigenvalue of the angular momentum operator $\widehat{L}_z = (x\widehat{p}_y - y\widehat{p}_x) = -i\hbar(\partial/\partial\phi)$. In terms of Cartesian coordinates these ground state wave functions are

$$\psi_{0,m_\phi}(x,y) = \mathcal{N}(x+iy)^{m_\phi} e^{-\frac{m\omega_c}{4\hbar}(x^2+y^2)}, \qquad m_\phi = 0,1,2,\ldots, \tag{3.416}$$

where $\mathcal{N}$ is the normalization constant. It is seen that we have an infinity of degenerate ground states for $\widehat{H}_{xy}$ all with the same energy $E_{0,m_\phi} = \hbar\omega_c/2$ independent of m_ϕ, which labels the angular momentum state (for two-dimensional motion, there is only one angular momentum $(x\widehat{p}_y - y\widehat{p}_x)$). The excited states of $\widehat{H}_{xy}$ are obtained by the repeated action of $\widehat{A}^\dagger$ on each of the ground states. We get

$$\psi_{n,m_\phi}(x,y) = \frac{1}{\sqrt{n!}}\left(\widehat{A}^\dagger\right)^n \psi_{0,m_\phi}(x,y), \quad n = 0,1,2,\ldots, \ m_\phi = 0,1,2,\ldots. \tag{3.417}$$

Thus, for a particle of charge q and mass m moving in a constant magnetic field $\vec{B} = B\vec{k}$, the complete set of orthonormal eigenstates are given by

$$\psi_{n,m_\phi,p_z}(x,y,z) = \frac{1}{\sqrt{2\pi\hbar n!}} e^{\frac{i}{\hbar}p_z z} \left(\widehat{A}^\dagger\right)^n \psi_{0,m_\phi}(x,y),$$

$$\psi_{0,m_\phi}(x,y) = \mathcal{N}(x+iy)^{m_\phi} e^{-\frac{m\omega_c}{4\hbar}(x^2+y^2)},$$

$$n = 0,1,2,\ldots, \ m_\phi = 0,1,2,\ldots, \ -\infty < p_z < \infty,$$

$$\tag{3.418}$$

corresponding to the energy spectrum, called the Landau levels,

$$E_{n,m_\phi,p_z} = \left(n + \frac{1}{2}\right)\hbar\omega_c + \frac{p_z^2}{2m}, \tag{3.419}$$

where $\mathcal{N}$ is the normalization constant and $\omega_c = qB/m$ is the nonrelativistic cyclotron frequency. Each eigenvalue is infinitely degenerate, and each degenerate eigenstate has distinct angular momentum eigenvalue $m_\phi\hbar$ on which the energy eigenvalue does not depend.

3.3.1.5 Scattering States

Let us now consider a particle of mass m moving along the x-axis in which it experiences a potential well

$$V(x) = \begin{cases} 0 & \text{for} \quad -\infty < x < -L \quad \text{region I} \\ -V & \text{for} \quad -L \leq x \leq L \quad \text{region II} \\ 0 & \text{for} \quad L < x < \infty \quad \text{region III} \end{cases} \quad (3.420)$$

where V is a positive constant. Since the corresponding Hamiltonian is time-independent, let us look at the time-independent Schrödinger equation for the energy eigenvalues of the system:

$$-\frac{\hbar^2}{2m}\frac{d^2\psi(x)}{dx^2} = E\psi(x), \quad \text{in region I}$$

$$\left(-\frac{\hbar^2}{2m}\frac{d^2}{dx^2} - V\right)\psi(x) = E\psi(x), \quad \text{in region II}$$

$$-\frac{\hbar^2}{2m}\frac{d^2\psi(x)}{dx^2} = E\psi(x), \quad \text{in region III} \quad (3.421)$$

First, let us take $-V < E < 0$. Classically, in region II, the particle will have positive kinetic energy and will move as a free particle between the turning points $x = -L$ and $x = L$, and it cannot enter the regions I and III since it cannot have positive kinetic energy in these regions. Let us look at the quantum mechanics in this case. The physically acceptable solution of the Schrödinger equation (3.421) is

Region I : $\psi(x) = \psi_I(x) = Ae^{\kappa x}$, with $\hbar\kappa = \sqrt{2m|E|}$,

Region II : $\psi(x) = \psi_{II}(x) = Ce^{ikx} + De^{-ikx}$,

with $\hbar k = \sqrt{2m(V - |E|)}$,

Region III : $\psi(x) = \psi_{III}(x) = Ge^{-\kappa x}$, with $\hbar\kappa = \sqrt{2m|E|}$, (3.422)

where A, C, D, and G are constants, and positive values of the square roots are taken. A solution $Be^{-\kappa x}$, with B as a constant, has been dropped for region I since $\lim_{x\to-\infty} e^{-\kappa x} \longrightarrow \infty$, and similarly a solution $Fe^{\kappa x}$, with F as a constant, has been dropped for region III as physically unacceptable since $\lim_{x\to\infty} e^{\kappa x} \longrightarrow \infty$.

The constants A, C, D, and G have to be fixed using the normalization condition, and the boundary conditions on $\psi(x)$ at $x = -L$ and $x = L$ where the potential changes discontinuously. The first condition on $\psi(x)$ is that it should be continuous at the boundary. By this condition, we ensure that the position probability $|\psi(x)|^2$ is continuous, as should be. The second condition on $\psi(x)$ is that its first derivative $d\psi/dx$ should be continuous at the boundary. The reason for this condition is as follows. If we integrate the Schrödinger equation (3.421) from $L - \varepsilon$ to $L + \varepsilon$, we get

$$-\frac{\hbar^2}{2m}\int_{L-\varepsilon}^{L+\varepsilon} dx \left(\frac{d^2\psi(x)}{dx^2}\right) = \int_{L-\varepsilon}^{L+\varepsilon} dx\, (E - V(x))\psi(x). \quad (3.423)$$

As $\varepsilon \longrightarrow 0$, the integral on the right is zero so that

$$\lim_{\varepsilon \to 0}\left(\left.\frac{d\psi(x)}{dx}\right|_{L+\varepsilon} - \left.\frac{d\psi(x)}{dx}\right|_{L-\varepsilon}\right) \longrightarrow 0, \qquad (3.424)$$

showing that $d\psi(x)/dx$ is continuous at the boundary $x = L$. Similarly, we have that $d\psi(x)/dx$ is continuous at the boundary $x = -L$.

Before applying the above boundary conditions, let us rewrite the solution in region II as

$$\psi_{II}(x) = \bar{C}\sin(kx) + \bar{D}\cos(kx), \qquad (3.425)$$

with $\bar{C} = i(C-D)$ and $\bar{D} = C+D$. Now, applying the boundary conditions, we have

$$Ae^{-\kappa L} = -\bar{C}\sin(kL) + \bar{D}\cos(kL),$$
$$A\kappa e^{-\kappa L} = k\left[\bar{C}\cos(kL) + \bar{D}\sin(kL)\right],$$
$$Ge^{-\kappa L} = \bar{C}\sin(kL) + \bar{D}\cos(kL),$$
$$-G\kappa e^{-\kappa L} = k\left[\bar{C}\cos(kL) - \bar{D}\sin(kL)\right]. \qquad (3.426)$$

Dividing the first equation by the second, on both sides, one gets an expression for $1/\kappa$ and dividing the third equation by the fourth, on both sides, one gets an expression for $-1/\kappa$. These two expressions are seen to be consistent only if either $\bar{C} = 0$ and $\bar{D} \neq 0$, or $\bar{C} \neq 0$ and $\bar{D} = 0$. For the solution corresponding to the first case, $\bar{C} = 0$ and $\bar{D} \neq 0$, we find that κ and k, which are functions of E, must satisfy a relation

$$\kappa = k\tan(kL). \qquad (3.427)$$

For the second case, $\bar{C} \neq 0$ and $\bar{D} = 0$, the corresponding relation is

$$\kappa = -k\cot(kL). \qquad (3.428)$$

By solving these relations for E, numerically or graphically, one gets the discrete set of allowed energy eigenvalues for the case $-V < E < 0$. For a chosen energy eigenvalue, the constants $\bar{C}$, $\bar{D}$, and G can be fixed in terms of A, using the relations in (3.426). Then, the normalization of $\psi(x)$, namely,

$$\int_{-\infty}^{\infty} dx\,|\psi(x)|^2 = \int_{-\infty}^{-L} dx\,|\psi_I(x)|^2 + \int_{-L}^{L} dx\,|\psi_{II}(x)|^2$$
$$+ \int_{L}^{\infty} dx\,|\psi_{III}(x)|^2 = 1, \qquad (3.429)$$

fixes A, and the time-dependent wave function is given by $\Psi(x,t) = \psi(x)e^{-iEt/\hbar}$. Note that classically the motion of the particle with energy in the range $(-V, 0)$ will be within the two walls of the potential well, at $x = -L$ and $x = L$, and any value of energy varying continuously in the range from $-V$ to 0 will be allowed. But, the quantum mechanics of the particle allows its energy to be only an eigenvalue from a discrete energy spectrum in the range $(-V, 0)$, and allows it to tunnel through the potential walls into the classically forbidden regions I and III, as seen from the exponentially decaying parts of the solution $\psi(x)$ in these regions.

Let us consider $E > 0$. Now, the solution of the Schrödinger equation (3.421) is

Region I: $\psi(x) = \psi_I(x) = Ae^{i\kappa x} + Be^{-i\kappa x}$, with $\hbar\kappa = \sqrt{2mE}$,

Region II: $\psi(x) = \psi_{II}(x) = Ce^{ikx} + De^{-ikx}$, with $\hbar k = \sqrt{2m(E+V)}$,

Region III: $\psi(x) = \psi_{III}(x) = Fe^{i\kappa x}$, with $\hbar\kappa = \sqrt{2mE}$, (3.430)

where A, B, C, D, and F, are constants. Here, in region I the time-dependent wave function $\left(Ae^{i\kappa x} + Be^{-i\kappa x}\right)e^{-iEt/\hbar}$ is a linear combination of wave functions of the incoming (from $-\infty$) particle of momentum $\hbar\kappa$ in the $+x$-direction and the reflected (from $x = -L$) particle of momentum $\hbar\kappa$ in the $-x$-direction. Similarly, in region II the time-dependent wave function $\left(Ce^{ikx} + De^{-ikx}\right)e^{-iEt/\hbar}$ is a linear combination of wave functions of the incoming (from $x = -L$, or transmitted at $x = -L$) particle of momentum $\hbar k$ and the reflected (from $x = L$) particle of momentum $-\hbar k$. In region III, the time-dependent wave function $Fe^{i[\kappa x - (Et/\hbar)]}$ is the wave function of the outgoing, or the transmitted, particle (from $x = L$) of momentum $\hbar\kappa$. Since the particle is not expected to be reflected from $+\infty$, there is no component of wave function $\sim e^{-i[\kappa x + (Et/\hbar)]}$ in region III.

To fix the constants we have to apply the boundary conditions at $x = -L$ and $x = L$ as earlier. Before that, let us rewrite the solution in region II as

$$\psi_{II}(x) = \bar{C}\sin(kx) + \bar{D}\cos(kx). \quad (3.431)$$

Then, the boundary conditions lead to the relations

$$Ae^{-i\kappa L} + Be^{i\kappa L} = -\bar{C}\sin(kL) + \bar{D}\cos(kL),$$
$$i\kappa\left(Ae^{-i\kappa L} - Be^{i\kappa L}\right) = k\left[\bar{C}\cos(kL) + \bar{D}\sin(kL)\right],$$
$$Fe^{i\kappa L} = \bar{C}\sin(kL) + \bar{D}\cos(kL),$$
$$i\kappa Fe^{i\kappa kL} = k\left[\bar{C}\cos(kL) - \bar{D}\sin(kL)\right]. \quad (3.432)$$

We are interested in solving for B and F in terms of A since A is the amplitude of the incident wave, and B and F are the amplitudes of the reflected and transmitted waves. To this end, we shall first solve for C and D in terms of F using the last pair of equations and then substitute these values of C and D in the first pair of equations to get B and F in terms of A. The algebra is straightforward, and the result is

$$B = i\frac{\sin(2kL)}{2\kappa k}\left(k^2 - \kappa^2\right)F,$$
$$F = \frac{e^{-2i\kappa L}}{\cos(2kL) - i\frac{(\kappa^2+k^2)}{2\kappa k}\sin(2kL)}A. \quad (3.433)$$

A particle obeying classical mechanics would pass through the potential well region without any reflection at any point since, with $E > 0$, its kinetic energy is positive throughout. The above result implies that this is not so for a particle obeying quantum mechanics. It can get reflected from the boundaries where the potential energy changes abruptly.

This example presents a quantum system for which there are both discrete and continuous energy eigenvalues. The discrete energy levels correspond to bound states with normalizable wave functions that vanish at $\pm\infty$. The eigenstates belonging to the continuous energy spectrum correspond to the particle incident on the potential well from, say, $-\infty$ and getting scattered (reflected/transmitted) to $\mp\infty$. The eigenfunctions of these states are not normalizable and do not vanish at $\pm\infty$. Such states are called scattering states.

Let us now consider a particle of mass m moving along the x-axis in which it encounters a potential barrier

$$V(x) = \begin{cases} 0 & \text{for} \quad -\infty < x < -L \quad \text{region I} \\ V > 0 & \text{for} \quad -L \leq x \leq L \quad \text{region II} \\ 0 & \text{for} \quad L < x < \infty \quad \text{region III} \end{cases} \quad (3.434)$$

Since the corresponding Hamiltonian is time-independent let us look at the time-independent Schrödinger equation for the energy eigenvalues of the system:

$$-\frac{\hbar^2}{2m}\frac{d^2\psi(x)}{dx^2} = E\psi(x), \quad \text{in region I}$$

$$\left(-\frac{\hbar^2}{2m}\frac{d^2}{dx^2} + V\right)\psi(x) = E\psi(x), \quad \text{in region II}$$

$$-\frac{\hbar^2}{2m}\frac{d^2\psi(x)}{dx^2} = E\psi(x), \quad \text{in region III} \quad (3.435)$$

First, let us take $E < V$. Classically, the particle hitting the barrier from region I with energy $E \leq V$ will be reflected back since it cannot enter region II without positive kinetic energy. Let us look at the quantum mechanics of the particle. The solution for the wave function in this case is

Region I: $\psi(x) = \psi_{\text{I}}(x) = Ae^{ikx} + Be^{-ikx}$, with $\hbar k = \sqrt{2mE}$,

Region II: $\psi(x) = \psi_{\text{II}}(x) = Ce^{\kappa x} + De^{-\kappa x}$, with $\hbar\kappa = \sqrt{2m(V-E)}$,

Region III: $\psi(x) = \psi_{\text{III}}(x) = Fe^{ikx}$, with $\hbar k = \sqrt{2mE}$, (3.436)

where the constants A, B, C, D, and F are to be determined using the same boundary conditions as earlier at the barrier walls at $x = -L$ and $x = L$. The result for F is

$$F = \frac{2k\kappa e^{-2ika}}{2k\kappa \cosh(2\kappa a) - i(k^2 - \kappa^2)\sinh(2\kappa a)} A. \quad (3.437)$$

In this case, the presence of B, amplitude for reflection, is natural. The surprising aspect of quantum mechanics is the presence of F, amplitude for transmission through the barrier. This quantum mechanical tunneling effect is the basic principle of scanning tunneling microscopy (see *e.g.*, Chen [21]).

When $E > V$, the solution for the wave function is

Region I : $\psi(x) = \psi_I(x) = Ae^{ikx} + Be^{-ikx}$, with $\hbar k = \sqrt{2mE}$,

Region II : $\psi(x) = \psi_{II}(x) = Ce^{i\kappa x} + De^{-i\kappa x}$, with $\hbar\kappa = \sqrt{2m(E-V)}$,

Region III : $\psi(x) = \psi_{III}(x) = Fe^{ikx}$, with $\hbar k = \sqrt{2mE}$. (3.438)

Applying the boundary conditions at $x = -L$ and $x = L$, we can determine the coefficients A, B, C, D, and F. In this case, what is surprising is the presence of B and D which represent the amplitudes for reflections from the boundaries $x = -L$ and $x = L$. Classically, the particle with $E > V$ will go through the potential barrier region with reduced momentum, but it will not be reflected from anywhere since it has positive kinetic energy everywhere. The particle obeying quantum mechanics is seen to be reflected from the boundaries of the potential barrier where the potential changes abruptly.

3.3.1.6 Approximation Methods, Time-Dependent Systems, and the Interaction Picture

There are very few quantum systems, like free particle, harmonic oscillator, and hydrogen-like atoms, for which the Schrödinger equation can be solved exactly. Therefore, approximation methods become necessary for many practical applications. For example, to study the one-dimensional anharmonic oscillator with the Hamiltonian $\widehat{H} = (\widehat{p}^2/2m) + (kx^2/2) + \varepsilon x^4$, where $\varepsilon > 0$ is a small parameter, time-independent perturbation theory is used treating εx^4 as a small perturbation to the harmonic oscillator Hamiltonian. Time-independent perturbation theory is a systematic procedure of getting approximate solutions to the perturbed problem in terms of the known exact solutions to the unperturbed problem. Variational method can be used to obtain approximately the bound-state energies and wave functions of a time-independent Hamiltonian. The Jeffreys–Wentzel–Kramers–Brillouin (JWKB) approximation technique is useful in analyzing any one-dimensional time-independent Schrödinger equation with a slowly varying potential. We will not be concerned here with these time-independent perturbation theory techniques.

We shall now consider time-dependent systems. The time evolution of a system with the time-dependent Hamiltonian $\widehat{H}(t)$ is given by the Schrödinger equation

$$i\hbar \frac{\partial |\Psi(t)\rangle}{\partial t} = \widehat{H}(t)|\Psi(t)\rangle. \quad (3.439)$$

Equivalently, we can write, as seen already (3.154–3.163),

$$|\Psi(t)\rangle = \widehat{U}(t,t_i)|\Psi(t_i)\rangle, \qquad t \geq t_i, \quad (3.440)$$

with t_i as the initial time and the time-evolution operator $\widehat{U}(t,t_i)$ satisfying the relations

$$i\hbar \frac{\partial}{\partial t}\widehat{U}(t,t_i) = \widehat{H}(t)\widehat{U}(t,t_i), \quad (3.441)$$

$$\widehat{U}(t,t_i)^\dagger \widehat{U}(t,t_i) = I, \qquad \widehat{U}(t_i,t_i) = I. \tag{3.442}$$

If the Hamiltonian is time-independent, we can integrate (3.441) exactly to write $\widehat{U}(t,t_i) = e^{-i\widehat{H}(t-t_i)/\hbar}$. But, when $\widehat{H}(t)$ is time-dependent, this cannot be done. This can be seen as follows. Let $t - t_i = N\Delta t$, where N is a large number and Δt is an infinitesimally small time interval. We can assume that the Hamiltonian does not change within the time interval Δt. Then, we can write

$$\begin{aligned}|\Psi(t)\rangle = \lim_{N\to\infty}\lim_{\Delta t\to 0}\Big\{ & e^{-\frac{i}{\hbar}\Delta t\widehat{H}(t_i+N\Delta t)} e^{-\frac{i}{\hbar}\Delta t\widehat{H}(t_i+(N-1)\Delta t)} e^{-\frac{i}{\hbar}\Delta t\widehat{H}(t_i+(N-2)\Delta t)} \cdots \\ & \cdots e^{-\frac{i}{\hbar}\Delta t\widehat{H}(t_i+2\Delta t)} e^{-\frac{i}{\hbar}\Delta t\widehat{H}(t_i+\Delta t)} \Big\} |\Psi(t_i)\rangle. \end{aligned} \tag{3.443}$$

If $\widehat{H}(t)$ is time-independent, then we have $\widehat{H}(t_i+N\Delta t) = \widehat{H}(t_i+(N-1)\Delta t) = \widehat{H}(t_i+(N-2)\Delta t) = \cdots = \widehat{H}(t_i+2\Delta t) = \widehat{H}(t_i+\Delta t) = \widehat{H}$ and hence the above equation can be written as

$$\begin{aligned}|\Psi(t)\rangle &= \lim_{\Delta t\to 0} e^{-\frac{i}{\hbar}\Sigma \Delta t \widehat{H}}|\Psi(t_i)\rangle = e^{-\frac{i}{\hbar}\int_{t_i}^{t} dt \widehat{H}}|\Psi(t_i)\rangle \\ &= e^{-\frac{i}{\hbar}\widehat{H}(t-t_i)}|\Psi(t_i)\rangle = \widehat{U}(t,t_i)|\Psi(t_i)\rangle. \end{aligned} \tag{3.444}$$

When the Hamiltonian is time-dependent, $\widehat{H}(t)$s at different times do not commute with each other and the exponents cannot be added, as done earlier, in taking the product of the exponentials since $e^{\widehat{A}}e^{\widehat{B}} \neq e^{\widehat{A}+\widehat{B}}$ if $\widehat{A}\widehat{B} \neq \widehat{B}\widehat{A}$. The Baker–Campbell–Hausdorff (BCH) formula for the product $e^{\widehat{A}}e^{\widehat{B}}$, when $\widehat{A}\widehat{B} \neq \widehat{B}\widehat{A}$, is

$$e^{\widehat{A}}e^{\widehat{B}} = e^{\widehat{A}+\widehat{B}+\frac{1}{2}[\widehat{A},\widehat{B}]+\frac{1}{12}([\widehat{A},[\widehat{A},\widehat{B}]]+[[\widehat{A},\widehat{B}],\widehat{B}])+\cdots}. \tag{3.445}$$

For a proof of the BCH formula, and for more operator techniques useful in physics, see, *e.g.*, Bellman and Vasudevan [10], and Wilcox [188].

When the Hamiltonian is time-dependent, we can proceed as follows to find $\widehat{U}(t,t_i)$. From (3.441), we have

$$\int_{t_i}^{t} dt \left(\frac{\partial}{\partial t}\widehat{U}(t,t_i)\right) = -\frac{i}{\hbar}\int_{t_i}^{t} dt_1 \widehat{H}(t_1)\widehat{U}(t_1,t_i), \tag{3.446}$$

or

$$\widehat{U}(t,t_i)\Big|_{t_i}^{t} = \widehat{U}(t,t_i) - I = -\frac{i}{\hbar}\int_{t_i}^{t} dt_1 \widehat{H}(t_1)\widehat{U}(t_1,t_i), \tag{3.447}$$

leading to the formal solution

$$\widehat{U}(t,t_i) = I - \frac{i}{\hbar}\int_{t_i}^{t} dt_1 \widehat{H}(t_1)\widehat{U}(t_1,t_i). \tag{3.448}$$

A Review of Quantum Mechanics

Iterating this formal solution, we get

$$\widehat{U}(t,t_i) = I - \frac{i}{\hbar}\int_{t_i}^{t} dt_1 \widehat{H}(t_1) + \left(-\frac{i}{\hbar}\right)^2 \int_{t_i}^{t} dt_2 \int_{t_i}^{t_2} dt_1 \widehat{H}(t_2)\widehat{H}(t_1)$$
$$+ \left(-\frac{i}{\hbar}\right)^3 \int_{t_i}^{t} dt_3 \int_{t_i}^{t_3} dt_2 \int_{t_i}^{t_2} dt_1 \widehat{H}(t_3)\widehat{H}(t_2)\widehat{H}(t_1)$$
$$+ \cdots,$$
$$= I + \sum_{n=1}^{\infty}\left(-\frac{i}{\hbar}\right)^n \int\cdots\int_{t>t_n>t_{n-1}>\cdots>t_2>t_1>t_i} dt_n dt_{n-1}\cdots dt_2 dt_1$$
$$\times \widehat{H}(t_n)\widehat{H}(t_{n-1})\cdots\widehat{H}(t_1). \qquad (3.449)$$

This series expression can be written in a compact form, symbolically, by introducing the time ordering operator:

$$\mathsf{T}[A(t_1)B(t_2)] = \begin{cases} A(t_1)B(t_2), & \text{if } t_1 > t_2 \\ B(t_2)A(t_1), & \text{if } t_2 > t_1 \end{cases} \qquad (3.450)$$

Now, note that

$$\mathsf{T}\left[\left(\int_{t_i}^{t} dt\, A(t)\right)^2\right] = \mathsf{T}\left[\left(\int_{t_i}^{t} dt_2\, A(t_2)\right)\left(\int_{t_i}^{t} dt_1\, A(t_1)\right)\right]$$
$$= \int_{t_i}^{t} dt_2 \int_{t_i}^{t} dt_1\, \mathsf{T}[A(t_2)A(t_1)]$$
$$= \iint_{t>t_2>t_1>t_i} dt_2 dt_1\, A(t_2)A(t_1)$$
$$+ \iint_{t>t_1>t_2>t_i} dt_2 dt_1\, A(t_1)A(t_2). \qquad (3.451)$$

The $t_1 \longleftrightarrow t_2$ symmetry between the two integrals on the right-hand side shows that the two integrals must be equal. Hence, we have

$$\iint_{t>t_2>t_1>t_i} dt_2 dt_1\, A(t_2)A(t_1) = \frac{1}{2}\mathsf{T}\left[\left(\int_{t_i}^{t} dt\, A(t)\right)^2\right]. \qquad (3.452)$$

Extending this argument leads easily to the result that

$$\int\cdots\int_{t>t_n>t_{n-1}>\cdots>t_2>t_1>t_i} dt_n dt_{n-1}\cdots dt_2 dt_1\, A(t_n)A(t_{n-1})\cdots A(t_1)$$
$$= \frac{1}{n!}\mathsf{T}\left[\left(\int_{t_i}^{t} dt\, A(t)\right)^n\right]. \qquad (3.453)$$

Thus, we can write (3.449) as

$$\widehat{U}(t,t_i) = I + \sum_{n=1}^{\infty} \frac{1}{n!} \left(-\frac{i}{\hbar}\right)^n \mathsf{T}\left[\left(\int_{t_i}^{t} dt\, \widehat{H}(t)\right)^n\right]$$

$$= \mathsf{T}\left[\sum_{n=0}^{\infty} \frac{1}{n!} \left(-\frac{i}{\hbar}\right)^n \left(\int_{t_i}^{t} dt\, \widehat{H}(t)\right)^n\right]$$

$$= \mathsf{T}\left(e^{-\frac{i}{\hbar}\int_{t_i}^{t} dt\, \widehat{H}(t)}\right), \tag{3.454}$$

where the expression on the right-hand side is known as the Dyson time-ordered exponential. It should be noted that the time-ordered exponential is not a true exponential and is only a notation for the series that has to be computed by truncating it up to any desired order.

An alternate expression for the time-evolution operator $\widehat{U}(t,t_i)$ is the Magnus formula (Magnus [130]):

$$\widehat{U}(t,t_i) = e^{-\frac{i}{\hbar}\widehat{T}(t,t_i)}, \tag{3.455}$$

where $\widehat{T}(t,t_i)$ is an infinite series with the first few terms given by

$$\widehat{T}(t,t_i) = \int_{t_i}^{t} dt_1\, \widehat{H}(t_1) + \frac{1}{2}\left(-\frac{i}{\hbar}\right)\int_{t_i}^{t} dt_2 \int_{t_i}^{t_2} dt_1 \left[\widehat{H}(t_2),\widehat{H}(t_1)\right]$$

$$+ \frac{1}{6}\left(-\frac{i}{\hbar}\right)^2 \int_{t_i}^{t} dt_3 \int_{t_i}^{t_3} dt_2 \int_{t_i}^{t_2} dt_1 \left\{\left[\left[\widehat{H}(t_3),\widehat{H}(t_2)\right],\widehat{H}(t_1)\right]\right.$$

$$\left. + \left[\left[\widehat{H}(t_1),\widehat{H}(t_2)\right],\widehat{H}(t_3)\right]\right\} + \cdots. \tag{3.456}$$

Note that $\widehat{T}(t,t_i)$ is Hermitian such that $\widehat{U}(t,t_i)$ is unitary. In the Magnus formula (3.455 and 3.456), $\widehat{U}(t,t_i)$ is a true exponential (see the Appendix at the end of this chapter for derivation). The Magnus formula is completely equivalent to the Dyson time-ordered exponential formula.

Let us now consider a system for which the Hamiltonian is time-dependent and can be written as

$$\widehat{H}(t) = \widehat{H}_0 + \widehat{H}'(t), \tag{3.457}$$

where $\widehat{H}_0$ is the unperturbed Hamiltonian and $\widehat{H}'(t)$ is a time-dependent perturbation Hamiltonian. Usually $\widehat{H}_0$ is chosen to be time-independent. With

$$i\hbar \frac{\partial |\Psi(t)\rangle}{\partial t} = \widehat{H}(t)|\Psi(t)\rangle, \tag{3.458}$$

define

$$|\Psi_\mathrm{i}(t)\rangle = \widehat{U}_{\widehat{H}_0}^{\dagger}(t,t_i)|\Psi(t)\rangle, \tag{3.459}$$

where $\widehat{U}_{\widehat{H}_0}(t,t_i)$ is the unitary time-evolution operator corresponding to the unperturbed Hamiltonian $\widehat{H}_0$ such that

$$i\hbar \frac{\partial \widehat{U}_{\widehat{H}_0}(t,t_i)}{\partial t} = \widehat{H}_0 \widehat{U}_{\widehat{H}_0}(t,t_i), \quad i\hbar \frac{\partial \widehat{U}_{\widehat{H}_0}^{\dagger}(t,t_i)}{\partial t} = -\widehat{U}_{\widehat{H}_0}^{\dagger}(t,t_i)\widehat{H}_0. \tag{3.460}$$

A Review of Quantum Mechanics

For the Hermitian operator $\hat{O}$ corresponding to an observable O, define

$$\hat{O}_{\text{i}}(t) = \hat{U}_{\hat{H}_0}^\dagger(t,t_i)\,\hat{O}\,\hat{U}_{\hat{H}_0}(t,t_i). \tag{3.461}$$

From (3.459–3.461) we have

$$i\hbar\frac{\partial |\Psi_{\text{i}}(t)\rangle}{\partial t} = i\hbar\frac{\partial \hat{U}_{\hat{H}_0}^\dagger(t,t_i)}{\partial t}|\Psi(t)\rangle + \hat{U}_{\hat{H}_0}^\dagger(t,t_i)\hat{H}(t)|\Psi(t)\rangle$$

$$= -\hat{U}_{\hat{H}_0}^\dagger(t,t_i)\hat{H}_0|\Psi(t)\rangle + \hat{U}_{\hat{H}_0}^\dagger(t,t_i)\hat{H}(t)|\Psi(t)\rangle$$

$$= \hat{U}_{\hat{H}_0}^\dagger(t,t_i)\left(\hat{H}(t) - \hat{H}_0\right)|\Psi(t)\rangle = \hat{U}_{\hat{H}_0}^\dagger(t,t_i)\hat{H}'(t)|\Psi(t)\rangle$$

$$= \hat{U}_{\hat{H}_0}^\dagger(t,t_i)\hat{H}'(t)\hat{U}_{\hat{H}_0}(t,t_i)\hat{U}_{\hat{H}_0}^\dagger(t,t_i)|\Psi(t)\rangle$$

$$= \hat{H}'_{\text{i}}(t)|\Psi_{\text{i}}(t)\rangle. \tag{3.462}$$

From (3.460) and (3.461) we have

$$i\hbar\frac{d\hat{O}_{\text{i}}}{dt} = \left(i\hbar\frac{\partial \hat{U}_{\hat{H}_0}^\dagger(t,t_i)}{\partial t}\right)\hat{O}\hat{U}_{\hat{H}_0}(t,t_i) + \hat{U}_{\hat{H}_0}^\dagger(t,t_i)\left(i\hbar\frac{\partial \hat{O}}{\partial t}\right)\hat{U}_{\hat{H}_0}(t,t_i)$$

$$+ \hat{U}_{\hat{H}_0}^\dagger(t,t_i)\hat{O}\left(i\hbar\frac{\partial \hat{U}_{\hat{H}_0}(t,t_i)}{\partial t}\right)$$

$$= -\hat{U}_{\hat{H}_0}^\dagger(t,t_i)\hat{H}_0\hat{O}\hat{U}_{\hat{H}_0}(t,t_i) + \hat{U}_{\hat{H}_0}^\dagger(t,t_i)\left(i\hbar\frac{\partial \hat{O}}{\partial t}\right)\hat{U}_{\hat{H}_0}(t,t_i)$$

$$+ \hat{U}_{\hat{H}_0}^\dagger(t,t_i)\hat{O}\hat{H}_0\hat{U}_{\hat{H}_0}(t,t_i)$$

$$= i\hbar\left(\frac{\partial \hat{O}}{\partial t}\right)_{\text{i}} + \left(\hat{U}_{\hat{H}_0}^\dagger(t,t_i)\hat{O}\hat{U}_{\hat{H}_0}(t,t_i)\right)\left(\hat{U}_{\hat{H}_0}^\dagger(t,t_i)\hat{H}_0\hat{U}_{\hat{H}_0}(t,t_i)\right)$$

$$- \left(\hat{U}_{\hat{H}_0}^\dagger(t,t_i)\hat{H}_0\hat{U}_{\hat{H}_0}(t,t_i)\right)\left(\hat{U}_{\hat{H}_0}^\dagger(t,t_i)\hat{O}\hat{U}_{\hat{H}_0}(t,t_i)\right)$$

$$= \left[\hat{O}_{\text{i}},\left(\hat{H}_0\right)_{\text{i}}\right] + i\hbar\left(\frac{\partial \hat{O}}{\partial t}\right)_{\text{i}}$$

$$= \left[-\left(\hat{H}_0\right)_{\text{i}},\hat{O}_{\text{i}}\right] + i\hbar\left(\frac{\partial \hat{O}}{\partial t}\right)_{\text{i}}, \tag{3.463}$$

or

$$\frac{d\hat{O}_{\text{i}}}{dt} = \frac{1}{i\hbar}\left[-\left(\hat{H}_0\right)_{\text{i}},\hat{O}_{\text{i}}\right] + \left(\frac{\partial \hat{O}}{\partial t}\right)_{\text{i}}. \tag{3.464}$$

Equations (3.462) and (3.464) constitute the interaction picture, introduced by Dirac. It is seen to be intermediate between the Schrödinger and the Heisenberg pictures. It becomes the Schrödinger picture when $\hat{H}_0 \longrightarrow 0$ and $\hat{H}'(t) \longrightarrow \hat{H}(t)$. It becomes the Heisenberg picture when $\widehat{H}_0 \longrightarrow \widehat{H}(t)$ and $\widehat{H}'(t) \longrightarrow 0$. Note that the equation of motion for the interaction picture observables (3.464) is exactly like the equation of motion for the Heisenberg picture observables (3.188).

The interaction picture is designed for use in time-dependent perturbation theory, where the goal is to study the time evolution of a system with a small time-dependent perturbation Hamiltonian. The time-evolution equation for $|\Psi_i(t)\rangle$, (3.462), can be formally integrated to give

$$|\Psi_i(t)\rangle = \widehat{U}_{\widehat{H}'_i}(t,t_i)|\Psi_i(t_i)\rangle, \qquad (3.465)$$

where

$$\widehat{U}_{\widehat{H}'_i}(t,t_i) = \mathsf{T}\left(e^{-\frac{i}{\hbar}\int_{t_i}^{t} dt\, \widehat{H}'_i(t)}\right). \qquad (3.466)$$

From the definition of $|\Psi_i(t)\rangle$, (3.459), we get

$$\begin{aligned}|\Psi(t)\rangle &= \widehat{U}_{\widehat{H}_0}(t,t_i)|\Psi_i(t)\rangle = \widehat{U}_{\widehat{H}_0}(t,t_i)\widehat{U}_{\widehat{H}'_i}(t,t_i)|\Psi_i(t_i)\rangle \\ &= \widehat{U}_{\widehat{H}_0}(t,t_i)\widehat{U}_{\widehat{H}'_i}(t,t_i)|\Psi(t_i)\rangle \\ &= \widehat{U}(t,t_i)|\Psi(t_i)\rangle, \end{aligned} \qquad (3.467)$$

relating the state of the system at t to its initial state at t_i. In the second step above, we have used the relation $|\Psi_i(t_i)\rangle = |\Psi(t_i)\rangle$ following directly from the definition (3.459). When the unperturbed Hamiltonian $\widehat{H}_0$ is chosen to be time-independent $\widehat{U}_{\widehat{H}_0}(t,t_i) = e^{-\frac{i}{\hbar}\widehat{H}_0(t-t_i)}$ and, when the time-dependent perturbation Hamiltonian $\widehat{H}'(t)$ is small, to compute $\widehat{U}_{\widehat{H}'_i}(t,t_i)$ only the first few terms of the infinite series in the Dyson, or the Magnus, formula need to be taken. We can directly verify that $\widehat{U}_{\widehat{H}_0}(t,t_i)\widehat{U}_{\widehat{H}'_i}(t,t_i) = \widehat{U}(t,t_i)$, the time-evolution operator corresponding to $\widehat{H}(t) = \widehat{H}_0 + \widehat{H}'(t)$:

$$\begin{aligned}&i\hbar\frac{\partial}{\partial t}\left(\widehat{U}_{\widehat{H}_0}(t,t_i)\widehat{U}_{\widehat{H}'_i}(t,t_i)\right) \\ &= \left[i\hbar\frac{\partial}{\partial t}\left(\widehat{U}_{\widehat{H}_0}(t,t_i)\right)\right]\widehat{U}_{\widehat{H}'_i}(t,t_i) + \widehat{U}_{\widehat{H}_0}(t,t_i)\left[i\hbar\frac{\partial}{\partial t}\left(\widehat{U}_{\widehat{H}'_i}(t,t_i)\right)\right] \\ &= \widehat{H}_0\widehat{U}_{\widehat{H}_0}(t,t_i)\widehat{U}_{\widehat{H}'_i}(t,t_i) + \widehat{U}_{\widehat{H}_0}(t,t_i)\widehat{H}'_i(t)\widehat{U}_{\widehat{H}'_i}(t,t_i) \\ &= \widehat{H}_0\widehat{U}_{\widehat{H}_0}(t,t_i)\widehat{U}_{\widehat{H}'_i}(t,t_i) + \widehat{U}_{\widehat{H}_0}(t,t_i)\widehat{U}^{\dagger}_{\widehat{H}_0}(t,t_i)\widehat{H}'(t)\widehat{U}_{\widehat{H}_0}(t,t_i)\widehat{U}_{\widehat{H}'_i}(t,t_i) \\ &= \left(\widehat{H}_0 + \widehat{H}'(t)\right)\left(\widehat{U}_{\widehat{H}_0}(t,t_i)\widehat{U}_{\widehat{H}'_i}(t,t_i)\right) \\ &= \widehat{H}(t)\left(\widehat{U}_{\widehat{H}_0}(t,t_i)\widehat{U}_{\widehat{H}'_i}(t,t_i)\right). \end{aligned} \qquad (3.468)$$

3.3.1.7 Schrödinger–Pauli Equation for the Electron

Electron has the nonclassical property, spin, the intrinsic angular momentum, with two values, $\pm\hbar/2$, for its z-component. In the presence of a magnetic field, it aligns itself either parallel or antiparallel to the field. As already discussed, the spin-$\frac{1}{2}$ operators, components of the vector $\vec{S}$, are to be represented in terms of the Pauli matrices as

$$S_x = \frac{\hbar}{2}\sigma_x, \quad S_y = \frac{\hbar}{2}\sigma_y, \quad S_z = \frac{\hbar}{2}\sigma_z. \tag{3.469}$$

The spin magnetic dipole moment, or simply called the spin magnetic moment, of the electron, with charge $-e$, is found to be

$$\vec{\mu}_e = -\mu_B \vec{\sigma} = -\frac{e\hbar}{2m_e}\vec{\sigma} = -\frac{e\vec{S}}{m_e}, \tag{3.470}$$

where $\mu_B = e\hbar/2m_e$ is the Bohr magneton, the unit of magnetic moment, and m_e is the mass of the electron. It is seen that $\vec{\mu}_e$ is twice the value of the magnetic moment it would have if the spin $\vec{S}$ is like a classical orbital angular momentum (see (3.402)). This factor of 2, known as the Landé g-factor, will be seen to be a direct consequence of Dirac's relativistic equation for the electron to be discussed later. Actually, the electron is observed to have an anomalous magnetic moment corresponding to a value of g slightly higher than 2, and the successful explanation of this $g - 2 \approx 0.00232$ is provided by quantum electrodynamics.

The nonrelativistic Hamiltonian of an electron in an electromagnetic field associated with the vector and scalar potentials $\vec{A}$ and ϕ, respectively, is, including spin,

$$\widehat{H} = \left[\frac{\vec{\pi}^2}{2m_e} - e\phi\right] I + \mu_B \vec{\sigma}\cdot\vec{B}, \quad \text{with } \vec{\pi} = \vec{p} + e\vec{A}, \tag{3.471}$$

where the last term, the Pauli term, is the interaction energy of the electron spin magnetic moment in the magnetic field and I is the 2×2 unit matrix inserted to be consistent with the last 2×2 matrix term. The Schrödinger equation now becomes

$$i\hbar\frac{\partial|\Psi(t)\rangle}{\partial t} = \left\{\left[\frac{\vec{\pi}^2}{2m_e} - e\phi\right] I + \mu_B \vec{\sigma}\cdot\vec{B}\right\}|\Psi(t)\rangle, \tag{3.472}$$

the Schrödinger–Pauli equation, or the Pauli equation, for the electron, where the state vector is a two-component column vector,

$$|\Psi(t)\rangle = \begin{pmatrix} |\Psi_1(t)\rangle \\ |\Psi_2(t)\rangle \end{pmatrix}, \tag{3.473}$$

since the 2×2 matrix Hamiltonian has to act on a two-component column vector. The nonrelativistic electron state vector $|\Psi(t)\rangle$ is called a two-component spinor. Note that we can write the Pauli equation (3.472) also as

$$i\hbar\frac{\partial|\Psi(t)\rangle}{\partial t} = \left\{\frac{1}{2m_e}\left(\vec{\sigma}\cdot\vec{\pi}\right)^2 - e\phi I\right\}|\Psi(t)\rangle \tag{3.474}$$

To derive (3.472) from (3.474), one has to proceed as follows. First, use the algebra of the Pauli matrices

$$\sigma_x^2 = I, \quad \sigma_y^2 = I, \quad \sigma_z^2 = I,$$
$$\sigma_x\sigma_y = -\sigma_y\sigma_x, \quad \sigma_y\sigma_z = -\sigma_z\sigma_y, \quad \sigma_z\sigma_x = -\sigma_x\sigma_z, \quad (3.475)$$

to observe that

$$\left(\vec{\sigma}\cdot\vec{\widehat{\pi}}\right)^2 = \vec{\widehat{\pi}}^2 + \sigma_x\sigma_y[\widehat{\pi}_x,\widehat{\pi}_y] + \sigma_y\sigma_z[\widehat{\pi}_y,\widehat{\pi}_z] + \sigma_z\sigma_x[\widehat{\pi}_z,\widehat{\pi}_x]. \quad (3.476)$$

Then, use the relations between the Pauli matrices

$$\sigma_x\sigma_y = i\sigma_z, \quad \sigma_y\sigma_z = i\sigma_x, \quad \sigma_z\sigma_x = i\sigma_y, \quad (3.477)$$

and the commutation relations

$$[\widehat{\pi}_x,\widehat{\pi}_y] = -ie\hbar B_z, \quad [\widehat{\pi}_y,\widehat{\pi}_z] = -ie\hbar B_x, \quad [\widehat{\pi}_z,\widehat{\pi}_x] = -ie\hbar B_y. \quad (3.478)$$

We can now write

$$|\underline{\Psi}(t)\rangle = |\Psi_1(t)\rangle \begin{pmatrix} 1 \\ 0 \end{pmatrix} + |\Psi_2(t)\rangle \begin{pmatrix} 0 \\ 1 \end{pmatrix}$$
$$= |\Psi_1(t)\rangle\,|\uparrow\rangle + |\Psi_2(t)\rangle\,|\downarrow\rangle. \quad (3.479)$$

As already discussed, $|\langle\vec{r}|\Psi_1(t)\rangle|^2$ gives the probability for the electron to be found at $\vec{r}$ with spin up and $|\langle\vec{r}|\Psi_2(t)\rangle|^2$ gives the probability for the electron to be found at $\vec{r}$ with spin down.

For an electron in a constant magnetic field $\vec{B} = B\vec{k}$, the Schrödinger–Pauli equation is

$$i\hbar\frac{\partial|\underline{\Psi}(t)\rangle}{\partial t} = \left\{\frac{1}{2m_e}\left[\left(\widehat{p}_x - \frac{1}{2}eBy\right)^2 + \left(\widehat{p}_y + \frac{1}{2}eBx\right)^2 + \widehat{p}_z^2\right]I + \mu_B B\sigma_z\right\}|\underline{\Psi}(t)\rangle. \quad (3.480)$$

For the stationary states, we can write $|\underline{\Psi}(t)\rangle = e^{-iEt/\hbar}|\underline{\psi}\rangle$, where $|\underline{\psi}\rangle$ satisfies the time-independent Pauli equation

$$\left\{\frac{1}{2m_e}\left[\left(\widehat{p}_x - \frac{1}{2}eBy\right)^2 + \left(\widehat{p}_y + \frac{1}{2}eBx\right)^2 + \widehat{p}_z^2\right]I + \mu_B B\sigma_z\right\}|\underline{\psi}\rangle = E|\underline{\psi}\rangle. \quad (3.481)$$

From the earlier discussion on the stationary states of a particle of mass m and charge q moving in a constant magnetic field, without considering spin, we know that the Hamiltonian

$$\widehat{H}_{\text{NRL}} = \frac{1}{2m}\left[\left(\widehat{p}_x + \frac{1}{2}qBy\right)^2 + \left(\widehat{p}_y - \frac{1}{2}qBx\right)^2 + \widehat{p}_z^2\right] \quad (3.482)$$

has the energy spectrum $E_{n,p_z} = [n+(1/2)]\hbar\omega_c$, where $\omega_c = |q|B/m$, $n = 0,1,2,\ldots$, and $-\infty < p_z < \infty$. The corresponding eigenfunctions are given by $\left\{\psi_{n,m_\phi,p_z}(\vec{r})\right\}$, as in (3.418), where m_ϕ $(= 0,1,2,\ldots,)$ represents an angular momentum quantum number with respect to which each energy level is infinitely degenerate. The subscript NRL in this Hamiltonian is to indicate that its energy spectrum corresponds to nonrelativistic Landau levels. Now, it is clear that equation (3.481) can be written as

$$\begin{pmatrix} \hat{H}_{NRL} + \mu_B B & 0 \\ 0 & \hat{H}_{NRL}I - \mu_B B \end{pmatrix} \begin{pmatrix} \psi_1(\vec{r}) \\ \psi_2(\vec{r}) \end{pmatrix} = E \begin{pmatrix} \psi_1(\vec{r}) \\ \psi_2(\vec{r}) \end{pmatrix}. \tag{3.483}$$

The general solution of this equation is

$$\langle \vec{r} | \psi \rangle = \underline{\psi}(\vec{r}) = \begin{pmatrix} \psi_1(\vec{r}) \\ \psi_2(\vec{r}) \end{pmatrix} = \begin{pmatrix} c_1 \psi_{n-1,m_\phi,p_z}(\vec{r}) \\ c_2 \psi_{n,m_\phi,p_z}(\vec{r}) \end{pmatrix}, \tag{3.484}$$

with $|c_1|^2 + |c_2|^2 = 1$ and $n = 1, 2, \ldots$. The energy eigenvalue is, with $\omega_c = eB/m_e$,

$$\begin{aligned} E_{n,p_z} &= \frac{p_z^2}{2m_e} + \left[(n-1) + \frac{1}{2}\right]\hbar\omega_c + \frac{\hbar eB}{2m_e} \\ &= \frac{p_z^2}{2m_e} + \left[n + \frac{1}{2}\right]\hbar\omega_c - \frac{\hbar eB}{2m_e} \\ &= \frac{p_z^2}{2m_e} + n\hbar\omega_c, \end{aligned} \tag{3.485}$$

with infinite degeneracy due to the angular momentum quantum number $m_\phi = 0, 1, 2, \ldots$. Note that in the energy spectrum (3.485) all levels are shifted up by $\hbar\omega_c/2$ compared to the Landau level spectrum (3.419) obtained ignoring spin. The probability for the particle to be found at $\vec{r}$ with spin up is given by $|c_1\psi_{n-1,m_\phi,p_z}(\vec{r})|^2$, and the probability for the particle to be found at $\vec{r}$ with spin down is given by $|c_2\psi_{n,m_\phi,p_z}(\vec{r})|^2$. Note that

$$\int d^3r \left(|c_1\psi_{n-1,m_\phi,p_z}(\vec{r})|^2 + |c_2\psi_{n,m_\phi,p_z}(\vec{r})|^2\right)$$
$$= \int d^3r \, \underline{\psi}^\dagger(\vec{r})\underline{\psi}(\vec{r}) = |c_1|^2 + |c_2|^2 = 1. \tag{3.486}$$

3.3.2 QUANTUM MECHANICS OF A SYSTEM OF IDENTICAL PARTICLES

When we have a quantum system of several, say N, particles moving in a common external field of force and interacting with each other, the Hamiltonian of the system and the state vectors will be functions of the dynamical variables of all the particles. We take the quantum operators of the corresponding observables belonging to two different particles to be commuting with each other. For example, the position operator of j-th particle $(\vec{r}_j)$ will commute with the momentum operator of the k-th particle $(\vec{p}_k = -i\hbar\vec{\nabla}_{\vec{r}_k})$, and any component of the spin operator of the j-th particle (say $S_x^{(j)}$)

will commute with any component of the spin operator of the k-th particle (say $S_y^{(k)}$). Just to fix the notations, let us consider the Hamiltonian of two electrons moving in the electromagnetic field of vector potential $\vec{A}$ and scalar potential ϕ given by

$$\widehat{H}\left(\left(\vec{r}_1,\vec{p}_1,\vec{S}^{(1)}\right),\left(\vec{r}_2,\vec{p}_2,\vec{S}^{(2)}\right),t\right) = \left[\frac{1}{2m_e}\left(\vec{p}_1 + e\vec{A}(\vec{r}_1,t)\right)^2 - e\phi(\vec{r}_1,t)\right.$$

$$\left. + \frac{1}{2m_e}\left(\vec{p}_2 + e\vec{A}(\vec{r}_2,t)\right)^2 - e\phi(\vec{r}_2,t)\right]I$$

$$+ \mu_B\vec{\sigma}^{(1)} \cdot \vec{B}(\vec{r}_1,t) + \mu_B\vec{\sigma}^{(2)} \cdot \vec{B}(\vec{r}_2,t),$$
(3.487)

where we have ignored completely the electron–electron interaction. The corresponding Schrödinger equation will be

$$i\hbar\frac{\partial|\Psi(t)\rangle}{\partial t} = \widehat{H}\left(\left(\vec{r}_1,\vec{p}_1,\vec{S}^{(1)}\right),\left(\vec{r}_2,\vec{p}_2,\vec{S}^{(2)}\right),t\right)|\Psi(t)\rangle.$$
(3.488)

The components of spin matrix operators $\vec{\sigma}^{(1)}$ and $\vec{\sigma}^{(2)}$ commute with each other. Hence, their matrix representations have to be

$$\sigma_x^{(1)} = \sigma_x \otimes I, \qquad \sigma_y^{(1)} = \sigma_y \otimes I, \qquad \sigma_z^{(1)} = \sigma_z \otimes I,$$

$$\sigma_x^{(2)} = I \otimes \sigma_x, \qquad \sigma_x^{(2)} = I \otimes \sigma_y, \qquad \sigma_x^{(2)} = I \otimes \sigma_z,$$
(3.489)

where I is the 2×2 identity matrix and $\otimes$ denotes the direct product.

The direct product of two 2×2 matrices, say A and B, is the 4×4 matrix defined by

$$A \otimes B = \begin{pmatrix} a_{11} & a_{12} \\ a_{21} & a_{22} \end{pmatrix} \otimes \begin{pmatrix} b_{11} & b_{12} \\ b_{21} & b_{22} \end{pmatrix}$$

$$= \begin{pmatrix} a_{11}B & a_{12}B \\ a_{21}B & a_{22}B \end{pmatrix}$$

$$= \begin{pmatrix} a_{11}b_{11} & a_{11}b_{12} & a_{12}b_{11} & a_{12}b_{12} \\ a_{11}b_{21} & a_{11}b_{22} & a_{12}b_{21} & a_{12}b_{22} \\ a_{21}b_{11} & a_{21}b_{12} & a_{22}b_{11} & a_{22}b_{12} \\ a_{21}b_{21} & a_{21}b_{22} & a_{22}b_{21} & a_{22}b_{22} \end{pmatrix}.$$
(3.490)

In general, the direct product of two $n \times n$ matrices, A and B, is the $n^2 \times n^2$ matrix with the matrix elements given by

$$(A \otimes B)_{jk,lm} = A_{jk}B_{lm}, \qquad j,k,l,m = 1,2,\ldots,n.$$
(3.491)

A Review of Quantum Mechanics

The direct product of matrices has the property

$$(A \otimes B)(C \otimes D) = AC \otimes BD. \tag{3.492}$$

Hence, if $AC = CA$ and $BD = DB$, then $(A \otimes B)(C \otimes D) = (C \otimes D)(A \otimes B)$. From this, it follows that

$$\sigma_j^{(1)} \sigma_k^{(2)} = \sigma_k^{(2)} \sigma_j^{(1)}, \qquad j,k = x,y,z, \tag{3.493}$$

as required. The direct product of two 2-dimensional vectors is the 4-dimensional vector defined by

$$\left|V^{(1)}\right\rangle \otimes \left|V^{(2)}\right\rangle = \begin{pmatrix} V_1^{(1)} \\ V_2^{(1)} \end{pmatrix} \otimes \begin{pmatrix} V_1^{(2)} \\ V_2^{(2)} \end{pmatrix} = \begin{pmatrix} V_1^{(1)} V_1^{(2)} \\ V_1^{(1)} V_2^{(2)} \\ V_2^{(1)} V_1^{(2)} \\ V_2^{(1)} V_2^{(2)} \end{pmatrix}. \tag{3.494}$$

Similarly, the direct product of two n-dimensional vectors will be an n^2-dimensional vector with the components

$$\left(\left|V^{(1)}\right\rangle \otimes \left|V^{(2)}\right\rangle\right)_{jk} = V_j^{(1)} V_k^{(2)}, \qquad j,k = 1,2,\ldots,n. \tag{3.495}$$

Now, if A and B are two $n \times n$ matrices and $V^{(1)}$ and $V^{(2)}$ are two n-dimensional vectors, then

$$(A \otimes B)\left(\left|V^{(1)}\right\rangle \otimes \left|V^{(2)}\right\rangle\right) = A\left|V^{(1)}\right\rangle \otimes B\left|V^{(2)}\right\rangle. \tag{3.496}$$

For more details on the direct product, or the Kronecker product, see, *e.g.*, Arfken, Weber, and Harris [3], and Byron and Fuller [18].

Let us now consider the direct products of two spin-$\frac{1}{2}$ vectors:

$$|\uparrow\rangle \otimes |\uparrow\rangle = \begin{pmatrix} 1 \\ 0 \end{pmatrix} \otimes \begin{pmatrix} 1 \\ 0 \end{pmatrix} = \begin{pmatrix} 1 \\ 0 \\ 0 \\ 0 \end{pmatrix},$$

$$|\uparrow\rangle \otimes |\downarrow\rangle = \begin{pmatrix} 1 \\ 0 \end{pmatrix} \otimes \begin{pmatrix} 0 \\ 1 \end{pmatrix} = \begin{pmatrix} 0 \\ 1 \\ 0 \\ 0 \end{pmatrix},$$

$$|\downarrow\rangle \otimes |\uparrow\rangle = \begin{pmatrix} 0 \\ 1 \end{pmatrix} \otimes \begin{pmatrix} 1 \\ 0 \end{pmatrix} = \begin{pmatrix} 0 \\ 0 \\ 1 \\ 0 \end{pmatrix},$$

$$|\downarrow\rangle \otimes |\downarrow\rangle = \begin{pmatrix} 0 \\ 1 \end{pmatrix} \otimes \begin{pmatrix} 0 \\ 1 \end{pmatrix} = \begin{pmatrix} 0 \\ 0 \\ 0 \\ 1 \end{pmatrix}. \tag{3.497}$$

Then, it follows from (3.496) that

$$S_z^{(1)}(|\uparrow\rangle \otimes |\uparrow\rangle) = \frac{\hbar}{2}\sigma_z^{(1)}(|\uparrow\rangle \otimes |\uparrow\rangle) = \frac{\hbar}{2}(\sigma_z \otimes I)(|\uparrow\rangle \otimes |\uparrow\rangle)$$
$$= \frac{\hbar}{2}(|\uparrow\rangle \otimes |\uparrow\rangle),$$
$$S_z^{(2)}(|\uparrow\rangle \otimes |\uparrow\rangle) = \frac{\hbar}{2}\sigma_z^{(2)}(|\uparrow\rangle \otimes |\uparrow\rangle) = \frac{\hbar}{2}(I \otimes \sigma_z)(|\uparrow\rangle \otimes |\uparrow\rangle)$$
$$= \frac{\hbar}{2}(|\uparrow\rangle \otimes |\uparrow\rangle), \quad (3.498)$$

Similarly, one can verify that

$$S_z^{(1)}(|\uparrow\rangle \otimes |\downarrow\rangle) = \frac{\hbar}{2}(|\uparrow\rangle \otimes |\downarrow\rangle), \quad S_z^{(2)}(|\uparrow\rangle \otimes |\downarrow\rangle) = -\frac{\hbar}{2}(|\uparrow\rangle \otimes |\downarrow\rangle),$$
$$S_z^{(1)}(|\downarrow\rangle \otimes |\uparrow\rangle) = -\frac{\hbar}{2}(|\downarrow\rangle \otimes |\uparrow\rangle), \quad S_z^{(2)}(|\downarrow\rangle \otimes |\uparrow\rangle) = \frac{\hbar}{2}(|\downarrow\rangle \otimes |\uparrow\rangle),$$
$$S_z^{(1)}(|\downarrow\rangle \otimes |\downarrow\rangle) = -\frac{\hbar}{2}(|\downarrow\rangle \otimes |\downarrow\rangle), \quad S_z^{(2)}(|\downarrow\rangle \otimes |\downarrow\rangle) = -\frac{\hbar}{2}(|\downarrow\rangle \otimes |\downarrow\rangle). \quad (3.499)$$

This means that the spin state of a system of two spin-$\frac{1}{2}$ particles will be represented by $|\uparrow\rangle \otimes |\uparrow\rangle$ if both the particles have their spins up. If the first particle has its spin up and the second particle has its spin down, then the state is $|\uparrow\rangle \otimes |\downarrow\rangle$. If the first particle has its spin down and the second particle has its spin up, then the state is $|\downarrow\rangle \otimes |\uparrow\rangle$. If both the particles have their spins down, then the state is $|\downarrow\rangle \otimes |\downarrow\rangle$. We shall write $|V^{(1)}\rangle \otimes |V^{(2)}\rangle$ simply as $|V^{(1)}\rangle|V^{(2)}\rangle$, as if $|V^{(1)}\rangle$ and $|V^{(2)}\rangle$ are one dimensional, to be understood as a direct product from the context.

It is now clear that the Hamiltonian in (3.487) is a 4-dimensional matrix with operators as elements. Hence, the state vector $|\Psi(t)\rangle$ in the Schrödinger equation (3.488) should be a 4-dimensional vector. Let

$$|\Psi(t)\rangle = \begin{pmatrix} |\Psi_{11}(t)\rangle \\ |\Psi_{12}(t)\rangle \\ |\Psi_{21}(t)\rangle \\ |\Psi_{22}(t)\rangle \end{pmatrix}. \quad (3.500)$$

From (3.497), we have

$$|\Psi(t)\rangle = |\Psi_{11}(t)\rangle|\uparrow\rangle|\uparrow\rangle + |\Psi_{12}(t)\rangle|\uparrow\rangle|\downarrow\rangle$$
$$+ |\Psi_{21}(t)\rangle|\downarrow\rangle|\uparrow\rangle + |\Psi_{22}(t)\rangle|\downarrow\rangle|\downarrow\rangle. \quad (3.501)$$

Thus, $|\langle \vec{r}_1, \vec{r}_2 |\Psi_{11}(t)\rangle|^2 = |\Psi_{11}(\vec{r}_1, \vec{r}_2, t)|^2$ gives the probability for the two particles to be found at positions $\vec{r}_1$ and $\vec{r}_2$ with both their spins up. Similarly, the other components of $|\Psi(t)\rangle$ can be interpreted. It is straightforward to extend this formalism to a system of several particles. When the spin of the particle is ignored, $|\Psi(t)\rangle$ is treated only as a single-component state vector.

When we have a system of identical particles, the Hamiltonian of the system should not change under the interchange of position coordinates, and other variables like spin, of any pair of particles. Let $\widehat{\xi}_j$ denote the set of operators $\left(\vec{\hat{r}}_j, \vec{\hat{p}}_j, \vec{S}^{(j)}\right)$ belonging to the j-th particle. Then, the invariance of the Hamiltonian under the interchange of any pair of particles means

$$P_{jk}\widehat{H}\left(\widehat{\xi}_1, \widehat{\xi}_2, \ldots, \widehat{\xi}_j, \ldots, \widehat{\xi}_k, \ldots, \widehat{\xi}_N, t\right) = \widehat{H}\left(\widehat{\xi}_1, \widehat{\xi}_2, \ldots, \widehat{\xi}_k, \ldots, \widehat{\xi}_j, \ldots, \widehat{\xi}_N, t\right)$$
$$= \widehat{H}\left(\widehat{\xi}_1, \widehat{\xi}_2, \ldots, \widehat{\xi}_j, \ldots, \widehat{\xi}_k, \ldots, \widehat{\xi}_N, t\right), \quad (3.502)$$

where P_{jk} is the exchange operator that interchanges the particles j and k. Let $|1, 2, \ldots, N\rangle$ be an eigenstate of a time-independent Hamiltonian of a system of N particles, say $\widehat{H}\left(\widehat{\xi}_1, \widehat{\xi}_2, \ldots, \widehat{\xi}_N\right)$. Since the Hamiltonian $\widehat{H}\left(\widehat{\xi}_1, \widehat{\xi}_2, \ldots, \widehat{\xi}_N\right)$ is invariant under the interchange of any pair of particles, it is found that $P_{jk}|1, 2, \ldots, j, \ldots, k, \ldots, N\rangle = |1, 2, \ldots, k, \ldots, j, \ldots, N\rangle$ differs only by a multiplicative factor ± 1 from $|1, 2, \ldots, j, \ldots, k, \ldots, N\rangle$. This multiplicative factor is ± 1 since $P_{jk}^2 = I$, the identity operator. Since the particles are indistinguishable, the multiplicative factor, $+1$ or -1, cannot depend on which pair of particles are interchanged. In other words, the eigenstate $|1, 2, \ldots, N\rangle$ must be either totally symmetric or totally antisymmetric under the interchange of any pair of particles. This should be true for all eigenstates of any time-independent Hamiltonian of the system. Since any state vector $|\Psi(t)\rangle$ of the system evolving under a time-dependent or time-independent Hamiltonian can be expressed as a linear combination of eigenstates of a time-independent Hamiltonian of the system, the state vector of the system has to be either totally symmetric or totally antisymmetric. It is found that for integer spin particles, bosons, the state vector is totally symmetric and for half integer spin particles, fermions, the state vector is totally antisymmetric.

Let us look at an example. Consider a system of N identical particles with the time-independent Hamiltonian

$$\widehat{\mathbb{H}}\left(\widehat{\xi}_1, \widehat{\xi}_2, \ldots, \widehat{\xi}_N\right) = \sum_{j=1}^{N} \widehat{H}\left(\widehat{\xi}_j\right), \quad (3.503)$$

where the single-particle Hamiltonians $\left\{\widehat{H}\left(\widehat{\xi}_j\right) \big| j = 1, 2, \ldots, N\right\}$ are all identical, except for the labeling of the variables. The N-particle Hamiltonian $\mathbb{H}$ is obviously invariant under any permutation of the particles. The form of $\mathbb{H}$ shows that the system is such that each of the N particles may be in a common external force field and the interparticle interactions have been ignored. For example, take the electronic Hamiltonian of an Z-electron atom,

$$\widehat{\mathbb{H}}_A = \sum_{j=1}^{Z}\left(\frac{\vec{\hat{p}}_j^{\,2}}{2m_e} - \frac{Ze^2}{4\pi\varepsilon_0|\vec{r}_j|}\right) + \sum_{j<k=1}^{Z}\frac{e^2}{4\pi\varepsilon_0\left|\vec{r}_j - \vec{r}_k\right|^2}, \quad (3.504)$$

where $\vec{r}_j$ is the position vector of the j-th electron with respect to the nucleus. Since the corresponding time-independent Schrödinger equation can be solved only by approximation methods, one takes

$$\widehat{\mathbb{H}}_A = \widehat{H}_C + \widehat{H}', \tag{3.505}$$

with

$$\widehat{H}_C = \sum_{j=1}^{Z} \left(\frac{\vec{p}_j^{\,2}}{2m_e} + V(|\vec{r}_j|) \right),$$

$$\widehat{H}' = \sum_{j<k=1}^{Z} \frac{e^2}{4\pi\varepsilon_0 |\vec{r}_j - \vec{r}_k|^2} - \sum_{j=1}^{Z} \left(\frac{Ze^2}{4\pi\varepsilon_0 |\vec{r}_j|} + V(|\vec{r}_j|) \right), \tag{3.506}$$

where $V(|\vec{r}_j|)$ is an effective central potential representing the attraction of the nucleus and the average effect of the repulsive interactions between the j-th electron and the other electrons. Then, the electronic structure of the atom is understood in terms of time-independent perturbation theory with $\widehat{H}_C$ as the unperturbed Hamiltonian and $\widehat{H}'$ as the perturbation. In this perturbation theory, the eigenstates of $\widehat{H}_C$ form the basis states. Note that $\widehat{H}_C$ is of the same type as $\widehat{\mathbb{H}}$ in (3.503). Hence, we shall analyse the general structure of the eigenstates of $\widehat{\mathbb{H}}$.

Let $\{|j\rangle\}$ be the complete set of orthonormal eigenstates of the single particle Hamiltonian $\widehat{H}\left(\widehat{\vec{\xi}}\right)$ corresponding to the energy eigenvalues $\{\varepsilon_j\}$, respectively, such that

$$\widehat{H}\left(\widehat{\vec{\xi}}\right)|j\rangle = \varepsilon_j|j\rangle. \tag{3.507}$$

Let the N particles of the system occupy the single particle eigenstates $\{|j_1\rangle, |j_2\rangle, \cdots, |j_N\rangle\}$ where

$$\widehat{H}|j_k\rangle = \varepsilon_{j_k}|j_k\rangle. \tag{3.508}$$

If we label the particle in the state $|j_1\rangle$ as 1, the particle in the state $|j_2\rangle$ as 2, ..., and the particle in the state $|j_N\rangle$ as N, we can represent the state of the N-particle system as $|j_1\rangle_1 |j_2\rangle_2 \cdots |j_N\rangle_N$. Then, it is easy to see that

$$\widehat{\mathbb{H}}(|j_1\rangle_1 |j_2\rangle_2 \cdots |j_N\rangle_N) = \left(\sum_{k=1}^{N} \varepsilon_{j_k} \right) (|j_1\rangle_1 |j_2\rangle_2 \cdots |j_N\rangle_N). \tag{3.509}$$

But, the indistinguishability of the particles does not allow us to label the particles as 1, 2, ..., and hence any eigenfunction of $\widehat{\mathbb{H}}$ should be symmetrized in such a way that it is totally symmetric for bosons and totally antisymmetric for fermions under the interchange of any pair of particles. We can construct such totally symmetric or antisymmetric wave functions starting with $|j_1\rangle_1 |j_2\rangle_2 \cdots |j_N\rangle_N$.

First, let us consider the case of fermions. Let

$$|1,2,\ldots,N;j_1,j_2,\ldots,j_N\rangle_A = \frac{1}{\sqrt{N!}}\det\begin{vmatrix} |j_1\rangle_1 & |j_2\rangle_1 & \cdots & |j_N\rangle_1 \\ |j_1\rangle_2 & |j_2\rangle_2 & \cdots & |j_N\rangle_2 \\ \vdots & \vdots & & \vdots \\ |j_1\rangle_N & |j_2\rangle_N & \cdots & |j_N\rangle_N \end{vmatrix},$$

(3.510)

where det stands for determinant. This expression is known as the Slater determinant. It can be verified directly that

$$\widehat{\mathbb{H}}|1,2,\ldots,N;j_1,j_2,\ldots,j_N\rangle_A = \left(\sum_{k=1}^{N}\varepsilon_{j_k}\right)|1,2,\ldots,N;j_1,j_2,\ldots,j_N\rangle_A. \quad (3.511)$$

The eigenfunction $|1,2,\ldots,N;j_1,j_2,\ldots,j_N\rangle_A$ gets multiplied by -1 when any pair of particles is exchanged, as required for total antisymmetry, since the corresponding rows of the above determinant are interchanged. It represents the state of the system of N identical fermions in which one particle is in the state $|j_1\rangle$, one particle is in the state $|j_2\rangle$, $\ldots$, and one particle is in the state $|j_N\rangle$, and the factor $1/\sqrt{N!}$ ensures its normalization. Note that more than one fermion cannot occupy any state, i.e, in the eigenfunction $|1,2,\ldots,N;j_1,j_2,\ldots,j_N\rangle_A$ more than one of $\{j_1,j_2,\ldots,j_N\}$ cannot have the same value. If two or more of $\{j_1,j_2,\ldots,j_N\}$ have the same value, then the corresponding columns of the determinant become identical making it vanish. Thus, the Pauli exclusion principle for fermions is built into the antisymmetry of the wave functions.

Let us now consider the case of bosons. Let a permutation of $\{1,2,\ldots,N\}$ give $\{p(1),p(2),\cdots p(N)\}$. We shall write this symbolically as $P\{1,2,\ldots,N\} = \{p(1),p(2),\cdots p(N)\}$. Now, let

$$|1,2,\ldots,N;j_1,j_2,\ldots,j_N\rangle_S = \frac{1}{\sqrt{N!}}\sum_{P}|j_1\rangle_{p(1)}|j_2\rangle_{p(2)}\cdots|j_N\rangle_{p(N)}, \quad (3.512)$$

where $\sum_P$ stands for sum over all $N!$ permutations of $\{1,2,\ldots,N\}$. It is seen that

$$\widehat{\mathbb{H}}|1,2,\ldots,N;j_1,j_2,\ldots,j_N\rangle_S = \left(\sum_{k=1}^{N}\varepsilon_{j_k}\right)|1,2,\ldots,N;j_1,j_2,\ldots,j_N\rangle_S. \quad (3.513)$$

The eigenfunction $|1,2,\ldots,N;j_1,j_2,\ldots,j_N\rangle_S$ is, by construction, totally symmetric under the exchange of any pair of particles, as required for bosons, and the factor $1/\sqrt{N!}$ ensures its normalization. It represents the state of the system of N identical bosons in which one particle is in the state $|j_1\rangle$, one particle is in the state $|j_2\rangle$, $\ldots$, and one particle is in the state $|j_N\rangle$. Note that more than one boson can occupy any state, i.e, the eigenfunction $|1,2,\ldots,N;j_1,j_2,\ldots,j_N\rangle_S$ exists when more than one of $\{j_1,j_2,\ldots,j_N\}$ have the same value, unlike in the case of fermions.

3.3.3 PURE AND MIXED STATES: DENSITY OPERATOR

It is not always possible to associate a state vector $|\Psi\rangle$ with a quantum system. If a state vector can be associated with a system, then it is said to be a pure state. In general, the state of a system can be a statistical mixture of pure states with classical probability weights representing our ignorance about the exact state in which the system is. In order to describe such mixed states, the concept of density operator, or density matrix, is introduced.

A pure state $|\Psi(t)\rangle$ is associated with the density operator

$$\widehat{\rho}(t) = |\Psi(t)\rangle\langle\Psi(t)|, \qquad (3.514)$$

which is a Hermitian operator. In a basis with $\{|\varphi_j\rangle\}$ as the basis vectors, the matrix elements of $\widehat{\rho}(t)$ are

$$\langle j|\widehat{\rho}(t)|k\rangle = \langle\varphi_j|\Psi(t)\rangle\langle\Psi(t)|\varphi_k\rangle. \qquad (3.515)$$

In the position representation

$$\langle\vec{r}|\widehat{\rho}(t)|\vec{r}'\rangle = \Psi(\vec{r},t)\Psi^*(\vec{r}',t). \qquad (3.516)$$

Note that the trace of this density matrix is

$$\operatorname{Tr}(\widehat{\rho}(t)) = \sum_j \langle j|\widehat{\rho}(t)|j\rangle = \sum_j \langle\varphi_j|\Psi(t)\rangle\langle\Psi(t)|\varphi_j\rangle$$
$$= \sum_j \langle\Psi(t)|\varphi_j\rangle\langle\varphi_j|\Psi(t)\rangle = \langle\Psi(t)|\Psi(t)\rangle = 1. \qquad (3.517)$$

Further, a pure state density operator has the property

$$\widehat{\rho}(t)^2 = (|\Psi(t)\rangle\langle\Psi(t)|)(|\Psi(t)\rangle\langle\Psi(t)|) = |\Psi(t)\rangle\langle\Psi(t)| = \widehat{\rho}(t). \qquad (3.518)$$

The time-evolution equation for the density operator, the von Neumann equation, follows from the Schrödinger equation (3.144) and its adjoint (3.145):

$$i\hbar\frac{\partial\widehat{\rho}(t)}{\partial t} = i\hbar\frac{\partial}{\partial t}(|\Psi(t)\rangle\langle\Psi(t)|)$$
$$= \widehat{H}(t)|\Psi(t)\rangle\langle\Psi(t)| - |\Psi(t)\rangle\langle\Psi(t)|\widehat{H}(t)$$
$$= \left[\widehat{H}(t),\widehat{\rho}(t)\right]. \qquad (3.519)$$

For any observable of the system, associated with the quantum Hermitian operator $\widehat{O}$, the average can be written as

$$\left\langle\widehat{O}\right\rangle_{|\Psi(t)\rangle} = \langle\Psi(t)|\widehat{O}|\Psi(t)\rangle$$
$$= \sum_j\sum_k \left\langle\Psi(t)|\varphi_j\right\rangle\left\langle\varphi_j\left|\widehat{O}\right|\varphi_k\right\rangle\left\langle\varphi_k|\Psi(t)\right\rangle$$

$$= \sum_j \sum_k \langle \varphi_k | \Psi(t) \rangle \langle \Psi(t) | \varphi_j \rangle \langle \varphi_j | \hat{O} | \varphi_k \rangle$$

$$= \sum_k \langle \varphi_k | (|\Psi(t)\rangle \langle \Psi(t)|) \hat{O} | \varphi_k \rangle$$

$$= \sum_k \langle \varphi_k | \hat{\rho}(t) \hat{O} | \varphi_k \rangle = \text{Tr}\left(\hat{\rho}(t)\hat{O}\right). \tag{3.520}$$

Using the time-evolution equation for $\hat{\rho}(t)$ (3.519), the equation of motion for the average (3.194) follows

$$\frac{d\langle \hat{O} \rangle_{\Psi(t)}}{dt} = \frac{d}{dt} \text{Tr}\left(\hat{\rho}(t)\hat{O}\right)$$

$$= \text{Tr}\left(\frac{\partial \hat{\rho}(t)}{\partial t}\hat{O} + \hat{\rho}(t)\frac{\partial \hat{O}}{\partial t}\right)$$

$$= \text{Tr}\left(-\frac{i}{\hbar}\left[\hat{H}(t), \hat{\rho}(t)\right]\hat{O} + \hat{\rho}(t)\frac{\partial \hat{O}}{\partial t}\right)$$

$$= \frac{i}{\hbar}\text{Tr}\left(\hat{\rho}(t)\left[\hat{H}(t), \hat{O}\right]\right) + \text{Tr}\left(\hat{\rho}(t)\frac{\partial \hat{O}}{\partial t}\right)$$

$$= \frac{i}{\hbar}\left\langle \left[\hat{H}, \hat{O}\right]\right\rangle(t) + \left\langle \frac{\partial \hat{O}}{\partial t}\right\rangle(t). \tag{3.521}$$

In this derivation, we have used the cyclic property of the trace operator, namely,

$$\text{Tr}\left(\hat{O}_1 \hat{O}_2\right) = \text{Tr}\left(\hat{O}_2 \hat{O}_1\right). \tag{3.522}$$

This follows from the observation that

$$\text{Tr}\left(\hat{O}_1 \hat{O}_2\right) = \sum_j \sum_k \langle \varphi_j | \hat{O}_1 | \varphi_k \rangle \langle \varphi_k | \hat{O}_2 | \varphi_j \rangle$$

$$= \sum_k \sum_j \langle \varphi_k | \hat{O}_2 | \varphi_j \rangle \langle \varphi_j | \hat{O}_1 | \varphi_k \rangle$$

$$= \text{Tr}\left(\hat{O}_2 \hat{O}_1\right). \tag{3.523}$$

The specification of the pure state by a time-dependent density operator as $\hat{\rho}(t) = |\Psi(t)\rangle\langle\Psi(t)|$ corresponds to the Schrödinger picture. In the Heisenberg picture, the system will be associated with the constant, initial, density operator

$$\hat{\rho}(t_i) = |\Psi(t_i)\rangle\langle\Psi(t_i)| = \hat{U}^\dagger(t, t_i) |\Psi(t)\rangle\langle\Psi(t)| \hat{U}(t, t_i)$$

$$= \hat{U}^\dagger(t, t_i) \hat{\rho}(t) \hat{U}(t, t_i). \tag{3.524}$$

Average of any observable will be given by

$$\begin{aligned}\left\langle \hat{O} \right\rangle_{|\Psi\rangle}(t) &= \text{Tr}\left(\hat{\rho}(t)\hat{O}\right) = \text{Tr}\left(\hat{U}(t,t_i)\hat{\rho}(t_i)\hat{U}^\dagger(t,t_i)\hat{O}\right) \\ &= \text{Tr}\left(\hat{\rho}(t_i)\hat{U}^\dagger(t,t_i)\hat{O}\hat{U}(t,t_i)\right) \\ &= \text{Tr}\left(\hat{\rho}(t_i)\hat{O}_{\text{H}}(t)\right). \end{aligned} \qquad (3.525)$$

The equation of motion for the average (3.194) follows from this:

$$\begin{aligned}\frac{d\left\langle \hat{O}\right\rangle(t)}{dt} &= \frac{d}{dt}\left\{\text{Tr}\left(\hat{\rho}(t_i)\hat{O}_{\text{H}}(t)\right)\right\} \\ &= \frac{d}{dt}\left\{\sum_j \left\langle \varphi_j \left| \hat{\rho}(t_i)\hat{O}_{\text{H}}(t)\right| \varphi_j\right\rangle\right\} \\ &= \sum_j \left\langle \varphi_j \left| \hat{\rho}(t_i)\left(\frac{d}{dt}\hat{O}_{\text{H}}(t)\right)\right| \varphi_j\right\rangle \\ &= \sum_j \left\langle \varphi_j \left| \hat{\rho}(t_i)\left\{\frac{i}{\hbar}\left(\left[\hat{H},\hat{O}\right]\right)_{\text{H}} + \left(\frac{\partial \hat{O}}{\partial t}\right)_{\text{H}}\right\}\right| \varphi_j\right\rangle \\ &= \frac{i}{\hbar}\text{Tr}\left[\hat{\rho}(t_i)\left(\left[\hat{H},\hat{O}\right]\right)_{\text{H}}\right] + \text{Tr}\left[\hat{\rho}(t_i)\left(\frac{\partial \hat{O}}{\partial t}\right)_{\text{H}}\right] \\ &= \frac{i}{\hbar}\left\langle \left[\hat{H},\hat{O}\right]\right\rangle(t) + \left\langle \frac{\partial \hat{O}}{\partial t}\right\rangle(t). \end{aligned} \qquad (3.526)$$

In general, the state of a quantum system can be represented, in the Schrödinger picture, by the density operator

$$\hat{\rho}(t) = \sum_{j=1}^{N} p_j \left|\Psi_j(t)\right\rangle\left\langle \Psi_j(t)\right|, \qquad (3.527)$$

where p_j is the classical probability for the system to be in the pure state $\left|\Psi_j(t)\right\rangle\left\langle \Psi_j(t)\right|$. Thus, the positive probability weight factors must satisfy the condition

$$\sum_{j=1}^{N} p_j = 1. \qquad (3.528)$$

Note that $\{\left|\Psi_j(t)\right\rangle \mid j=1,2,\ldots,N\}$ need not be orthogonal to each other. A pure state corresponds to $N=1$. As for the pure state, we find

$$\text{Tr}(\hat{\rho}(t)) = \sum_{j=1}^{N} p_j \text{Tr}\left(\left|\Psi_j(t)\right\rangle\left\langle \Psi_j(t)\right|\right) = \sum_{j=1}^{N} p_j = 1. \qquad (3.529)$$

A Review of Quantum Mechanics

As seen in (3.518), for a pure state $\widehat{\rho}(t)^2 = \widehat{\rho}(t)$. But, in the case of a mixed state we have

$$\widehat{\rho}(t)^2 = \left(\sum_{j=1}^{N} p_j |\Psi_j(t)\rangle \langle \Psi_j(t)|\right)^2 \neq \widehat{\rho}(t). \tag{3.530}$$

This is what distinguishes a mixed state from a pure state. Since each pure state component $|\Psi_j(t)\rangle \langle \Psi_j(t)|$ of the mixed state (3.527) evolves in time under the same Hamiltonian $\widehat{H}(t)$, the time evolution of $\widehat{\rho}(t)$, the von Neumann equation, becomes

$$\begin{aligned}
i\hbar \frac{\partial \widehat{\rho}(t)}{\partial t} &= \sum_{j=1}^{N} p_j i\hbar \frac{\partial}{\partial t} \left(|\Psi_j(t)\rangle \langle \Psi_j(t)|\right) \\
&= \sum_{j=1}^{N} p_j \left[\widehat{H}(t), |\Psi_j(t)\rangle \langle \Psi_j(t)|\right] \\
&= \left[\widehat{H}(t), \widehat{\rho}(t)\right],
\end{aligned} \tag{3.531}$$

same as in the case of pure state. Further, since

$$|\Psi_j(t)\rangle \langle \Psi_j(t)| = \widehat{U}(t,t_i) |\Psi_j(t_i)\rangle \langle \Psi_j(t_i)| \widehat{U}^\dagger(t,t_i), \tag{3.532}$$

we have

$$\widehat{\rho}(t) = \widehat{U}(t,t_i) \widehat{\rho}(t_i) \widehat{U}^\dagger(t,t_i), \tag{3.533}$$

which is the integral version of the equation of motion (3.531), as can be seen by direct differentiation,

$$\begin{aligned}
i\hbar \frac{\partial \widehat{\rho}(t)}{\partial t} &= \left(i\hbar \frac{\partial}{\partial t} \widehat{U}(t,t_i)\right) \widehat{\rho}(t_i) \widehat{U}^\dagger(t,t_i) + \widehat{U}(t,t_i) \widehat{\rho}(t_i) \left(i\hbar \frac{\partial}{\partial t} \widehat{U}^\dagger(t,t_i)\right) \\
&= \widehat{H}(t) \widehat{U}(t,t_i) \widehat{\rho}(t_i) \widehat{U}^\dagger(t,t_i) - \widehat{U}(t,t_i) \widehat{\rho}(t_i) \widehat{U}^\dagger(t,t_i) \widehat{H}(t) \\
&= \left[\widehat{H}(t), \widehat{\rho}(t)\right].
\end{aligned} \tag{3.534}$$

Now, for any observable O, the expression for the average is

$$\begin{aligned}
\langle \widehat{O} \rangle_{\widehat{\rho}(t)} &= \sum_{j=1}^{N} p_j \langle \widehat{O} \rangle_{|\Psi_j(t)\rangle} \\
&= \sum_{j=1}^{N} p_j \text{Tr}\left(|\Psi_j(t)\rangle \langle \Psi_j(t)| \widehat{O}\right) \\
&= \text{Tr}\left(\widehat{\rho}(t) \widehat{O}\right),
\end{aligned} \tag{3.535}$$

same as in the case of pure state. Exactly as in the case of pure state, we can derive the equation of motion for the average (3.194) by following the derivation of (3.521) starting from (3.519).

As in the case of a pure state, the density operator of a mixed state in the Heisenberg picture is given by the constant initial time density operator

$$\begin{aligned}\widehat{\rho}(t_i) &= \sum_{j=1}^{N} p_j |\Psi_j(t_i)\rangle \langle \Psi_j(t_i)| \\ &= \sum_{j=1}^{N} p_j \widehat{U}^\dagger(t,t_i) |\Psi_j(t)\rangle \langle \Psi_j(t)| \widehat{U}(t,t_i) \\ &= \widehat{U}^\dagger(t,t_i) \widehat{\rho}(t) \widehat{U}(t,t_i).\end{aligned} \qquad (3.536)$$

Then, as can be easily verified, the average value of any observable can be written as

$$\langle \widehat{O} \rangle(t) = \text{Tr}\left(\widehat{\rho}(t_i) \widehat{O}_\text{H}(t) \right), \qquad (3.537)$$

where $\widehat{O}_\text{H}(t)$ is the Heisenberg picture operator for the observable O.

When the wave function of a system has multiple components, the definition of the density operator is easily extended. For example, a nonrelativistic spin-$\frac{1}{2}$ particle obeying the Schrödinger–Pauli equation has a two-component wave function and a relativistic spin-$\frac{1}{2}$ particle obeying the Dirac equation has a four-component wave function. In the case of a system with two-component wave function, the pure state density operator is given by

$$\begin{aligned}\widehat{\rho}(t) &= |\underline{\Psi}(t)\rangle \langle \underline{\Psi}(t)| \\ &= \begin{pmatrix} |\Psi_1(t)\rangle \\ |\Psi_2(t)\rangle \end{pmatrix} \begin{pmatrix} \langle \Psi_1(t)| & \langle \Psi_2(t)| \end{pmatrix} \\ &= \begin{pmatrix} |\Psi_1(t)\rangle\langle \Psi_1(t)| & |\Psi_1(t)\rangle\langle \Psi_2(t)| \\ |\Psi_2(t)\rangle\langle \Psi_1(t)| & |\Psi_2(t)\rangle\langle \Psi_2(t)| \end{pmatrix} \\ &= \begin{pmatrix} \widehat{\rho}_{11}(t) & \widehat{\rho}_{12}(t) \\ \widehat{\rho}_{21}(t) & \widehat{\rho}_{22}(t) \end{pmatrix}.\end{aligned} \qquad (3.538)$$

The two-component wave function is normalized as

$$\langle \underline{\Psi}(t)| \underline{\Psi}(t)\rangle = \sum_{i=1}^{2} \langle \Psi_i(t)| \Psi_i(t)\rangle = \sum_{i=1}^{2} \int d^3 r\, |\Psi_i(\vec{r},t)|^2 = 1. \qquad (3.539)$$

Thus, the density operator is normalized as

$$\text{Tr}(\widehat{\rho}(t)) = \sum_{i=1}^{2} \int d^3 r\, \langle \vec{r}|\widehat{\rho}_{ii}(t)|\vec{r}\rangle = 1. \qquad (3.540)$$

For a mixed state, the density operator in the case of a system with two-component wave function will become

A Review of Quantum Mechanics

$$\widehat{\rho}(t) = \sum_{j=1}^{N} p_j |\Psi^j(t)\rangle\langle\Psi^j(t)|$$

$$= \sum_{j=1}^{N} p_j \begin{pmatrix} |\Psi_1^j(t)\rangle\langle\Psi_1^j(t)| & |\Psi_1^j(t)\rangle\langle\Psi_2^j(t)| \\ |\Psi_2^j(t)\rangle\langle\Psi_1^j(t)| & |\Psi_2^j(t)\rangle\langle\Psi_2^j(t)| \end{pmatrix}$$

$$= \sum_{j=1}^{N} p_j \begin{pmatrix} \widehat{\rho}_{11}^j(t) & \widehat{\rho}_{12}^j(t) \\ \widehat{\rho}_{21}^j(t) & \widehat{\rho}_{22}^j(t) \end{pmatrix}$$

$$= \sum_{j=1}^{N} p_j \widehat{\rho}^j(t) = \begin{pmatrix} \widehat{\rho}_{11}(t) & \widehat{\rho}_{12}(t) \\ \widehat{\rho}_{21}(t) & \widehat{\rho}_{22}(t) \end{pmatrix},$$

$$\text{with } \sum_{j=1}^{N} p_j = 1. \tag{3.541}$$

For the mixed state, the normalization condition is

$$\sum_{j=1}^{N} p_j \left(\sum_{i=1}^{2} \int d^3 r \left|\Psi_i^j(\vec{r},t)\right|^2 \right) = 1. \tag{3.542}$$

Then, the normalization for the mixed state density operator becomes

$$\sum_{j=1}^{N} p_j \left(\sum_{i=1}^{2} \int d^3 r \left|\Psi_i^j(\vec{r},t)\right|^2 \right) = \sum_{j=1}^{N} p_j \text{Tr}\left(\widehat{\rho}^j(t)\right)$$

$$= \text{Tr}\left(\widehat{\rho}(t)\right) = 1. \tag{3.543}$$

It is straightforward to extend the equations of motion for the density operator and the average of an observable to the case of systems with multicomponent wave functions. The results are the same:

$$i\hbar \frac{\partial \widehat{\rho}(t)}{\partial t} = \left[\widehat{H}(t), \widehat{\rho}(t)\right],$$

$$\frac{d\langle \widehat{O}\rangle(t)}{dt} = \frac{i}{\hbar} \left\langle \left[\widehat{H}, \widehat{O}\right] \right\rangle (t) + \left\langle \frac{\partial \widehat{O}}{\partial t} \right\rangle (t). \tag{3.544}$$

Let $\widehat{O}$ be a Hermitian 2×2 matrix operator associated with some observable of a system with a two-component wave function. For example, $\vec{\sigma} \cdot \vec{p}$ is such an operator associated with a spin-$\frac{1}{2}$ particle. An observable like $\widehat{p}^2/2m$ can also be considered as a 2×2 matrix operator $\left(\widehat{p}^2/2m\right) I$. For such an observable, the average value in a pure state is

$$\langle \widehat{O} \rangle (t) = \left\langle \underline{\Psi}(t) \Big| \widehat{O} \Big| \underline{\Psi}(t) \right\rangle = \sum_{i,j=1}^{2} \left\langle \Psi_i(t) \Big| \widehat{O}_{ij} \Big| \Psi_j(t) \right\rangle$$

$$= \sum_{i,j=1}^{2} \int\int d^3r d^3r' \, \langle \Psi_i(t)|\vec{r}\rangle \left\langle \vec{r} \Big| \widehat{O}_{ij} \Big| \vec{r}' \right\rangle \langle \vec{r}'|\Psi_j(t)\rangle$$

$$= \sum_{i,j=1}^{2} \int\int d^3r d^3r' \, \langle \vec{r}'|\Psi_j(t)\rangle \langle \Psi_i(t)|\vec{r}\rangle \left\langle \vec{r} \Big| \widehat{O}_{ij} \Big| \vec{r}' \right\rangle$$

$$= \sum_{i,j=1}^{2} \int\int d^3r d^3r' \, \langle \vec{r}'|\widehat{\rho}_{ji}|\vec{r}\rangle \left\langle \vec{r} \Big| \widehat{O}_{ij} \Big| \vec{r}' \right\rangle$$

$$= \mathrm{Tr}\left(\widehat{\rho}(t)\widehat{O}\right). \tag{3.545}$$

In the case of the mixed state, we have

$$\langle \widehat{O} \rangle (t) = \sum_{j=1}^{N} p_j \left\langle \underline{\Psi}^j(t) \Big| \widehat{O} \Big| \underline{\Psi}^j(t) \right\rangle = \sum_{j=1}^{N} p_j \mathrm{Tr}\left(\widehat{\rho}^j(t)\widehat{O}\right)$$

$$= \mathrm{Tr}\left[\left(\sum_{j=1}^{N} p_j \widehat{\rho}^j(t)\right)\widehat{O}\right] = \mathrm{Tr}\left(\widehat{\rho}(t)\widehat{O}\right). \tag{3.546}$$

For a system of N identical particles, we can define the density matrix as

$$\widehat{\rho}_N(t) = \sum_j p_j |\Psi_j(1,2,\ldots,N,t)\rangle \langle \Psi_j(1,2,\ldots,N,t)|, \tag{3.547}$$

by generalizing (3.527), where the classical probability weight factors $\{p_j\}$ satisfy the condition $\sum_j p_j = 1$, and the N-particle state vectors $\{|\Psi_j(1,2,\ldots,N,t)\rangle\}$ are properly symmetrized (totally symmetric for bosons and totally antisymmetric for fermions under the exchange of any pair of particles). In the application of the density matrix formalism for many-particle systems, one further introduces reduced density matrices of different orders. We shall not pursue this topic further.

3.4 RELATIVISTIC QUANTUM MECHANICS

3.4.1 KLEIN–GORDON EQUATION

3.4.1.1 Free-Particle Equation and Difficulties in Interpretation

To generalize the nonrelativistic Schrödinger equation for a free particle of mass m, the starting point is, naturally, to convert the classical relativistic Hamiltonian

$$H = \sqrt{m^2 c^4 + c^2 p^2}, \tag{3.548}$$

to a quantum Hamiltonian $\widehat{H}$ by replacing $\vec{p}$ by $\widehat{\vec{p}}$, and write down

$$i\hbar \frac{\partial |\Psi(t)\rangle}{\partial t} = \widehat{H}|\Psi(t)\rangle. \tag{3.549}$$

A Review of Quantum Mechanics

The resulting equation,

$$i\hbar \frac{\partial |\Psi(t)\rangle}{\partial t} = \sqrt{m^2 c^4 + c^2 \widehat{p}^2}\, |\Psi(t)\rangle, \qquad (3.550)$$

involves the obvious difficulty of taking the square root of the differential operator. A Taylor series expansion of the square root would lead to an infinite series of spatial derivatives, and time and space would not occur symmetrically contrary to what is expected in a basic relativistic equation. In (3.550) if we expand the square root and retain only the first two terms, we get

$$i\hbar \frac{\partial |\Psi(t)\rangle}{\partial t} \approx \left[mc^2 \left(1 + \frac{\widehat{p}^2}{2m^2 c^2} \right) \right] |\Psi(t)\rangle$$

$$= \left(mc^2 + \frac{\widehat{p}^2}{2m} \right) |\Psi(t)\rangle, \qquad (3.551)$$

which becomes the nonrelativistic free-particle Schrödinger equation when the constant rest energy term, mc^2, is dropped inconsequentially.

Instead of this approach in which the square root creates a problem, let us note that for a time-independent $\widehat{H}$, we get, from (3.549),

$$\left(i\hbar \frac{\partial}{\partial t} \right)^2 |\Psi(t)\rangle = \widehat{H}^2 |\Psi(t)\rangle. \qquad (3.552)$$

Now, substituting $m^2 c^4 + c^2 \widehat{p}^2$ for $\widehat{H}^2$ leads to the equation

$$\left(-\hbar^2 \frac{\partial^2}{\partial t^2} - c^2 \widehat{p}^2 \right) |\Psi(t)\rangle = m^2 c^4 |\Psi(t)\rangle, \qquad (3.553)$$

the Klein–Gordon equation for the free particle. It is also sometimes called the Klein–Gordon–Fock equation. In position representation, we can write this equation as

$$\left(\nabla^2 - \frac{1}{c^2} \frac{\partial^2}{\partial t^2} \right) \Psi(\vec{r},t) = \left(\frac{mc}{\hbar} \right)^2 \Psi(\vec{r},t). \qquad (3.554)$$

Schrödinger also had considered this equation and rejected it as unsatisfactory before proposing his nonrelativistic equation. A plane wave,

$$\Psi_{\vec{p}}(\vec{r},t) \sim e^{\frac{i}{\hbar}(\vec{p}\cdot\vec{r} - Et)}, \qquad (3.555)$$

apart from a normalization constant, is seen to be a solution of the free-particle Klein–Gordon equation (3.554) when the momentum $\vec{p}$ (note $\vec{\widehat{p}}\Psi_{\vec{p}}(\vec{r},t) = \vec{p}\Psi_{\vec{p}}(\vec{r},t)$) is related to E by the relation $E^2 = (m^2 c^4 + c^2 p^2)$. It should be noted that the particle associated with the plane wave $\Psi_{\vec{p}}(\vec{r},t)$ can have its energy as $\pm (m^2 c^4 + c^2 p^2)$. Thus, the free-particle energy spectrum has two branches, a positive-energy branch ($E \geq mc^2$) and a negative-energy branch ($E \leq -mc^2$) with a gap of energy $2mc^2$. Unlike in the classical case, one cannot discard the negative-energy solutions as unphysical since all the solutions form a complete set and are needed for representing a general state of the particle.

Let us recall the nonrelativistic free-particle Schrödinger equation:

$$i\hbar \frac{\partial \Psi_{NR}}{\partial t} = -\frac{\hbar^2}{2m}\nabla^2 \Psi_{NR}. \qquad (3.556)$$

The complex conjugate of this equation is

$$-i\hbar \frac{\partial \Psi^*_{NR}}{\partial t} = -\frac{\hbar^2}{2m}\nabla^2 \Psi^*_{NR}. \qquad (3.557)$$

Multiplying both sides of the first equation (3.556) on the left by Ψ^*_{NR}, the second equation (3.557) by Ψ_{NR}, and subtracting the second from the first, we get

$$\begin{aligned} i\hbar \left(\Psi^*_{NR} \frac{\partial \Psi_{NR}}{\partial t} + \Psi_{NR} \frac{\partial \Psi^*_{NR}}{\partial t} \right) &= -\frac{\hbar^2}{2m} \left[\Psi^*_{NR} \nabla^2 \Psi_{NR} - \Psi_{NR} \nabla^2 \Psi^*_{NR} \right] \\ &= -\frac{\hbar^2}{2m} \left[\vec{\nabla} \cdot \left(\Psi^*_{NR} \vec{\nabla} \Psi_{NR} - \Psi_{NR} \vec{\nabla} \Psi^*_{NR} \right) \right]. \end{aligned} \qquad (3.558)$$

We can rewrite this equation as

$$\frac{\partial \rho_{NR}}{\partial t} + \vec{\nabla} \cdot \vec{j}_{NR} = 0, \qquad (3.559)$$

where

$$\rho_{NR}(\vec{r},t) = \Psi^*_{NR}(\vec{r},t) \Psi_{NR}(\vec{r},t) = |\Psi_{NR}(\vec{r},t)|^2 \qquad (3.560)$$

can be identified with the nonrelativistic probability density and

$$\begin{aligned} \vec{j}_{NR}(\vec{r},t) &= \frac{\hbar}{2mi} \left(\Psi^*_{NR}(\vec{r},t) \vec{\nabla} \Psi_{NR}(\vec{r},t) - \Psi_{NR}(\vec{r},t) \vec{\nabla} \Psi^*_{NR}(\vec{r},t) \right) \\ &= \frac{1}{2m} \left(\Psi^*_{NR}(\vec{r},t) \vec{p} \Psi_{NR}(\vec{r},t) - \Psi_{NR}(\vec{r},t) \vec{p} \Psi^*_{NR}(\vec{r},t) \right), \end{aligned} \qquad (3.561)$$

defines the nonrelativistic probability current density. Equation (3.559) is the continuity equation for the probability density. As we know, $\int_V d^3r\, \rho_{NR}(\vec{r},t)$ gives the probability of finding the particle within a specified volume V. The rate of change of this probability is given by

$$\frac{\partial}{\partial t} \int_V d^3r\, \rho_{NR} = \int_V d^3r\, \frac{\partial \rho_{NR}}{\partial t} = -\int_V d^3r\, \vec{\nabla} \cdot \vec{j}_{NR} = -\int_S \vec{j}_{NR} \cdot d\vec{S}, \qquad (3.562)$$

where S denotes the closed surface boundary of the volume V, and we have used the continuity equation (3.559) and the Gauss theorem of vector calculus. This result shows that the rate at which the probability of finding the particle within a specified volume changes is equal to the flux of the probability current through the surface boundary of the volume. In other words, the continuity equation (3.559) represents the conservation of probability. When the volume considered is the entire space, the probability current density and the surface integral vanish since $\Psi_{NR}(\vec{r},t) \longrightarrow 0$ as $|\vec{r}| \longrightarrow \infty$. This leads to the result that the normalization of $\Psi_{NR}(\vec{r},t)$, i.e., $\int d^3r\, |\Psi_{NR}(\vec{r},t)|^2 = 1$, with integration over the whole space, is preserved in time. We know already that this is implied by the unitary time evolution of $|\Psi(t)\rangle$. Let us note that for a free particle of mass m with momentum $\vec{p}$ associated with the plane wave

$$\Psi(\vec{r},t) = \frac{1}{(2\pi\hbar)^{3/2}} e^{\frac{i}{\hbar}\left(\vec{p}\cdot\vec{r} - \frac{p^2}{2m}t\right)}, \qquad (3.563)$$

the nonrelativistic probability density and the probability current density satisfy the relation

$$\begin{aligned}\vec{j}_{NR} &= \frac{1}{2m}\left(\Psi^*\vec{p}\Psi - \Psi\vec{p}\Psi^*\right) \\ &= \frac{\vec{p}}{(2\pi\hbar)^3 m} = \Psi^*\Psi\vec{v} = \rho\vec{v},\end{aligned} \qquad (3.564)$$

as expected.

Let us now return to the free-particle Klein–Gordon equation (3.554). Its complex conjugate equation is

$$\left(\nabla^2 - \frac{1}{c^2}\frac{\partial^2}{\partial t^2}\right)\Psi^*(\vec{r},t) = \left(\frac{mc}{\hbar}\right)^2 \Psi^*(\vec{r},t). \qquad (3.565)$$

Multiplying both sides of (3.554) on the left by Ψ^*, both sides of (3.565) on the left by Ψ, and subtracting we get

$$-\frac{1}{c^2}\left(\Psi^*\frac{\partial^2\Psi}{\partial t^2} - \Psi\frac{\partial^2\Psi^*}{\partial t^2}\right) + \left(\Psi^*\nabla^2\Psi - \Psi\nabla^2\Psi^*\right) = 0. \qquad (3.566)$$

Multiplying throughout by $\hbar/2mi$, we can rewrite this equation as a continuity equation

$$\frac{\partial \rho_{KG}}{\partial t} + \vec{\nabla}\cdot\vec{j}_{KG} = 0, \qquad (3.567)$$

in which we can take

$$\rho_{KG}(\vec{r},t) = \frac{i\hbar}{2mc^2}\left(\Psi^*(\vec{r},t)\frac{\partial\Psi(\vec{r},t)}{\partial t} - \Psi(\vec{r},t)\frac{\partial\Psi^*(\vec{r},t)}{\partial t}\right) \qquad (3.568)$$

as the Klein–Gordon probability density and

$$\vec{j}_{KG}(\vec{r},t) = \frac{\hbar}{2mi}\left(\Psi^*(\vec{r},t)\vec{\nabla}\Psi(\vec{r},t) - \Psi(\vec{r},t)\vec{\nabla}\Psi^*(\vec{r},t)\right) \qquad (3.569)$$

as the Klein–Gordon probability current density. It is clear from (3.567) that $\int d^3r\, \rho_{KG}(\vec{r},t)$ is preserved in time.

Let us look at the nonrelativistic limit of the free-particle Klein–Gordon equation. In the nonrelativistic limit, we can consider to take

$$\Psi(\vec{r},t) = \Psi_{NR}(\vec{r},t)e^{-\frac{i}{\hbar}mc^2 t}, \qquad (3.570)$$

with

$$\frac{\partial \Psi_{NR}(\vec{r},t)}{\partial t} \approx \left(\frac{\varepsilon}{i\hbar}\right)\Psi_{NR}(\vec{r},t), \qquad 0 < \varepsilon \ll mc^2, \qquad (3.571)$$

where $(mc^2 + \varepsilon)$ is the total energy of the nonrelativistic particle including its rest energy. Substituting this ansatz in (3.554), and considering $\varepsilon/\hbar c \ll mc/\hbar$, we arrive at

$$\left(\frac{mc}{\hbar}\right)^2 \Psi_{NR} e^{-\frac{i}{\hbar}mc^2 t}$$
$$= \left(\nabla^2 \Psi_{NR}\right) e^{-\frac{i}{\hbar}mc^2 t} - \frac{1}{c^2}\frac{\partial^2}{\partial t^2}\left(\Psi_{NR} e^{-\frac{i}{\hbar}mc^2 t}\right)$$
$$= \left[\nabla^2 \Psi_{NR} - \frac{1}{c^2}\frac{\partial^2 \Psi_{NR}}{\partial t^2} + \left(\frac{2im}{\hbar}\right)\frac{\partial \Psi_{NR}}{\partial t} + \left(\frac{mc}{\hbar}\right)^2 \Psi_{NR}\right] e^{-\frac{i}{\hbar}mc^2 t}$$
$$\approx \left[\nabla^2 \Psi_{NR} + \left(\frac{\varepsilon}{\hbar c}\right)^2 \Psi_{NR} + \left(\frac{2im}{\hbar}\right)\frac{\partial \Psi_{NR}}{\partial t} + \left(\frac{mc}{\hbar}\right)^2 \Psi_{NR}\right] e^{-\frac{i}{\hbar}mc^2 t}$$
$$\approx \left[\nabla^2 \Psi_{NR} + \left(\frac{2im}{\hbar}\right)\frac{\partial \Psi_{NR}}{\partial t} + \left(\frac{mc}{\hbar}\right)^2 \Psi_{NR}\right] e^{-\frac{i}{\hbar}mc^2 t}. \qquad (3.572)$$

From this, we have

$$\nabla^2 \Psi_{NR} + \left(\frac{2im}{\hbar}\right)\frac{\partial \Psi_{NR}}{\partial t} \approx 0. \qquad (3.573)$$

Multiplying by $-\hbar^2/2m$, we can rewrite this equation as

$$i\hbar \frac{\partial \Psi_{NR}}{\partial t} \approx -\frac{\hbar^2}{2m}\nabla^2 \Psi_{NR}. \qquad (3.574)$$

Thus, we see that, in the nonrelativistic limit, the free-particle nonrelativistic Schrödinger equation is the approximation of the free-particle Klein–Gordon equation.

In the nonrelativistic limit, taking $\Psi(\vec{r},t) = \Psi_{NR}(\vec{r},t)e^{-imc^2 t/\hbar}$,

$$\rho_{KG} = \frac{i\hbar}{2mc^2}\left(\Psi^* \frac{\partial \Psi}{\partial t} - \Psi \frac{\partial \Psi^*}{\partial t}\right)$$
$$= \frac{i\hbar}{2mc^2}\left(\Psi_{NR}^* \frac{\partial \Psi_{NR}}{\partial t} - \Psi_{NR}\frac{\partial \Psi_{NR}^*}{\partial t}\right) + \Psi_{NR}^* \Psi_{NR}$$
$$\approx \left(1 + \frac{\varepsilon}{mc^2}\right)\Psi_{NR}^* \Psi_{NR} \approx |\Psi_{NR}|^2 = \rho_{NR}, \qquad (3.575)$$

where we have used (3.571) to make the approximations. For the nonrelativistic limit of the Klein–Gordon probability current density, we have

$$\vec{j}_{KG} = \frac{\hbar}{2mi}\left(\Psi^*\vec{\nabla}\Psi - \Psi\vec{\nabla}\Psi^*\right)$$
$$= \frac{\hbar}{2mi}\left(\Psi_{NR}^*\vec{\nabla}\Psi_{NR} - \Psi_{NR}\vec{\nabla}\Psi_{NR}^*\right) = \vec{j}_{NR}. \quad (3.576)$$

Thus, in the nonrelativistic limit, ρ_{KG} and $\vec{j}_{KG}$ become the corresponding nonrelativistic probability density and probability current density. However, ρ_{KG} is not acceptable as a probability density. A probability density $\rho(\vec{r},t)$ has to be nonnegative for all $\vec{r}$ and t. When $\Psi(\vec{r},t)$ is real, $\rho_{KG}(\vec{r},t)$ vanishes identically at any position and time, meaning that the particle exists nowhere. When $\Psi(\vec{r},t)$ is complex $\rho_{KG}(\vec{r},t)$ can become negative for some choice of $\partial\Psi/\partial t$. Note that since the Klein–Gordon equation (3.554) is of second order in time derivative, both Ψ and $\partial\Psi/\partial t$ have to be specified independently at the initial time for studying the time evolution.

It is only to be expected that in relativistic quantum mechanics there should be no one-particle time-evolution equation, or wave equation, since in relativistic mechanics, particle number is not necessarily conserved. Relativistic quantum theory should be able to describe processes in which particles are created and annihilated. This is achieved in the relativistic quantum field theory where $\Psi(\vec{r},t)$ becomes a quantum field operator $\widehat{\Psi}(\vec{r},t)$ in which both the space coordinates $\vec{r}$ and time t are parameters with equal status, unlike in nonrelativistic quantum mechanics where the space coordinates are dynamical variables and time is a parameter. Then, $\widehat{\Psi}(\vec{r},t)$ satisfying the Klein–Gordon equation is associated with spin-0 particles. The negative-energy states of a Klein–Gordon particle are the positive-energy states of the corresponding antiparticle.

3.4.1.2 Feshbach–Villars Representation

Now, let us look at a two-component representation of the Klein–Gordon equation introduced by Feshbach and Villars [50]. To this end, let us write

$$i\hbar\frac{\partial\Psi}{\partial t} = mc^2\left(\frac{i\hbar}{mc^2}\frac{\partial\Psi}{\partial t}\right),$$

$$i\hbar\frac{\partial}{\partial t}\left(\frac{i\hbar}{mc^2}\frac{\partial\Psi}{\partial t}\right) = -\frac{\hbar^2}{m}\left(\frac{1}{c^2}\frac{\partial^2\Psi}{\partial t^2}\right)$$

$$= -\frac{\hbar^2}{m}\left[\nabla^2 - \left(\frac{mc}{\hbar}\right)^2\right]\Psi$$

$$= \left(-\frac{\hbar^2}{m}\nabla^2 + mc^2\right)\Psi, \quad (3.577)$$

and define

$$\Psi_+ = \frac{1}{2}\left[\Psi + \frac{1}{mc^2}\left(i\hbar\frac{\partial\Psi}{\partial t}\right)\right],$$

$$\Psi_- = \frac{1}{2}\left[\Psi - \frac{1}{mc^2}\left(i\hbar\frac{\partial\Psi}{\partial t}\right)\right]. \quad (3.578)$$

Adding and subtracting both sides of the two equations of (3.577), and multiplying the resulting equations by $1/2$ on both sides, we get

$$i\hbar \frac{\partial \Psi_+}{\partial t} = -\frac{\hbar^2}{2m} \nabla^2 (\Psi_+ + \Psi_-) + mc^2 \Psi_+,$$

$$i\hbar \frac{\partial \Psi_-}{\partial t} = \frac{\hbar^2}{2m} \nabla^2 (\Psi_+ + \Psi_-) - mc^2 \Psi_-. \quad (3.579)$$

We shall rewrite this set of coupled equations for Ψ_+ and Ψ_- as

$$i\hbar \frac{\partial \underline{\Psi}}{\partial t} = \widehat{H}_{\mathrm{FV}} \underline{\Psi}, \quad (3.580)$$

where

$$\underline{\Psi} = \begin{pmatrix} \Psi_+ \\ \Psi_- \end{pmatrix}, \quad (3.581)$$

and

$$\widehat{H}_{\mathrm{FV}} = \begin{pmatrix} mc^2 - \frac{\hbar^2 \nabla^2}{2m} & -\frac{\hbar^2 \nabla^2}{2m} \\ \frac{\hbar^2 \nabla^2}{2m} & -mc^2 + \frac{\hbar^2 \nabla^2}{2m} \end{pmatrix}$$

$$= \left(mc^2 + \frac{\widehat{p}^2}{2m} \right) \sigma_z + \frac{\widehat{p}^2}{2m} (i\sigma_y)$$

$$= mc^2 \sigma_z + \frac{\widehat{p}^2}{2m} (\sigma_z + i\sigma_y) \quad (3.582)$$

is the Feshbach–Villars Hamiltonian. Since $\sigma_z^2 = I$, $(i\sigma_y)^2 = -I$, $\sigma_z \sigma_y + \sigma_y \sigma_z = 0$, and the constant mc^2 commutes with $\widehat{p}^2/2m$, we get

$$\widehat{H}_{\mathrm{FV}}^2 = \left[\left(mc^2 + \frac{\widehat{p}^2}{2m} \right) \sigma_z + \frac{\widehat{p}^2}{2m} (i\sigma_y) \right]^2$$

$$= \left(mc^2 + \frac{\widehat{p}^2}{2m} \right)^2 - \left(\frac{\widehat{p}^2}{2m} \right)^2 = m^2 c^4 + c^2 \widehat{p}^2, \quad (3.583)$$

as if $\widehat{H}_{\mathrm{FV}}$ is the desired square root of $m^2 c^4 + c^2 \widehat{p}^2$. However, the time and space derivatives occur differently in the Feshbach–Villars representation of the Klein–Gordon equation, which is not acceptable in a relativistic theory. Further, $\widehat{H}_{\mathrm{FV}}$ is seen to be non-Hermitian. The reason for this is as follows.

In the Feshbach–Villars representation, we find

$$\rho_{\mathrm{KG}} = \frac{i\hbar}{2mc^2} \left(\Psi^* \frac{\partial \Psi}{\partial t} - \Psi \frac{\partial \Psi^*}{\partial t} \right)$$

$$= \frac{i\hbar}{2mc^2} \left\{ (\Psi_+^* + \Psi_-^*) \left[\frac{mc^2}{i\hbar} (\Psi_+ - \Psi_-) \right] \right.$$

$$-(\Psi_+ + \Psi_-)\left[-\frac{mc^2}{i\hbar}\left(\Psi_+^* - \Psi_-^*\right)\right]\Big\}$$
$$= |\Psi_+|^2 - |\Psi_-|^2. \tag{3.584}$$

We saw earlier that $\int d^3 r\, \rho_{KG}$ is conserved in time, i.e., $\int d^3 r \left(|\Psi_+|^2 - |\Psi_-|^2\right)$ is conserved in time. This means that the Feshbach–Villars state $\underline{\Psi}$ does not have a unitary evolution with a Hermitian Hamiltonian. For unitary evolution of $\underline{\Psi}$, we should have $\int d^3 r\, \underline{\Psi}(\vec{r},t)^\dagger \underline{\Psi}(\vec{r},t) = \int d^3 r \left(|\Psi_+(\vec{r},t)|^2 + |\Psi_-(\vec{r},t)|^2\right)$ conserved in time. Anyhow, one can use the Feshbach–Villars equation (3.580) to study the time evolution of $\underline{\Psi}(\vec{r},t)$.

3.4.1.3 Charged Klein–Gordon Particle in a Constant Magnetic Field

For a particle of mass m and charge q moving in an electromagnetic field, the relativistic classical Hamiltonian is, as we know,

$$H(\vec{r},\vec{p},t) = \sqrt{m^2 c^4 + c^2 \vec{\pi}^2} + q\phi, \tag{3.585}$$

where ϕ and $\vec{A}$ are, respectively, the scalar and vector potentials of the field. Now, the Klein–Gordon equation for the particle becomes

$$\left(i\hbar\frac{\partial}{\partial t} - q\phi\right)^2 \Psi = \left(m^2 c^4 + c^2 \vec{\pi}^2\right)\Psi, \tag{3.586}$$

or

$$\left[\left(i\hbar\frac{\partial}{\partial t} - q\phi\right)^2 - c^2(-i\hbar\vec{\nabla} - q\vec{A})^2\right]\Psi = m^2 c^4 \Psi. \tag{3.587}$$

We can rewrite this equation as

$$\left[\left(\vec{\nabla} - \frac{i}{\hbar}q\vec{A}\right)^2 - \frac{1}{c^2}\left(\frac{\partial}{\partial t} + \frac{i}{\hbar}q\phi\right)^2\right]\Psi = \left(\frac{mc}{\hbar}\right)^2 \Psi, \tag{3.588}$$

which becomes the free-particle Klein–Gordon equation (3.554) in the absence of electromagnetic field. Note that equation (3.586) is related to the free-particle equation through minimal electromagnetic coupling.

The Feshbach–Villars equation corresponding to (3.587) is given by

$$i\hbar\frac{\partial \underline{\Psi}}{\partial t} = \left[mc^2 \sigma_z + \frac{\vec{\pi}^2}{2m}(\sigma_z + i\sigma_y) + q\phi\right]\underline{\Psi}, \tag{3.589}$$

with

$$\underline{\Psi} = \begin{pmatrix} \frac{1}{2}\left[\Psi + \frac{1}{mc^2}\left(i\hbar\frac{\partial}{\partial t} - q\phi\right)\Psi\right] \\ \frac{1}{2}\left[\Psi - \frac{1}{mc^2}\left(i\hbar\frac{\partial}{\partial t} - q\phi\right)\Psi\right] \end{pmatrix}. \tag{3.590}$$

To see this, let us write

$$\left(i\hbar\frac{\partial}{\partial t} - q\phi\right)\Psi = mc^2\left[\frac{1}{mc^2}\left(i\hbar\frac{\partial}{\partial t} - q\phi\right)\Psi\right], \quad (3.591)$$

and

$$\left(i\hbar\frac{\partial}{\partial t} - q\phi\right)\left[\frac{1}{mc^2}\left(i\hbar\frac{\partial}{\partial t} - q\phi\right)\Psi\right] = \frac{1}{mc^2}\left(i\hbar\frac{\partial}{\partial t} - q\phi\right)^2\Psi$$

$$= \frac{1}{mc^2}\left(m^2c^4 + c^2\vec{\pi}^2\right)\Psi = \left(mc^2 + \frac{\vec{\pi}^2}{m}\right)\Psi. \quad (3.592)$$

Adding and subtracting both sides of these two equations, and multiplying the resulting equations by $1/2$ on both sides, we get

$$\left(i\hbar\frac{\partial}{\partial t} - q\phi\right)\underline{\Psi} = \left[mc^2\sigma_z + \frac{\vec{\pi}^2}{2m}(\sigma_z + i\sigma_y)\right]\underline{\Psi}, \quad (3.593)$$

which can be rewritten as (3.589).

When we are dealing with spin-0 particles with positive energy and the processes studied do not involve any particle creation and annihilation, it should be appropriate to use the Klein–Gordon equation as if it is a one-particle relativistic equation. It can also be used in this way for treating the relativistic quantum mechanics of spinless particles, *i.e.*, when the spin is ignored, or treated as a spectator variable. As an example, let us consider a spinless, or spin-0, Klein–Gordon particle of mass m and charge q moving in a constant magnetic field in the z-direction. It will obey the equation

$$\left(i\hbar\frac{\partial}{\partial t}\right)^2\Psi = \left\{m^2c^4 + c^2\left[\left(\widehat{p}_x + \frac{1}{2}qBy\right)^2 + \left(\widehat{p}_y - \frac{1}{2}qBx\right)^2 + \widehat{p}_z^2\right]\right\}\Psi, \quad (3.594)$$

obtained by taking in (3.586) $\phi = 0$ and $\vec{A} = (-By/2, Bx/2, 0)$ corresponding to the magnetic field $\vec{B} = B\vec{k}$. Let us now write this equation as

$$\left(i\hbar\frac{\partial}{\partial t}\right)^2\Psi = \left(m^2c^4 + 2mc^2\widehat{H}_{\text{NRL}}\right)\Psi, \quad (3.595)$$

where

$$\widehat{H}_{\text{NRL}} = \frac{1}{2m}\left[\left(\widehat{p}_x + \frac{1}{2}qBy\right)^2 + \left(\widehat{p}_y - \frac{1}{2}qBx\right)^2 + \widehat{p}_z^2\right] \quad (3.596)$$

is the Hamiltonian of a nonrelativistic particle of mass m and charge q moving in the field $\vec{B} = B\vec{k}$ which has the Landau levels (3.419) as its spectrum. Since the system is time-independent, there will be stationary eigenstates for which we can

A Review of Quantum Mechanics

take $\Psi(\vec{r},t) = \psi(\vec{r})e^{-iEt/\hbar}$, where E is the energy of the eigenstate. Substituting this Ψ in (3.594), we get

$$E^2 \psi e^{-iEt/\hbar} = \left(m^2c^4 + 2mc^2\widehat{H}_{\mathrm{NRL}}\right)\psi e^{-iEt/\hbar}, \qquad (3.597)$$

leading to the eigenvalue equation

$$\left(m^2c^4 + 2mc^2\widehat{H}_{\mathrm{NRL}}\right)\psi = E^2\psi. \qquad (3.598)$$

Let us take $\psi(\vec{r})$ as $\psi_{n,m_\phi,p_z}(\vec{r})$, defined in (3.418), which is an eigenfunction of $\widehat{H}_{\mathrm{NRL}}$ corresponding to the eigenvalue $E_{n,m_\phi,p_z} = (2n+1)\hbar qB/2m + p_z^2/2m$. Then, we find that

$$\left(m^2c^4 + 2mc^2\widehat{H}_{\mathrm{NRL}}\right)\psi_{n,m_\phi,p_z}(\vec{r}) = \left[m^2c^4 + (2n+1)c^2\hbar qB + c^2 p_z^2\right]\psi_{n,m_\phi,p_z}(\vec{r})$$
$$= E^2 \psi_{n,m_\phi,p_z}(\vec{r}), \qquad (3.599)$$

showing that the energy eigenvalues of the system are given by

$$E_{n,m_\phi,p_z} = \sqrt{m^2c^4 + (2n+1)c^2\hbar qB + c^2 p_z^2}$$
$$= mc^2\sqrt{1 + \left(\frac{p_z}{mc}\right)^2 + (2n+1)\frac{\hbar qB}{m^2c^2}}, \quad n = 0,1,2,\ldots, \qquad (3.600)$$

where we are considering only the positive square root. As in the nonrelativistic case, each energy eigenvalue has infinite degeneracy characterized by the angular momentum quantum number $m_\phi = 0, 1, 2, \ldots$. In the nonrelativistic limit, when the rest energy of the particle is larger than other energies of the particle, we have

$$E_{n,m_\phi,p_z} \approx mc^2\left[1 + \frac{1}{2}\left(\frac{p_z}{mc}\right)^2 + \left(n + \frac{1}{2}\right)\frac{\hbar qB}{m^2c^2}\right]$$
$$= mc^2 + \frac{p_z^2}{2m} + \left(n + \frac{1}{2}\right)\frac{\hbar qB}{m}, \quad n = 0,1,2,\ldots, \qquad (3.601)$$

which gives the nonrelativistic Landau energy spectrum of spinless particle added to the rest energy.

3.4.2 DIRAC EQUATION

3.4.2.1 Free-Particle Equation

The correct relativistic equation with all the desired properties (linear in time, nonnegative probability density, etc.) was discovered by Dirac. It describes the electron, or any spin-$\frac{1}{2}$ particle, in a natural way. He wrote down, for a free particle of mass m,

$$\widehat{H} = \sqrt{m^2c^4 + c^2\widehat{\vec{p}}^{\,2}} = \beta mc^2 + c\left(\alpha_x\widehat{p}_x + \alpha_y\widehat{p}_y + \alpha_z\widehat{p}_z\right), \qquad (3.602)$$

and demanded that

$$\widehat{H}^2 = \left[\beta mc^2 + c\left(\alpha_x \widehat{p}_x + \alpha_y \widehat{p}_y + \alpha_z \widehat{p}_z\right)\right]^2 = m^2c^4 + c^2\widehat{p}^2. \tag{3.603}$$

This implies, with mc^2, $\widehat{p}_x$, $\widehat{p}_y$, and $\widehat{p}_z$ commuting with each other,

$$\begin{aligned}
\beta^2 &= \alpha_x^2 = \alpha_y^2 = \alpha_z^2 = 1, \\
\beta\alpha_x &= -\alpha_x\beta, \quad \beta\alpha_y = -\alpha_y\beta, \quad \beta\alpha_z = -\alpha_z\beta, \\
\alpha_x\alpha_y &= -\alpha_y\alpha_x, \quad \alpha_y\alpha_z = -\alpha_z\alpha_y, \quad \alpha_z\alpha_x = -\alpha_x\alpha_z.
\end{aligned} \tag{3.604}$$

It is obvious that β, α_x, α_y, and α_z cannot be ordinary, real or complex, numbers. The relations $\beta^2 = 1$, $\alpha_x^2 = 1$, etc., show that, if we represent them by square matrices, all their eigenvalues must be ± 1. The relations $\beta\alpha_x = -\alpha_x\beta$ and $\alpha_x^2 = 1$ imply that $\alpha_x\beta\alpha_x = -\beta$. Since $\text{Tr}(ABC) = \text{Tr}(BCA)$, we get $\text{Tr}(\alpha_x\beta\alpha_x) = \text{Tr}(\beta) = \text{Tr}(-\beta) = -\text{Tr}(\beta)$. Thus, $\text{Tr}(\beta) = 0$. By the same argument, we get $\text{Tr}(\alpha_x) = \text{Tr}(\alpha_y) = \text{Tr}(\alpha_z) = 0$. Recall that for any finite-dimensional matrix, the sum of its eigenvalues is equal to its trace. This means that for each of the matrices β, α_x, α_y, and α_z, the spectrum of eigenvalues should contain an equal number of $+1$s and -1s such that the trace is zero. This leads to the result that the dimension of these matrices β, α_x, α_y, and α_z must be even. We know already that the set of three two-dimensional Pauli matrices, $(\sigma_x, \sigma_y, \sigma_z)$, obey the same algebra as the αs and β obey (3.604), i.e.,

$$\begin{aligned}
\sigma_x^2 &= \sigma_y^2 = \sigma_z^2 = 1, \\
\sigma_x\sigma_y &= -\sigma_y\sigma_z, \quad \sigma_x\sigma_y = -\sigma_y\sigma_z, \quad \sigma_x\sigma_y = -\sigma_y\sigma_z.
\end{aligned} \tag{3.605}$$

Now, one may try to construct a fourth two-dimensional matrix, say M, which anticommutes with all the three Pauli matrices by taking it to be

$$M = \begin{pmatrix} M_{11} & M_{12} \\ M_{21} & M_{22} \end{pmatrix} \tag{3.606}$$

and demanding it to satisfy the relations

$$M\sigma_x = -\sigma_x M, \quad M\sigma_y = -\sigma_y M, \quad M\sigma_z = -\sigma_z M. \tag{3.607}$$

The result will be that M has to be O, the 2×2 null matrix. Thus, there does not exist a fourth two-dimensional matrix that anticommutes with all the three Pauli matrices. Hence, the matrices β, α_x, α_y, and α_z have to be of minimum dimension four. Further, using group theory it can be shown that the Dirac algebra (3.604) has only one inequivalent irreducible faithful representation of dimension four, i.e., all four-dimensional representations of the Dirac algebra would be equivalent to each other related by similarity transformations. In other words, if $(\beta, \alpha_x, \alpha_y, \alpha_z)$ and $(\beta', \alpha_x', \alpha_y', \alpha_z')$ are two sets of four-dimensional matrices obeying the Dirac algebra, then there will exist a nonsingular matrix $\mathscr{S}$ such that

$$\beta' = \mathscr{S}\beta\mathscr{S}^{-1}, \quad \alpha_x' = \mathscr{S}\alpha_x\mathscr{S}^{-1}, \quad \alpha_y' = \mathscr{S}\alpha_y\mathscr{S}^{-1}, \quad \alpha_z' = \mathscr{S}\alpha_z\mathscr{S}^{-1}. \tag{3.608}$$

A Review of Quantum Mechanics

The Pauli algebra and the Dirac algebra are the lowest order special cases of the Clifford algebra (for more details, see, *e.g.*, Ramakrishnan [155], Jagannathan [86] and references therein).

Dirac constructed the required four-dimensional matrices as

$$\beta = \sigma_z \otimes I = \begin{pmatrix} 1 & 0 \\ 0 & -1 \end{pmatrix} \otimes \begin{pmatrix} 1 & 0 \\ 0 & 1 \end{pmatrix} = \begin{pmatrix} I & 0 \\ 0 & -I \end{pmatrix}$$

$$= \begin{pmatrix} 1 & 0 & 0 & 0 \\ 0 & 1 & 0 & 0 \\ 0 & 0 & -1 & 0 \\ 0 & 0 & 0 & -1 \end{pmatrix},$$

$$\alpha_x = \sigma_x \otimes \sigma_x = \begin{pmatrix} 0 & 1 \\ 1 & 0 \end{pmatrix} \otimes \begin{pmatrix} 0 & 1 \\ 1 & 0 \end{pmatrix} = \begin{pmatrix} 0 & \sigma_x \\ \sigma_x & 0 \end{pmatrix}$$

$$= \begin{pmatrix} 0 & 0 & 0 & 1 \\ 0 & 0 & 1 & 0 \\ 0 & 1 & 0 & 0 \\ 1 & 0 & 0 & 0 \end{pmatrix},$$

$$\alpha_y = \sigma_x \otimes \sigma_y = \begin{pmatrix} 0 & 1 \\ 1 & 0 \end{pmatrix} \otimes \begin{pmatrix} 0 & -i \\ i & 0 \end{pmatrix} = \begin{pmatrix} 0 & \sigma_y \\ \sigma_y & 0 \end{pmatrix}$$

$$= \begin{pmatrix} 0 & 0 & 0 & -i \\ 0 & 0 & i & 0 \\ 0 & -i & 0 & 0 \\ i & 0 & 0 & 0 \end{pmatrix},$$

$$\alpha_z = \sigma_x \otimes \sigma_z = \begin{pmatrix} 0 & 1 \\ 1 & 0 \end{pmatrix} \otimes \begin{pmatrix} 1 & 0 \\ 0 & -1 \end{pmatrix} = \begin{pmatrix} 0 & \sigma_z \\ \sigma_z & 0 \end{pmatrix}$$

$$= \begin{pmatrix} 0 & 0 & 1 & 0 \\ 0 & 0 & 0 & -1 \\ 1 & 0 & 0 & 0 \\ 0 & -1 & 0 & 0 \end{pmatrix}. \tag{3.609}$$

As mentioned earlier, any other four-dimensional representation of the Dirac algebra (3.604) would be equivalent to the standard Dirac representation (3.609). It is straightforward to verify, using (3.492) or directly, that these matrices satisfy the relations (3.604) as required. They are all Hermitian. Let us call the Hamiltonian $\widehat{H}$ defined by (3.602) and (3.609) as the free-particle Dirac Hamiltonian $\widehat{H}_\mathrm{D}$:

$$\widehat{H}_\mathrm{D} = mc^2 \beta + c\vec{\alpha} \cdot \vec{p} = \begin{pmatrix} mc^2 I & c\vec{\sigma} \cdot \vec{p} \\ c\vec{\sigma} \cdot \vec{p} & -mc^2 I \end{pmatrix}$$

$$= \begin{pmatrix} mc^2 & 0 & c\widehat{p}_z & c(\widehat{p}_x - i\widehat{p}_y) \\ 0 & mc^2 & c(\widehat{p}_x + i\widehat{p}_y) & -c\widehat{p}_z \\ c\widehat{p}_z & c(\widehat{p}_x - i\widehat{p}_y) & -mc^2 & 0 \\ c(\widehat{p}_x + i\widehat{p}_y) & -c\widehat{p}_z & 0 & -mc^2 \end{pmatrix}. \tag{3.610}$$

Then, the time-dependent Dirac equation

$$i\hbar \frac{\partial |\underline{\Psi}(t)\rangle}{\partial t} = \widehat{H}_\text{D} |\underline{\Psi}(t)\rangle \tag{3.611}$$

represents the time evolution of a relativistic free particle of mass m. Note that $\widehat{H}_\text{D}$ is Hermitian: $\widehat{H}_\text{D}^\dagger = \widehat{H}_\text{D}$. Since $\widehat{H}_\text{D}$ is a four-dimensional matrix operator, the state vector $|\underline{\Psi}(t)\rangle$ has to be a four-dimensional column vector, say,

$$|\underline{\Psi}(t)\rangle = \begin{pmatrix} |\Psi_1(t)\rangle \\ |\Psi_2(t)\rangle \\ |\Psi_3(t)\rangle \\ |\Psi_4(t)\rangle \end{pmatrix}, \tag{3.612}$$

known as the Dirac spinor.

In position representation, the Dirac equation (3.611) becomes

$$i\hbar \frac{\partial \underline{\Psi}(\vec{r},t)}{\partial t} = \left(mc^2 \beta - i\hbar c \vec{\alpha} \cdot \vec{\nabla} \right) \underline{\Psi}(\vec{r},t), \tag{3.613}$$

where

$$\underline{\Psi}(\vec{r},t) = \begin{pmatrix} \Psi_1(\vec{r},t) \\ \Psi_2(\vec{r},t) \\ \Psi_3(\vec{r},t) \\ \Psi_4(\vec{r},t) \end{pmatrix}. \tag{3.614}$$

Correspondingly,

$$\underline{\Psi}^\dagger(\vec{r},t) = (\Psi_1^*(\vec{r},t)\ \Psi_2^*(\vec{r},t)\ \Psi_3^*(\vec{r},t)\ \Psi_4^*(\vec{r},t)) = \langle \underline{\Psi}(t) | \vec{r} \rangle. \tag{3.615}$$

Hermiticity of $\widehat{H}_\text{D}$ implies that the time evolution of $|\underline{\Psi}(t)\rangle$ is unitary, i.e., $\langle \underline{\Psi}(t) | \underline{\Psi}(t) \rangle = \sum_{j=1}^4 \langle \Psi_j(t) | \Psi_j(t) \rangle = \int d^3 r \left(\sum_{j=1}^4 |\Psi_j(\vec{r},t)|^2 \right)$ is conserved in time. This suggests that $\sum_{j=1}^4 |\Psi_j(\vec{r},t)|^2 = \underline{\Psi}^\dagger(\vec{r},t) \underline{\Psi}(\vec{r},t)$ should be the probability density for the Dirac particle. Let us look for the corresponding probability current density and the continuity equation. To this end, we proceed as follows. Let us write (3.613) as

$$i\hbar \frac{\partial \underline{\Psi}}{\partial t} = mc^2 \beta \underline{\Psi} - i\hbar c \vec{\alpha} \cdot \vec{\nabla} \underline{\Psi}. \tag{3.616}$$

The Hermitian conjugate of this equation is

$$-i\hbar \frac{\partial \underline{\Psi}^\dagger}{\partial t} = mc^2 \beta \underline{\Psi}^\dagger + i\hbar c \vec{\nabla} \underline{\Psi}^\dagger \cdot \vec{\alpha}. \tag{3.617}$$

From these two equations, we get

A Review of Quantum Mechanics

$$\frac{\partial}{\partial t}\left(\underline{\Psi}^\dagger \underline{\Psi}\right) = \frac{\partial \underline{\Psi}^\dagger}{\partial t}\underline{\Psi} + \underline{\Psi}^\dagger \frac{\partial \underline{\Psi}}{\partial t}$$

$$= \left(\frac{imc^2}{\hbar}\beta\underline{\Psi}^\dagger - c\vec{\nabla}\underline{\Psi}^\dagger \cdot \vec{\alpha}\right)\underline{\Psi} + \underline{\Psi}^\dagger\left(-\frac{imc^2}{\hbar}\beta\underline{\Psi} - c\vec{\alpha}\cdot\vec{\nabla}\underline{\Psi}\right)$$

$$= -c\left(\vec{\nabla}\underline{\Psi}^\dagger \cdot \vec{\alpha}\underline{\Psi} + \underline{\Psi}^\dagger \vec{\alpha}\cdot\vec{\nabla}\underline{\Psi}\right) = -c\vec{\nabla}\cdot\left(\underline{\Psi}^\dagger\vec{\alpha}\underline{\Psi}\right). \tag{3.618}$$

Thus, we get the continuity equation

$$\frac{\partial \rho_\mathrm{D}}{\partial t} + \vec{\nabla}\cdot\vec{j}_\mathrm{D} = 0, \tag{3.619}$$

where the Dirac probability density and the Dirac probability current density are given, respectively, by

$$\rho_\mathrm{D}(\vec{r},t) = \underline{\Psi}^\dagger(\vec{r},t)\underline{\Psi}(\vec{r},t),$$
$$\vec{j}_\mathrm{D}(\vec{r},t) = c\underline{\Psi}^\dagger(\vec{r},t)\vec{\alpha}\underline{\Psi}(\vec{r},t). \tag{3.620}$$

Note that $\rho_\mathrm{D}(\vec{r},t)$ is nonnegative for all $\vec{r}$ and t.

Note that starting with (3.613), and noting that $\widehat{H}_\mathrm{D}$ is time-independent, we can write

$$\left(i\hbar\frac{\partial}{\partial t}\right)^2 \underline{\Psi}(\vec{r},t) = \left(mc^2\beta - i\hbar c\vec{\alpha}\cdot\vec{\nabla}\right)\left(i\hbar\frac{\partial}{\partial t}\right)\underline{\Psi}(\vec{r},t)$$

$$= \left(mc^2\beta - i\hbar c\vec{\alpha}\cdot\vec{\nabla}\right)^2 \underline{\Psi}(\vec{r},t), \tag{3.621}$$

which, after dividing by $\hbar^2 c^2$ and rearranging, simplifies to

$$\left(\nabla^2 - \frac{1}{c^2}\frac{\partial^2}{\partial t^2}\right)\underline{\Psi}(\vec{r},t) = \left(\frac{mc}{\hbar}\right)^2 \underline{\Psi}(\vec{r},t), \tag{3.622}$$

the free-particle Klein–Gordon equation for each component of the free-particle Dirac spinor. This shows that the solutions of the free-particle Dirac equation satisfy the free-particle Klein–Gordon equation. But, the converse is not true.

Let us now find the solutions of the free-particle Dirac equation. For a free particle of mass m with momentum $\vec{p}$, let us take

$$\underline{\Psi}(\vec{r},t) = \underline{\psi}e^{\frac{i}{\hbar}(\vec{p}\cdot\vec{r}-Et)} = \begin{pmatrix} \psi_1 \\ \psi_2 \\ \psi_3 \\ \psi_4 \end{pmatrix} e^{\frac{i}{\hbar}(\vec{p}\cdot\vec{r}-Et)} \tag{3.623}$$

as the general form of the Dirac plane wave spinor. Substituting this solution in (3.613), we get

$$\left(mc^2\beta + c\vec{\alpha}\cdot\vec{p}\right)\underline{\psi}e^{\frac{i}{\hbar}(\vec{p}\cdot\vec{r}-Et)} = E\underline{\psi}e^{\frac{i}{\hbar}(\vec{p}\cdot\vec{r}-Et)}. \tag{3.624}$$

This leads to the eigenvalue equation

$$H\underline{\psi} = E\underline{\psi}, \quad \text{where } H = \left(mc^2\beta + c\vec{\alpha}\cdot\vec{p}\right), \qquad (3.625)$$

which is the time-independent Dirac equation for the free particle. Let us now take

$$\underline{\psi} = \begin{pmatrix} \underline{\psi}_u \\ \underline{\psi}_\ell \end{pmatrix}, \qquad (3.626)$$

where

$$\underline{\psi}_u = \begin{pmatrix} \psi_1 \\ \psi_2 \end{pmatrix}, \quad \underline{\psi}_\ell = \begin{pmatrix} \psi_3 \\ \psi_4 \end{pmatrix}. \qquad (3.627)$$

are the upper and lower pairs of components of $\underline{\psi}$. Now, the equation (3.625) becomes

$$\begin{pmatrix} mc^2 & c\vec{\sigma}\cdot\vec{p} \\ c\vec{\sigma}\cdot\vec{p} & -mc^2 \end{pmatrix} \begin{pmatrix} \underline{\psi}_u \\ \underline{\psi}_\ell \end{pmatrix} = E \begin{pmatrix} \underline{\psi}_u \\ \underline{\psi}_\ell \end{pmatrix}, \qquad (3.628)$$

or

$$\begin{aligned}(E - mc^2)\underline{\psi}_u &= c\vec{\sigma}\cdot\vec{p}\,\underline{\psi}_\ell \\ (E + mc^2)\underline{\psi}_\ell &= c\vec{\sigma}\cdot\vec{p}\,\underline{\psi}_u.\end{aligned} \qquad (3.629)$$

Solving these equations, one can express $\underline{\psi}_\ell$ or $\underline{\psi}_u$ in terms of the other:

$$\begin{aligned}\underline{\psi}_\ell &= \frac{c}{E + mc^2}\vec{\sigma}\cdot\vec{p}\,\underline{\psi}_u \\ \underline{\psi}_u &= \frac{c}{E - mc^2}\vec{\sigma}\cdot\vec{p}\,\underline{\psi}_\ell.\end{aligned} \qquad (3.630)$$

Substituting the above expression for $\underline{\psi}_\ell$ in terms of $\underline{\psi}_u$ on the right-hand side of the first equation of (3.629), we get

$$(E^2 - m^2c^4)\underline{\psi}_u = c^2p^2\underline{\psi}_u. \qquad (3.631)$$

Note that $(\vec{\sigma}\cdot\vec{p})^2 = p^2$. Similarly, substituting the above expression for $\underline{\psi}_u$ in terms of $\underline{\psi}_\ell$ on the right-hand side of the second equation of (3.629), we get

$$(E^2 - m^2c^4)\underline{\psi}_\ell = c^2p^2\underline{\psi}_\ell. \qquad (3.632)$$

This shows that we must have

$$E = \pm\sqrt{m^2c^4 + c^2p^2} \qquad (3.633)$$

as the eigenvalues of H in (3.625). Since H is a four-dimensional Hermitian traceless matrix, its four real eigenvalues should be $\{E(p), E(p), -E(p), -E(p)\}$, where $E(p) = +\sqrt{m^2c^4 + c^2p^2}$, the positive square root of $m^2c^4 + c^2p^2$.

A Review of Quantum Mechanics

Now, following the first equation of (3.630) and taking $E = \mathsf{E}(p)$, let us write

$$\underline{\psi}_{\mathsf{E}(p)} = \begin{pmatrix} \underline{\psi}_u \\ \frac{c}{\mathsf{E}(p)+mc^2}\vec{\sigma}\cdot\vec{p}\,\underline{\psi}_u \end{pmatrix}, \tag{3.634}$$

where $\underline{\psi}_u$ is arbitrary. It can be verified directly, with simple algebra, that $\underline{\psi}_{\mathsf{E}(p)}$ is an eigenvector of H corresponding to the energy eigenvalue $\mathsf{E}(p)$ for any nontrivial choice of $\underline{\psi}_u$, i.e.,

$$\begin{pmatrix} mc^2 & c\vec{\sigma}\cdot\vec{p} \\ c\vec{\sigma}\cdot\vec{p} & -mc^2 \end{pmatrix} \begin{pmatrix} \underline{\psi}_u \\ \frac{c}{\mathsf{E}(p)+mc^2}\vec{\sigma}\cdot\vec{p}\,\underline{\psi}_u \end{pmatrix} = \mathsf{E}(p) \begin{pmatrix} \underline{\psi}_u \\ \frac{c}{\mathsf{E}(p)+mc^2}\vec{\sigma}\cdot\vec{p}\,\underline{\psi}_u \end{pmatrix}. \tag{3.635}$$

To construct the positive-energy eigenvector, we have chosen the first equation of (3.630) since the second equation becomes indeterminate when the particle is at its rest frame, i.e., when $p = 0$ and hence $\mathsf{E}(p) = mc^2$. There are two linearly independent choices for $\underline{\psi}_u$. Let us choose

$$\underline{\psi}_u^{(1)} = \begin{pmatrix} 1 \\ 0 \end{pmatrix}, \quad \underline{\psi}_u^{(2)} = \begin{pmatrix} 0 \\ 1 \end{pmatrix}, \tag{3.636}$$

in terms of which any two-dimensional column vector can be written as a linear combination. Substituting these two choices for $\underline{\psi}_u$ in (3.634), we write the two linearly independent positive-energy eigenvectors of H, apart from normalization factors, as

$$\underline{\psi}^{(+)}_{(\mathsf{E}(p))} = \begin{pmatrix} 1 \\ 0 \\ \frac{cp_z}{\mathsf{E}(p)+mc^2} \\ \frac{cp_+}{\mathsf{E}(p)+mc^2} \end{pmatrix}, \quad \underline{\psi}^{(-)}_{(\mathsf{E}(p))} = \begin{pmatrix} 0 \\ 1 \\ \frac{cp_-}{\mathsf{E}(p)+mc^2} \\ \frac{-cp_z}{\mathsf{E}(p)+mc^2} \end{pmatrix}, \tag{3.637}$$

where $p_+ = p_x + ip_y$ and $p_- = p_x - ip_y$. When the particle moves nonrelativistically, i.e., $p \ll mc$, the lower pair of components of $\underline{\psi}^{(\pm)}_{(\mathsf{E}(p))}$ are seen to be much smaller than the upper pair of components.

Similarly, starting from the second equation of (3.630) and substituting $E = -\mathsf{E}(p)$, we can construct the negative-energy eigenvectors of H. It can be verified directly that

$$\begin{pmatrix} mc^2 & c\vec{\sigma}\cdot\vec{p} \\ c\vec{\sigma}\cdot\vec{p} & -mc^2 \end{pmatrix} \begin{pmatrix} \frac{-c}{\mathsf{E}(p)+mc^2}\vec{\sigma}\cdot\vec{p}\,\underline{\psi}_\ell \\ \underline{\psi}_\ell \end{pmatrix} = -\mathsf{E}(p) \begin{pmatrix} \frac{-c}{\mathsf{E}(p)+mc^2}\vec{\sigma}\cdot\vec{p}\,\underline{\psi}_\ell \\ \underline{\psi}_\ell \end{pmatrix}. \tag{3.638}$$

From this it follows that, taking $\underline{\psi}_\ell$ to be $\begin{pmatrix} 1 \\ 0 \end{pmatrix}$ and $\begin{pmatrix} 0 \\ 1 \end{pmatrix}$, we can write down the two linearly independent negative energy eigenvectors of H, apart from normalization factors, as

$$\underline{\psi}^{(+)}_{(-E(p))} = \begin{pmatrix} \frac{-cp_z}{E(p)+mc^2} \\ \frac{-cp_+}{E(p)+mc^2} \\ 1 \\ 0 \end{pmatrix}, \quad \underline{\psi}^{(-)}_{(-E(p))} = \begin{pmatrix} \frac{-cp_-}{E(p)+mc^2} \\ \frac{cp_z}{E(p)+mc^2} \\ 0 \\ 1 \end{pmatrix}. \quad (3.639)$$

Note that the two negative-energy eigenvectors are linearly independent of the two positive-energy eigenvectors since a negative-energy eigenvector cannot be a linear combination of positive-energy eigenvectors. It is seen that when the particle moves nonrelativistically the upper pair of components of the aforementioned negative-energy eigenvectors are much smaller than the lower pair of components.

Let us write

$$\underline{\psi}^{(+)}_{(E(p))} = \underline{\psi}^{(1)}, \quad \underline{\psi}^{(-)}_{(E(p))} = \underline{\psi}^{(2)},$$
$$\underline{\psi}^{(+)}_{(-E(p))} = \underline{\psi}^{(3)}, \quad \underline{\psi}^{(-)}_{(-E(p))} = \underline{\psi}^{(4)}. \quad (3.640)$$

It can be verified that they are orthogonal to each other, *i.e.*, $\left(\underline{\psi}^{(j)}\right)^\dagger \underline{\psi}^{(k)} = 0$, for $j \neq k$. Then, the four linearly independent orthonormal eigensolutions of the time-dependent Dirac equation for a free particle can be written as

$$\underline{\Psi}^1_{\vec{p}}(\vec{r},t) = N\underline{\psi}^{(1)} e^{\frac{i}{\hbar}(\vec{p}\cdot\vec{r}-E(p)t)},$$
$$\underline{\Psi}^2_{\vec{p}}(\vec{r},t) = N\underline{\psi}^{(2)} e^{\frac{i}{\hbar}(\vec{p}\cdot\vec{r}-E(p)t)},$$
$$\underline{\Psi}^3_{\vec{p}}(\vec{r},t) = N\underline{\psi}^{(3)} e^{\frac{i}{\hbar}(\vec{p}\cdot\vec{r}+E(p)t)},$$
$$\underline{\Psi}^4_{\vec{p}}(\vec{r},t) = N\underline{\psi}^{(4)} e^{\frac{i}{\hbar}(\vec{p}\cdot\vec{r}+E(p)t)}, \quad (3.641)$$

where N is a normalization factor. Note that N is common for all the four solutions since $\left(\underline{\psi}^{(j)}\right)^\dagger \underline{\psi}^{(j)} = (2E(p)/E(p)+mc^2)$, for all $j = 1,2,3,4$.

It can be checked that for both the positive-energy eigenstates $\underline{\Psi}^1_{\vec{p}}$ and $\underline{\Psi}^2_{\vec{p}}$ the probability density $\rho_D = \underline{\Psi}^\dagger \underline{\Psi}$ is $2|N|^2 E(p)/(E(p)+mc^2)$ and the probability current density $\vec{j}_D = c\underline{\Psi}^\dagger \vec{\alpha} \underline{\Psi}$ is $\rho_D(c^2\vec{p}/E(p)) = \rho_D\vec{v}$, where $\vec{v}$ is the velocity of the particle. For the negative-energy eigenstates, $\underline{\Psi}^3_{\vec{p}}$ and $\underline{\Psi}^4_{\vec{p}}$, the probability density is $2|N|^2 E(p)/(E(p)+mc^2)$ and the current density turns out to be $-\rho_D(c^2\vec{p}/E(p))$. Since for a relativistic negative-energy particle, with $-m$ as the rest mass, the momentum is $\vec{p} = -m\vec{v}/\sqrt{1-(v^2/c^2)} = -E(p)\vec{v}/c^2$, the current density is $\rho_D\vec{v}$ as expected.

3.4.2.2 Zitterbewegung

In the Heisenberg picture, we have

$$\vec{p}(t) = e^{\frac{it}{\hbar}\hat{H}_D} \vec{p} e^{-\frac{it}{\hbar}\hat{H}_D} = \vec{p}(0), \tag{3.642}$$

since $\left[\hat{H}_D, \vec{p}\right] = 0$. In other words, momentum is conserved for a free Dirac particle. For position, we get

$$\begin{aligned}\vec{r}(t) &= e^{\frac{it}{\hbar}\hat{H}_D} \vec{r} e^{-\frac{it}{\hbar}\hat{H}_D} \\ &= \vec{r}(0) + \frac{it}{\hbar}\left[\hat{H}_D, \vec{r}\right] + \frac{1}{2!}\left(\frac{it}{\hbar}\right)^2 \left[\hat{H}_D, \left[\hat{H}_D, \vec{r}\right]\right] \\ &\quad + \frac{1}{3!}\left(\frac{it}{\hbar}\right)^3 \left[\hat{H}_D, \left[\hat{H}_D, \left[\hat{H}_D, \vec{r}\right]\right]\right] + \cdots \\ &= \vec{r}(0) + ct\vec{\alpha} + \frac{c^2 t^2}{\hbar}\left(imc\beta\vec{\alpha} + \vec{p}\times\vec{\Sigma}\right) + \cdots, \end{aligned} \tag{3.643}$$

where

$$\vec{\Sigma} = I \otimes \vec{\sigma} = \begin{pmatrix} \vec{\sigma} & 0 \\ 0 & \vec{\sigma} \end{pmatrix}, \tag{3.644}$$

a very complicated result. So, we have to follow a different approach as is usually done. In the Heisenberg picture, we have for the time-independent free particle,

$$\begin{aligned}\frac{d\vec{r}(t)}{dt} &= \frac{i}{\hbar}\left[\hat{H}_D, \vec{r}(t)\right] = \frac{i}{\hbar} e^{\frac{it}{\hbar}\hat{H}_D} \left[\hat{H}_D, \vec{r}\right] e^{-\frac{it}{\hbar}\hat{H}_D} \\ &= c e^{\frac{it}{\hbar}\hat{H}_D} \vec{\alpha} e^{-\frac{it}{\hbar}\hat{H}_D} = c\vec{\alpha}(t).\end{aligned} \tag{3.645}$$

Then,

$$\begin{aligned}\frac{d^2\vec{r}(t)}{dt^2} &= c\frac{d\vec{\alpha}(t)}{dt} = \frac{ic}{\hbar}\left[\hat{H}_D, \vec{\alpha}(t)\right] \\ &= \frac{ic}{\hbar} e^{\frac{it}{\hbar}\hat{H}_D} \left[\hat{H}_D, \vec{\alpha}\right] e^{-\frac{it}{\hbar}\hat{H}_D} \\ &= \frac{ic}{\hbar} e^{\frac{it}{\hbar}\hat{H}_D} \left[\left(\hat{H}_D\vec{\alpha} + \vec{\alpha}\hat{H}_D\right) - 2\vec{\alpha}\hat{H}_D\right] e^{-\frac{it}{\hbar}\hat{H}_D} \\ &= \frac{ic}{\hbar} e^{\frac{it}{\hbar}\hat{H}_D} \left[2\left(c\vec{p} - \vec{\alpha}\hat{H}_D\right)\right] e^{-\frac{it}{\hbar}\hat{H}_D} \\ &= \frac{2ic}{\hbar}\left(c\vec{p}(0) - \vec{\alpha}(t)\hat{H}_D\right)\end{aligned} \tag{3.646}$$

Taking the Hermitian conjugate of this equation and dividing throughout by c, we get

$$\frac{d\vec{\alpha}(t)}{dt} = \frac{2i}{\hbar}\left(\hat{H}_D \vec{\alpha}(t) - c\vec{p}(0)\right). \tag{3.647}$$

From this, we find

$$\frac{d}{dt}\left(e^{-\frac{2it}{\hbar}\widehat{H}_D}\vec{\alpha}(t)\right) = -\frac{2ic}{\hbar}e^{-\frac{2it}{\hbar}\widehat{H}_D}\vec{\widehat{p}}(0). \tag{3.648}$$

It is straightforward to integrate this equation to get

$$e^{-\frac{2it}{\hbar}\widehat{H}_D}\vec{\alpha}(t) - \vec{\alpha} = \left(e^{-\frac{2it}{\hbar}\widehat{H}_D} - 1\right)\frac{c\vec{\widehat{p}}(0)}{\widehat{H}_D}. \tag{3.649}$$

Solving for $\vec{\alpha}(t)$ and substituting in (3.645), we get

$$\frac{d\vec{r}(t)}{dt} = \frac{c^2\vec{\widehat{p}}(0)}{\widehat{H}_D} + e^{\frac{2it}{\hbar}\widehat{H}_D}\left(c\vec{\alpha} - \frac{c^2\vec{\widehat{p}}(0)}{\widehat{H}_D}\right). \tag{3.650}$$

Integrating this equation, we have

$$\vec{r}(t) = \vec{r}(0) + \frac{c^2\vec{\widehat{p}}(0)}{\widehat{H}_D}t - \frac{i\hbar c}{2\widehat{H}_D}e^{\frac{2it}{\hbar}\widehat{H}_D}\left(\vec{\alpha} - \frac{c\vec{\widehat{p}}(0)}{\widehat{H}_D}\right). \tag{3.651}$$

The first term gives the initial value of $\vec{r}$. The second term gives correctly the position at time t corresponding to the constant velocity $c^2\vec{\widehat{p}}(0)/\widehat{H}_D$, or $c^2\vec{p}(0)/E(p)$. The last term indicates a complicated oscillatory motion with an extremely small amplitude $\hbar c/2E \sim \hbar/2mc$ and extremely high frequency $2E/\hbar \sim 2mc^2/\hbar$. This rapid oscillation is called Zitterbewegung (German word meaning trembling motion). The amplitude of this oscillation is $\sim h/mc$, the Compton wavelength of the electron, and hence it is meaningless to specify the position of an electron to an accuracy less than its Compton wavelength.

3.4.2.3 Spin and Helicity of the Dirac Particle

We have found already that all the components of the orbital angular momentum operator $\vec{\widehat{L}}$ commute with $\widehat{p}^2$ (see (3.205)). Hence, $\vec{\widehat{L}}$ commutes with the nonrelativistic free-particle Hamiltonian $\widehat{p}^2/2m$. Then, from the Heisenberg equation of motion, it follows that the orbital angular momentum is conserved for a nonrelativistic free particle. Let us check whether $\vec{\widehat{L}}$ is conserved for a free Dirac particle. To this end, we find the commutator of the free-particle Dirac Hamiltonian with $\vec{\widehat{L}}$:

$$\left[\widehat{H}_D, \widehat{L}_z\right] = \left[mc^2\beta + c\vec{\alpha}\cdot\vec{\widehat{p}}, \widehat{L}_z\right] = c\left[\alpha_x\widehat{p}_x + \alpha_y\widehat{p}_y, \widehat{L}_z\right] = -i\hbar c\left(\alpha_x\widehat{p}_y - \alpha_y\widehat{p}_x\right), \tag{3.652}$$

using the commutation relations (3.205). Similarly, we have

$$\left[\widehat{H}_D, \widehat{L}_x\right] = -i\hbar c\left(\alpha_y\widehat{p}_z - \alpha_z\widehat{p}_y\right), \quad \left[\widehat{H}_D, \widehat{L}_y\right] = -i\hbar c\left(\alpha_z\widehat{p}_x - \alpha_x\widehat{p}_z\right). \tag{3.653}$$

These three commutation relations, which can be written compactly as

$$\left[\widehat{H}_D, \vec{\widehat{L}}\right] = -i\hbar c\vec{\alpha}\times\vec{\widehat{p}}, \tag{3.654}$$

A Review of Quantum Mechanics

show that the orbital angular momentum is not conserved for the free Dirac particle. Now, define

$$\vec{\mathbb{S}} = \frac{\hbar}{2}\vec{\Sigma}. \tag{3.655}$$

It is straightforward to see, using the Pauli algebra (3.475), that

$$\left[\widehat{H}_D, \vec{\mathbb{S}}\right] = \frac{\hbar}{2}\left[\begin{pmatrix} mc^2 I & c\vec{\sigma}\cdot\vec{p} \\ c\vec{\sigma}\cdot\vec{p} & -mc^2 I \end{pmatrix}, \begin{pmatrix} \vec{\sigma} & 0 \\ 0 & \vec{\sigma} \end{pmatrix}\right]$$

$$= i\hbar c\vec{\alpha}\times\vec{p}, \tag{3.656}$$

exactly opposite of $\left[\widehat{H}_D, \vec{L}\right]$! Thus,

$$\left[\widehat{H}_D, \vec{L}+\vec{\mathbb{S}}\right] = 0. \tag{3.657}$$

So, we can take

$$\vec{J}_D = \vec{L}+\vec{\mathbb{S}} = \left(\vec{r}\times\vec{p}\right) + \frac{\hbar}{2}\vec{\Sigma} \tag{3.658}$$

as the conserved total angular momentum of the Dirac particle and recognize $\vec{\mathbb{S}}$ as its intrinsic angular momentum, or spin. Since $\mathbb{S}_z$ has eigenvalues $\pm\frac{\hbar}{2}$ and $\mathbb{S}^2 = \mathbb{S}_x^2 + \mathbb{S}_y^2 + \mathbb{S}_z^2 = \frac{3}{4}\hbar^2 I = \frac{1}{2}\left(\left(\frac{1}{2}+1\right)\right)\hbar^2 I$, we find that the Dirac particle has spin $\frac{\hbar}{2}$.

From (3.637) to (3.641), it is seen that when the Dirac particle is at rest $E = \pm mc^2$ and its eigenstates are

$$\underline{\Psi}_0^1(\vec{r},t) = N\begin{pmatrix} 1 \\ 0 \\ 0 \\ 0 \end{pmatrix} e^{-imc^2 t/\hbar},$$

$$\underline{\Psi}_0^2(\vec{r},t) = N\begin{pmatrix} 0 \\ 1 \\ 0 \\ 0 \end{pmatrix} e^{-imc^2 t/\hbar},$$

$$\underline{\Psi}_0^3(\vec{r},t) = N\begin{pmatrix} 0 \\ 0 \\ 1 \\ 0 \end{pmatrix} e^{imc^2 t/\hbar},$$

$$\underline{\Psi}_0^4(\vec{r},t) = N\begin{pmatrix} 0 \\ 0 \\ 0 \\ 1 \end{pmatrix} e^{imc^2 t/\hbar}. \tag{3.659}$$

Eigenstates $\underline{\Psi}_0^1$ and $\underline{\Psi}_0^2$ correspond to positive energy mc^2, and $\underline{\Psi}_0^3$ and $\underline{\Psi}_0^4$ correspond to negative energy $-mc^2$. The first two states are also eigenstates of $\mathbb{S}_z$ corresponding to eigenvalues $\frac{\hbar}{2}$ and $-\frac{\hbar}{2}$, respectively:

$$\mathbb{S}_z \underline{\Psi}_0^1 = \frac{\hbar}{2}\Sigma_z \underline{\Psi}_0^1 = \frac{\hbar}{2}\underline{\Psi}_0^1,$$
$$\mathbb{S}_z \underline{\Psi}_0^2 = \frac{\hbar}{2}\Sigma_z \underline{\Psi}_0^2 = -\frac{\hbar}{2}\underline{\Psi}_0^2. \tag{3.660}$$

Similarly, the last two states are seen to be eigenstates of $\mathbb{S}_z$ corresponding to eigenvalues $\frac{\hbar}{2}$ and $-\frac{\hbar}{2}$, respectively:

$$\mathbb{S}_z \underline{\Psi}_0^3 = \frac{\hbar}{2}\Sigma_z \underline{\Psi}_0^3 = \frac{\hbar}{2}\underline{\Psi}_0^3,$$
$$\mathbb{S}_z \underline{\Psi}_0^4 = \frac{\hbar}{2}\Sigma_z \underline{\Psi}_0^4 = -\frac{\hbar}{2}\underline{\Psi}_0^4. \tag{3.661}$$

All the four states $\underline{\Psi}_0^j$, $j = 1, 2, 3, 4$, have $\mathbb{S}^2 = \frac{3}{4}\hbar^2$. Thus, when the particle is at rest, the two degenerate energy (positive or negative) eigenstates correspond to the two eigenstates of the z-component of spin, $\mathbb{S}_z$, namely, $+$ and $-$, or up and down.

In general, when the particle is moving, the four eigenstates (3.641) are not eigenstates of spin. To understand what distinguishes the two degenerate energy (positive or negative) eigenstates corresponding to the particle in motion, let us introduce the helicity operator

$$\widehat{h} = \begin{pmatrix} \frac{\vec{\sigma}\cdot\vec{p}}{|\vec{p}|} & 0 \\ 0 & \frac{\vec{\sigma}\cdot\vec{p}}{|\vec{p}|} \end{pmatrix} = \frac{\vec{\Sigma}\cdot\vec{p}}{|\vec{p}|}. \tag{3.662}$$

Helicity gives the component of spin in the direction of the momentum of the particle in units of $\frac{\hbar}{2}$ and has eigenvalues ± 1 since $\widehat{h}^2 = I$. Note that the helicity operator commutes with the free-particle Dirac Hamiltonian, i.e.,

$$\left[\widehat{H}_D, \widehat{h}\right] = \left[\begin{pmatrix} mc^2 I & c\vec{\sigma}\cdot\vec{p} \\ c\vec{\sigma}\cdot\vec{p} & -mc^2 I \end{pmatrix}, \begin{pmatrix} \frac{\vec{\sigma}\cdot\vec{p}}{|\vec{p}|} & 0 \\ 0 & \frac{\vec{\sigma}\cdot\vec{p}}{|\vec{p}|} \end{pmatrix}\right] = 0. \tag{3.663}$$

For a free Dirac particle, the momentum is a constant of motion since $\vec{p}$ commutes with $\widehat{H}_D$. When the particle is moving with a constant momentum $\vec{p}$, if the z-axis is aligned with the direction of motion of the particle, the two positive-energy eigenstates (3.641) become, with $p = |\vec{p}|$,

$$\underline{\Psi}_{\vec{p}}^1 = \begin{pmatrix} 1 \\ 0 \\ \frac{cp}{E(p)+mc^2} \\ 0 \end{pmatrix} e^{\frac{i}{\hbar}(pz - E(p)t)},$$

$$\underline{\Psi}_{\vec{p}}^2 = \begin{pmatrix} 0 \\ 1 \\ 0 \\ \frac{-cp}{E(p)+mc^2} \end{pmatrix} e^{\frac{i}{\hbar}(pz - E(p)t)}, \tag{3.664}$$

ns of helicity:
$$\widehat{h}\,\underline{\Psi}^1_{\vec{p}} = \underline{\Psi}^1_{\vec{p}}, \qquad \widehat{h}\,\underline{\Psi}^2_{\vec{p}} = -\underline{\Psi}^2_{\vec{p}}. \tag{3.665}$$

In other words, while in the case of the particle being at rest, a measurement of the component of spin in any direction will give a definite result $\pm\hbar/2$, and in the case of the particle in motion, the component of spin in the direction of its momentum will give a definite result $\pm\hbar/2$. If the helicity is $+1$, the particle is called right handed, and if the helicity is -1, the particle is said to be left handed. In a similar way we can construct the two helicity eigenstates corresponding to negative energy. With the z-axis aligned in the direction of the momentum $\vec{p}$, the negative-energy eigenstates $\underline{\Psi}^3_{\vec{p}}$ and $\underline{\Psi}^4_{\vec{p}}$ are eigenstates of helicity with eigenvalues $+1$ and -1, respectively.

Negative-energy states of a Dirac particle are actually positive-energy states of the corresponding antiparticle. The Dirac equation is also not a single particle equation. However, when we are dealing with the dynamics of a spin-$\frac{1}{2}$ particle with positive energy and the process studied does not involve any particle creation and annihilation, it should be appropriate to use the Dirac equation as if it is a one-particle relativistic equation.

3.4.2.4 Spin Magnetic Moment of the Electron and the Dirac–Pauli Equation

In presence of an electromagnetic field, with ϕ and $\vec{A}$ as the scalar and vector potentials, the Dirac equation for the electron is

$$\left(i\hbar\frac{\partial}{\partial t} + e\phi\right)\underline{\Psi}(\vec{r},t) = \left(m_e c^2 \beta + c\vec{\alpha}\cdot\vec{\pi}\right)\underline{\Psi}(\vec{r},t), \quad \text{with } \vec{\pi} = \vec{p} + e\vec{A}, \tag{3.666}$$

obtained from the free-particle equation by the replacement $i\hbar\partial/\partial t \longrightarrow (i\hbar\partial/\partial t) + e\phi$ and $\vec{p} \longrightarrow \vec{\pi}$ following the principle of minimal electromagnetic coupling. Let us now consider the state of a positive-energy electron to be given by

$$\underline{\Psi} = \begin{pmatrix} \underline{\Psi}_u \\ \underline{\Psi}_\ell \end{pmatrix} = e^{-im_e c^2 t/\hbar} \begin{pmatrix} \underline{\Psi}_u \\ \underline{\Psi}_l \end{pmatrix}. \tag{3.667}$$

Now, the time-dependent Dirac equation (3.666) becomes

$$\left(m_e c^2 + i\hbar\frac{\partial}{\partial t}\right)\begin{pmatrix} \underline{\Psi}_u \\ \underline{\Psi}_l \end{pmatrix}$$
$$= \begin{pmatrix} (m_e c^2 - e\phi)I & c\vec{\sigma}\cdot\vec{\pi} \\ c\vec{\sigma}\cdot\vec{\pi} & -(m_e c^2 + e\phi)I \end{pmatrix}\begin{pmatrix} \underline{\Psi}_u \\ \underline{\Psi}_l \end{pmatrix}. \tag{3.668}$$

Rewriting this as a pair of coupled equations for $\underline{\Psi}_u$ and $\underline{\Psi}_l$, we have

$$i\hbar\frac{\partial \underline{\Psi}_u}{\partial t} = c\vec{\sigma}\cdot\vec{\pi}\underline{\Psi}_l - e\phi\underline{\Psi}_u$$
$$i\hbar\frac{\partial \underline{\Psi}_l}{\partial t} = c\vec{\sigma}\cdot\vec{\pi}\underline{\Psi}_u - (2m_e c^2 + e\phi)\underline{\Psi}_l. \tag{3.669}$$

Let us now consider the electron to be nonrelativistic such that the rest energy $m_e c^2$ is very much larger than its kinetic and potential energies. Because of this in the second of the above pair of equations, we can replace $(2m_e c^2 + e\phi)$ by $2m_e c^2$. Further, we can take $i\hbar \partial \underline{\Psi}_{\text{l}}/\partial t$ to be negligible compared to the term $2m_e c^2 \underline{\Psi}_{\text{l}}$, since the former term will be $\sim$ (kinetic energy)$\underline{\Psi}_{\text{l}}$ after having pulled out the factor $e^{-im_e c^2 t/\hbar}$ from $\underline{\Psi}_\ell$. With this approximation, the second of the above pair of equations can be written as

$$2m_e c^2 \underline{\Psi}_{\text{l}} = c\vec{\sigma} \cdot \vec{\pi} \underline{\Psi}_{\text{u}}. \qquad (3.670)$$

This shows that in the nonrelativistic limit the upper pair of components are large compared to the lower pair of components in the positive-energy Dirac spinors even in the presence of an electromagnetic field. We saw this behavior in the case of free-particle solutions (3.641) earlier. Using the relation (3.670) to replace $\underline{\Psi}_{\text{l}}$ in the first of the pair of equations (3.669), and renaming $\underline{\Psi}_{\text{u}}$ as $\underline{\Psi}_{\text{NRP}}$, we get

$$i\hbar \frac{\partial \underline{\Psi}_{\text{NRP}}}{\partial t} = \left[\frac{1}{2m_e} \left(\vec{\sigma} \cdot \vec{\pi} \right)^2 - e\phi \right] \underline{\Psi}_{\text{NRP}}, \qquad (3.671)$$

which is same as the nonrelativistic equation for the electron including spin (3.474). As seen already, on expanding $\left(\vec{\sigma} \cdot \vec{\pi} \right)^2$, this equation becomes

$$i\hbar \frac{\partial \underline{\Psi}_{\text{NRP}}}{\partial t} = \left[\left(\frac{\vec{\pi}^2}{2m_e} - e\phi \right) I + \mu_B \vec{\sigma} \cdot \vec{B} \right] \underline{\Psi}_{\text{NRP}}, \qquad (3.672)$$

where the last term represents the interaction energy of the electron spin magnetic moment with the magnetic field. While in the Schrödinger–Pauli equation (3.472) this last term was added by hand, of course necessitated by experimental results, the Dirac equation leads to it, the correct spin magnetic moment term, naturally.

As already noted, electron has a magnetic moment slightly greater than the magnetic moment predicted by the Dirac equation. When this anomalous magnetic moment μ_a is taken into account, the time-evolution equation for the electron in an electromagnetic field becomes

$$i\hbar \frac{\partial \underline{\Psi}}{\partial t} = \left(m_e c^2 \beta + c\vec{\alpha} \cdot \vec{\pi} - e\phi - \mu_a \beta \vec{\Sigma} \cdot \vec{B} \right) \underline{\Psi}, \qquad (3.673)$$

the Pauli–Dirac equation in which the Pauli term, the last term, has been added to the Dirac Hamiltonian.

3.4.2.5 Electron in a Constant Magnetic Field

Let us consider an electron moving in a constant magnetic field $\vec{B}$. We can take $\phi = 0$ and $\vec{A} = \vec{A}(\vec{r})$ independent of time. Let us ignore the anomalous magnetic moment. Then, the time-independent Dirac equation for the stationary state is

$$\left(m_e c^2 \beta + c\vec{\alpha} \cdot \vec{\pi} \right) \underline{\psi}(\vec{r}) = E \underline{\psi}(\vec{r}), \qquad (3.674)$$

A Review of Quantum Mechanics

obtained by taking $\underline{\Psi}(\vec{r},t) = e^{-iEt/\hbar}\underline{\psi}(\vec{r})$ in (3.666). Taking $\underline{\psi} = \begin{pmatrix} \underline{\psi}_u \\ \underline{\psi}_\ell \end{pmatrix}$, we have

$$\begin{pmatrix} m_e c^2 I & c\vec{\sigma}\cdot\vec{\pi} \\ c\vec{\sigma}\cdot\vec{\pi} & -m_e c^2 I \end{pmatrix} \begin{pmatrix} \underline{\psi}_u \\ \underline{\psi}_\ell \end{pmatrix} = E \begin{pmatrix} \underline{\psi}_u \\ \underline{\psi}_\ell \end{pmatrix}. \tag{3.675}$$

This leads to

$$c\vec{\sigma}\cdot\vec{\pi}\,\underline{\psi}_\ell = (E - m_e c^2)\,\underline{\psi}_u,$$
$$c\vec{\sigma}\cdot\vec{\pi}\,\underline{\psi}_u = (E + m_e c^2)\,\underline{\psi}_\ell. \tag{3.676}$$

Solving for $\underline{\psi}_\ell$ from the second equation and substituting it in the first equation, we get

$$c^2\left(\vec{\sigma}\cdot\vec{\pi}\right)^2 \underline{\psi}_u = (E^2 - m_e^2 c^4)\,\underline{\psi}_u \tag{3.677}$$

leading to an eigenvalue equation for E^2

$$\left[m_e^2 c^4 + c^2\left(\vec{\sigma}\cdot\vec{\pi}\right)^2\right]\underline{\psi}_u = E^2 \underline{\psi}_u. \tag{3.678}$$

If we solve for $\underline{\psi}_u$ from the first equation and substitute it in the second, it will lead to the same eigenvalue equation for E^2 with just $\underline{\psi}_u$ replaced by $\underline{\psi}_\ell$. Expanding the above equation, we get

$$\left(m_e^2 c^4 + c^2 \vec{\pi}^2 + c^2 e\hbar\vec{\sigma}\cdot\vec{B}\right)\underline{\psi}_u = E^2 \underline{\psi}_u. \tag{3.679}$$

For the magnetic field $\vec{B} = B\vec{k}$, we can take $\vec{A} = \vec{B}\times\vec{r}/2$ so that

$$\left\{c^2\left[\left(\hat{p}_x - \frac{1}{2}eBy\right)^2 + \left(\hat{p}_y + \frac{1}{2}eBx\right)^2 + \hat{p}_z^2\right] + m_e^2 c^4 + c^2 e\hbar\sigma_z B\right\}\underline{\psi}_u = E^2 \underline{\psi}_u. \tag{3.680}$$

Following what we did in the case of the Klein–Gordon equation, we shall write the above equation as

$$\left(2m_e c^2 \widehat{H}_{\text{NRL}} + m^2 c^4 + c^2 e\hbar\sigma_z B\right)\underline{\psi}_u$$

$$= \begin{pmatrix} 2m_e c^2 \widehat{H}_{\text{NRL}} + m_e^2 c^4 \\ + c^2 \hbar e B & 0 \\ 0 & 2m_e c^2 \widehat{H}_{\text{NRL}} + m_e^2 c^4 \\ & -c^2 \hbar e B \end{pmatrix} \begin{pmatrix} \psi_1 \\ \psi_2 \end{pmatrix}$$

$$= E^2 \begin{pmatrix} \psi_1 \\ \psi_2 \end{pmatrix}. \tag{3.681}$$

The general solution of this equation is

$$\underline{\Psi}_u(\vec{r}) = \begin{pmatrix} \psi_1(\vec{r}) \\ \psi_2(\vec{r}) \end{pmatrix} = \begin{pmatrix} c_1 \Psi_{n-1,m_\phi,p_z}(\vec{r}) \\ c_2 \Psi_{n,m_\phi,p_z}(\vec{r}) \end{pmatrix}, \qquad (3.682)$$

and

$$\begin{aligned} E_{n,p_z}^2 &= 2m_e c^2 \left[\frac{p_z^2}{2m_e} + \left(n - \frac{1}{2}\right) \frac{\hbar e B}{m_e} \right] + m_e^2 c^4 + c^2 \hbar e B \\ &= 2m_e c^2 \left[\frac{p_z^2}{2m_e} + \left(n + \frac{1}{2}\right) \frac{\hbar e B}{m_e} \right] + m_e^2 c^4 - c^2 \hbar e B \\ &= m_e^2 c^4 + c^2 p_z^2 + 2m_e c^2 \left[\left(n \mp \frac{1}{2}\right) \frac{\hbar e B}{m_e} \right] \pm c^2 \hbar e B \\ &= m_e^2 c^4 + c^2 p_z^2 + 2n c^2 \hbar e B \\ &= m_e^2 c^4 \left[1 + \left(\frac{p_z}{m_e c}\right)^2 + \frac{2n \hbar e B}{m_e^2 c^2} \right], \end{aligned} \qquad (3.683)$$

with infinite degeneracy due to the angular momentum quantum number $m_\phi = 0, 1, 2, \ldots$. We shall consider the positive square root and write

$$E_{n,p_z} = m_e c^2 \sqrt{\left[1 + \left(\frac{p_z}{m_e c}\right)^2 + \frac{2n \hbar e B}{m_e^2 c^2} \right]}. \qquad (3.684)$$

In the nonrelativistic limit, when $p_z \ll m_e c$ and $\hbar e B \ll m_e^2 c^2$, we get

$$\begin{aligned} E_{n,p_z} &\approx m_e c^2 \left[1 + \frac{1}{2} \left(\frac{p_z}{m_e c}\right)^2 + \frac{n \hbar e B}{m_e^2 c^2} \right] \\ &= m_e c^2 + \frac{p_z^2}{2m_e} + n \left(\frac{\hbar e B}{m_e}\right), \quad n = 1, 2, 3, \ldots, \end{aligned} \qquad (3.685)$$

which coincides with the Landau energy levels (3.485) for a nonrelativistic electron obeying the Schrödinger–Pauli equation. Once $\underline{\Psi}_u$ is known, $\underline{\Psi}_\ell$ can be found from the second equation in (3.676) (For more details on the problem of Dirac electron in a constant magnetic field, see, *e.g.*, Johnson and Lippmann [88]).

3.4.3 FOLDY–WOUTHUYSEN TRANSFORMATION

3.4.3.1 Foldy–Wouthuysen Representation of the Dirac Equation

We obtained the Schrödinger–Pauli equation by taking the nonrelativistic limit of the Dirac equation. In this approximation we found the upper pair of components to be larger than the lower pair of components in the positive-energy Dirac spinors. We have seen already that, in the negative-energy free-particle spinors, the lower pair of components are large compared to the upper pair of components. This indicates that if we could somehow transform the Hamiltonian to a direct sum of two-dimensional blocks, then we would have separate equations for two-component positive-energy and negative-energy spinors. It is not possible to achieve this through a similarity transformation using a unitary operator, because the minimum dimension of irreducible faithful representation of the algebra of β and α matrices is four. So, at best, what we can hope for is to transform to a new representation in which the Hamiltonian has the off-diagonal two-dimensional blocks as small as we desire, in terms of some order parameter, compared to the diagonal blocks. Then, the new Hamiltonian can be approximated by its block diagonal form dropping the off-diagonal blocks. This is achieved through a transformation technique, the Foldy–Wouthuysen transformation technique (Foldy and Wouthuysen [53]). Pryce [151] and Tani [177] had also anticipated this transformation, and hence it is sometimes called the Pryce–Tani–Foldy–Wouthuysen transformation (see, *e.g.*, Bell [8]; see Acharya and Sudarshan [1] for a general discussion of the role of Foldy–Wouthuysen-type transformations in particle interpretation of relativistic wave equations).

In general, if we make a unitary transformation of the state vector evolving in time according to a Hamiltonian $\widehat{H}$, *i.e.*,

$$i\hbar \frac{\partial |\Psi(t)\rangle}{\partial t} = \widehat{H} |\Psi(t)\rangle, \qquad (3.686)$$

then the transformed state vector, say $|\Psi'(t)\rangle = e^{i\widehat{S}(t)}|\Psi(t)\rangle$, obeys the equation

$$i\hbar \frac{\partial}{\partial t}\left(e^{-i\widehat{S}(t)}|\Psi'(t)\rangle\right) = i\hbar \left(\frac{\partial e^{-i\widehat{S}(t)}}{\partial t}\right)|\Psi'(t)\rangle + i\hbar e^{-i\widehat{S}(t)}\frac{\partial |\Psi'(t)\rangle}{\partial t}$$
$$= \widehat{H} e^{-i\widehat{S}(t)}|\Psi'(t)\rangle. \qquad (3.687)$$

Operating from left by $e^{i\widehat{S}(t)}$ and rearranging the terms, this equation becomes

$$i\hbar \frac{\partial |\Psi'(t)\rangle}{\partial t} = \left[e^{i\widehat{S}(t)} \widehat{H} e^{-i\widehat{S}(t)} - i\hbar \left(e^{i\widehat{S}(t)} \frac{\partial e^{-i\widehat{S}(t)}}{\partial t}\right)\right] |\Psi'(t)\rangle. \qquad (3.688)$$

In other words, the transformed state vector $|\Psi'(t)\rangle$ evolves in time according to the equation

$$i\hbar \frac{\partial |\Psi'(t)\rangle}{\partial t} = \widehat{H}' |\Psi'(t)\rangle,$$
$$\widehat{H}' = e^{i\widehat{S}(t)} \widehat{H} e^{-i\widehat{S}(t)} - i\hbar \left(e^{i\widehat{S}(t)} \frac{\partial e^{-i\widehat{S}(t)}}{\partial t}\right), \qquad (3.689)$$

with $\widehat{H}'$ as the transformed Hamiltonian. To calculate $\widehat{H}'$ from $\widehat{H}$ and $\widehat{S}(t)$, we will need, besides (3.138), the following identity:

$$e^{\widehat{A}(t)}\frac{\partial}{\partial t}\left(e^{-\widehat{A}(t)}\right) = \left(I+\widehat{A}(t)+\frac{1}{2!}\widehat{A}(t)^2+\cdots\right)$$
$$\times \frac{\partial}{\partial t}\left(I-\widehat{A}(t)+\frac{1}{2!}\widehat{A}(t)^2-\cdots\right)$$
$$= \left(I+\widehat{A}(t)+\frac{1}{2!}\widehat{A}(t)^2+\cdots\right)\left(-\frac{\partial \widehat{A}(t)}{\partial t}\right.$$
$$\left. +\frac{1}{2!}\left\{\frac{\partial \widehat{A}(t)}{\partial t}\widehat{A}(t)+\widehat{A}(t)\frac{\partial \widehat{A}(t)}{\partial t}\right\}-\cdots\right)$$
$$= -\frac{\partial \widehat{A}(t)}{\partial t}-\frac{1}{2!}\left[\widehat{A}(t),\frac{\partial \widehat{A}(t)}{\partial t}\right]$$
$$-\frac{1}{3!}\left[\widehat{A}(t),\left[\widehat{A}(t),\frac{\partial \widehat{A}(t)}{\partial t}\right]\right]$$
$$-\frac{1}{4!}\left[\widehat{A}(t),\left[\widehat{A}(t),\left[\widehat{A}(t),\frac{\partial \widehat{A}(t)}{\partial t}\right]\right]\right]-\cdots. \quad (3.690)$$

For a guidance to arrive at a new representation in which the Dirac Hamiltonian in the presence of an electromagnetic field would be block diagonalized to a desired order of approximation, let us look at the free-particle Hamiltonian for which we have found the exact eigenvectors, or in other words, the exact diagonalizing transformation. For the free-particle Hamiltonian matrix $H = mc^2\beta + c\vec{\alpha}\cdot\vec{p}$ in (3.625), we have found four eigenvectors $\left\{\underline{\psi}^{(j)}\,|\,j=1,2,3,4\right\}$ such that

$$H\underline{\psi}^{(j)} = \mathsf{E}(p)\underline{\psi}^{(j)}, \quad \text{for } j=1,2,$$
$$H\underline{\psi}^{(j)} = -\mathsf{E}(p)\underline{\psi}^{(j)}, \quad \text{for } j=3,4, \quad (3.691)$$

which obey the orthogonality relations

$$\underline{\psi}^{(j)\dagger}\underline{\psi}^{(k)} = \frac{2\mathsf{E}(p)}{\mathsf{E}(p)+mc^2}\delta_{jk}, \quad j,k=1,2,3,4. \quad (3.692)$$

Let us form the matrix

$$U = \sqrt{\frac{\mathsf{E}(p)+mc^2}{2\mathsf{E}(p)}}\begin{pmatrix}\underline{\psi}^{(1)\dagger}\\ \underline{\psi}^{(2)\dagger}\\ \underline{\psi}^{(3)\dagger}\\ \underline{\psi}^{(4)\dagger}\end{pmatrix}. \quad (3.693)$$

It follows that, with

A Review of Quantum Mechanics

$$
\begin{aligned}
U^\dagger &= \frac{1}{\sqrt{2\mathrm{E}(p)\,(\mathrm{E}(p)+mc^2)}} \left(\underline{\psi}^{(1)} \ \underline{\psi}^{(2)} \ \underline{\psi}^{(3)} \ \underline{\psi}^{(4)} \right) \\
&= \frac{1}{\sqrt{2\mathrm{E}(p)\,(\mathrm{E}(p)+mc^2)}} \times \\
&\quad \begin{pmatrix} \mathrm{E}(p)+mc^2 & 0 & -cp_z & -cp_- \\ 0 & \mathrm{E}(p)+mc^2 & -cp_+ & cp_z \\ cp_z & cp_- & \mathrm{E}(p)+mc^2 & 0 \\ cp_+ & -cp_z & 0 & \mathrm{E}(p)+mc^2 \end{pmatrix} \\
&= \sqrt{\frac{\mathrm{E}(p)+mc^2}{2\mathrm{E}(p)}} I - \frac{cp}{\sqrt{2\mathrm{E}(p)\,(\mathrm{E}(p)+mc^2)}} \beta \frac{c\vec{\alpha}\cdot\vec{p}}{cp},
\end{aligned}
\tag{3.694}
$$

we have

$$
UHU^\dagger = \mathrm{E}(p)\beta,
\tag{3.695}
$$

where $\mathrm{E}(p)\beta$ is the diagonal form of H. Note that

$$
\left(\sqrt{\frac{\mathrm{E}(p)+mc^2}{2\mathrm{E}(p)}}\right)^2 + \left(\frac{cp}{\sqrt{2\mathrm{E}(p)\,(\mathrm{E}(p)+mc^2)}}\right)^2 = 1.
\tag{3.696}
$$

Hence, taking

$$
\sqrt{\frac{\mathrm{E}(p)+mc^2}{2\mathrm{E}(p)}} = \cos(cp\theta), \quad \frac{cp}{\sqrt{2\mathrm{E}(p)\,(\mathrm{E}(p)+mc^2)}} = \sin(cp\theta),
\tag{3.697}
$$

let us write

$$
U^\dagger = \cos(cp\theta)I - \sin(cp\theta)\beta\frac{c\vec{\alpha}\cdot\vec{p}}{cp},
\tag{3.698}
$$

and note that

$$
\begin{aligned}
UU^\dagger &= \left(\cos(cp\theta)I - \sin(cp\theta)\frac{c\vec{\alpha}\cdot\vec{p}}{cp}\beta\right)\left(\cos(cp\theta)I - \sin(cp\theta)\beta\frac{c\vec{\alpha}\cdot\vec{p}}{cp}\right) \\
&= \cos^2(cp\theta)I - \cos(cp\theta)\sin(cp\theta)\left(\beta\frac{c\vec{\alpha}\cdot\vec{p}}{cp} + \frac{c\vec{\alpha}\cdot\vec{p}}{cp}\beta\right) \\
&\quad + \sin^2(cp\theta)\left(\frac{c\vec{\alpha}\cdot\vec{p}}{cp}\beta^2\frac{c\vec{\alpha}\cdot\vec{p}}{cp}\right) \\
&= \left(\cos^2(cp)\theta\right) + \sin^2(cp\theta)\right)I = I,
\end{aligned}
\tag{3.699}
$$

where we have used the hermiticity of β and αs and their anticommutation relations (3.604). It can be checked from (3.697) that $\tan(2cp\theta) = p/mc$. Since $(\beta c\vec{\alpha}\cdot\vec{p})^2 = -c^2p^2I$, we have

$$e^{-i(-i\theta\beta c\vec{\alpha}\cdot\vec{p})} = e^{-\theta\beta c\vec{\alpha}\cdot\vec{p}}$$

$$= I - \theta\beta c\vec{\alpha}\cdot\vec{p} - \frac{(cp\theta)^2}{2!}I + \frac{(cp)^2\theta^3}{3!}\beta c\vec{\alpha}\cdot\vec{p} + \cdots$$

$$= \left(1 - \frac{(cp\theta)^2}{2!} + \frac{(cp\theta)^4}{4!} - \cdots\right)I$$

$$- \left(cp\theta - \frac{(cp\theta)^3}{3!} + \frac{(cp\theta)^5}{5!} - \cdots\right)\beta\frac{c\vec{\alpha}\cdot\vec{p}}{cp}$$

$$= \cos(cp\theta)I - \sin(cp\theta)\beta\frac{c\vec{\alpha}\cdot\vec{p}}{cp} = U^\dagger. \tag{3.700}$$

Now, if we take

$$S = -i\theta\beta c\vec{\alpha}\cdot\vec{p}, \tag{3.701}$$

then

$$S^\dagger = S, \quad U^\dagger = e^{-iS}, \quad U = e^{iS}. \tag{3.702}$$

Thus, we can write (3.695) as

$$e^{iS}(mc^2\beta + c\vec{\alpha}\cdot\vec{p})e^{-iS} = \sqrt{m^2c^4 + c^2p^2}\,\beta,$$

$$\text{with } S = -i\theta\beta c\vec{\alpha}\cdot\vec{p}, \quad \text{and } \tan(2cp\theta) = \frac{p}{mc}. \tag{3.703}$$

Since the components of $\vec{p}$ commute with each other, we can replace $\vec{p}$ in the above equation by $\vec{\hat{p}}$ so that we have

$$e^{i\hat{S}}(mc^2\beta + c\vec{\alpha}\cdot\vec{\hat{p}})e^{-i\hat{S}} = \sqrt{m^2c^4 + c^2\vec{\hat{p}}^2}\,\beta,$$

$$\text{with } \hat{S} = -i\theta\beta c\vec{\alpha}\cdot\vec{\hat{p}}, \quad \text{and } 2c|\vec{\hat{p}}|\theta = \tan^{-1}\left(|\vec{\hat{p}}|/mc\right). \tag{3.704}$$

Now, it is clear from the above discussion that under the change of representation $|\Psi\rangle \longrightarrow |\Psi'\rangle = e^{i\hat{S}}|\Psi\rangle = e^{\theta\beta c\vec{\alpha}\cdot\vec{\hat{p}}}|\Psi\rangle$ the free-particle Dirac Hamiltonian $\hat{H}_D = mc^2\beta + c\vec{\alpha}\cdot\vec{\hat{p}}$ is reduced to its diagonal form $\hat{H}' = e^{i\hat{S}}\hat{H}e^{-i\hat{S}} = \sqrt{m^2c^4 + c^2p^2}\,\beta$. Note that in this case of the Foldy–Wouthuysen transformation for the free particle, $e^{i\hat{S}}$ is time-independent, and hence the resulting transformation of the Hamiltonian (3.689) becomes $\hat{H}' = e^{i\hat{S}}\hat{H}e^{-i\hat{S}}$. Now, note that we can write $U^\dagger$ in (3.694) as

$$U^\dagger = (H + E(p)\beta)\beta. \tag{3.705}$$

Thus, the eigenvectors of the free-particle Dirac Hamiltonian matrix H are given simply by the columns of $(H + E(p)\beta)\beta$, or $(H + E(p)\beta)$ apart from an overall minus sign for the negative-energy eigenvectors (see Ramakrishnan [155]). Further, we note that in the free-particle case, the Foldy–Wouthuysen transformation matrix is given simply by $U = \beta(H + E(p)\beta)$ with its row vectors as the Hermitian conjugates of the eigenvectors of H apart from an overall minus sign for the negative-energy eigenvectors (see Tekumalla [178]), and Tekumalla and Santhanam [179] for related work).

A Review of Quantum Mechanics

To develop the Foldy–Wouthuysen transformation technique for the case of the Dirac particle in the presence of an electromagnetic field, we shall proceed as follows. First let us have a closer look at the free-particle case. Using the identities (3.138) and (3.690), in general, for $\hat{H}'$ in (3.689), we have

$$\begin{aligned}\hat{H}' &= e^{i\hat{S}}\hat{H}e^{-i\hat{S}} - i\hbar\left(e^{i\hat{S}}\frac{\partial e^{-i\hat{S}}}{\partial t}\right) \\ &= \hat{H} + i\left[\hat{S},\hat{H}\right] + \frac{i^2}{2!}\left[\hat{S},\left[\hat{S},\hat{H}\right]\right] + \frac{i^3}{3!}\left[\hat{S},\left[\hat{S},\left[\hat{S},\hat{H}\right]\right]\right] \\ &\quad + \frac{i^4}{4!}\left[\hat{S},\left[\hat{S},\left[\hat{S},\left[\hat{S},\hat{H}\right]\right]\right]\right] + \cdots \\ &\quad - i\hbar\left(-i\frac{\partial\hat{S}}{\partial t} - \frac{i^2}{2!}\left[\hat{S},\frac{\partial\hat{S}}{\partial t}\right] - \frac{i^3}{3!}\left[\hat{S},\left[\hat{S},\frac{\partial\hat{S}}{\partial t}\right]\right]\right. \\ &\quad \left. - \frac{i^4}{4!}\left[\hat{S},\left[\hat{S},\left[\hat{S},\frac{\partial\hat{S}}{\partial t}\right]\right]\right] + \cdots\right) \\ &= \hat{H} - \hbar\frac{\partial\hat{S}}{\partial t} + i\left[\hat{S},\hat{H} - \frac{\hbar}{2}\frac{\partial\hat{S}}{\partial t}\right] + \frac{i^2}{2!}\left[\hat{S},\left[\hat{S},\hat{H} - \frac{\hbar}{3}\frac{\partial\hat{S}}{\partial t}\right]\right] \\ &\quad + \frac{i^3}{3!}\left[\hat{S},\left[\hat{S},\left[\hat{S},\hat{H} - \frac{\hbar}{4}\frac{\partial\hat{S}}{\partial t}\right]\right]\right] + \cdots.\end{aligned} \quad (3.706)$$

In the case of the free particle, $\hat{H} = mc^2\beta + c\vec{\alpha}\cdot\vec{p}$ and $\hat{S} = -i\theta\beta c\vec{\alpha}\cdot\vec{p}$ with $2c|\vec{p}|\theta = \tan^{-1}\left(|\vec{p}|/mc\right)$. Noting that $\hat{S}$ is independent of time, we have

$$\begin{aligned}\hat{H}' &= mc^2\beta + c\vec{\alpha}\cdot\vec{p} - 2\theta mc^3\vec{\alpha}\cdot\vec{p} + 2\theta c^2\vec{p}^2\beta \\ &\quad - 2\theta^2 mc^4\vec{p}^2\beta - 2\theta^2 c^3\vec{p}^2\vec{\alpha}\cdot\vec{p} + \cdots.\end{aligned} \quad (3.707)$$

Let us consider the nonrelativistic case when $|\vec{p}|/mc \ll 1$. Then, $\theta \approx 1/2mc^2$ and we get

$$\begin{aligned}\hat{H}' &\approx mc^2\beta + c\vec{\alpha}\cdot\vec{p} - c\vec{\alpha}\cdot\vec{p} + \frac{\vec{p}^2}{m}\beta - \frac{\vec{p}^2}{2m}\beta - \frac{\vec{p}^2}{2m^2c^2}c\vec{\alpha}\cdot\vec{p} \\ &= \left(mc^2 + \frac{\vec{p}^2}{2m}\right)\beta - \frac{1}{2}\left(\frac{\vec{p}}{mc}\right)^2 c\vec{\alpha}\cdot\vec{p}.\end{aligned} \quad (3.708)$$

In this expression for $\hat{H}'$, the first two terms are the first two terms of the nonrelativistic approximation of $\sqrt{m^2c^4 + c^2\vec{p}^2}\,\beta$, representing the rest energy and the nonrelativistic kinetic energy, respectively. This term, like $mc^2\beta$ in $\hat{H}$, containing nonzero 2×2 matrix blocks only along the diagonal, is known as an even term that does not couple the upper and lower pairs of the Dirac spinor, or the larger and smaller components of the Dirac spinor in the nonrelativistic case. The third, last, term of $\widehat{H}'$ is like the second, last, term of $\widehat{H}$, an odd term that contains nonzero 2×2 matrix blocks only along the off-diagonal and hence couples the upper and lower pairs of the Dirac spinor. It should be noted that the odd term of $\widehat{H}'$ is much weaker, being of second order in $|\vec{p}|/mc$ ($\ll 1$), compared to the odd term of $\widehat{H}$ which is of zeroth order in $|\vec{p}|/mc$.

It is clear from the derivation of $\widehat{H}'$ from $\widehat{H}$ that the weakening of the odd term in $\widehat{H}'$ is the result of the anticommutation relation between β and the odd term $c\vec{\alpha}\cdot\vec{p}$. In general, one can see that β anticommutes with any odd term:

$$\begin{pmatrix} I & O \\ O & -I \end{pmatrix} \begin{pmatrix} O & A \\ B & O \end{pmatrix} = - \begin{pmatrix} O & A \\ B & O \end{pmatrix} \begin{pmatrix} I & O \\ O & -I \end{pmatrix}, \qquad (3.709)$$

where A and B are arbitrary nonzero 2×2 matrices, i.e., if $\widehat{\mathcal{O}}$ is an odd term in a Dirac Hamiltonian, with nonzero 2×2 block matrices only along the off-diagonal, then

$$\beta \widehat{\mathcal{O}} = -\widehat{\mathcal{O}} \beta. \qquad (3.710)$$

Any even term having nonzero 2×2 block matrices only along the diagonal commutes with β. This relation between β and the odd and even terms is the basis for the systematic approximation of the Dirac equation using the Foldy–Wouthuysen transformation.

Let us now consider the Dirac equation for a particle of mass m and charge q in an electromagnetic field:

$$i\hbar \frac{\partial |\underline{\Psi}(t)\rangle}{\partial t} = \widehat{H} |\underline{\Psi}(t)\rangle,$$

$$\widehat{H} = mc^2 \beta + \widehat{\mathcal{E}} + \widehat{\mathcal{O}},$$

$$\widehat{\mathcal{E}} = q\phi I = q\phi \begin{pmatrix} I & O \\ O & I \end{pmatrix},$$

$$\widehat{\mathcal{O}} = c\vec{\alpha} \cdot \vec{\pi} = c \begin{pmatrix} O & \vec{\sigma}\cdot\vec{\pi} \\ \vec{\sigma}\cdot\vec{\pi} & O \end{pmatrix}, \qquad (3.711)$$

where $\vec{\pi} = \vec{p} - q\vec{A}$, and $\widehat{\mathcal{E}}$ and $\widehat{\mathcal{O}}$ are even and odd terms, respectively. Let us now make the transformation

$$\left|\underline{\Psi}^{(1)}(t)\right\rangle = e^{i\widehat{S}_1} |\underline{\Psi}(t)\rangle, \qquad (3.712)$$

with

$$\widehat{S}_1 = -\frac{i}{2mc^2} \beta \widehat{\mathcal{O}}. \qquad (3.713)$$

This will transform (3.711) into

$$i\hbar \frac{\partial}{\partial t} \left|\underline{\Psi}^{(1)}(t)\right\rangle = \widehat{H}^{(1)} \left|\underline{\Psi}^{(1)}(t)\right\rangle, \qquad (3.714)$$

where

$$\widehat{H}^{(1)} = e^{i\widehat{S}_1}\widehat{H}e^{-i\widehat{S}_1} - i\hbar\left(e^{i\widehat{S}_1}\frac{\partial e^{-i\widehat{S}_1}}{\partial t}\right). \tag{3.715}$$

It is straightforward to calculate $\widehat{H}^{(1)}$ using (3.706), (3.713), and the relations $\beta\widehat{\mathcal{O}} = -\widehat{\mathcal{O}}\beta$ and $\beta\widehat{\mathcal{E}} = \widehat{\mathcal{E}}\beta$. The result is independent of the particular forms of $\widehat{\mathcal{E}}$ and $\widehat{\mathcal{O}}$. The result is

$$\widehat{H}^{(1)} \approx mc^2\beta + \widehat{\mathcal{E}}^{(1)} + \widehat{\mathcal{O}}^{(1)}, \tag{3.716}$$

where

$$\widehat{\mathcal{E}}^{(1)} = \widehat{\mathcal{E}} + \frac{1}{2mc^2}\beta\widehat{\mathcal{O}}^2 - \frac{1}{8m^2c^4}\left[\widehat{\mathcal{O}},\left(\left[\widehat{\mathcal{O}},\widehat{\mathcal{E}}\right] + i\hbar\frac{\partial\widehat{\mathcal{O}}}{\partial t}\right)\right] - \frac{1}{8m^3c^6}\beta\widehat{\mathcal{O}}^4,$$

$$\widehat{\mathcal{O}}^{(1)} = \frac{1}{2mc^2}\beta\left(\left[\widehat{\mathcal{O}},\widehat{\mathcal{E}}\right] + i\hbar\frac{\partial\widehat{\mathcal{O}}}{\partial t}\right) - \frac{1}{3m^2c^4}\widehat{\mathcal{O}}^3. \tag{3.717}$$

The even and odd terms of $\widehat{H}^{(1)}$, namely $\widehat{\mathcal{E}}^{(1)}$ and $\widehat{\mathcal{O}}^{(1)}$, obey the same relations with β like $\widehat{\mathcal{E}}$ and $\widehat{\mathcal{O}}$ of $\widehat{H}$, i.e.,

$$\beta\widehat{\mathcal{E}}^{(1)} = \widehat{\mathcal{E}}^{(1)}\beta, \qquad \beta\widehat{\mathcal{O}}^{(1)} = -\widehat{\mathcal{O}}^{(1)}\beta. \tag{3.718}$$

Note that we can regard $1/mc^2$ as an expansion parameter for the expression (3.716 and 3.717) for $\widehat{H}^{(1)}$. It is seen that while the odd term $\widehat{\mathcal{O}}$ in $\widehat{H}$ is of order zero with respect to $1/mc^2$ the odd term in $\widehat{H}^{(1)}$, $\widehat{\mathcal{O}}^{(1)}$, contains only terms of order $1/mc^2$ and higher powers of $1/mc^2$.

Let us now apply another transformation with the same prescription:

$$\left|\Psi^{(2)}(t)\right\rangle = e^{i\widehat{S}_2}\left|\Psi^{(1)}(t)\right\rangle, \tag{3.719}$$

with

$$\widehat{S}_2 = -\frac{i}{2mc^2}\beta\widehat{\mathcal{O}}^{(1)}$$

$$= -\frac{i}{2mc^2}\beta\left\{\frac{1}{2mc^2}\beta\left(\left[\widehat{\mathcal{O}},\widehat{\mathcal{E}}\right] + i\hbar\frac{\partial\widehat{\mathcal{O}}}{\partial t}\right) - \frac{1}{3m^2c^4}\widehat{\mathcal{O}}^3\right\}. \tag{3.720}$$

After this transformation, we have

$$i\hbar\frac{\partial}{\partial t}\left|\Psi^{(2)}(t)\right\rangle = \widehat{H}^{(2)}\left|\Psi^{(2)}(t)\right\rangle, \tag{3.721}$$

with

$$\widehat{H}^{(2)} = mc^2\beta + \widehat{\mathscr{E}}^{(2)} + \widehat{\mathscr{O}}^{(2)}, \tag{3.722}$$

where

$$\widehat{\mathscr{E}}^{(2)} \approx \widehat{\mathscr{E}}^{(1)},$$

$$\widehat{\mathscr{O}}^{(2)} \approx \frac{1}{2mc^2}\beta\left(\left[\widehat{\mathscr{O}}^{(1)}, \widehat{\mathscr{E}}^{(1)}\right] + i\hbar\frac{\partial \widehat{\mathscr{O}}^{(1)}}{\partial t}\right). \tag{3.723}$$

Note that $\widehat{\mathscr{O}}^{(2)}$ is of order $(1/mc^2)^2$.

A third transformation

$$\left|\Psi^{(3)}(t)\right\rangle = e^{i\widehat{S}_3}\left|\Psi^{(2)}(t)\right\rangle, \tag{3.724}$$

with

$$\widehat{S}_3 = -\frac{i}{2mc^2}\beta\widehat{\mathscr{O}}^{(2)}$$

$$= -\frac{i}{2mc^2}\beta\left\{\frac{1}{2mc^2}\beta\left(\left[\widehat{\mathscr{O}}^{(1)}, \widehat{\mathscr{E}}^{(1)}\right] + i\hbar\frac{\partial \widehat{\mathscr{O}}^{(1)}}{\partial t}\right)\right\}, \tag{3.725}$$

takes us to

$$i\hbar\frac{\partial}{\partial t}\left|\Psi^{(3)}(t)\right\rangle = \widehat{H}^{(3)}\left|\Psi^{(3)}(t)\right\rangle, \tag{3.726}$$

with

$$\widehat{H}^{(3)} = mc^2\beta + \widehat{\mathscr{E}}^{(3)} + \widehat{\mathscr{O}}^{(3)}, \tag{3.727}$$

where

$$\widehat{\mathscr{E}}^{(3)} \approx \widehat{\mathscr{E}}^{(2)} \approx \widehat{\mathscr{E}}^{(1)}$$

$$\widehat{\mathscr{O}}^{(3)} \approx \frac{1}{2mc^2}\beta\left(\left[\widehat{\mathscr{O}}^{(2)}, \widehat{\mathscr{E}}^{(2)}\right] + i\hbar\frac{\partial \widehat{\mathscr{O}}^{(2)}}{\partial t}\right). \tag{3.728}$$

Note that $\widehat{\mathscr{O}}^{(3)}$ is of order $(1/mc^2)^3$. Further, it may be noted that starting with the second transformation successive even and odd terms can be obtained recursively using the rule

$$\widehat{\mathscr{E}}^{(j)} = \widehat{\mathscr{E}}^{(1)}\left(\widehat{\mathscr{E}} \longrightarrow \widehat{\mathscr{E}}^{(j-1)}, \widehat{\mathscr{O}} \longrightarrow \widehat{\mathscr{O}}^{(j-1)}\right),$$

$$\widehat{\mathscr{O}}^{(j)} = \widehat{\mathscr{O}}^{(1)}\left(\widehat{\mathscr{E}} \longrightarrow \widehat{\mathscr{E}}^{(j-1)}, \widehat{\mathscr{O}} \longrightarrow \widehat{\mathscr{O}}^{(j-1)}\right), \quad j \geq 2, \tag{3.729}$$

and retaining only the relevant terms of required order at each step.

Stopping with the third transformation and neglecting $\widehat{\mathscr{O}}^{(3)}$, we have

$$\widehat{H}^{(3)} = mc^2\beta + \widehat{\mathscr{E}} + \frac{1}{2mc^2}\beta\widehat{\mathscr{O}}^2$$
$$- \frac{1}{8m^2c^4}\left[\widehat{\mathscr{O}},\left(\left[\widehat{\mathscr{O}},\widehat{\mathscr{E}}\right] + i\hbar\frac{\partial\widehat{\mathscr{O}}}{\partial t}\right)\right] - \frac{1}{8m^3c^6}\beta\widehat{\mathscr{O}}^4. \quad (3.730)$$

Let us call $\left|\Psi^{(3)}(t)\right\rangle$ as $|\Psi_{\text{FW}}(t)\rangle$ and $\widehat{H}^{(3)}$ as $\widehat{H}_{\text{FW}}$. We can calculate $\widehat{H}_{\text{FW}}$ by substituting $\widehat{\mathscr{E}} = q\phi$ and $\widehat{\mathscr{O}} = c\vec{\alpha}\cdot\vec{\pi}$ in (3.730). The resulting Dirac equation in the Foldy–Wouthuysen representation is

$$i\hbar\frac{\partial}{\partial t}|\Psi_{\text{FW}}(t)\rangle = \widehat{H}_{\text{FW}}|\Psi_{\text{FW}}(t)\rangle, \quad (3.731)$$

with

$$\widehat{H}_{\text{FW}} \approx \left(mc^2 + \frac{\vec{\pi}^2}{2m} - \frac{\vec{\pi}^4}{8m^3c^2}\right)\beta + q\phi - \frac{q\hbar}{2mc}\beta\vec{\Sigma}\cdot\vec{B}$$
$$- \frac{iq\hbar^2}{8m^2c^2}\vec{\Sigma}\cdot\left(\vec{\nabla}\times\vec{E}\right) - \frac{q\hbar}{4m^2c^2}\vec{\Sigma}\cdot\left(\vec{E}\times\vec{p}\right)$$
$$- \frac{q\hbar^2}{8m^2c^2}\vec{\nabla}\cdot\vec{E}. \quad (3.732)$$

In this Foldy–Wouthuysen representation of the Dirac Hamiltonian, $\widehat{H}_{\text{FW}}$, the three terms in the first parenthesis, corresponding to the rest energy, nonrelativistic kinetic energy, and the lowest order relativistic correction to the kinetic energy, result from the binomial expansion of $\sqrt{m^2c^4 + c^2\vec{\pi}^2}$. The second and the third terms are the electrostatic energy and the magnetic dipole energy, respectively. The next two terms, taken together (for hermiticity), contain the spin-orbit interaction. The last term, called the Darwin term, is attributed to the zitterbewegung, resulting in the particle being under the influence of a somewhat smeared-out electrical potential. Thus, it is clear that the Foldy–Wouthuysen transformation technique expands the Dirac Hamiltonian as a power series in the parameter $1/mc^2$, leading to a systematic approximation procedure for studying the deviations from the nonrelativistic situation. Further, with $\widehat{H}_{\text{FW}}$ having nonzero 2×2 blocks only along its diagonal, we have actually two uncoupled equations for the two two-component spinors, one for positive energy and the other for negative energy. In (3.731), the upper two-component spinor corresponds to positive energy and the lower two-component spinor corresponds to negative energy. This can be seen by noting that, in the field-free case, the upper two-component spinor evolves with the Hamiltonian $\sqrt{m^2c^4 + c^2\vec{p}^2}$ and the lower two-component spinor evolves with the Hamiltonian $-\sqrt{m^2c^4 + c^2\vec{p}^2}$. Thus, we can use the Foldy–Wouthuysen representation to develop a two-component spinor theory for positive-energy spin-$\frac{1}{2}$ particles.

We have found that for the Dirac particle, the position operator, $\vec{r}$, undergoes a complicated motion, zitterbewegung, as found in the Heisenberg picture. Following the discussion on change of representation (see (3.130)), in the Foldy–Wouthuysen representation, the corresponding position operator would be

$$\vec{r}_{FW} = \hat{U}_{FW} \vec{r} \hat{U}_{FW}^\dagger, \qquad (3.733)$$

where $\hat{U}_{FW}$ is the transformation relating the Foldy–Wouthuysen and the Dirac representations as

$$|\Psi_{FW}\rangle = \hat{U}_{FW} |\Psi_D\rangle. \qquad (3.734)$$

with

$$\hat{U}_{FW} = e^{i\hat{S}_3} e^{i\hat{S}_2} e^{i\hat{S}_1}, \qquad (3.735)$$

and $\hat{S}_1$, $\hat{S}_2$, and $\hat{S}_3$, given, respectively, by (3.713), (3.720), and (3.725). Let us consider the case of the free particle with the Hamiltonian

$$\hat{H}_D = mc^2 \beta + c\vec{\alpha} \cdot \vec{p}. \qquad (3.736)$$

We know that

$$\hat{U}_{FW} \hat{H}_D \hat{U}_{FW}^\dagger = \hat{H}_{FW} = \sqrt{m^2 c^4 + c^2 \vec{p}^2}\, \beta, \qquad (3.737)$$

where

$$\hat{U}_{FW} = e^{\theta c \beta \vec{\alpha} \cdot \vec{p}}, \quad \text{with} \quad \tan\left(2c\theta \left|\vec{p}\right|\right) = \frac{\left|\vec{p}\right|}{mc}. \qquad (3.738)$$

Since unitary transformations do not change the commutation relations, in the Foldy–Wouthuysen representation the position operator $\hat{r}_{FW}$ would exhibit the zitterbewegung phenomenon. However, if we take $\vec{r}$ itself as the position operator in the Foldy–Wouthuysen representation, then we get, with

$$\vec{r}(t) = e^{\frac{it}{\hbar} \hat{H}_{FW}} \vec{r} e^{-\frac{it}{\hbar} \hat{H}_{FW}}, \qquad (3.739)$$

in the Heisenberg picture,

$$\frac{d\vec{r}}{dt} = \frac{i}{\hbar} \left[\hat{H}_{FW}, \vec{r}(t) \right] = \frac{i}{\hbar} e^{\frac{it}{\hbar} \hat{H}_{FW}} \left[\hat{H}_{FW}, \vec{r} \right] e^{-\frac{it}{\hbar} \hat{H}_{FW}}. \qquad (3.740)$$

For the free particle, we have

$$\frac{d\vec{r}}{dt} = \frac{i}{\hbar} e^{\frac{it}{\hbar} \hat{H}_{FW}} \left(\frac{c^2 \vec{p}}{\sqrt{m^2 c^4 + c^2 \vec{p}^2}} \beta \right) e^{-\frac{it}{\hbar} \hat{H}_{FW}} = \frac{c^2 \vec{p}}{\hat{H}_{FW}}. \qquad (3.741)$$

We shall take the canonical momentum operator in the Foldy–Wouthuysen representation to be $\vec{p} = -i\hbar \vec{\nabla}$. For a free particle, there is no difference between the canonical momentum and the kinetic momentum. Then, we can interpret $c^2 \vec{p}/\hat{H}_{FW}$ as the operator for the velocity of the particle since $c^2 \vec{p}/E(\vec{p})$ is the velocity of a relativistic particle. Thus, there is no zitterbewegung for the operator $\vec{r}$ in the Foldy–Wouthuysen representation. Following Foldy and Wouthuysen [53], we can call $\vec{r}$ as the mean position operator in the Foldy–Wouthuysen representation. It would correspond to the operator

$$\vec{r}_{NW} = \widehat{U}_{FW}^{\dagger} \vec{r} \widehat{U}_{FW}$$

$$= \vec{r} + \frac{i\beta}{2E(\vec{p})} \left[c\vec{\alpha} - \frac{c^3}{E(\vec{p})\left(E(\vec{p}) + mc^2\right)} (\vec{\alpha} \cdot \vec{p})\vec{p} \right]$$

$$- \frac{c^2 \hbar}{2E(\vec{p})\left(E(\vec{p}) + mc^2\right)} \vec{\Sigma} \times \vec{p} \qquad (3.742)$$

representing the mean position in the Dirac representation which would not have zitterbewegung. The notation $\vec{r}_{NW}$ is because of the fact that precisely this operator was introduced, prior to the Foldy–Wouthuysen work, by Newton and Wigner [139] as the position operator for the Dirac particle from general considerations of localizability. We have taken the canonical momentum operator in the Foldy–Wouthuysen representation to be $\vec{p}$. In the case of the free particle, the corresponding operator for the canonical momentum in the Dirac representation is seen to be the same, as we have

$$\widehat{U}_{FW}^{\dagger} \vec{p} \widehat{U}_{FW} = \vec{p}, \qquad (3.743)$$

since $\widehat{U}_{FW}$ commutes with $\vec{p}$. Thus, for a free particle, we can take $\vec{r}$ and $\vec{p} = -i\hbar \vec{\nabla}$ as the operators corresponding to the mean position and momentum in the Foldy–Wouthuysen representation.

Now, it may be noted that for the free particle

$$\left[\widehat{H}_{FW}, \vec{L}\right] = \left[\sqrt{m^2 c^4 + c^2 \vec{p}^{\,2}} \beta, \vec{L}\right] = 0,$$

$$\left[\widehat{H}_{FW}, \vec{\Sigma}\right] = \left[\sqrt{m^2 c^4 + c^2 \vec{p}^{\,2}} \beta, \vec{\Sigma}\right] = 0, \qquad (3.744)$$

where $\vec{L} = \vec{r} \times \vec{p}$. This means that in the Foldy–Wouthuysen representation the orbital angular momentum $\vec{L}$ and the spin $\vec{S} = \frac{\hbar}{2}\vec{\Sigma}$ are conserved separately, unlike in the case of the Dirac representation, in which only their sum is conserved. As the position operator, we have to regard $\vec{L}$ and $\vec{S}$ as mean orbital angular momentum and mean spin operators in the Foldy–Wouthuysen representation. It should be that the experimentally measured values of these observables (position, orbital angular momentum, spin, etc.) correspond to these mean operators in the Foldy–Wouthuysen representation.

3.4.3.2 Foldy–Wouthuysen Representation of the Feshbach–Villars form of the Klein–Gordon Equation

We have found that the Feshbach–Villars form of the Klein–Gordon equation (3.589) is given by

$$i\hbar \frac{\partial |\Psi(t)\rangle}{\partial t} = \left[mc^2 \sigma_z + \frac{\vec{\pi}^2}{2m}(\sigma_z + i\sigma_y) + q\phi \right] |\Psi(t)\rangle, \quad (3.745)$$

where $|\Psi\rangle$ is a two-component state vector. This equation is in the Schrödinger form with the Hamiltonian, the Feshbach–Villars Hamiltonian, given by

$$\widehat{H}_{\text{FV}} = mc^2 \sigma_z + \frac{\vec{\pi}^2}{2m}(\sigma_z + i\sigma_y) + q\phi. \quad (3.746)$$

Let us note that $\widehat{H}_{\text{FV}}$ has the same structure as the Dirac Hamiltonian:

$$\widehat{H}_{\text{FV}} = mc^2 \sigma_z + \widehat{\mathscr{E}} + \widehat{\mathscr{O}}, \quad (3.747)$$

where σ_z is the two-dimensional β, and

$$\widehat{\mathscr{E}} = \frac{\vec{\pi}^2}{2m}\sigma_z + q\phi, \qquad \widehat{\mathscr{O}} = \frac{\vec{\pi}^2}{2m}i\sigma_y, \quad (3.748)$$

are, respectively, even and odd operators such that

$$\sigma_z \widehat{\mathscr{E}} = \widehat{\mathscr{E}} \sigma_z, \qquad \sigma_z \widehat{\mathscr{O}} = -\widehat{\mathscr{O}} \sigma_z. \quad (3.749)$$

This suggests exploring the application of the Foldy–Wouthuysen transformation. Let us consider the case of static fields, *i.e.*, the potentials ϕ and $\vec{A}$ are time-independent. When we make the first Foldy–Wouthuysen transformation

$$\widehat{H}^{(1)} = e^{i\widehat{S}_1} \widehat{H}_{\text{FV}} e^{-i\widehat{S}_1}, \qquad \text{with } \widehat{S}_1 = -\frac{i}{2mc^2} \sigma_z \widehat{\mathscr{O}}, \quad (3.750)$$

we get

$$\widehat{H}^{(1)} \approx mc^2 \sigma_z + \widehat{\mathscr{E}}^{(1)} + \widehat{\mathscr{O}}^{(1)}, \quad (3.751)$$

where

$$\widehat{\mathscr{E}}^{(1)} = \widehat{\mathscr{E}} + \frac{1}{2mc^2} \sigma_z \widehat{\mathscr{O}}^2 - \frac{1}{8m^2c^4}\left[\widehat{\mathscr{O}}, \left[\widehat{\mathscr{O}}, \widehat{\mathscr{E}}\right]\right] - \frac{1}{8m^3c^6} \sigma_z \widehat{\mathscr{O}}^4,$$

$$\widehat{\mathscr{O}}^{(1)} = \frac{1}{2mc^2} \sigma_z \left[\widehat{\mathscr{O}}, \widehat{\mathscr{E}}\right] - \frac{1}{3m^2c^4} \widehat{\mathscr{O}}^3. \quad (3.752)$$

The second transformation

$$\widehat{H}^{(2)} = e^{i\widehat{S}_2} \widehat{H}^{(1)} e^{-i\widehat{S}_2}, \qquad \text{with } \widehat{S}_2 = -\frac{i}{2mc^2} \sigma_z \widehat{\mathscr{O}}^{(1)}, \quad (3.753)$$

A Review of Quantum Mechanics

leads to

$$\widehat{H}^{(2)} = mc^2 \sigma_z + \widehat{\mathcal{E}}^{(2)} + \widehat{\mathcal{O}}^{(2)}, \tag{3.754}$$

where

$$\widehat{\mathcal{E}}^{(2)} \approx \widehat{\mathcal{E}}^{(1)},$$
$$\widehat{\mathcal{O}}^{(2)} \approx \frac{1}{2mc^2} \sigma_z \left[\widehat{\mathcal{O}}^{(1)}, \widehat{\mathcal{E}}^{(1)} \right]. \tag{3.755}$$

Stopping with the third transformation

$$\widehat{H}^{(3)} = e^{i\widehat{S}_3} \widehat{H}^{(2)} e^{-i\widehat{S}_3}, \quad \text{with } \widehat{S}_3 = -\frac{i}{2mc^2} \sigma_z \widehat{\mathcal{O}}^{(2)}, \tag{3.756}$$

we arrive at

$$\widehat{H}^{(3)} = mc^2 \sigma_z + \widehat{\mathcal{E}}^{(3)} + \widehat{\mathcal{O}}^{(3)}, \tag{3.757}$$

where

$$\widehat{\mathcal{E}}^{(3)} \approx \widehat{\mathcal{E}}^{(2)} \approx \widehat{\mathcal{E}}^{(1)},$$
$$\widehat{\mathcal{O}}^{(3)} \approx \frac{1}{2mc^2} \sigma_z \left[\widehat{\mathcal{O}}^{(2)}, \widehat{\mathcal{E}}^{(2)} \right]. \tag{3.758}$$

Let us call $\widehat{H}^{(3)}$, after dropping $\widehat{\mathcal{O}}^{(3)}$ from it, as $\widehat{H}_{\text{KGFW}}$. Then, we have

$$\widehat{H}_{\text{KGFW}} = mc^2 \sigma_z + \widehat{\mathcal{E}} + \frac{1}{2mc^2} \sigma_z \widehat{\mathcal{O}}^2$$
$$- \frac{1}{8m^2 c^4} \left[\widehat{\mathcal{O}}, \left[\widehat{\mathcal{O}}, \widehat{\mathcal{E}} \right] \right] - \frac{1}{8m^3 c^6} \sigma_z \widehat{\mathcal{O}}^4. \tag{3.759}$$

Substituting $\widehat{\mathcal{E}} = \left(\vec{\pi}^2 / 2m \right) \sigma_z + q\phi$ and $\widehat{\mathcal{O}} = \left(\vec{\pi}^2 / 2m \right) i\sigma_y$, we have

$$\widehat{H}_{\text{KGFW}} \approx \left(mc^2 + \frac{\vec{\pi}^2}{2m} - \frac{\vec{\pi}^4}{8m^3 c^2} \right) \sigma_z$$
$$+ q\phi + \frac{1}{32m^4 c^4} \left[\vec{\pi}^2, \left[\vec{\pi}^2, q\phi \right] \right]. \tag{3.760}$$

For the free particle, we would have

$$\widehat{H}_{\text{KGFW}} \approx \left(mc^2 + \frac{\vec{p}^2}{2m} - \frac{\vec{p}^4}{8m^3 c^2} \right) \sigma_z, \tag{3.761}$$

resulting from the binomial expansion of $\sqrt{m^2 c^4 + c^2 \vec{p}^2} \sigma_z$. In this Foldy–Wouthuysen representation of the Feshbach–Villars equation, the two components of the state vector have been decoupled, with the upper component representing the positive-energy state and the lower component representing the negative-energy state. This is clear from the fact that in the free-particle case the Hamiltonian $\widehat{H}_{\text{KGFW}}$ in (3.761) is diagonal such that the upper component of the state vector evolves with the Hamiltonian $\sqrt{m^2c^4 + c^2\vec{\hat{p}}^2}$ and the lower component of the state vector evolves with the Hamiltonian $-\sqrt{m^2c^4 + c^2\vec{\hat{p}}^2}$. Note that the operator $\vec{r}$ has the equation of motion,

$$\frac{d\vec{r}}{dt} = \frac{c^2\vec{\hat{p}}}{\widehat{H}_{\text{KGFW}}}, \qquad (3.762)$$

in the Heisenberg picture of the Foldy–Wouthuysen representation, as expected for a free relativistic particle. Thus, the position operator can be taken as $\vec{r}$, in the Foldy–Wouthuysen representation of the Feshbach–Villars formalism of the Klein-Gordon equation, corresponding to experimental observation. Similarly, the operators corresponding to linear momentum and angular momentum are given, respectively, by $\vec{\hat{p}}$ and $\vec{\hat{L}}$.

3.5 APPENDIX: THE MAGNUS FORMULA FOR THE EXPONENTIAL SOLUTION OF A LINEAR DIFFERENTIAL EQUATION

Suppose we have to solve the differential equation

$$\frac{\partial \psi(t)}{\partial t} = \mathscr{L}(t)\psi(t), \qquad (3.763)$$

to get $\psi(t)$ for any $t > t_0$, where $\mathscr{L}(t)$ is a linear operator and $\psi(t_0)$ is given. For an infinitesimal Δt, we can write

$$\psi(t_0 + \Delta t) = e^{\Delta t \mathscr{L}(t_0)} \psi(t_0). \qquad (3.764)$$

Iterating this solution, we get

$$\psi(t_0 + 2\Delta t) = e^{\Delta t \mathscr{L}(t_0 + \Delta t)} e^{\Delta t \mathscr{L}(t_0)} \psi(t_0),$$
$$\psi(t_0 + 3\Delta t) = e^{\Delta t \mathscr{L}(t_0 + 2\Delta t)} e^{\Delta t \mathscr{L}(t_0 + \Delta t)} e^{\Delta t \mathscr{L}(t_0)} \psi(t_0),$$
$$\text{and so on.} \qquad (3.765)$$

Let $t = t_0 + N\Delta t$. Then, we have

$$\psi(t) = \left(\Pi_{n=0}^{N-1} e^{\Delta t \mathscr{L}(t_0 + n\Delta t)}\right) \psi(t_0). \qquad (3.766)$$

Thus, $\psi(t)$ is given by computing the product in (3.766) using successively the BCH formula (3.445),

$$e^{\widehat{A}} e^{\widehat{B}} = e^{\widehat{A} + \widehat{B} + \frac{1}{2}[\widehat{A}, \widehat{B}] + \frac{1}{12}([\widehat{A}, [\widehat{A}, \widehat{B}]] + [[\widehat{A}, \widehat{B}], \widehat{B}]) + \cdots}, \qquad (3.767)$$

A Review of Quantum Mechanics

and considering the limit $\Delta t \longrightarrow 0$, $N \longrightarrow \infty$, such that $N\Delta t = t - t_0$. The resulting expression is the Magnus formula (see Magnus [130]):

$$\psi(t) = \mathbb{T}(t,t_0)\,\psi(t_0),$$

$$\mathbb{T}(t,t_0) = \exp\left\{\int_{t_0}^{t} dt_1\,\mathscr{L}(t_1) + \frac{1}{2}\int_{t_0}^{t} dt_2 \int_{t_0}^{t_2} dt_1\,[\mathscr{L}(t_2),\mathscr{L}(t_1)]\right.$$
$$+ \frac{1}{6}\int_{t_0}^{t} dt_3 \int_{t_0}^{t_3} dt_2 \int_{t_0}^{t_2} dt_1 \left([[\mathscr{L}(t_3),\mathscr{L}(t_2)],\mathscr{L}(t_1)]\right.$$
$$\left.\left. + [[\mathscr{L}(t_1),\mathscr{L}(t_2)],\mathscr{L}(t_3)]\right) + \cdots \right\}. \tag{3.768}$$

To see how the Magnus formula (3.768) is obtained, let us substitute the solution $\psi(t) = \mathbb{T}(t,t_0)\,\psi(t_0)$ in (3.763). Then, we find that the t-evolution, or t-transfer, operator $\mathbb{T}(t,t_0)$ has to satisfy the equation

$$\frac{\partial \mathbb{T}(t,t_0)}{\partial t} = \mathscr{L}(t)\mathbb{T}(t,t_0), \qquad \mathbb{T}(t_0,t_0) = I, \tag{3.769}$$

where I is the identity operator. Introducing an iteration parameter λ, let us write

$$\frac{\partial \mathbb{T}(t,t_0;\lambda)}{\partial t} = \lambda \mathscr{L}(t)\mathbb{T}(t,t_0;\lambda),$$
$$\mathbb{T}(t_0,t_0;\lambda) = I, \qquad \mathbb{T}(t,t_0;1) = \mathbb{T}(t,t_0). \tag{3.770}$$

Assume a solution of the form

$$\mathbb{T}(t,t_0;\lambda) = e^{\tau(t,t_0;\lambda)}, \tag{3.771}$$

with

$$\tau(t,t_0;\lambda) = \sum_{n=0}^{\infty} \lambda^n \Delta_n(t,t_0), \qquad \Delta_n(t_0,t_0) = 0, \text{ for all } n. \tag{3.772}$$

Now, using the identity (see Wilcox [188])

$$\frac{\partial}{\partial t} e^{\tau(t,t_0;\lambda)} = \left(\int_0^1 ds\, e^{s\tau(t,t_0;\lambda)} \frac{\partial}{\partial t}\tau(t,t_0;\lambda)\, e^{-s\tau(t,t_0;\lambda)}\right) e^{\tau(t,t_0;\lambda)}, \tag{3.773}$$

one has

$$\left(\int_0^1 ds\, e^{s\tau(t,t_0;\lambda)} \frac{\partial}{\partial t}\tau(t,t_0;\lambda)\, e^{-s\tau(t,t_0;\lambda)}\right) = \lambda \mathscr{L}(t). \tag{3.774}$$

Substituting the series expression for $\tau(t,t_0;\lambda)$, expanding the left-hand side using (3.138), integrating, and equating the coefficients of powers of λ on both sides, we get, recursively, the equations for $\Delta_1(t,t_0)$, $\Delta_2(t,t_0)$, etc. For the coefficients of λ, we have

$$\frac{\partial \Delta_1(t,t_0)}{\partial t} = \mathscr{L}(t), \qquad \Delta_1(t_0,t_0) = 0, \tag{3.775}$$

and hence
$$\Delta_1(t,t_0) = \int_{t_0}^{t} dt_1 \mathscr{L}(t_1). \tag{3.776}$$

For the coefficients of λ^2, we have
$$\frac{\partial \Delta_2(t,t_0)}{\partial t} + \frac{1}{2}\left[\Delta_1(t,t_0), \frac{\partial \Delta_1(t,t_0)}{\partial t}\right] = 0, \quad \Delta_1(t_0,t_0) = 0, \tag{3.777}$$

and hence
$$\Delta_2(t_0,t_0) = \frac{1}{2}\int_{t_0}^{t} dt_2 \int_{t_0}^{t_2} dt_1 \left[\mathscr{L}(t_2), \mathscr{L}(t_1)\right]. \tag{3.778}$$

Similarly, we get
$$\Delta_3(t,t_0) = \frac{1}{6}\int_{t_0}^{t} dt_3 \int_{t_0}^{t_3} dt_2 \int_{t_0}^{t_2} dt_1 \left([[\mathscr{L}(t_3), \mathscr{L}(t_2)], \mathscr{L}(t_1)]\right.$$
$$\left. + [[\mathscr{L}(t_1), \mathscr{L}(t_2)], \mathscr{L}(t_3)]\right). \tag{3.779}$$

Then, the Magnus formula (3.768) follows from (3.771) and (3.772) with $\lambda = 1$.

In the time-evolution problems of quantum mechanics, the linear operator $\mathscr{L}$ is the Hamiltonian of the system. As we shall see later, in the z-evolution problems of classical charged particle beam optics, $\mathscr{L}$ is : $-\mathscr{H}_o :_z$, the Poisson bracket operation, and in the z-evolution problems of quantum charged particle beam optics, $\mathscr{L}$ is the quantum beam optical Hamiltonian $\widehat{\mathscr{H}_o}$ in the case of the Schrödinger picture, and is $\frac{1}{i\hbar} : -\widehat{\mathscr{H}_o} :$, the commutator bracket operation, in the case of the Heisenberg picture. The Magnus formula is applicable to any form of the linear operator $\mathscr{L}$. For more details on the exponential solutions of linear differential equations, related operator techniques, and applications to physical problems, see, e.g., Wilcox [188], Bellman and Vasudevan [10], Dattoli, Reneiri, and Torre [31], and Blanes, Casas, Oteo, and Ros [14], and references therein.

4 An Introduction to Classical Charged Particle Beam Optics

4.1 INTRODUCTION: RELATIVISTIC CLASSICAL CHARGED PARTICLE BEAM OPTICS

Charged particle beam optics is the study of transport of charged particle beams through electromagnetic optical systems. Such systems, like different types of electromagnetic lenses used for focusing or defocusing the beams, bending magnets, etc., are the main components of charged particle beam devices—low-energy electron microscopes to high-energy particle accelerators. In this chapter, we shall consider the classical mechanics of a few examples of such optical elements with a view to demonstrate how we can understand, in later chapters, the quantum mechanics underlying their behaviors. We are not concerned about the technical aspects of these optical elements. We shall consider only the theoretical aspects of the perfect, or idealized, versions of these optical elements. We are considering the beam particles to be noninteracting and independent, and hence we are treating the beam propagation on the basis of single particle dynamics. For more details on any aspect of classical charged particle beam optics, see *e.g.*, Hawkes and Kasper [70, 71], Orloff [141], Groves [68], Lubk [129], Pozzi [150], Berz, Makino, and Wan [12], Conte and MacKay [27], Lee [127], Reiser [158], Rosenzweig [159], Seryi [168], Weidemann [187], Wolski [192], Chao, Mess, Tigner, and Zimmermann [20], and references therein.

Any charged particle beam device has an optic axis. The trajectory of the charged particle along this axis is the design, or reference, trajectory. This optic axis can be straight or curved depending on the purpose of the device. For example, an axially symmetric, or round, magnetic lens used in an electron microscope to focus the beam will have a straight optic axis, and a dipole magnet used in devices for bending the beam will have a curved optic axis. First, we shall consider optical elements with straight optic axis chosen to be along the z-direction. We shall always take the beam to be moving forward along the $+z$-direction. We have already derived, in (2.132), the z-evolution Hamiltonian for a particle moving relativistically in an electromagnetic field as

$$\mathscr{H} = -\frac{1}{c}\sqrt{(E - q\phi)^2 - m^2 c^4 - c^2 \pi_\perp^2} - qA_z, \qquad (4.1)$$

where $\pi_\perp^2 = \pi_x^2 + \pi_y^2$, m is the rest mass of the particle with charge q. We shall consider the electromagnetic fields of the optical elements to be static, *i.e.*, time-independent. Hence, the total energy of the particle, E, is always conserved along its trajectory. In the free space outside the optical system, from where the particle enters the system, $E = \sqrt{m^2c^4 + c^2p_0^2}$, where p_0 is the design momentum, and this E is the total conserved energy of the particle throughout its trajectory. Further, for the optical elements, we consider $E \gg q\phi$. Thus, for the propagation of a paraxial, or quasiparaxial, beam through any electromagnetic optical system along its optic axis, we can take the classical charged particle beam optical Hamiltonian, or simply, the classical beam optical Hamiltonian, as

$$\begin{aligned}
\mathcal{H}_o &= -\frac{1}{c}\sqrt{(E-q\phi)^2 - m^2c^4 - c^2\pi_\perp^2} - qA_z \\
&= -\frac{1}{c}\sqrt{E^2 + q^2\phi^2 - 2Eq\phi - m^2c^4 - c^2\pi_\perp^2} - qA_z \\
&= -\frac{1}{c}\sqrt{E^2 - m^2c^4 + q^2\phi^2 - 2Eq\phi - c^2\pi_\perp^2} - qA_z \\
&= -\frac{1}{c}\sqrt{c^2p_0^2 - 2Eq\phi\left(1 - \frac{q\phi}{2E}\right) - c^2\pi_\perp^2} - qA_z \\
&\approx -\frac{1}{c}\sqrt{c^2\left(p_0^2 - \pi_\perp^2\right) - 2Eq\phi} - qA_z \\
&= -\sqrt{p_0^2 - \pi_\perp^2 - \frac{2Eq\phi}{c^2}} - qA_z.
\end{aligned} \qquad (4.2)$$

This is the relativistic classical beam optical Hamiltonian. We shall first study some examples of the optical elements based on this relativistic Hamiltonian and consider later the nonrelativistic classical charged particle beam optics as an approximation. Note that $\mathcal{H}_o$ is the optical Hamiltonian of the beam propagating through the optical element: it contains the parameters of both the beam $(E, p_0, \ldots)$ and the optical element $(\phi, \vec{A}, \ldots)$. However, often we will refer to $\mathcal{H}_o$ simply as the optical Hamiltonian of that particular optical element. For example, the paraxial optical Hamiltonian of a magnetic quadrupole will mean the optical Hamiltonian for the propagation of a paraxial beam through the magnetic quadrupole along its optic axis.

4.2 FREE PROPAGATION

When a beam propagates in free space, for example, between two optical elements of a device, motion of the beam particle is governed by the free-space optical Hamiltonian

$$\mathcal{H}_o = -\sqrt{p_0^2 - p_\perp^2}, \qquad (4.3)$$

as seen by taking $\phi = 0$ and $\vec{A} = (0,0,0)$ in (4.2). Let us now consider the propagation of the beam entering the xy-plane, the plane perpendicular to the optic axis, at $z(\text{in}) = z_i$ and exiting the xy-plane at z. Let a particle of the beam have $x(z_i)$ and $y(z_i)$ as its x and y coordinates, and $p_x(z_i)$ and $p_y(z_i)$ as its x and y components of momentum, in the xy-plane at z_i. Let it have $x(z)$ and $y(z)$ as its x and y coordinates, and $p_x(z)$ and $p_y(z)$ as its x and y components of momentum, when it arrives at the xy-plane at z. Hamilton's equations of motion for $(x(z), y(z))$ and $(p_x(z), p_y(z))$, the components of position and momentum of the particle in the plane perpendicular to the optic axis at the point z, become

$$\frac{dx}{dz} =: -\mathcal{H}_o :_z x =: \sqrt{p_0^2 - p_\perp^2} :_z x$$

$$= \frac{p_x}{\sqrt{p_0^2 - p_\perp^2}} = \frac{p_x}{p_z},$$

$$\frac{dy}{dz} =: -\mathcal{H}_o :_z y =: \sqrt{p_0^2 - p_\perp^2} :_z y$$

$$= \frac{p_y}{\sqrt{p_0^2 - p_\perp^2}} = \frac{p_y}{p_z},$$

$$\frac{dp_x}{dz} =: -\mathcal{H}_o :_z p_x =: \sqrt{p_0^2 - p_\perp^2} :_z p_x = 0,$$

$$\frac{dp_y}{dz} =: -\mathcal{H}_o :_z p_y =: \sqrt{p_0^2 - p_\perp^2} :_z p_y = 0, \qquad (4.4)$$

where we have used the definitions in (2.97) and (2.99). Note that when the z-evolution Hamiltonian $\mathcal{H}_o$ and any observable O are time-independent

$$: -\mathcal{H}_o :_z O = -\left\{ \left(\frac{\partial \mathcal{H}_o}{\partial x} \frac{\partial O}{\partial p_x} - \frac{\partial \mathcal{H}_o}{\partial p_x} \frac{\partial O}{\partial x} \right) \right.$$

$$\left. + \left(\frac{\partial \mathcal{H}_o}{\partial y} \frac{\partial O}{\partial p_y} - \frac{\partial \mathcal{H}_o}{\partial p_y} \frac{\partial O}{\partial y} \right) \right\}. \qquad (4.5)$$

We can write the equation (4.4) as

$$\frac{d}{dz} \begin{pmatrix} x(z) \\ y(z) \\ \frac{p_x(z)}{p_z} \\ \frac{p_y(z)}{p_z} \end{pmatrix} = \begin{pmatrix} 0 & 0 & 1 & 0 \\ 0 & 0 & 0 & 1 \\ 0 & 0 & 0 & 0 \\ 0 & 0 & 0 & 0 \end{pmatrix} \begin{pmatrix} x(z) \\ y(z) \\ \frac{p_x(z)}{p_z} \\ \frac{p_y(z)}{p_z} \end{pmatrix} = \mu \begin{pmatrix} x(z) \\ y(z) \\ \frac{p_x(z)}{p_z} \\ \frac{p_y(z)}{p_z} \end{pmatrix}. \qquad (4.6)$$

Since μ is z-independent, we can readily integrate this equation to get

$$\begin{pmatrix} x(z) \\ y(z) \\ \frac{p_x(z)}{p_z} \\ \frac{p_y(z)}{p_z} \end{pmatrix} = e^{\int_{z_i}^{z} dz\, \mu} \begin{pmatrix} x(z_i) \\ y(z_i) \\ \frac{p_x(z_i)}{p_z} \\ \frac{p_y(z_i)}{p_z} \end{pmatrix}$$

$$= \begin{pmatrix} 1 & 0 & z-z_i & 0 \\ 0 & 1 & 0 & z-z_i \\ 0 & 0 & 1 & 0 \\ 0 & 0 & 0 & 1 \end{pmatrix} \begin{pmatrix} x(z_i) \\ y(z_i) \\ \frac{p_x(z_i)}{p_z} \\ \frac{p_y(z_i)}{p_z} \end{pmatrix},$$

$$= \mathscr{M}_D(z, z_i) \begin{pmatrix} x(z_i) \\ y(z_i) \\ \frac{p_x(z_i)}{p_z} \\ \frac{p_y(z_i)}{p_z} \end{pmatrix} = \mathscr{M}_D(\Delta z) \begin{pmatrix} x(z_i) \\ y(z_i) \\ \frac{p_x(z_i)}{p_z} \\ \frac{p_y(z_i)}{p_z} \end{pmatrix}, \quad (4.7)$$

where $\Delta z = z - z_i$, and the subscript D stands for drift or free propagation.

Instead of the above derivation of (4.7), we can use (2.100) directly to get,

$$x(z) = \left[e^{\{\Delta z : -\mathscr{H}_o : z\}} x \right](z_i) = \left[e^{\{\Delta z : \sqrt{p_0^2 - p_\perp^2} : z\}} x \right](z_i)$$

$$= \left[\sum_{n=0}^{\infty} \frac{(\Delta z)^n}{n!} \left(: \sqrt{p_0^2 - p_\perp^2} :_z \right)^n x \right](z_i)$$

$$= x(z_i) + \Delta z \frac{p_x}{\sqrt{p_0^2 - p_\perp^2}} = x(z_i) + \Delta z \frac{p_x(z_i)}{p_z}. \quad (4.8)$$

For p_x we have

$$p_x(z) = \left[e^{\{\Delta z : -\mathscr{H}_o : z\}} p_x \right](z_i) = \left[e^{\{\Delta z : \sqrt{p_0^2 - p_\perp^2} : z\}} p_x \right](z_i)$$

$$= \left[\sum_{n=0}^{\infty} \frac{(\Delta z)^n}{n!} \left(: \sqrt{p_0^2 - p_\perp^2} :_z \right)^n p_x \right](z_i) = p_x(z_i), \quad (4.9)$$

showing that p_x is conserved, as should be for a free particle. Similarly, for y and p_y, we get

$$y(z) = y(z_i) + \Delta z \frac{p_y(z_i)}{p_z}, \quad p_y(z) = p_y(z_i). \quad (4.10)$$

Since $|\vec{p}_\perp| < p_0$, we have

$$\mathscr{H}_o = -p_0 \sqrt{1 - \frac{p_\perp^2}{p_0^2}} = -p_0 + \frac{p_\perp^2}{2p_0} + \frac{p_\perp^4}{8p_0^3} + \cdots . \quad (4.11)$$

Classical Charged Particle Beam Optics

When $|\vec{p}_\perp| \ll p_z$, we have a paraxial beam in which all the particles move along and very close to the $+z$-direction. Then, we have $p_z \approx p_0$ and $|\vec{p}_\perp| \ll p_0$ so that we can make the paraxial approximation

$$\mathcal{H}_o \approx -p_0 + \frac{p_\perp^2}{2p_0}, \qquad (4.12)$$

and get the same results (4.8–4.10). For example,

$$\begin{aligned}
x(z) &= \left[e^{\{\Delta z : -\mathcal{H}_o : z\}} x \right](z_i) = \left[e^{\left\{ \Delta z : \sqrt{p_0^2 - p_\perp^2} : z \right\}} x \right](z_i) \\
&\approx \left[\left(1 + \Delta z : p_0 - \frac{p_\perp^2}{2p_0} : _z \right) x \right](z_i) \\
&= x(z_i) + \Delta z \frac{p_x(z_i)}{p_0} \approx x(z_i) + \Delta z \frac{p_x(z_i)}{p_z}.
\end{aligned} \qquad (4.13)$$

If the beam is quasiparaxial, it would be required to take into account more terms, nonparaxial terms, in the Hamiltonian like

$$\mathcal{H}_o \approx -p_0 + \frac{p_\perp^2}{2p_0} + \frac{p_\perp^4}{8p_0^3}, \qquad (4.14)$$

to get the same results (4.8–4.10). For example,

$$\begin{aligned}
x(z) &= \left[e^{\{\Delta z : -\mathcal{H}_o : z\}} x \right](z_i) = \left[e^{\left\{ \Delta z : \sqrt{p_0^2 - p_\perp^2} : z \right\}} x \right](z_i) \\
&\approx \left[\left(1 + \Delta z : p_0 - \frac{p_\perp^2}{2p_0} - \frac{p_\perp^4}{8p_0^3} : _z \right) x \right](z_i) \\
&= x(z_i) + \Delta z \frac{p_x}{p_0} \left(1 + \frac{p_\perp^2}{2p_0^2} \right) \approx x(z_i) + \Delta z \frac{p_x(z_i)}{p_0 \sqrt{1 - \frac{p_\perp^2}{p_0^2}}} \\
&= x(z_i) + \Delta z \frac{p_x(z_i)}{\sqrt{p_0^2 - p_\perp^2}} = x(z_i) + \Delta z \frac{p_x(z_i)}{p_z}.
\end{aligned} \qquad (4.15)$$

Since we will be dealing only with paraxial or quasiparaxial beams with $p_z \approx p_0$, we can write the transverse phase space maps (4.8–4.10) across the free space between the xy-planes at z_i and z as

$$\begin{pmatrix} \vec{r}_\perp \\ \frac{\vec{p}_\perp}{p_0} \end{pmatrix}_z = \begin{pmatrix} 1 & \Delta z \\ 0 & 1 \end{pmatrix} \begin{pmatrix} \vec{r}_\perp \\ \frac{\vec{p}_\perp}{p_0} \end{pmatrix}_{z_i} = \mathscr{M}_D(\Delta z) \begin{pmatrix} \vec{r}_\perp \\ \frac{\vec{p}_\perp}{p_0} \end{pmatrix}_{z_i}. \qquad (4.16)$$

Note that $\mathscr{M}_D(\Delta z) = M_D(\Delta z) \otimes I$, where I is the 2×2 identity matrix. Observe that for a particle of paraxial or quasiparaxial beam

$$\frac{\vec{p}_\perp}{p_0} \approx \frac{\vec{p}_\perp}{p_z} = \frac{\frac{d\vec{r}_\perp}{dt}}{\frac{dz}{dt}} = \frac{d\vec{r}_\perp}{dz} = \vec{r}_\perp'. \qquad (4.17)$$

Thus, the map (4.16) is usually written as

$$\begin{pmatrix} \vec{r}_\perp \\ \vec{r}'_\perp \end{pmatrix}_z = M_D(\Delta z) \begin{pmatrix} \vec{r}_\perp \\ \vec{r}'_\perp \end{pmatrix}_{z_i}, \qquad (4.18)$$

in geometrical charged particle optics, or charged particle ray optics, and $\vec{r}_\perp$ and $\vec{r}'_\perp$ are called the ray coordinates representing the position and the slope of the ray intersecting the xy-plane at z. The beam is paraxial when $|\vec{r}'_\perp| \ll 1$.

4.3 OPTICAL ELEMENTS WITH STRAIGHT OPTIC AXIS

4.3.1 AXIALLY SYMMETRIC MAGNETIC LENS: IMAGING IN ELECTRON MICROSCOPY

Let us now consider a round magnetic lens symmetric about its straight optic axis. This is the central component of any electron microscope. It comprises an axially symmetric magnetic field. There is no electric field, *i.e.*, $\phi(\vec{r}) = 0$. Taking the z-axis as the optic axis, the vector potential can be taken, in general, as

$$\vec{A}(\vec{r}) = \left(-\frac{1}{2} y \Pi(\vec{r}_\perp, z), \frac{1}{2} x \Pi(\vec{r}_\perp, z), 0 \right), \qquad (4.19)$$

with

$$\begin{aligned}
\Pi(\vec{r}_\perp, z) &= \sum_{n=0}^{\infty} \frac{1}{n!(n+1)!} \left(-\frac{r_\perp^2}{4} \right)^n B^{(2n)}(z) \\
&= B(z) - \frac{1}{8} r_\perp^2 B''(z) + \frac{1}{192} r_\perp^4 B''''(z) - \cdots,
\end{aligned} \qquad (4.20)$$

where $r_\perp^2 = x^2 + y^2$, $B^{(0)}(z) = B(z)$, $B'(z) = dB(z)/dz$, $B''(z) = d^2 B(z)/dz^2$, $B'''(z) = d^3 B(z)/dz^3$, $B''''(z) = d^4 B(z)/dz^4$, ..., and $B^{(2n)}(z) = d^{2n} B(z)/dz^{2n}$. The corresponding magnetic field, $\vec{\nabla} \times \vec{A}$, is given by

$$\begin{aligned}
\vec{B}_\perp &= -\frac{1}{2} \left(B'(z) - \frac{1}{8} B'''(z) r_\perp^2 + \cdots \right) \vec{r}_\perp, \\
B_z &= B(z) - \frac{1}{4} B''(z) r_\perp^2 + \frac{1}{64} B''''(z) r_\perp^4 - \cdots.
\end{aligned} \qquad (4.21)$$

The field $\vec{B}$, characterized completely by the function $B(z)$, is seen to be rotationally symmetric about the z-axis, the optic axis of the system. The practical boundaries of the lens, say z_ℓ and z_r, are determined by where $B(z)$ becomes negligible, *i.e.*, $B(z < z_\ell) \approx 0$ and $B(z > z_r) \approx 0$.

We are concerned with a monoenergetic paraxial beam with design momentum p_0 such that $|\vec{p}| = p_0 \approx p_z$ for all its constituent particles. For a paraxial beam propagating through the round magnetic lens, along the optic axis, the vector potential $\vec{A}$ can be taken as

$$\vec{A} \approx \vec{A}_0 = \left(-\frac{1}{2} B(z) y, \frac{1}{2} B(z) x, 0 \right), \qquad (4.22)$$

Classical Charged Particle Beam Optics

corresponding to the field

$$\vec{B} \approx \vec{B}_0 = \left(-\frac{1}{2}B'(z)x, -\frac{1}{2}B'(z)y, B(z)\right). \quad (4.23)$$

In the paraxial approximation, only the lowest order terms in $\vec{r}_\perp$ are considered to contribute to the effective field felt by the particles moving close to the optic axis. Further, it entails approximating the z-evolution, or optical, Hamiltonian by dropping from it the terms of order higher than second in $|\vec{p}_\perp|/p_0$ in view of the condition $|\vec{p}_\perp|/p_0 \ll 1$. Thus, from (4.2) and (4.22), we can write down the classical beam optical Hamiltonian of the system as

$$\begin{aligned}
\mathcal{H}_o &= -\sqrt{p_0^2 - \pi_\perp^2} - qA_z \\
&\approx -p_0 + \frac{\pi_\perp^2}{2p_0} \\
&= -p_0 + \frac{1}{2p_0}\left[(p_x - qA_x)^2 + (p_y - qA_y)^2\right] \\
&= -p_0 + \frac{1}{2p_0}\left[\left(p_x + \frac{1}{2}qB(z)y\right)^2 + \left(p_y - \frac{1}{2}qB(z)x\right)^2\right] \\
&= -p_0 + \frac{1}{2p_0}\left(p_\perp^2 + \frac{1}{4}q^2B(z)^2 r_\perp^2 - qB(z)(xp_y - yp_x)\right) \\
&= -p_0 + \frac{1}{2p_0}\left(p_\perp^2 + \frac{1}{4}q^2B(z)^2 r_\perp^2 - qB(z)L_z\right),
\end{aligned} \quad (4.24)$$

where

$$B(z) \begin{cases} \neq 0 & \text{in the lens region } (z_\ell \leq z \leq z_r) \\ = 0 & \text{outside the lens region } (z < z_\ell, z > z_r), \end{cases} \quad (4.25)$$

and L_z is the z-component of the orbital angular momentum of the particle. With the notation

$$\alpha(z) = \frac{qB(z)}{2p_0}, \quad (4.26)$$

we shall write

$$\mathcal{H}_o = -p_0 + \frac{1}{2p_0}p_\perp^2 + \frac{1}{2}p_0\alpha^2(z)r_\perp^2 - \alpha(z)L_z. \quad (4.27)$$

Hamilton's equations of motion are

$$\frac{d}{dz}\begin{pmatrix} x(z) \\ y(z) \\ \frac{p_x(z)}{p_0} \\ \frac{p_y(z)}{p_0} \end{pmatrix} = \begin{pmatrix} 0 & \alpha(z) & 1 & 0 \\ -\alpha(z) & 0 & 0 & 1 \\ -\alpha^2(z) & 0 & 0 & \alpha(z) \\ 0 & -\alpha^2(z) & -\alpha(z) & 0 \end{pmatrix} \begin{pmatrix} x(z) \\ y(z) \\ \frac{p_x(z)}{p_0} \\ \frac{p_y(z)}{p_0} \end{pmatrix}. \quad (4.28)$$

It is seen that in the regions outside the lens, where $\alpha(z) = 0$, this equation of motion becomes (4.6) with $p_0 \approx p_z$ for a paraxial beam. Let us write (4.28) as

$$\frac{d}{dz}\begin{pmatrix} x(z) \\ y(z) \\ \frac{p_x(z)}{p_0} \\ \frac{p_y(z)}{p_0} \end{pmatrix} = [\mu(z) \otimes I + I \otimes \rho(z)] \begin{pmatrix} x(z) \\ y(z) \\ \frac{p_x(z)}{p_0} \\ \frac{p_y(z)}{p_0} \end{pmatrix}, \qquad (4.29)$$

where

$$\mu(z) = \begin{pmatrix} 0 & 1 \\ -\alpha^2(z) & 0 \end{pmatrix}, \qquad (4.30)$$

and

$$\rho(z) = \alpha(z)\begin{pmatrix} 0 & 1 \\ -1 & 0 \end{pmatrix}. \qquad (4.31)$$

We shall now integrate (4.29). Note that $\mu(z) \otimes I$ and $I \otimes \rho(z)$ commute with each other. Let

$$\begin{pmatrix} x(z) \\ y(z) \\ \frac{p_x(z)}{p_0} \\ \frac{p_y(z)}{p_0} \end{pmatrix} = \mathcal{M}(z,z_i)\mathcal{R}(z,z_i) \begin{pmatrix} x(z_i) \\ y(z_i) \\ \frac{p_x(z_i)}{p_0} \\ \frac{p_y(z_i)}{p_0} \end{pmatrix}, \quad \text{for any } z \geq z_i, \qquad (4.32)$$

where

$$\frac{d\mathcal{M}(z,z_i)}{dz} = (\mu(z) \otimes I)\mathcal{M}(z,z_i), \qquad \mathcal{M}(z_i,z_i) = I, \qquad (4.33)$$

$$\frac{d\mathcal{R}(z,z_i)}{dz} = (I \times \rho(z))\mathcal{R}(z,z_i), \qquad \mathcal{R}(z_i,z_i) = I. \qquad (4.34)$$

It is clear that we can write

$$\mathcal{M}(z,z_i) = M(z,z_i) \otimes I, \qquad \frac{d\mathcal{M}(z,z_i)}{dz} = \frac{dM(z,z_i)}{dz} \otimes I, \qquad (4.35)$$

with

$$\frac{dM(z,z_i)}{dz} = \mu(z)M(z,z_i), \qquad (4.36)$$

and

$$\mathcal{R}(z,z_i) = I \otimes R(z,z_i), \qquad \frac{d\mathcal{R}(z,z_i)}{dz} = I \otimes \frac{dR(z,z_i)}{dz}, \qquad (4.37)$$

with

$$\frac{dR(z,z_i)}{dz} = \rho(z)R(z,z_i). \qquad (4.38)$$

Note that $\mathcal{M}(z,z_i)$ and $\mathcal{R}(z,z_i)$ commute with each other, $d\mathcal{M}(z,z_i)/dz$ commutes with $\mathcal{R}(z,z_i)$, and $d\mathcal{R}(z,z_i)/dz$ commutes with $\mathcal{M}(z,z_i)$. Then,

$$\frac{d}{dz}\begin{pmatrix} x(z) \\ y(z) \\ \frac{p_x(z)}{p_0} \\ \frac{p_y(z)}{p_0} \end{pmatrix} = \left(\frac{d\mathcal{M}(z,z_i)}{dz}\mathcal{R}(z,z_i) + \mathcal{M}(z,z_i)\frac{d\mathcal{R}(z,z_i)}{dz}\right)\begin{pmatrix} x(z_i) \\ y(z_i) \\ \frac{p_x(z_i)}{p_0} \\ \frac{p_y(z_i)}{p_0} \end{pmatrix}$$

$$= [\mu(z)\otimes I + I \otimes \rho(z)]\mathcal{M}(z,z_i)\mathcal{R}(z,z_i)\begin{pmatrix} x(z_i) \\ y(z_i) \\ \frac{p_x(z_i)}{p_0} \\ \frac{p_y(z_i)}{p_0} \end{pmatrix}$$

$$= [\mu(z)\otimes I + I \otimes \rho(z)]\begin{pmatrix} x(z) \\ y(z) \\ \frac{p_x(z)}{p_0} \\ \frac{p_y(z)}{p_0} \end{pmatrix}, \quad (4.39)$$

as required by (4.29).

The equation for $\mathcal{R}(z,z_i)$, (4.34), can be readily integrated by ordinary exponentiation since $[\rho(z'),\rho(z'')]=0$ for any z' and z''. Thus, we have

$$\mathcal{R}(z,z_i) = e^{\int_{z_i}^{z} dz\, \rho(z)}$$

$$= \exp\left\{I \otimes \left[\theta(z,z_i)\begin{pmatrix} 0 & 1 \\ -1 & 0 \end{pmatrix}\right]\right\}$$

$$= I \otimes \begin{pmatrix} \cos\theta(z,z_i) & \sin\theta(z,z_i) \\ -\sin\theta(z,z_i) & \cos\theta(z,z_i) \end{pmatrix}$$

$$= I \otimes R(z,z_i), \quad (4.40)$$

with

$$\theta(z,z_i) = \int_{z_i}^{z} dz\, \alpha(z) = \frac{q}{2p_0}\int_{z_i}^{z} dz\, B(z). \quad (4.41)$$

Let us now introduce an XYz-coordinate system with its X and Y axes rotating along the z-axis at the rate $d\theta(z)/dz = \alpha(z) = qB(z)/2p_0$, such that we can write

$$\begin{pmatrix} x(z) \\ y(z) \end{pmatrix} = R(z,z_i)\begin{pmatrix} X(z) \\ Y(z) \end{pmatrix}. \quad (4.42)$$

or

$$\begin{pmatrix} X(z) \\ Y(z) \end{pmatrix} = R^{-1}(z,z_i)\begin{pmatrix} x(z) \\ y(z) \end{pmatrix}$$

$$= \begin{pmatrix} \cos\theta(z,z_i) & -\sin\theta(z,z_i) \\ \sin\theta(z,z_i) & \cos\theta(z,z_i) \end{pmatrix}\begin{pmatrix} x(z) \\ y(z) \end{pmatrix}. \quad (4.43)$$

Then, we have

$$\begin{pmatrix} p_x(z) \\ p_y(z) \end{pmatrix} = R(z, z_i) \begin{pmatrix} P_X(z) \\ P_Y(z) \end{pmatrix}, \qquad (4.44)$$

or

$$\begin{pmatrix} P_X(z) \\ P_Y(z) \end{pmatrix} = \begin{pmatrix} \cos\theta(z, z_i) & -\sin\theta(z, z_i) \\ \sin\theta(z, z_i) & \cos\theta(z, z_i) \end{pmatrix} \begin{pmatrix} p_x(z) \\ p_y(z) \end{pmatrix}, \qquad (4.45)$$

where P_X and P_Y are the components of momentum in the rotating coordinate system. Note that in the vertical plane at $z = z_i$, where the particle enters the system, XYz coordinate system coincides with the xyz coordinate system. Thus, we can write (4.32) as, for any $z \geq z_i$,

$$\mathscr{R}(z, z_i) \begin{pmatrix} X(z) \\ Y(z) \\ \frac{P_X(z)}{p_0} \\ \frac{P_Y(z)}{p_0} \end{pmatrix} = \mathscr{M}(z, z_i) \mathscr{R}(z, z_i) \begin{pmatrix} X(z_i) \\ Y(z_i) \\ \frac{P_X(z_i)}{p_0} \\ \frac{P_Y(z_i)}{p_0} \end{pmatrix}$$

$$= \mathscr{R}(z, z_i) \mathscr{M}(z, z_i) \begin{pmatrix} X(z_i) \\ Y(z_i) \\ \frac{P_X(z_i)}{p_0} \\ \frac{P_Y(z_i)}{p_0} \end{pmatrix}. \qquad (4.46)$$

In other words, the position and momentum of the particle with reference to the rotating coordinate system evolve along the optic axis according to

$$\begin{pmatrix} X(z) \\ Y(z) \\ \frac{P_X(z)}{p_0} \\ \frac{P_Y(z)}{p_0} \end{pmatrix} = \mathscr{M}(z, z_i) \begin{pmatrix} X(z_i) \\ Y(z_i) \\ \frac{P_X(z_i)}{p_0} \\ \frac{P_Y(z_i)}{p_0} \end{pmatrix}, \quad \text{for any } z \geq z_i. \qquad (4.47)$$

From (4.30) and (4.33), the corresponding equations of motion follow:

$$\frac{d}{dz} \begin{pmatrix} \vec{R}_\perp(z) \\ \frac{\vec{P}_\perp(z)}{p_0} \end{pmatrix} = \mu(z) \begin{pmatrix} \vec{R}_\perp(z) \\ \frac{\vec{P}_\perp(z)}{p_0} \end{pmatrix} = \begin{pmatrix} 0 & 1 \\ -\alpha^2(z) & 0 \end{pmatrix} \begin{pmatrix} \vec{R}_\perp(z) \\ \frac{\vec{P}_\perp(z)}{p_0} \end{pmatrix}, \qquad (4.48)$$

with

$$\vec{R}_\perp(z) = \begin{pmatrix} X(z) \\ Y(z) \end{pmatrix}, \qquad \vec{P}_\perp(z) = \begin{pmatrix} P_X(z) \\ P_Y(z) \end{pmatrix}. \qquad (4.49)$$

From this, it follows that

$$\frac{d^2}{dz^2} \begin{pmatrix} \vec{R}_\perp(z) \\ \frac{\vec{P}_\perp(z)}{p_0} \end{pmatrix} = \begin{pmatrix} -\alpha^2(z) & 0 \\ -2\alpha(z)\alpha'(z) & -\alpha^2(z) \end{pmatrix} \begin{pmatrix} \vec{R}_\perp(z) \\ \frac{\vec{P}_\perp(z)}{p_0} \end{pmatrix}, \qquad (4.50)$$

or

$$\vec{R}_\perp''(z) + \alpha^2(z)\vec{R}_\perp = 0, \tag{4.51}$$

$$\frac{1}{p_0}\vec{P}_\perp''(z) + 2\alpha(z)\alpha'(z)\vec{R}_\perp(z) + \frac{1}{p_0}\alpha^2(z)\vec{P}_\perp(z) = 0. \tag{4.52}$$

The first of the above equations (4.51) is the paraxial equation of motion in the rotating coordinate system. The second equation is not independent of the first equation since it is just a consequence of the relation $\vec{P}_\perp(z)/p_0 = d\vec{R}_\perp(z)/dz$.

Equation (4.48) cannot be integrated by ordinary exponentiation since $\mu(z')$ and $\mu(z'')$ do not commute when $z' \neq z''$ if $\alpha^2(z') \neq \alpha^2(z'')$. So, to integrate (4.48) we have to follow the same method used to integrate the time-dependent Schrödinger equation (3.439) as in (3.440) with $\widehat{U}(t, t_0)$ given by the series expression in (3.449). Thus, replacing t by z, and $-i\widehat{H}(t)/\hbar$ by $\mu(z)$, we get

$$\begin{pmatrix} \vec{R}_\perp(z) \\ \frac{\vec{P}_\perp(z)}{p_0} \end{pmatrix} = \begin{pmatrix} g(z, z_i) & h(z, z_i) \\ g'(z, z_i) & h'(z, z_i) \end{pmatrix} \begin{pmatrix} \vec{R}_\perp(z_i) \\ \frac{\vec{P}_\perp(z_i)}{p_0} \end{pmatrix}$$

$$= M(z, z_i) \begin{pmatrix} \vec{R}_\perp(z_i) \\ \frac{\vec{P}_\perp(z_i)}{p_0} \end{pmatrix}, \tag{4.53}$$

where

$$M(z, z_i) = I + \int_{z_i}^z dz_1 \mu(z_1) + \int_{z_i}^z dz_2 \int_{z_i}^{z_2} dz_1 \mu(z_2) \mu(z_1)$$
$$+ \int_{z_i}^z dz_3 \int_{z_i}^{z_3} dz_2 \int_{z_i}^{z_2} dz_1 \mu(z_3) \mu(z_2) \mu(z_1)$$
$$+ \cdots . \tag{4.54}$$

Note that $g(z, z_i)$ and $h(z, z_i)$ are two linearly independent solutions of the paraxial equation of motion (4.51) corresponding to the initial conditions

$$\begin{pmatrix} \vec{R}_\perp(z_i) \\ \frac{\vec{P}_\perp(z_i)}{p_0} \end{pmatrix} = \begin{pmatrix} 1 \\ 0 \end{pmatrix}, \quad \begin{pmatrix} \vec{R}_\perp(z_i) \\ \frac{\vec{P}_\perp(z_i)}{p_0} \end{pmatrix} = \begin{pmatrix} 0 \\ 1 \end{pmatrix}. \tag{4.55}$$

It can be directly verified that

$$\frac{dM(z, z_i)}{dz} = \mu(z) M(z, z_i), \tag{4.56}$$

and hence

$$\frac{d}{dz} \begin{pmatrix} \vec{R}_\perp(z) \\ \frac{\vec{P}_\perp(z)}{p_0} \end{pmatrix} = \frac{dM(z, z_i)}{dz} \begin{pmatrix} \vec{R}_\perp(z_i) \\ \frac{\vec{P}_\perp(z_i)}{p_0} \end{pmatrix} = \mu(z) \begin{pmatrix} \vec{R}_\perp(z_i) \\ \frac{\vec{P}_\perp(z_i)}{p_0} \end{pmatrix}, \tag{4.57}$$

as required in (4.48). Explicit expressions for the elements of M can be written down from (4.54) as follows:

$$g(z,z_i) = 1 - \int_{z_i}^{z} dz_2 \int_{z_i}^{z_2} dz_1 \, \alpha^2(z_1)$$

$$+ \int_{z_i}^{z} dz_4 \int_{z_i}^{z_4} dz_3 \, \alpha^2(z_3) \int_{z_i}^{z_3} dz_2 \int_{z_i}^{z_2} dz_1 \, \alpha^2(z_1) - \cdots,$$

$$h(z,z_i) = (z - z_i) - \int_{z_i}^{z} dz_2 \int_{z_i}^{z_2} dz_1 \, \alpha^2(z_1)(z_1 - z_i)$$

$$+ \int_{z_i}^{z} dz_4 \int_{z_i}^{z_4} dz_3 \, \alpha^2(z_3) \int_{z_i}^{z_3} dz_2 \int_{z_i}^{z_2} dz_1 \, \alpha^2(z_1)(z_1 - z_i) - \cdots,$$

$$g'(z,z_i) = -\int_{z_i}^{z} dz_1 \, \alpha^2(z_1) + \int_{z_i}^{z} dz_3 \, \alpha^2(z_3) \int_{z_i}^{z_3} dz_2 \int_{z_i}^{z_2} dz_1 \, \alpha^2(z_1) - \cdots,$$

$$h'(z,z_i) = 1 - \int_{z_i}^{z} dz_1 \, \alpha^2(z_1)(z_1 - z_i)$$

$$+ \int_{z_i}^{z} dz_3 \, \alpha^2(z_3) \int_{z_i}^{z_3} dz_2 \int_{z_i}^{z_2} dz_1 \, \alpha^2(z_1)(z_1 - z_i) - \cdots . \quad (4.58)$$

Note that

$$\frac{dg(z,z_i)}{dz} = g'(z,z_i), \qquad \frac{dh(z,z_i)}{dz} = h'(z,z_i), \quad (4.59)$$

consistent with the relation $d\vec{R}_\perp(z)/dz = \vec{P}_\perp(z)/p_0$. From the differential equation, (4.56) for M we know that we can write

$$M(z,z_i) = \lim_{N\to\infty} \lim_{\Delta z \to 0} \left\{ e^{\Delta z \mu(z_i + N\Delta z)} e^{\Delta z \mu(z_i + (N-1)\Delta z)} e^{\Delta z \mu(z_i + (N-2)\Delta z)} \cdots \right.$$

$$\left. \cdots e^{\Delta z \mu(z_i + 2\Delta z)} e^{\Delta z \mu(z_i + \Delta z)} \right\}, \quad (4.60)$$

with $N\Delta z = (z - z_i)$. Since the trace of $\mu(z)$ is zero, each of the factors in the above product expression, of the type $e^{\Delta z \mu(z_i + j\Delta z)}$, has unit determinant. Thus, the matrix M has an unit determinant, i.e.,

$$g(z,z_i) h'(z,z_i) - h(z,z_i) g'(z,z_i) = 1, \quad (4.61)$$

for any (z,z_i).

In a simplified picture of an electron microscope, let us consider the monoenergetic paraxial beam comprising of electrons scattered elastically from the specimen (the object to be imaged) being illuminated. The beam transmitted by the specimen carries the information about the structure of the specimen. This beam going through the magnetic lens gets magnified and is recorded at the image plane. Let us take the

Classical Charged Particle Beam Optics

positions of the object plane and the image plane in the z-axis, the optic axis, as z_{obj} and z_{img}, respectively. As already mentioned, we shall take the practical boundaries of the lens to be at z_ℓ and z_r. The magnetic field of the lens is practically confined to the region between z_ℓ and z_r and $z_{obj} < z_\ell < z_r < z_{img}$. The region between z_{obj} and z_ℓ and the region between z_r and z_{img} are free spaces. Thus, the electron beam transmitted by the specimen (object) travels first through the free space between z_{obj} and z_ℓ, then the lens between z_ℓ and z_r, and finally the free space between z_r and z_{img} where it is recorded (image). Hence, the z-evolution matrix, or the transfer matrix, $M(z_{img}, z_{obj})$ can be written as

$$M(z_{img}, z_{obj}) = M_D(z_{img}, z_r) M_L(z_r, z_\ell) M_D(z_\ell, z_{obj}), \tag{4.62}$$

where subscript D indicates the drift region and L indicates the lens region. Since $\alpha^2(z) = 0$ in the free space regions, or drift regions, we get from (4.58)

$$M_D(z_{img}, z_r) = \begin{pmatrix} 1 & (z_{img} - z_r) \\ 0 & 1 \end{pmatrix}, \quad M_D(z_\ell, z_{obj}) = \begin{pmatrix} 1 & (z_\ell - z_{obj}) \\ 0 & 1 \end{pmatrix}. \tag{4.63}$$

For the lens region, we have from (4.58)

$$M_L(z_r, z_\ell) = \begin{pmatrix} g(z_r, z_\ell) & h(z_r, z_\ell) \\ g'(z_r, z_\ell) & h'(z_r, z_\ell) \end{pmatrix} = \begin{pmatrix} g_L & h_L \\ g'_L & h'_L \end{pmatrix}. \tag{4.64}$$

Thus, we get

$$M(z_{img}, z_{obj}) = \begin{pmatrix} 1 & (z_{img} - z_r) \\ 0 & 1 \end{pmatrix} \begin{pmatrix} g_L & h_L \\ g'_L & h'_L \end{pmatrix} \begin{pmatrix} 1 & (z_\ell - z_{obj}) \\ 0 & 1 \end{pmatrix}$$

$$= \begin{pmatrix} g_L + g'_L(z_{img} - z_r) & \begin{aligned} & [g_L(z_\ell - z_{obj}) + h_L \\ & + g'_L(z_{img} - z_r)(z_\ell - z_{obj}) \\ & + h'_L(z_{img} - z_r)] \end{aligned} \\ g'_L & g'_L(z_\ell - z_{obj}) + h'_L \end{pmatrix}$$

$$= \begin{pmatrix} g(z_{img}, z_{obj}) & h(z_{img}, z_{obj}) \\ g'(z_{img}, z_{obj}) & h'(z_{img}, z_{obj}) \end{pmatrix}. \tag{4.65}$$

Note that $M_D(z_\ell, z_{obj})$ and $M_D(z_{img}, z_r)$ have unit determinant. From the general theory, we know that $M_L(z_r, z_\ell)$ should also have unit determinant, i.e.,

$$g_L h'_L - h_L g'_L = 1. \tag{4.66}$$

Since z_{img} is the position of the image plane $h(z_{img}, z_{obj})$, the 12-element of the matrix $M(z_{img}, z_{obj})$, should vanish such that

$$\vec{R}_\perp(z_{img}) \propto \vec{R}_\perp(z_{obj}). \tag{4.67}$$

This means that we should have

$$g_L(z_\ell - z_{obj}) + h_L + g'_L(z_{img} - z_r)(z_\ell - z_{obj}) + h'_L(z_{img} - z_r) = 0. \quad (4.68)$$

Now, let

$$u = (z_\ell - z_{obj}) + \frac{h'_L - 1}{g'_L}, \qquad v = (z_{img} - z_r) + \frac{g_L - 1}{g'_L}, \quad (4.69)$$

and

$$f = -\frac{1}{g'_L}. \quad (4.70)$$

We are to interpret u as the object distance, v as the image distance, and f as the focal length of the lens. Using straightforward algebra, and the relation (4.66), one can verify that the familiar lens equation,

$$\frac{1}{u} + \frac{1}{v} = \frac{1}{f}, \quad (4.71)$$

implies (4.68). From (4.69) it is seen that the principal planes from which u and v are to be measured in the case of a thick lens are situated at

$$z_{P_{obj}} = z_\ell + \frac{h'_L - 1}{g'_L} = z_\ell + f(1 - h'_L),$$

$$z_{P_{img}} = z_r - \frac{g_L - 1}{g'_L} = z_r - f(1 - g_L), \quad (4.72)$$

such that

$$u = z_{P_{obj}} - z_{obj}, \qquad v = z_{img} - z_{P_{img}}. \quad (4.73)$$

The explicit expression for the focal length f follows from (4.70) and (4.58):

$$\frac{1}{f} = \int_{z_\ell}^{z_r} dz\, \alpha^2(z) - \int_{z_\ell}^{z_r} dz_2\, \alpha^2(z_2) \int_{z_\ell}^{z_2} dz_1 \int_{z_\ell}^{z_1} dz\, \alpha^2(z) + \cdots$$

$$= \frac{q^2}{4p_0^2} \int_{z_\ell}^{z_r} dz\, B^2(z) - \frac{q^4}{16p_0^4} \int_{z_\ell}^{z_r} dz_2\, B^2(z_2) \int_{z_\ell}^{z_2} dz_1 \int_{z_\ell}^{z_1} dz\, B^2(z) + \cdots. \quad (4.74)$$

To understand the behavior of the lens, let us consider the idealized model in which $B(z) = B$ is a constant in the lens region and zero outside. Then, we get

$$\frac{1}{f} = \frac{qB}{2p_0} \sin\left(\frac{qBw}{2p_0}\right), \quad (4.75)$$

where $w = (z_r - z_\ell)$ is the width, or thickness, of the lens. This shows that the focal length is always positive to start with and is then periodic with respect to the variation of the field strength. Thus, the round magnetic lens is convergent up to a certain strength of the field. Round magnetic lenses commonly used in electron microscopy

Classical Charged Particle Beam Optics

are always convergent. Note that $\int_{z_\ell}^{z_r} dz\, \alpha^2(z)$ has the dimension of reciprocal length. The round magnetic lens is considered weak when

$$\int_{z_\ell}^{z_r} dz\, \alpha^2(z) \ll \frac{1}{w}. \tag{4.76}$$

For such a weak lens, the expression for the focal length (4.74) can be approximated as

$$\frac{1}{f} \approx \int_{z_\ell}^{z_r} dz\, \alpha^2(z) = \frac{q^2}{4p_0^2} \int_{z_\ell}^{z_r} dz\, B^2(z) = \frac{q^2}{4p_0^2} \int_{-\infty}^{\infty} dz\, B^2(z). \tag{4.77}$$

A weak lens is said to be thin since in this case $1/f \ll 1/w$, or $f \gg w$. The formula (4.77), known as the Busch formula, for the focal length of a thin axially symmetric magnetic lens was derived in 1927 by Busch [17].

The paraxial z-evolution matrix from the object plane to the image plane (4.65) becomes, taking into account (4.68),

$$M(z_{img}, z_{obj}) = \begin{pmatrix} g_L - \frac{z_{img}-z_r}{f} & 0 \\ -\frac{1}{f} & h'_L - \frac{z_\ell - z_{obj}}{f} \end{pmatrix}$$

$$= \begin{pmatrix} g_L - \frac{v+f(g_L-1)}{f} & 0 \\ -\frac{1}{f} & h'_L - \frac{u+f(h'_L-1)}{f} \end{pmatrix}$$

$$= \begin{pmatrix} 1 - \frac{v}{f} & 0 \\ -\frac{1}{f} & 1 - \frac{u}{f} \end{pmatrix} = \begin{pmatrix} -\frac{v}{u} & 0 \\ -\frac{1}{f} & -\frac{u}{v} \end{pmatrix}$$

$$= \begin{pmatrix} M & 0 \\ -\frac{1}{f} & \frac{1}{M} \end{pmatrix}, \tag{4.78}$$

where M denotes the magnification. In our notation, both u and v are positive, and hence M is negative, showing the inverted nature of the image, as should be in the case of imaging by a convergent lens.

For a thin lens with $f \gg w$, we can take $w \approx 0$, and the two principal planes collapse into a single plane at the center of the lens, $i.e.$, we can take $z_P = (z_\ell + z_r)/2$. Then, from (4.58), it is clear that for the thin lens we can make the approximation

$$M_{TL}(z_r, z_\ell) \approx \begin{pmatrix} 1 & 0 \\ -\frac{1}{f} & 1 \end{pmatrix}, \tag{4.79}$$

where the subscript TL stands for thin lens. Thus, corresponding to (4.65), we have for the thin lens

$$M\left(z_{img}, z_{obj}\right) \approx \begin{pmatrix} 1 & (z_{img} - z_P) \\ 0 & 1 \end{pmatrix} \begin{pmatrix} 1 & 0 \\ -\frac{1}{f} & 1 \end{pmatrix} \begin{pmatrix} 1 & (z_P - z_{obj}) \\ 0 & 1 \end{pmatrix}$$

$$= \begin{pmatrix} 1 - \frac{1}{f}(z_{img} - z_P) & \begin{array}{c} [(z_P - z_{obj}) \\ -\frac{1}{f}(z_{img} - z_P)(z_P - z_{obj}) \\ + (z_{img} - z_P)] \end{array} \\ -\frac{1}{f} & 1 - \frac{1}{f}(z_P - z_{obj}) \end{pmatrix}.$$

(4.80)

Since $(z_P - z_{obj}) = u$ and $(z_{img} - z_P) = v$, and $1/u + 1/v = 1/f$, we get

$$M\left(z_{img}, z_{obj}\right) = \begin{pmatrix} 1 - \frac{v}{f} & u - \frac{uv}{f} + v \\ -\frac{1}{f} & 1 - \frac{u}{f} \end{pmatrix}$$

$$= \begin{pmatrix} -\frac{v}{u} & 0 \\ -\frac{1}{f} & -\frac{u}{v} \end{pmatrix} = \begin{pmatrix} M & 0 \\ -\frac{1}{f} & \frac{1}{M} \end{pmatrix},$$

(4.81)

where $M = -v/u$ is the magnification.

So far, we have been describing the behavior of the lens in the rotated XYz-coordinate system. Now, if we return to the original xyz-coordinate system, we would have, as seen from (4.32), (4.35), and (4.37),

$$\begin{pmatrix} \vec{r}_\perp(z_{img}) \\ \frac{\vec{p}_\perp}{p_0}(z_{img}) \end{pmatrix} = M\left(z_{img}, z_{obj}\right) \otimes R\left(z_{img}, z_{obj}\right) \begin{pmatrix} \vec{r}_\perp(z_{obj}) \\ \frac{\vec{p}_\perp}{p_0}(z_{obj}) \end{pmatrix}.$$

(4.82)

Since the rotation of the image is the effect of only the lens magnetic field

$$R\left(z_{img}, z_{obj}\right) = R(z_\ell, z_r) = \begin{pmatrix} \cos\theta_L & \sin\theta_L \\ -\sin\theta_L & \cos\theta_L \end{pmatrix},$$

(4.83)

where

$$\theta_L = \theta(z_\ell, z_r) = \frac{q}{2p_0} \int_{z_\ell}^{z_r} dz\, B(z).$$

(4.84)

From (4.80) and (4.82), it is clear that apart from rotation and drifts through field-free regions in the front and back of the lens, the effect of a thin convergent lens is essentially described by the transfer matrix $\begin{pmatrix} 1 & 0 \\ -1/f & 1 \end{pmatrix}$. This can also be seen simply as follows. If a particle of a paraxial beam enters a thin lens in the transverse xy-plane with the coordinates $(x, 0)$ and momenta $(0, 0)$, $i.e.$, in the xz-plane and parallel to the optic (z) axis, then we have

$$\begin{pmatrix} 1 & f \\ 0 & 1 \end{pmatrix} \begin{pmatrix} 1 & 0 \\ -\frac{1}{f} & 1 \end{pmatrix} \begin{pmatrix} x \\ 0 \end{pmatrix} = \begin{pmatrix} 0 \\ -\frac{x}{f} \end{pmatrix}.$$

(4.85)

This means that all the particles of the paraxial beam hitting the lens parallel to the optic axis in the xz-plane meet the optic axis at a distance f from the lens independent of the x coordinate at which it enters the lens. This implies that the lens is a convergent lens with the focal length f. If the lens is a thin divergent lens of focal length $-f$, then correspondingly we would have

$$\begin{pmatrix} 1 & -f \\ 0 & 1 \end{pmatrix} \begin{pmatrix} 1 & 0 \\ \frac{1}{f} & 1 \end{pmatrix} \begin{pmatrix} x \\ 0 \end{pmatrix} = \begin{pmatrix} 0 \\ \frac{x}{f} \end{pmatrix}, \qquad (4.86)$$

implying that the particles appear virtually to meet behind the lens at a distance f. Thus, the effect of a thin divergent lens is essentially described by the transfer matrix $\begin{pmatrix} 1 & 0 \\ 1/f & 1 \end{pmatrix}$. Note that, in general, if the transfer matrix of a lens system is $\begin{pmatrix} A & B \\ C & D \end{pmatrix}$, then it will be focusing or defocusing depending on whether C is negative or positive, respectively. This is so because if C is negative, then p_x/p_0 decreases proportional to the x coordinate of the beam particle driving it towards the optic axis, and if C is positive, then p_x/p_0 increases proportional to the x coordinate of the beam particle driving it away from the optic axis.

In this discussion, we have considered the beam to be paraxial and approximated the optical Hamiltonian by keeping only terms up to second order in $\vec{r}_\perp$ and $\vec{p}_\perp$. This paraxial, or Gaussian, approximation has resulted in perfect imaging in which there is a one-to-one relationship between the object points and the image points. When the beam deviates from the ideal paraxial condition, as is usually the case in practice, we have to go beyond the paraxial approximation and retain in the Hamiltonian terms of order higher than second in $\vec{r}_\perp$ and $\vec{p}_\perp$. The resulting theory accounts for the image aberrations. We shall not deal with the classical theory of aberrations here. We shall treat aberrations in the quantum theory of electron optical imaging, and the classical theory of aberrations will emerge in the classical limit of the quantum theory.

The discussion so far is based on Hamilton's equations of motion. Let us now understand the results based on the Lie transfer operator method. Extensive techniques for applying the Lie transfer operator methods to light optics, charged particle beam optics, and accelerator optics have been developed by Dragt et al. (see Dragt [34], Dragt and Forest [35], Dragt et al. [36], Forest, Berz, and Irwin [54], Rangarajan, Dragt, and Neri [156], Forest and Hirata [55], Forest [56], Radlička [154], and references therein; see also Berz [11], Mondragon and Wolf [135], Wolf [191], Rangarajan and Sachidanand [157], Lakshminarayanan, Sridhar, and Jagannathan [124], and Wolski [192]). In general, from Hamilton's equations, we have

$$\frac{d}{dz} \begin{pmatrix} \vec{r}_\perp(z) \\ \frac{\vec{p}_\perp(z)}{p_0} \end{pmatrix} =: -\mathcal{H}_o(z) :_z \begin{pmatrix} \vec{r}_\perp(z) \\ \frac{\vec{p}_\perp(z)}{p_0} \end{pmatrix}. \qquad (4.87)$$

We shall now follow the treatment of the Schrödinger equation with time-dependent Hamiltonian in quantum mechanics. Let us write the solution of the above equation as

$$\begin{pmatrix} \vec{r}_\perp(z) \\ \frac{\vec{p}_\perp(z)}{p_0} \end{pmatrix} = \mathcal{T}(z, z_i) \begin{pmatrix} \vec{r}_\perp(z_i) \\ \frac{\vec{p}_\perp(z_i)}{p_0} \end{pmatrix}, \quad z \geq z_i, \tag{4.88}$$

where it is understood that

$$\mathcal{T}(z, z_i) \begin{pmatrix} \vec{r}_\perp(z_i) \\ \frac{\vec{p}_\perp(z_i)}{p_0} \end{pmatrix} = \left(\mathcal{T}(z, z_i) \begin{pmatrix} \vec{r}_\perp(z) \\ \frac{\vec{p}_\perp(z)}{p_0} \end{pmatrix} \right) \bigg|_{z=z_i}. \tag{4.89}$$

Then, we should have

$$\frac{\partial}{\partial z} \mathcal{T}(z, z_i) =: -\mathcal{H}_o(z) :_z \mathcal{T}(z, z_i), \quad \mathcal{T}(z_i, z_i) = 1. \tag{4.90}$$

Writing

$$\int_{z_i}^{z} dz \left(\frac{\partial}{\partial z} \mathcal{T}(z, z_i) \right) = \int_{z_i}^{z} dz_1 : -\mathcal{H}_o(z_1) :_z \mathcal{T}(z_1, z_i), \tag{4.91}$$

and integrating, we get

$$\mathcal{T}(z, z_i)|_{z_i}^{z} = \mathcal{T}(z, z_i) - 1 = \int_{z_i}^{z} dz_1 : -\mathcal{H}_o(z_1) :_z \mathcal{T}(z_1, z_i). \tag{4.92}$$

This leads to the formal solution

$$\mathcal{T}(z, z_i) = 1 - \int_{z_i}^{z} dz_1 : -\mathcal{H}_o(z_1) :_z \mathcal{T}(z_1, z_i). \tag{4.93}$$

Iterating this formal solution, we get

$$\begin{aligned}
\mathcal{T}(z, z_i) = &\, 1 + \int_{z_i}^{z} dz_1 : -\mathcal{H}_o(z_1) :_z \\
&+ \int_{z_i}^{z} dz_2 \int_{z_i}^{z_2} dz_1 : -\mathcal{H}_o(z_2) :_z : -\mathcal{H}_o(z_1) :_z \\
&+ \int_{z_i}^{z} dz_3 \int_{z_i}^{z_3} dz_2 \int_{z_i}^{z_2} dz_1 : -\mathcal{H}_o(z_3) :_z : -\mathcal{H}_o(z_2) :_z : -\mathcal{H}_o(z_1) :_z \\
&+ \cdots.
\end{aligned} \tag{4.94}$$

Introducing an z-ordering operator, analogous to the time-ordering operator (3.450) in quantum mechanics, we can write

$$\begin{aligned}
\mathcal{T}(z, z_i) &= 1 + \sum_{n=1}^{\infty} \frac{1}{n!} \mathsf{P} \left[\left(\int_{z_i}^{z} dz : -\mathcal{H}_o(z) :_z \right)^n \right] \\
&= \mathsf{P} \left[\sum_{n=0}^{\infty} \frac{1}{n!} \left(\int_{z_i}^{z} dz : -\mathcal{H}_o(z) :_z \right)^n \right] \\
&= \mathsf{P} \left(e^{\int_{z_i}^{z} dz : -\mathcal{H}_o(z) :_z} \right),
\end{aligned} \tag{4.95}$$

Classical Charged Particle Beam Optics

which may be called the z-ordered, or path ordered, exponential of $:-\mathcal{H}_o(z):_z$. In exact analogy with (3.455 and 3.456) $\mathcal{T}(z,z_i)$ can also be written in the Magnus form as

$$\mathcal{T}(z,z_i) = e^{T(z,z_i)}, \qquad (4.96)$$

where $T(z,z_i)$ is an infinite series with the first few terms given by

$$\begin{aligned} T(z,z_i) &= \int_{z_i}^{z} dz \; :-\mathcal{H}_o:_z (z) \\ &+ \frac{1}{2} \int_{z_i}^{z} dz_2 \int_{z_i}^{z_2} dz_1 \, [:-\mathcal{H}_o:_z (z_2), :-\mathcal{H}_o:_z (z_1)] \\ &+ \frac{1}{6} \int_{z_i}^{z} dz_3 \int_{z_i}^{z_3} dz_2 \int_{z_i}^{z_2} dz_1 \\ &\quad \{[[:-\mathcal{H}_o:_z (z_3), :-\mathcal{H}_o:_z (z_2)], :-\mathcal{H}_o:_z (z_1)] \\ &\quad + [[:-\mathcal{H}_o:_z (z_1), :-\mathcal{H}_o:_z (z_2)], :-\mathcal{H}_o:_z (z_3)]\} \\ &+ \cdots, \end{aligned} \qquad (4.97)$$

with

$$\begin{aligned} &[:-\mathcal{H}_o:_z (z_1), :-\mathcal{H}_o:_z (z_2)] \\ &= :-\mathcal{H}_o:_z (z_1) :-\mathcal{H}_o:_z (z_2) - :-\mathcal{H}_o:_z (z_2) :-\mathcal{H}_o:_z (z_1) \\ &= :\{-\mathcal{H}_o(z_1), -\mathcal{H}_o(z_2)\}_z :, \end{aligned} \qquad (4.98)$$

as follows from (2.57). When $\mathcal{H}_o$ is z-independent, the commutators in (4.97) vanish making $\mathcal{T}(z,z_i)$ an ordinary exponential:

$$\mathcal{T}(z,z_i) = e^{\int_{z_i}^{z} dz \, :-\mathcal{H}_o(z):_z}. \qquad (4.99)$$

Note that we can write

$$\mathcal{T}(z,z_i) = \mathsf{P}\left(e^{\int_{z_i}^{z} dz \, :-\mathcal{H}_o(z):_z}\right) = e^{T(z,z_i)}. \qquad (4.100)$$

In general, for any observable O, we would have

$$O(z) = e^{T(z,z_i)} O(z_i). \qquad (4.101)$$

Analogous to the semigroup property of the time-evolution operator in quantum mechanics (3.167), the Lie transfer operator $\mathcal{T}(z,z_i)$ has the semigroup property,

$$\mathcal{T}(z,z_i) = \mathcal{T}(z,z') \, \mathcal{T}(z',z_i), \qquad (4.102)$$

which follows from the observation

$$\begin{pmatrix} \vec{r}_\perp(z) \\ \frac{\vec{p}_\perp(z)}{p_0} \end{pmatrix} = \mathscr{T}(z,z_i) \begin{pmatrix} \vec{r}_\perp(z_i) \\ \frac{\vec{p}_\perp(z_i)}{p_0} \end{pmatrix},$$

$$\begin{pmatrix} \vec{r}_\perp(z) \\ \frac{\vec{p}_\perp(z)}{p_0} \end{pmatrix} = \mathscr{T}(z,z') \begin{pmatrix} \vec{r}_\perp(z') \\ \frac{\vec{p}_\perp(z')}{p_0} \end{pmatrix},$$

$$\begin{pmatrix} \vec{r}_\perp(z') \\ \frac{\vec{p}_\perp(z')}{p_0} \end{pmatrix} = \mathscr{T}(z',z_i) \begin{pmatrix} \vec{r}_\perp(z_i) \\ \frac{\vec{p}_\perp(z_i)}{p_0} \end{pmatrix}, \quad (4.103)$$

In general, we have, for $z > z_n > z_{n-1} > \cdots > z_2 > z_1 > z_i$,

$$\mathscr{T}(z,z_i) = \mathscr{T}(z,z_n)\mathscr{T}(z_n,z_{n-1})\ldots\mathscr{T}(z_2,z_1)\mathscr{T}(z_1,z_i), \quad (4.104)$$

For the round magnetic lens, we have

$$\mathscr{H}_o = \mathscr{H}_{o,L} + \mathscr{H}_{o,R}, \quad (4.105)$$

where

$$\mathscr{H}_{o,L} = -p_0 + \frac{1}{2p_0}p_\perp^2 + \frac{1}{2}p_0\alpha^2(z)r_\perp^2,$$

$$\mathscr{H}_{o,R} = -\alpha(z)L_z. \quad (4.106)$$

Let us now note the following:

$$\{r_\perp^2, L_z\}_z = 0, \quad (4.107)$$

where for any two time-independent f and g

$$\{f,g\}_z = \left(\frac{\partial f}{\partial x}\frac{\partial g}{\partial p_x} - \frac{\partial f}{\partial p_x}\frac{\partial g}{\partial x}\right) + \left(\frac{\partial f}{\partial y}\frac{\partial g}{\partial p_y} - \frac{\partial f}{\partial p_y}\frac{\partial g}{\partial y}\right). \quad (4.108)$$

Proof of (4.107) is as follows:

$$\begin{aligned} \{r_\perp^2, L_z\}_z &= \{x^2+y^2, xp_y - yp_x\}_z \\ &= \{x^2, xp_y\}_z - \{x^2, yp_x\}_z + \{y^2, xp_y\}_z - \{y^2, yp_x\}_z \\ &= -y\{x^2, p_x\}_z + x\{y^2, p_y\}_z \\ &= -2yx + 2xy = 0. \end{aligned} \quad (4.109)$$

Similarly we have

$$\{p_\perp^2, L_z\}_z = 0, \quad (4.110)$$

as follows:

$$\begin{aligned} \{p_\perp^2, L_z\}_z &= \{p_x^2 + p_y^2, xp_y - yp_x\}_z \\ &= \{p_x^2, xp_y\}_z - \{p_x^2, yp_x\}_z + \{p_y^2, xp_y\}_z - \{p_y^2, yp_x\}_z \\ &= \{p_x^2, x\}_z p_y - \{p_y^2, y\}_z p_x \\ &= -2p_xp_y + 2p_yp_x = 0. \end{aligned} \quad (4.111)$$

Classical Charged Particle Beam Optics

In view of (4.107) and (4.110), we have

$$\{\mathcal{H}_{o,L}, \mathcal{H}_{o,R}\}_z = 0. \tag{4.112}$$

Then, the commutator terms between $\mathcal{H}_{o,L}$ and $\mathcal{H}_{o,R}$ in (4.97) vanish, leading to the result that for the round magnetic lens

$$\mathcal{T}(z, z_i) = \mathcal{M}(z, z_i) \mathcal{R}(z, z_i) \tag{4.113}$$

with

$$\mathcal{M}(z, z_i) = \mathsf{P}\left(e^{\int_{z_i}^z dz \, :-\mathcal{H}_{o,L}(z):_z}\right), \quad \mathcal{R}(z, z_i) = \mathsf{P}\left(e^{\int_{z_i}^z dz \, :-\mathcal{H}_{o,R}(z):_z}\right). \tag{4.114}$$

It is easy to see that $:-\mathcal{H}_{o,R}(z'):_z$ commutes with $:-\mathcal{H}_{o,R}(z''):_z$ for any two z' and z'', and hence we can write

$$\mathcal{R}(z, z_i) = e^{\int_{z_i}^z dz \, :-\mathcal{H}_{o,R}(z):_z} = e^{\frac{q}{2p_0}\left(\int_{z_i}^z dz B(z)\right) :L_z:_z} = e^{\theta(z,z_i):L_z:_z}. \tag{4.115}$$

Note that

$$\{\mathcal{H}_{o,L}(z'), \mathcal{H}_{o,L}(z'')\}_z$$
$$= \left\{\left(\frac{1}{2p_0}p_\perp^2 + \frac{1}{2}p_0\alpha^2(z')r_\perp^2\right), \left(\frac{1}{2p_0}p_\perp^2 + \frac{1}{2}p_0\alpha^2(z'')r_\perp^2\right)\right\}_z$$
$$= (\alpha^2(z') - \alpha^2(z''))(\vec{r}_\perp \cdot \vec{p}_\perp), \tag{4.116}$$

showing that $:-\mathcal{H}_{o,L}(z'):_z$ does not commute with $:-\mathcal{H}_{o,L}(z''):_z$ if $\alpha^2(z') \neq \alpha^2(z'')$. Thus, the expression for $\mathcal{M}(z, z_i)$ cannot be simplified further.

Let us first consider the effect of $\mathcal{R}(z, z_i)$ on $\vec{r}_\perp$ and $\frac{\vec{p}_\perp}{p_0}$. To this end, we proceed as follows:

$$e^{\theta:L_z:_z} x = \left(1 + \theta : L_z :_z + \frac{\theta^2}{2!} : L_z :_z^2 + \frac{\theta^3}{3!} : L_z :_z^3 + \cdots\right) x$$
$$= x + \theta \{L_z, x\}_z + \frac{\theta^2}{2!} \{L_z, \{L_z, x\}_z\}_z + \frac{\theta^3}{3!} \{L_z, \{L_z, \{L_z, x\}_z\}_z\}_z \cdots$$
$$= x\left(1 - \frac{\theta^2}{2!} + \frac{\theta^4}{4!} - \cdots\right) + y\left(\theta - \frac{\theta^3}{3!} + \frac{\theta^5}{5!} - \cdots\right)$$
$$= \cos\theta x + \sin\theta y. \tag{4.117}$$

Similarly, it is seen that

$$e^{\theta:L_z:_z} y = -\sin\theta x + \cos\theta y,$$
$$e^{\theta:L_z:_z} \frac{p_x}{p_0} = \cos\theta \frac{p_x}{p_0} + \sin\theta \frac{p_y}{p_0},$$
$$e^{\theta:L_z:_z} \frac{p_y}{p_0} = -\sin\theta \frac{p_x}{p_0} + \cos\theta \frac{p_y}{p_0}. \tag{4.118}$$

This shows that we can write

$$\mathscr{R}(z,z_i)\begin{pmatrix} x(z_i) \\ y(z_i) \\ \frac{p_x(z_i)}{p_0} \\ \frac{p_y(z_i)}{p_0} \end{pmatrix}$$

$$= \begin{pmatrix} \cos\theta(z,z_i) & \sin\theta(z,z_i) & 0 & 0 \\ -\sin\theta(z,z_i) & \cos\theta(z,z_i) & 0 & 0 \\ 0 & 0 & \cos\theta(z,z_i) & \sin\theta(z,z_i) \\ 0 & 0 & -\sin\theta(z,z_i) & \cos\theta(z,z_i) \end{pmatrix} \begin{pmatrix} x(z_i) \\ y(z_i) \\ \frac{p_x(z_i)}{p_0} \\ \frac{p_y(z_i)}{p_0} \end{pmatrix}$$

$$= [I \otimes R(z,z_i)] \begin{pmatrix} x(z_i) \\ y(z_i) \\ \frac{p_x(z_i)}{p_0} \\ \frac{p_y(z_i)}{p_0} \end{pmatrix}. \tag{4.119}$$

We can derive this result in the following way also. Observe that

$$:L_z:_z \begin{pmatrix} x \\ y \\ \frac{p_x}{p_0} \\ \frac{p_y}{p_0} \end{pmatrix} = \begin{pmatrix} \{L_z, x\}_z \\ \{L_z, y\}_z \\ \{L_z, \frac{p_x}{p_0}\}_z \\ \{L_z, \frac{p_y}{p_0}\}_z \end{pmatrix} = L \begin{pmatrix} x \\ y \\ \frac{p_x}{p_0} \\ \frac{p_y}{p_0} \end{pmatrix}, \tag{4.120}$$

where

$$L = \begin{pmatrix} 0 & 1 & 0 & 0 \\ -1 & 0 & 0 & 0 \\ 0 & 0 & 0 & 1 \\ 0 & 0 & -1 & 0 \end{pmatrix}. \tag{4.121}$$

It can be verified directly that

$$:L:_z^2 \begin{pmatrix} x \\ y \\ \frac{p_x}{p_0} \\ \frac{p_y}{p_0} \end{pmatrix} = L^2 \begin{pmatrix} x \\ y \\ \frac{p_x}{p_0} \\ \frac{p_y}{p_0} \end{pmatrix}, \quad :L:_z^3 \begin{pmatrix} x \\ y \\ \frac{p_x}{p_0} \\ \frac{p_y}{p_0} \end{pmatrix} = L^3 \begin{pmatrix} x \\ y \\ \frac{p_x}{p_0} \\ \frac{p_y}{p_0} \end{pmatrix}, \ldots \tag{4.122}$$

Hence we can write

$$\mathscr{R}(z,z_i) \begin{pmatrix} x(z_i) \\ y(z_i) \\ \frac{p_x(z_i)}{p_0} \\ \frac{p_y(z_i)}{p_0} \end{pmatrix} = e^{\theta(z,z_i):L_z:_z} \begin{pmatrix} x(z_i) \\ y(z_i) \\ \frac{p_x(z_i)}{p_0} \\ \frac{p_y(z_i)}{p_0} \end{pmatrix}$$

$$= \sum_{n=0}^{\infty} \frac{1}{n!} \theta(z,z_i)^n :L_z:_z^n \begin{pmatrix} x(z_i) \\ y(z_i) \\ \frac{p_x(z_i)}{p_0} \\ \frac{p_y(z_i)}{p_0} \end{pmatrix}$$

Classical Charged Particle Beam Optics

$$= \sum_{n=0}^{\infty} \frac{1}{n!} \theta(z,z_i)^n \mathrm{L}^n \begin{pmatrix} x(z_i) \\ y(z_i) \\ \frac{p_x(z_i)}{p_0} \\ \frac{p_y(z_i)}{p_0} \end{pmatrix}$$

$$= [I \otimes R(z,z_i)] \begin{pmatrix} x(z_i) \\ y(z_i) \\ \frac{p_x(z_i)}{p_0} \\ \frac{p_y(z_i)}{p_0} \end{pmatrix}, \qquad (4.123)$$

as found already in (4.119).

To find the effect of $\mathcal{M}(z,z_i)$ on $(x,y,p_x/p_0,p_y/p_0)$, let us proceed as follows. Observe that

$$\begin{aligned}
:-\mathcal{H}_{o,L}(z):_z x &= \{-\mathcal{H}_{o,L}(z),x\}_z \\
&= \left\{-\left(\frac{1}{2p_0}p_\perp^2 + \frac{1}{2}p_0\alpha^2(z)r_\perp^2\right),x\right\}_z \\
&= \left\{-\frac{p_x^2}{2p_0},x\right\}_z = \frac{p_x}{p_0}, \qquad (4.124)
\end{aligned}$$

and

$$\begin{aligned}
:-\mathcal{H}_{o,L}(z):_z \frac{p_x}{p_0} &= \left\{-\mathcal{H}_{o,L}(z),\frac{p_x}{p_0}\right\}_z \\
&= \left\{-\left(\frac{1}{2p_0}p_\perp^2 + \frac{1}{2}p_0\alpha^2(z)r_\perp^2\right),\frac{p_x}{p_0}\right\}_z \\
&= \left\{-\frac{1}{2}p_0\alpha^2(z)x^2,\frac{p_x}{p_0}\right\}_z = -\alpha^2(z)x. \qquad (4.125)
\end{aligned}$$

Similar equations follow for the Poisson brackets of $-\mathcal{H}_{o,L}(z)$ with y and p_y/p_0. Thus, we get

$$\begin{aligned}
:-\mathcal{H}_{o,L}(z):_z \begin{pmatrix} x \\ y \\ \frac{p_x}{p_0} \\ \frac{p_y}{p_0} \end{pmatrix} &= \begin{pmatrix} 0 & 0 & 1 & 0 \\ 0 & 0 & 0 & 1 \\ -\alpha^2(z) & 0 & 0 & 0 \\ 0 & -\alpha^2(z) & 0 & 0 \end{pmatrix} \begin{pmatrix} x \\ y \\ \frac{p_x}{p_0} \\ \frac{p_y}{p_0} \end{pmatrix} \\
&= \left[\begin{pmatrix} 0 & 1 \\ -\alpha^2(z) & 0 \end{pmatrix} \otimes I\right] \begin{pmatrix} x \\ y \\ \frac{p_x}{p_0} \\ \frac{p_y}{p_0} \end{pmatrix} \\
&= (\mu(z) \otimes I) \begin{pmatrix} x \\ y \\ \frac{p_x}{p_0} \\ \frac{p_y}{p_0} \end{pmatrix}. \qquad (4.126)
\end{aligned}$$

It can be verified directly that

$$: -\mathcal{H}_{o,L}(z_2) :_z: -\mathcal{H}_{o,L}(z_1) :_z \begin{pmatrix} x \\ y \\ p_x \\ p_0 \\ p_y \\ p_0 \end{pmatrix}$$

$$= (\mu(z_2) \otimes I)(\mu(z_1) \otimes I) \begin{pmatrix} x \\ y \\ p_x \\ p_0 \\ p_y \\ p_0 \end{pmatrix}$$

$$= (\mu(z_2) \mu(z_1) \otimes I) \begin{pmatrix} x \\ y \\ p_x \\ p_0 \\ p_y \\ p_0 \end{pmatrix}. \qquad (4.127)$$

From this, it follows that we can write

$$\mathcal{M}(z, z_i) \begin{pmatrix} x(z_i) \\ y(z_i) \\ p_x(z_i) \\ p_0 \\ p_y(z_i) \\ p_0 \end{pmatrix} = \mathsf{P}\left(e^{\int_{z_i}^{z} dz : -\mathcal{H}_{o,L}(z) :_z}\right) \begin{pmatrix} x(z_i) \\ y(z_i) \\ p_x(z_i) \\ p_0 \\ p_y(z_i) \\ p_0 \end{pmatrix}$$

$$= \mathsf{P}\left(e^{\int_{z_i}^{z} dz\, \mu(z) \otimes I}\right) \begin{pmatrix} x(z_i) \\ y(z_i) \\ p_x(z_i) \\ p_0 \\ p_y(z_i) \\ p_0 \end{pmatrix}$$

$$= \left[\mathsf{P}\left(e^{\int_{z_i}^{z} dz\, \mu(z)}\right) \otimes I\right] \begin{pmatrix} x(z_i) \\ y(z_i) \\ p_x(z_i) \\ p_0 \\ p_y(z_i) \\ p_0 \end{pmatrix}$$

$$= (M(z, z_i) \otimes I) \begin{pmatrix} x(z_i) \\ y(z_i) \\ p_x(z_i) \\ p_0 \\ p_y(z_i) \\ p_0 \end{pmatrix}, \qquad (4.128)$$

where $M(z, z_I)$ is given by (4.54). Now, from (4.113), we get the full phase space transfer map as

$$\begin{pmatrix} x(z) \\ y(z) \\ p_x(z) \\ p_0 \\ p_y(z) \\ p_0 \end{pmatrix} = \mathcal{T}(z, z_i) \begin{pmatrix} x(z_i) \\ y(z_i) \\ p_x(z_i) \\ p_0 \\ p_y(z_i) \\ p_0 \end{pmatrix} = (M(z, z_i) \otimes R(z, z_i)) \begin{pmatrix} x(z_i) \\ y(z_i) \\ p_x(z_i) \\ p_0 \\ p_y(z_i) \\ p_0 \end{pmatrix}. \qquad (4.129)$$

Thus, we have analyzed the performance of an axially symmetric magnetic lens using direct integration of Hamilton's equations and also using the Lie transfer operator method. The paraxial approximation is seen to lead to a linear relationship between the initial transverse coordinates and momenta to the final coordinates and momenta for a particle of a paraxial beam traveling through the system close to the optic axis. If the beam deviates from the ideal paraxial condition, the transfer map from the initial to the final transverse coordinates and momenta will become nonlinear leading to aberrations in the imaging systems. Then, one has to go beyond the paraxial approximation and use perturbation techniques to understand the performance of the system. As already mentioned, we shall not deal with the classical theory of aberrations here, and later we shall see that the classical theory of aberrations can be understood as an approximation of the quantum theory.

4.3.2 NORMAL MAGNETIC QUADRUPOLE

Let us now consider a paraxial beam propagating through a normal magnetic quadrupole lens. Let the optic axis be along the z-direction and the practical boundaries of the quadrupole be at z_ℓ and z_r with $z_\ell < z_r$. The field of the normal magnetic quadrupole is

$$\vec{B}(\vec{r}) = (-Q_n y, -Q_n x, 0), \qquad (4.130)$$

where

$$Q_n = \begin{cases} \text{constant in the lens region} & (z_\ell \leq z \leq z_r) \\ 0 \text{ outside the lens region} & (z < z_\ell, z > z_r) \end{cases} \qquad (4.131)$$

There is no electric field and hence $\phi(\vec{r}) = 0$. The vector potential of the magnetic field can be taken as

$$\vec{A}(\vec{r}) = \left(0, 0, \frac{1}{2} Q_n \left(x^2 - y^2\right)\right). \qquad (4.132)$$

Now, from (4.2), the classical beam optical Hamiltonian is seen to be, with $K_n = qQ_n/p_0$,

$$\mathcal{H}_o = -\sqrt{p_0^2 - \pi_\perp^2} - qA_z \approx -p_0 + \frac{p_\perp^2}{2p_0} - qA_z$$

$$= -p_0 + \frac{p_\perp^2}{2p_0} - \frac{1}{2} p_0 K_n \left(x^2 - y^2\right). \qquad (4.133)$$

Let us consider the propagation of the beam from the transverse xy-plane at $z = z_i < z_\ell$ to the transverse plane at $z > z_r$. The beam propagates in free space from z_i to z_ℓ, passes through the lens from z_ℓ to z_r, and propagates through free space again from z_r to z. Let us specify the transfer map for the transverse coordinates and momenta of a beam particle by

$$\begin{pmatrix} x(z) \\ \frac{p_x(z)}{p_0} \\ y(z) \\ \frac{p_y(z)}{p_0} \end{pmatrix} = \mathcal{T}(z, z_i) \begin{pmatrix} x(z_i) \\ \frac{p_x(z_i)}{p_0} \\ y(z_i) \\ \frac{p_y(z_i)}{p_0} \end{pmatrix}. \qquad (4.134)$$

In view of the semigroup property of the transfer operator (4.104), we can write

$$\begin{aligned}\mathcal{T}(z,z_i) &= \mathcal{T}(z,z_r)\,\mathcal{T}(z_r,z_\ell)\,\mathcal{T}(z_\ell,z_i) \\ &= \mathcal{T}_D(z,z_r)\,\mathcal{T}_L(z_r,z_\ell)\,\mathcal{T}_D(z_\ell,z_i) \\ &= e^{\int_{z_r}^{z} dz\,:-\frac{p_\perp^2}{2p_0}:_z}\,e^{\int_{z_\ell}^{z_r} dz\,:-\left(\frac{p_\perp^2}{2p_0}-\frac{1}{2}p_0 K_n(x^2-y^2)\right):_z}\,e^{\int_{z_i}^{z_\ell} dz\,:-\frac{p_\perp^2}{2p_0}:_z} \\ &= e^{(z-z_r):-\frac{p_\perp^2}{2p_0}:_z}\,e^{(z_r-z_\ell):-\left(\frac{p_\perp^2}{2p_0}-\frac{1}{2}p_0 K_n(x^2-y^2)\right):_z}\,e^{(z_\ell-z_i):-\frac{p_\perp^2}{2p_0}:_z},\quad (4.135)\end{aligned}$$

where the subscripts D and L denote drift and propagation through the lens. Note that $\mathcal{T}_D$ and $\mathcal{T}_L$ are expressible as ordinary exponentials since $\mathcal{H}_o$ is z-independent in the free and lens regions. Note that we have dropped the additive constant term $-p_0$ from $\mathcal{H}_o$ in the transfer operators, since its Poisson brackets with the coordinates and momenta vanish. As we already know, we have

$$:-\frac{p_\perp^2}{2p_0}:_z \begin{pmatrix} x \\ \frac{p_x}{p_0} \\ y \\ \frac{p_y}{p_0} \end{pmatrix} = \begin{pmatrix} 0 & 1 & 0 & 0 \\ 0 & 0 & 0 & 0 \\ 0 & 0 & 0 & 1 \\ 0 & 0 & 0 & 0 \end{pmatrix} \begin{pmatrix} x \\ \frac{p_x}{p_0} \\ y \\ \frac{p_y}{p_0} \end{pmatrix}, \quad (4.136)$$

and hence

$$\begin{pmatrix} x(z) \\ \frac{p_x(z)}{p_0} \\ y(z) \\ \frac{p_y(z)}{p_0} \end{pmatrix} = \mathcal{T}_D(z,z_r) \begin{pmatrix} x(z_r) \\ \frac{p_x(z_r)}{p_0} \\ y(z_r) \\ \frac{p_y(z_r)}{p_0} \end{pmatrix}$$

$$= \begin{pmatrix} 1 & (z-z_r) & 0 & 0 \\ 0 & 1 & 0 & 0 \\ 0 & 0 & 1 & (z-z_r) \\ 0 & 0 & 0 & 1 \end{pmatrix} \begin{pmatrix} x(z_r) \\ \frac{p_x(z_r)}{p_0} \\ y(z_r) \\ \frac{p_y(z_r)}{p_0} \end{pmatrix}. \quad (4.137)$$

Similarly, we have

$$\begin{pmatrix} x(z_\ell) \\ \frac{p_x(z_\ell)}{p_0} \\ y(z_\ell) \\ \frac{p_y(z_\ell)}{p_0} \end{pmatrix} = \mathcal{T}_D(z_\ell,z_i) \begin{pmatrix} x(z_i) \\ \frac{p_x(z_i)}{p_0} \\ y(z_i) \\ \frac{p_y(z_i)}{p_0} \end{pmatrix}$$

$$= \begin{pmatrix} 1 & (z_\ell-z_i) & 0 & 0 \\ 0 & 1 & 0 & 0 \\ 0 & 0 & 1 & (z_\ell-z_i) \\ 0 & 0 & 0 & 1 \end{pmatrix} \begin{pmatrix} x(z_i) \\ \frac{p_x(z_i)}{p_0} \\ y(z_i) \\ \frac{p_y(z_i)}{p_0} \end{pmatrix}. \quad (4.138)$$

Classical Charged Particle Beam Optics

To get $\mathscr{T}_L(z_r, z_\ell)$, let us observe that

$$:-\left(\frac{p_\perp^2}{2p_0} - \frac{1}{2}p_0 K_n(x^2 - y^2)\right): _z \begin{pmatrix} x \\ \frac{p_x}{p_0} \\ y \\ \frac{p_y}{p_0} \end{pmatrix} = \begin{pmatrix} 0 & 1 & 0 & 0 \\ K_n & 0 & 0 & 0 \\ 0 & 0 & 0 & 1 \\ 0 & 0 & -K_n & 0 \end{pmatrix} \begin{pmatrix} x \\ \frac{p_x}{p_0} \\ y \\ \frac{p_y}{p_0} \end{pmatrix}. \tag{4.139}$$

Then, we get

$$\begin{pmatrix} x(z_r) \\ \frac{p_x(z_r)}{p_0} \\ y(z_r) \\ \frac{p_y(z_r)}{p_0} \end{pmatrix} = \mathscr{T}_L(z_r, z_\ell) \begin{pmatrix} x(z_\ell) \\ \frac{p_x(z_\ell)}{p_0} \\ y(z_\ell) \\ \frac{p_y(z_\ell)}{p_0} \end{pmatrix}$$

$$= \exp\left[w \begin{pmatrix} 0 & 1 & 0 & 0 \\ K_n & 0 & 0 & 0 \\ 0 & 0 & 0 & 1 \\ 0 & 0 & -K_n & 0 \end{pmatrix}\right] \begin{pmatrix} x(z_\ell) \\ \frac{p_x(z_\ell)}{p_0} \\ y(z_\ell) \\ \frac{p_y(z_\ell)}{p_0} \end{pmatrix}$$

$$= \left(\begin{pmatrix} \cosh(w\sqrt{K_n}) & \frac{1}{\sqrt{K_n}}\sinh(w\sqrt{K_n}) \\ \sqrt{K_n}\sinh(w\sqrt{K_n}) & \cosh(w\sqrt{K_n}) \end{pmatrix} \oplus \right.$$

$$\left. \begin{pmatrix} \cos(w\sqrt{K_n}) & \frac{1}{\sqrt{K_n}}\sin(w\sqrt{K_n}) \\ -\sqrt{K_n}\sin(w\sqrt{K_n}) & \cos(w\sqrt{K_n}) \end{pmatrix}\right) \begin{pmatrix} x(z_\ell) \\ \frac{p_x(z_\ell)}{p_0} \\ y(z_\ell) \\ \frac{p_y(z_\ell)}{p_0} \end{pmatrix},$$

$$\text{with } w = (z_r - z_\ell). \tag{4.140}$$

Here,

$$(A \oplus B) = \begin{pmatrix} A & O \\ O & B \end{pmatrix} \tag{4.141}$$

is the direct sum of the matrices A and B with O as the 2×2 null matrix with all its entries as zero. Thus, we have

$$\begin{pmatrix} x(z) \\ \frac{p_x(z)}{p_0} \\ y(z) \\ \frac{p_y(z)}{p_0} \end{pmatrix} = \mathscr{T}_D(z, z_r)\, \mathscr{T}_L(z_r, z_\ell)\, \mathscr{T}_D(z_\ell, z_i) \begin{pmatrix} x(z_i) \\ \frac{p_x(z_i)}{p_0} \\ y(z_i) \\ \frac{p_y(z_i)}{p_0} \end{pmatrix},$$

$$\mathscr{T}_D(z, z_r) = \begin{pmatrix} 1 & (z - z_r) & 0 & 0 \\ 0 & 1 & 0 & 0 \\ 0 & 0 & 1 & (z - z_r) \\ 0 & 0 & 0 & 1 \end{pmatrix},$$

$$\mathscr{T}_L(z_r, z_\ell) = \left(\begin{pmatrix} \cosh(w\sqrt{K_n}) & \frac{1}{\sqrt{K_n}}\sinh(w\sqrt{K_n}) \\ \sqrt{K_n}\sinh(w\sqrt{K_n}) & \cosh(w\sqrt{K_n}) \end{pmatrix} \oplus \right.$$

$$\left. \begin{pmatrix} \cos\left(w\sqrt{K_n}\right) & \frac{1}{\sqrt{K_n}}\sin\left(w\sqrt{K_n}\right) \\ -\sqrt{K_n}\sin\left(w\sqrt{K_n}\right) & \cos\left(w\sqrt{K_n}\right) \end{pmatrix} \right)$$

$$\mathscr{T}_D(z_\ell, z_i) = \begin{pmatrix} 1 & (z_\ell - z_i) & 0 & 0 \\ 0 & 1 & 0 & 0 \\ 0 & 0 & 1 & (z_\ell - z_i) \\ 0 & 0 & 0 & 1 \end{pmatrix}. \tag{4.142}$$

It is readily seen from this map that when $K_n > 0$ the lens is divergent in the xz-plane and convergent in the yz-plane. In other words, the normal magnetic quadrupole lens produces a line focus. When $K_n < 0$ the lens is convergent in the xz-plane and divergent in the yz-plane. Note that K_n has the dimensions of length^{-2}. In the weak field case, when $w\sqrt{|K_n|} \ll 1$, the aforementioned transfer map becomes

$$\begin{pmatrix} x(z) \\ \frac{p_x(z)}{p_0} \\ y(z) \\ \frac{p_y(z)}{p_0} \end{pmatrix} \approx \begin{pmatrix} 1 & (z - z_r) & 0 & 0 \\ 0 & 1 & 0 & 0 \\ 0 & 0 & 1 & (z - z_r) \\ 0 & 0 & 0 & 1 \end{pmatrix} \times$$

$$\begin{pmatrix} 1 & 0 & 0 & 0 \\ wK_n & 1 & 0 & 0 \\ 0 & 0 & 1 & 0 \\ 0 & 0 & -wK_n & 1 \end{pmatrix} \times$$

$$\begin{pmatrix} 1 & (z_\ell - z_i) & 0 & 0 \\ 0 & 1 & 0 & 0 \\ 0 & 0 & 1 & (z_\ell - z_i) \\ 0 & 0 & 0 & 1 \end{pmatrix} \begin{pmatrix} x(z_i) \\ \frac{p_x(z_i)}{p_0} \\ y(z_i) \\ \frac{p_y(z_i)}{p_0} \end{pmatrix}, \tag{4.143}$$

showing that the lens can be considered a thin lens ($w \approx 0$) with the focal lengths

$$\frac{1}{f^{(x)}} = -wK_n, \qquad \frac{1}{f^{(y)}} = wK_n. \tag{4.144}$$

When the beam is not ideally paraxial, more terms will have to be included in the optical Hamiltonian, and this will lead to nonlinearities in the transfer map.

Let us consider two thin lenses of focal lengths f_1 and f_2 kept separated by a distance d. The transfer matrix from the entrance of the lens with focal length f_1 to the exit of the lens with focal length f_2 will be

$$\begin{pmatrix} 1 & 0 \\ -\frac{1}{f_2} & 1 \end{pmatrix} \begin{pmatrix} 1 & d \\ 0 & 1 \end{pmatrix} \begin{pmatrix} 1 & 0 \\ -\frac{1}{f_1} & 1 \end{pmatrix} = \begin{pmatrix} 1 - \frac{d}{f_1} & d \\ -\left(\frac{1}{f_1} + \frac{1}{f_2} - \frac{d}{f_1 f_2}\right) & 1 - \frac{d}{f_2} \end{pmatrix}, \tag{4.145}$$

showing that the combination has the equivalent focal length given by

$$\frac{1}{f} = \frac{1}{f_1} + \frac{1}{f_2} - \frac{d}{f_1 f_2}. \tag{4.146}$$

This shows that a lens of focal length $\pm f$ can be considered effectively to be two lenses of focal length $\pm 2f$, respectively, joined together, i.e., $f_1 = f_2 = \pm 2f$, and $d = 0$ in the formula (4.146). This also shows that a doublet of focusing and defocusing lenses of focal lengths f and $-f$ with a distance d between them would have the effective focal length $f^2/d > 0$, and hence would behave as a focusing lens.

Let us consider a triplet of identical quadrupoles, with focal lengths f and $-f$ in the two transverse planes, arranged as follows. The first and the third quadrupoles, situated symmetrically about the middle quadrupole at a distance d from them, are focusing in the xz-plane and defocusing in the yz-plane. The middle quadrupole is rotated by $90°$ with respect to first (and the third) so that it is defocusing in the xz-plane and focusing in the yz-plane. The transfer matrix from the center of the first lens to the center of the third lens is given by

$$\begin{pmatrix} 1 & 0 & 0 & 0 \\ -\frac{1}{2f} & 1 & 0 & 0 \\ 0 & 0 & 1 & 0 \\ 0 & 0 & \frac{1}{2f} & 1 \end{pmatrix} \begin{pmatrix} 1 & d & 0 & 0 \\ 0 & 1 & 0 & 0 \\ 0 & 0 & 1 & d \\ 0 & 0 & 0 & 1 \end{pmatrix} \begin{pmatrix} 1 & 0 & 0 & 0 \\ \frac{1}{f} & 1 & 0 & 0 \\ 0 & 0 & 1 & 0 \\ 0 & 0 & -\frac{1}{f} & 1 \end{pmatrix} \times$$

$$\times \begin{pmatrix} 1 & d & 0 & 0 \\ 0 & 1 & 0 & 0 \\ 0 & 0 & 1 & d \\ 0 & 0 & 0 & 1 \end{pmatrix} \begin{pmatrix} 1 & 0 & 0 & 0 \\ -\frac{1}{2f} & 1 & 0 & 0 \\ 0 & 0 & 1 & 0 \\ 0 & 0 & \frac{1}{2f} & 1 \end{pmatrix}$$

$$= \begin{pmatrix} 1 - \frac{d}{2f^2} & 2d\left(1 + \frac{d}{2f}\right) & 0 & 0 \\ -\frac{d}{2f^2}\left(1 - \frac{d}{2f}\right) & 1 - \frac{d}{2f^2} & 0 & 0 \\ 0 & 0 & 1 - \frac{d}{2f^2} & 2d\left(1 - \frac{d}{2f}\right) \\ 0 & 0 & -\frac{d}{2f^2}\left(1 + \frac{d}{2f}\right) & 1 - \frac{d}{2f^2} \end{pmatrix}, \quad (4.147)$$

corresponding to net focusing effect in both the transverse planes when $d < 2f$.

In electron optical technology, for electron energies in the range of tens or hundreds of kilovolts to a few megavolts, magnetic quadrupole lenses are used, if at all, as components in aberration-correcting units for round lenses and in devices required to produce a line focus. It is mainly at higher energies, where round lenses are too weak, quadrupole lenses are utilized to provide the principal focusing field. As seen above, a triplet of quadrupoles can be arranged to have net focusing effect in both the transverse planes. Series of such quadrupole triplets are the main design elements of long beam transport lines or circular accelerators to provide a periodic focusing structure called a FODO-channel. FODO stands for F(ocusing)O(nonfocusing)D(efocusing)O(nonfocusing), where O can be a drift space or a bending magnet.

4.3.3 SKEW MAGNETIC QUADRUPOLE

A magnetic quadrupole associated with the magnetic field

$$\vec{B}(\vec{r}) = (-Q_s x, Q_s y, 0), \qquad (4.148)$$

corresponding to the vector potential

$$\vec{A}(\vec{r}) = (0, 0, -Q_s xy), \qquad (4.149)$$

is known as a skew magnetic quadrupole. If z_ℓ and $z_r > z_\ell$ are the boundaries of the quadrupole, then Q_s is a nonzero constant for $z_\ell \leq z \leq z_r$ and $Q_s = 0$ outside the quadrupole ($z < z_\ell, z > z_r$). For a paraxial beam propagating through this skew magnetic quadrupole along its optic axis, z-axis, the classical beam optical Hamiltonian will be

$$\mathcal{H}_o = \begin{cases} -p_0 + \frac{p_\perp^2}{2p_0}, & \text{for } z < z_\ell,\, z > z_r, \\ -p_0 + \frac{p_\perp^2}{2p_0} + p_0 K_s xy, & \text{for } z_\ell \leq z \leq z_r, \end{cases} \qquad (4.150)$$

where $K_s = qQ_s/p_0$. Now, if we make the transformation

$$\begin{pmatrix} x \\ p_x \\ y \\ p_y \end{pmatrix} = \frac{1}{\sqrt{2}} \begin{pmatrix} 1 & 0 & 1 & 0 \\ 0 & 1 & 0 & 1 \\ -1 & 0 & 1 & 0 \\ 0 & -1 & 0 & 1 \end{pmatrix} \begin{pmatrix} x' \\ p'_x \\ y' \\ p'_y \end{pmatrix} = \mathbb{R} \begin{pmatrix} x' \\ p'_x \\ y' \\ p'_y \end{pmatrix}, \qquad (4.151)$$

the Hamiltonian is seen to become

$$\mathcal{H}'_o = \begin{cases} -p_0 + \frac{p'^{\,2}_\perp}{2p_0}, & \text{for } z < z_\ell,\, z > z_r, \\ -p_0 + \frac{p'^{\,2}_\perp}{2p_0} - \frac{1}{2} p_0 K_s \left(x'^2 - y'^2 \right), & \text{for } z_\ell \leq z \leq z_r, \end{cases} \qquad (4.152)$$

same as for the propagation of the paraxial beam through a normal magnetic quadrupole (4.133) in the new (x', y') coordinate system, with K_n replaced by K_s. The above transformation corresponds to a clockwise rotation of the (x, y) coordinate axes by $\frac{\pi}{4}$ about the optic axis. Hence, the transfer map for the skew magnetic quadrupole can be obtained from the transfer map for the normal magnetic quadrupole as

$$\mathcal{T}_{sq} = \mathbb{R} \mathcal{T}_{nq} \mathbb{R}^{-1}, \qquad (4.153)$$

where the subscripts sq and nq stand for the skew magnetic quadrupole and the normal magnetic quadrupole, respectively. The beam optical Hamiltonian for propagation through free space is axially symmetric. Thus, the free space transfer map through any distance, say z' to z'', $\mathcal{T}_D(z'', z')$, is seen to satisfy the relation

$$\mathbb{R}(\theta) \mathcal{T}_D(z'', z') \mathbb{R}(\theta)^{-1} = \mathcal{T}_D(z'', z'), \quad \text{for any } \theta, \qquad (4.154)$$

where

$$\mathbb{R}(\theta) = \begin{pmatrix} \cos\theta & 0 & \sin\theta & 0 \\ 0 & \cos\theta & 0 & \sin\theta \\ -\sin\theta & 0 & \cos\theta & 0 \\ 0 & -\sin\theta & 0 & \cos\theta \end{pmatrix},$$

$$\mathscr{T}_D\left(z'',z'\right) = \begin{pmatrix} 1 & z''-z' & 0 & 0 \\ 0 & 1 & 0 & 0 \\ 0 & 0 & 1 & z''-z' \\ 0 & 0 & 0 & 1 \end{pmatrix}. \tag{4.155}$$

Note that $\mathbb{R}(\theta)^{-1} = \mathbb{R}(-\theta)$ and $\mathbb{R}$ in (4.151) is $\mathbb{R}\left(\frac{\pi}{4}\right)$. Then, from the relations (4.153) and (4.154), we get

$$\mathscr{T}_{sq}(z,z_i) = \mathscr{T}_D(z,z_r)\,\mathscr{T}_{sq,L}(z_r,z_\ell)\,\mathscr{T}_D(z_\ell,z_i)$$

$$= \mathscr{T}_D(z,z_r)\,\mathbb{R}\,\mathscr{T}_{nq,L}(z_r,z_\ell)\,\mathbb{R}^{-1}\,\mathscr{T}_D(z_\ell,z_i), \tag{4.156}$$

where $\mathbb{R}$ is as in (4.151), and $\mathscr{T}_{nq,L}(z_r,z_\ell)$ is the same as $\mathscr{T}_L(z_r,z_\ell)$ in (4.142) with K_n replaced by K_s. One can get $\mathscr{T}_{sq,L}(z_r,z_\ell)$ directly as follows. Observe that

$$:-\left(\frac{p_\perp^2}{2p_0}+p_0K_sxy\right):_z \begin{pmatrix} x \\ \frac{p_x}{p_0} \\ y \\ \frac{p_y}{p_0} \end{pmatrix} = \begin{pmatrix} 0 & 1 & 0 & 0 \\ 0 & 0 & -K_s & 0 \\ 0 & 0 & 0 & 1 \\ -K_s & 0 & 0 & 0 \end{pmatrix} \begin{pmatrix} x \\ \frac{p_x}{p_0} \\ y \\ \frac{p_y}{p_0} \end{pmatrix}. \tag{4.157}$$

Then, one gets

$$\begin{pmatrix} x(z_r) \\ \frac{p_x(z_r)}{p_0} \\ y(z_r) \\ \frac{p_y(z_r)}{p_0} \end{pmatrix} = \mathscr{T}_{sq,L}(z_r,z_\ell) \begin{pmatrix} x(z_\ell) \\ \frac{p_x(z_\ell)}{p_0} \\ y(z_\ell) \\ \frac{p_y(z_\ell)}{p_0} \end{pmatrix}$$

$$= \exp\left[w\begin{pmatrix} 0 & 1 & 0 & 0 \\ 0 & 0 & -K_s & 0 \\ 0 & 0 & 0 & 1 \\ -K_s & 0 & 0 & 0 \end{pmatrix}\right] \begin{pmatrix} x(z_\ell) \\ \frac{p_x(z_\ell)}{p_0} \\ y(z_\ell) \\ \frac{p_y(z_\ell)}{p_0} \end{pmatrix}$$

$$= \frac{1}{2}\begin{pmatrix} C^+ & \frac{1}{\sqrt{K_s}}S^+ & C^- & \frac{1}{\sqrt{K_s}}S^- \\ -\sqrt{K_s}S^- & C^+ & -\sqrt{K_s}S^+ & C^- \\ C^- & \frac{1}{\sqrt{K_s}}S^- & C^+ & \frac{1}{\sqrt{K_s}}S^+ \\ -\sqrt{K_s}S^+ & C^- & -\sqrt{K_s}S^- & C^+ \end{pmatrix} \begin{pmatrix} x(z_\ell) \\ \frac{p_x(z_\ell)}{p_0} \\ y(z_\ell) \\ \frac{p_y(z_\ell)}{p_0} \end{pmatrix},$$

with $w = (z_r - z_\ell)$. \hfill (4.158)

where
$$C^{\pm} = \cos(w\sqrt{K_s}) \pm \cosh(w\sqrt{K_s}),$$
$$S^{\pm} = \sin(w\sqrt{K_s}) \pm \sinh(w\sqrt{K_s}). \qquad (4.159)$$

It can be verified that $\mathscr{T}_{sq,L} = \mathbb{R}\mathscr{T}_{nq,L}(z_r, z_\ell)\mathbb{R}^{-1}$.

4.3.4 AXIALLY SYMMETRIC ELECTROSTATIC LENS

An axially symmetric, round, electrostatic lens with the axis along the z-direction consists of the electric field corresponding to the potential

$$\begin{aligned}\phi(\vec{r}_\perp, z) &= \sum_{n=0}^{\infty} \frac{(-1)^n}{(n!)^2 4^n} \phi^{(2n)}(z) r_\perp^{2n} \\ &= \phi(z) - \frac{1}{4}\phi''(z)r_\perp^2 + \frac{1}{64}\phi''''(z)r_\perp^4 - \cdots, \end{aligned} \qquad (4.160)$$

inside the lens region ($z_\ell < z < z_r$). Outside the lens, *i.e.*, ($z < z_\ell$, $z > z_r$), $\phi(z) = 0$. And, there is no magnetic field, *i.e.*, $\vec{A}(\vec{r}) = (0, 0, 0)$.

Starting with (4.2), the classical paraxial beam optical Hamiltonian of a general electrostatic lens, in the lens region, is obtained as follows:

$$\begin{aligned}\mathscr{H}_o &= -\sqrt{p_0^2 - \pi_\perp^2 - \frac{1}{c^2}2qE\phi} - qA_z \\ &= -\sqrt{p_0^2 - p_\perp^2 - \frac{1}{c^2}2qE\phi} \\ &= -p_0\sqrt{1 - \left(\frac{p_\perp^2}{p_0^2} + \frac{2qE\phi}{c^2 p_0^2}\right)} \\ &\approx -p_0\left\{1 - \frac{1}{2}\left(\frac{p_\perp^2}{p_0^2} + \frac{2qE\phi}{c^2 p_0^2}\right) - \frac{1}{8}\left(\frac{p_\perp^2}{p_0^2} + \frac{2qE\phi}{c^2 p_0^2}\right)^2\right\} \\ &\approx -p_0 + \frac{p_\perp^2}{2p_0} + \frac{qE\phi}{c^2 p_0} + \frac{qE\phi p_\perp^2}{2c^2 p_0^3} + \frac{q^2 E^2 \phi^2}{2c^4 p_0^3}. \end{aligned} \qquad (4.161)$$

For the round electrostatic lens, we get the classical beam optical Hamiltonian in the lens region, in the paraxial approximation, by using (4.160) in (4.161) as follows:

$$\begin{aligned}\mathscr{H}_o \approx &-p_0 + \frac{p_\perp^2}{2p_0} + \frac{qE}{c^2 p_0}\left(\phi(z) - \frac{1}{4}\phi''(z)r_\perp^2\right) + \frac{qE p_\perp^2}{2c^2 p_0^3}\left(\phi(z) - \frac{1}{4}\phi''(z)r_\perp^2\right) \\ &+ \frac{q^2 E^2}{2c^4 p_0^3}\left(\phi(z)^2 - \frac{1}{2}\phi(z)\phi''(z)r_\perp^2\right)\end{aligned}$$

$$\approx -\left(1 - \frac{qE}{c^2 p_0^2}\phi(z) - \frac{q^2 E^2}{2c^4 p_0^4}\phi^2(z)\right) p_0 + \left(1 + \frac{qE}{c^2 p_0^2}\phi(z)\right) \frac{p_\perp^2}{2p_0}$$
$$- \frac{qE}{4c^2 p_0}\left(1 + \frac{qE}{c^2 p_0^2}\phi(z)\right) \phi''(z) r_\perp^2. \tag{4.162}$$

In the field-free regions outside the lens, the optical Hamiltonian will be $-p_0 + (p_\perp^2/2p_0)$.

The aforementioned optical Hamiltonian of the round electrostatic lens is seen to be similar to the optical Hamiltonian of the round magnetic lens (4.24), except for the absence of the rotation term and the presence of an z-dependent coefficient for the drift term $p_\perp^2/2p_0$. The first constant term, though z-dependent, can be ignored as before, since it has vanishing Poisson bracket with the transverse coordinates and momenta. The paraxial Hamiltonian in (4.162), being quadratic in $r_\perp$ and $p_\perp$, leads to a linear transfer map for the transverse coordinates and momenta, and it is straightforward to calculate the map following the same procedure used in the case of the round magnetic lens. When the beam is not paraxial, there will be aberrations. We shall not pursue the performance of this lens further. Some uses of electrostatic lenses are in the extraction, preparation, and initial acceleration of electron and ion beams in a variety of applications.

4.3.5 ELECTROSTATIC QUADRUPOLE

The electric field of an ideal electrostatic quadrupole lens with the optic axis along the z-direction corresponds to the potential

$$\phi(\vec{r}_\perp, z) = \begin{cases} \frac{1}{2} Q_e (x^2 - y^2) & \text{in the lens region } (z_\ell < z < z_r), \\ 0 & \text{outside the lens } (z < z_\ell, z > z_r), \end{cases} \tag{4.163}$$

where Q_e is a constant and $z_r - z_\ell = w$ is the width of the lens. Substituting this expression for ϕ and $\vec{A} = (0,0,0)$, since there is no magnetic field, in the general form of the optical Hamiltonian of an electrostatic lens (4.161), we get

$$\mathscr{H}_o \approx -p_0 + \frac{p_\perp^2}{2p_0} + \frac{qEQ_e}{2c^2 p_0}(x^2 - y^2), \tag{4.164}$$

as the classical beam optical Hamiltonian of the electrostatic quadrupole lens in the paraxial approximation.

Simply by comparing the Hamiltonian of the normal magnetic quadrupole lens (4.133) with the Hamiltonian of the electrostatic quadrupole lens (4.164), it is readily seen that the electrostatic quadrupole lens is convergent in the xz-plane and divergent in the yz-plane when

$$K_e = \frac{qEQ_e}{c^2 p_0^2} > 0. \tag{4.165}$$

When $K_e < 0$, this lens is divergent in the xz-plane and convergent in the yz-plane. Note that K_e has the dimension of length^{-2}. In the weak field case, when $w^2 \ll 1/|K_e|$, the lens can be considered as thin and has focal lengths

$$\frac{1}{f^{(x)}} = wK_e, \qquad \frac{1}{f^{(y)}} = -wK_e. \tag{4.166}$$

Deviations from the ideal behavior result from nonparaxial conditions of the beam. We shall not pursue the performance of this lens further here.

4.4 BENDING MAGNET: AN OPTICAL ELEMENT WITH A CURVED OPTIC AXIS

Circular accelerators and storage rings require dipole magnets for bending the beams to guide them along the desired curved paths. By applying a constant magnetic field in the vertical direction using a dipole magnet, the beam is bent along a circular arc in the horizontal plane. The circular arc of radius of curvature ρ, the design trajectory of the particle, is the optic axis of the bending magnet. It is natural to use the arclength, say S, measured along the optic axis from some reference point as the independent coordinate, instead of z. Let the reference particle moving along the design trajectory carry an orthonormal XY-coordinate frame with it. The X-axis is taken to be perpendicular to the tangent to the design orbit and in the same horizontal plane as the trajectory, and the Y-axis is taken to be in the vertical direction perpendicular to both the X-axis and the trajectory. The curved S-axis is along the design trajectory and perpendicular to both the X and Y axes at any point on the design trajectory. The instantaneous position of the reference particle in the design trajectory at an arclength S from the reference point corresponds to $X = 0$ and $Y = 0$. Let any particle of the beam have coordinates (x, y, z) with respect to a fixed right-handed Cartesian coordinate frame, with its origin at the reference point on the design trajectory from which the arclength S is measured. Then, the two sets of coordinates of any particle of the beam, (X, Y, S) and (x, y, z), will be related as follows:

$$x = (\rho + X)\cos\left(\frac{S}{\rho}\right) - \rho, \quad z = (\rho + X)\sin\left(\frac{S}{\rho}\right), \quad y = Y \tag{4.167}$$

To study the motion of the beam particles through the dipole magnet with a circular arc as the optic axis, we have to transform the optical Hamiltonian to the (X, Y, S)-coordinate system. This can be done through a canonical transformation with the generating function chosen as

$$F_3(p_x, p_y, p_z, X, Y, S) = -\left[(\rho + X)\cos\left(\frac{S}{\rho}\right) - \rho\right] p_x$$
$$- \left[(\rho + X)\sin\left(\frac{S}{\rho}\right)\right] p_z - Y p_y, \tag{4.168}$$

such that

$$x_j = -\frac{\partial F_3}{\partial p_j}, \qquad P_j = -\frac{\partial F_3}{\partial X_j}. \tag{4.169}$$

The first part of the above equation just reproduces (4.167) giving the relation between (x,y,z) and (X,Y,S). The second part gives the new momentum components conjugate to (X,Y,S): with $1/\rho = \kappa$, the curvature, and $\zeta = 1 + \kappa X$,

$$\begin{pmatrix} P_X \\ \frac{1}{\zeta} P_S \\ P_Y \end{pmatrix} = \begin{pmatrix} \cos(\kappa S) & \sin(\kappa S) & 0 \\ -\sin(\kappa S) & \cos(\kappa S) & 0 \\ 0 & 0 & 1 \end{pmatrix} \begin{pmatrix} p_x \\ p_z \\ p_y \end{pmatrix} \qquad (4.170)$$

Since X and S axes are rotated clockwise through the angle S/ρ relative to the x and z axes, the components of the vector potential in the two frames are related as

$$\begin{pmatrix} A_X \\ A_S \\ A_Y \end{pmatrix} = \begin{pmatrix} \cos(\kappa S) & \sin(\kappa S) & 0 \\ -\sin(\kappa S) & \cos(\kappa S) & 0 \\ 0 & 0 & 1 \end{pmatrix} \begin{pmatrix} A_x \\ A_z \\ A_y \end{pmatrix}. \qquad (4.171)$$

For more details, see, e.g., Wolski [192].

The general Hamiltonian of the particle in an electromagnetic field (2.37) is

$$H(x,y,z,p_x,p_y,p_z,t) = \sqrt{m^2 c^4 + c^2 (\vec{p} - q\vec{A})^2} + q\phi, \qquad (4.172)$$

in the Cartesian coordinate system. If we change it to (X,Y,S) coordinate system using the relations (4.170) and (4.171), we get

$$\begin{aligned} H(X,Y,S,P_X,P_Y,P_S,t) &= \left\{ m^2 c^4 + c^2 \left[(P_X - qA_X)^2 + (P_Y - qA_Y)^2 \right.\right. \\ &\left.\left. + \frac{1}{\zeta^2} (P_S - q\zeta A_S)^2 \right] \right\}^{1/2} + q\phi, \end{aligned} \qquad (4.173)$$

in which (x,y,z) are expressed in terms of (X,Y,S) using (4.167). Now, we have to change the independent variable in the above Hamiltonian from t to S, and the desired classical beam optical Hamiltonian in the chosen curvilinear coordinate system is given as the solution for $-P_S$:

$$\begin{aligned} \widetilde{\mathscr{H}_o} &= -P_S \\ &= -\frac{1}{c} \zeta \left(\sqrt{(E - q\phi)^2 - m^2 c^4 - c^2 \left(\vec{P}_\perp - q\vec{A}_\perp \right)^2} \right) - \zeta q A_S. \end{aligned} \qquad (4.174)$$

From (4.167), we find that ds, the distance between two infinitesimally close points, is given by

$$ds^2 = dx^2 + dy^2 + dz^2 = dX^2 + dY^2 + \zeta^2 dS^2. \qquad (4.175)$$

In general, in an orthogonal curvilinear (u,v,w)-coordinate system with the line element ds given by

$$ds^2 = h_u^2 du^2 + h_v^2 dv^2 + h_w^2 dw^2, \qquad (4.176)$$

the curl of any vector $\vec{A}$ has the following expression (see, *e.g.*, Arfken, Weber, and Harris [3]):

$$(\vec{\nabla} \times \vec{A})_u = \frac{1}{h_v h_w} \left(\frac{\partial (h_w A_w)}{\partial v} - \frac{\partial (h_v A_v)}{\partial w} \right),$$

$$(\vec{\nabla} \times \vec{A})_v = \frac{1}{h_w h_u} \left(\frac{\partial (h_u A_u)}{\partial w} - \frac{\partial (h_w A_w)}{\partial u} \right),$$

$$(\vec{\nabla} \times \vec{A})_w = \frac{1}{h_u h_v} \left(\frac{\partial (h_v A_v)}{\partial u} - \frac{\partial (h_u A_u)}{\partial v} \right). \tag{4.177}$$

Thus, in the (X, Y, S)-coordinate system, we have

$$B_X = (\vec{\nabla} \times \vec{A})_X = \frac{1}{\zeta} \left(\frac{\partial (\zeta A_S)}{\partial Y} - \frac{\partial A_Y}{\partial S} \right),$$

$$B_Y = (\vec{\nabla} \times \vec{A})_Y = \frac{1}{\zeta} \left(\frac{\partial A_X}{\partial S} - \frac{\partial (\zeta A_S)}{\partial X} \right),$$

$$B_S = (\vec{\nabla} \times \vec{A})_S = \frac{\partial A_Y}{\partial X} - \frac{\partial A_X}{\partial Y}. \tag{4.178}$$

The dipole magnetic field is a uniform field perpendicular to the design trajectory such that the beam particle is bent along the design trajectory. So, we should have

$$B_X = 0, \quad B_Y = B_0, \quad B_S = 0. \tag{4.179}$$

Using (4.178), we find that the corresponding vector potential in the curved coordinate system can be taken as

$$A_X = 0, \quad A_Y = 0, \quad A_S = -B_0 \left(X - \frac{\kappa X^2}{2\zeta} \right). \tag{4.180}$$

Let us now find the optical Hamiltonian obtained by substituting in (4.174) $\phi = 0$ and choosing the vector potential as in (4.180). The resulting classical beam optical Hamiltonian for the dipole magnet, in the paraxial approximation, is

$$\begin{aligned}\widetilde{\mathcal{H}}_o &= -\frac{1}{c}\zeta\sqrt{E^2 - m^2c^4 - c^2 P_\perp^2} + \zeta q B_0 \left(X - \frac{\kappa X^2}{2\zeta} \right) \\ &= -\zeta\sqrt{p_0^2 - P_\perp^2} + qB_0 \left(X\zeta - \frac{1}{2}\kappa X^2 \right) \\ &\approx \zeta \left(-p_0 + \frac{P_\perp^2}{2p_0} \right) + qB_0 \left(X + \frac{1}{2}\kappa X^2 \right) \\ &\approx -p_0 + \frac{P_\perp^2}{2p_0} + (qB_0 - \kappa p_0) X + \frac{1}{2} q B_0 \kappa X^2. \end{aligned} \tag{4.181}$$

Let us now find what would be the effect of the term $\propto X$ in the above Hamiltonian: with $\Delta S = S - S_i$,

$$e^{\int_{S_i}^{S} ds\ :-(qB_0-\kappa p_0)X:_S} \begin{pmatrix} X(S_i) \\ \frac{P_X(S_i)}{p_0} \\ Y(S_i) \\ P_Y(S_i) \end{pmatrix}$$

$$= e^{:-\Delta S(qB_0-\kappa p_0)X:_S} \begin{pmatrix} X(S_i) \\ \frac{P_X(S_i)}{p_0} \\ Y(S_i) \\ P_Y(S_i) \end{pmatrix}$$

$$= \sum_{0}^{\infty} \frac{1}{n!} : -\Delta S (qB_0 - \kappa p_0) X :_S^n \begin{pmatrix} X(S_i) \\ \frac{P_X(S_i)}{p_0} \\ Y(S_i) \\ P_Y(S_i) \end{pmatrix}$$

$$= \begin{pmatrix} X(S_i) \\ \frac{1}{p_0}(P_X(S_i) + \Delta S(qB_0 - \kappa p_0)) \\ Y(S_i) \\ P_Y(S_i) \end{pmatrix}, \qquad (4.182)$$

where S_i and $S > S_i$ are two points on the curved S-axis or the design trajectory. This shows that the presence of the term $\propto X$ in the Hamiltonian causes a change in P_X and hence a deflection of the particle in the horizontal plane perpendicular to the design trajectory. If the curvature of the design trajectory is properly matched to the dipole magnetic field, *i.e.*,

$$qB_0 = \kappa p_0 \quad \text{or,} \quad B_0\rho = \frac{p_0}{q}, \qquad (4.183)$$

then the term $(qB_0 - \kappa p_0) X$ in the Hamiltonian vanishes, and if a particle is initially on the design trajectory, it will remain on the design trajectory through the dipole. The condition (4.183) follows simply from the fact that if a constant magnetic field of magnitude B bends the path of a particle of charge q moving with a speed v perpendicular to it in a circular arc of radius ρ, then we must have

$$\frac{\gamma m v^2}{\rho} = qBv, \quad \text{or,} \quad \frac{\gamma m v}{q} = \frac{p}{q} = B\rho. \qquad (4.184)$$

The quantity $B\rho$ is called magnetic rigidity. Thus, if a beam of design momentum p_0 is to be bent by a circular arc of radius ρ by a dipole magnet, the magnetic field should have the magnitude $B_0 = p_0/(q\rho)$. With the condition (4.183) satisfied, the classical beam optical Hamiltonian of the dipole magnet becomes

$$\widetilde{\mathscr{H}}_{o,d} = -p_0 + \frac{P_\perp^2}{2p_0} + \frac{1}{2}p_0\kappa^2 X^2$$

$$= -p_0 + \left(\frac{P_X^2}{2p_0} + \frac{1}{2}p_0\kappa^2 X^2\right) + \frac{P_Y^2}{2p_0}. \quad (4.185)$$

which is independent of S. Then, the transfer map for the dipole is seen to be

$$\begin{pmatrix} X(S) \\ \frac{P_X(S)}{p_0} \\ Y(S) \\ \frac{P_Y(S)}{p_0} \end{pmatrix} = e^{\int_{S_i}^{S} d\mathbf{s} \, : -\widetilde{\mathscr{H}}_{o,d}(\mathbf{s}): S} \begin{pmatrix} X(S_i) \\ \frac{P_X(S_i)}{p_0} \\ Y(S_i) \\ \frac{P_Y(S_i)}{p_0} \end{pmatrix}$$

$$= \exp\left[\Delta S \begin{pmatrix} 0 & 1 & 0 & 0 \\ -\kappa^2 & 0 & 0 & 0 \\ 0 & 0 & 0 & 1 \\ 0 & 0 & 0 & 0 \end{pmatrix}\right] \begin{pmatrix} X(S_i) \\ \frac{P_X(S_i)}{p_0} \\ Y(S_i) \\ \frac{P_Y(S_i)}{p_0} \end{pmatrix}$$

$$= \begin{pmatrix} \cos(\kappa\Delta S) & \frac{1}{\kappa}\sin(\kappa\Delta S) & 0 & 0 \\ -\kappa\sin(\kappa\Delta S) & \cos(\kappa\Delta S) & 0 & 0 \\ 0 & 0 & 1 & \Delta S \\ 0 & 0 & 0 & 1 \end{pmatrix} \begin{pmatrix} X(S_i) \\ \frac{P_X(S_i)}{p_0} \\ Y(S_i) \\ \frac{P_Y(S_i)}{p_0} \end{pmatrix}.$$

(4.186)

This shows that a particle entering the dipole magnet along the design trajectory ($X = 0, Y = 0, P_X = 0, P_Y = 0$) will follow exactly the curved design trajectory. Such a steering of the beam is the purpose of the dipole magnet. Any other particle of the paraxial beam making a small angle with the design trajectory will have a small oscillation in the X-direction about the design trajectory and will have free propagation along the Y-direction.

4.5 NONRELATIVISTIC CLASSICAL CHARGED PARTICLE BEAM OPTICS

We shall now see how the nonrelativistic classical charged particle beam optics becomes an approximation of the relativistic classical charged particle beam optics. We have been so far using the exact relativistic expression for the energy of a charged particle in an electromagnetic field,

$$E = \sqrt{m^2c^4 + c^2\vec{\pi}^2} + q\phi, \quad (4.187)$$

in deriving the classical beam optical Hamiltonians for the various charged particle optical elements. The expression (4.187) for energy is exact and valid for all values of particle momentum irrespective of its velocity. In high-energy particle accelerators, the velocities of the particles being accelerated are very high, and so one has to use the relativistic expression (4.187). However, in applications like the common electron microscopy, or some ion accelerators, where the velocities of the beam particles are not so high it would be appropriate to use the nonrelativistic approximation. So, let us try to understand the nonrelativistic approximation of the above formalism.

In nonrelativistic situations when the velocities involved are $\ll c$, the expression for energy can be approximated as

$$\begin{aligned} E &= mc^2 \sqrt{1 + \frac{\vec{\pi}^2}{m^2 c^2}} + q\phi \\ &\approx mc^2 \left(1 + \frac{\vec{\pi}^2}{2m^2 c^2}\right) + q\phi \qquad (\because |\vec{\pi}| \ll mc) \\ &= mc^2 + \frac{\vec{\pi}^2}{2m} + q\phi, \end{aligned} \qquad (4.188)$$

which is the sum of rest energy, kinetic energy, and potential energy of the particle. Since the rest energy is not involved in any nonrelativistic interactions, one can consider the sum of kinetic energy and potential energy as the total energy of the particle. Thus, we take

$$E = \frac{\vec{\pi}^2}{2m} + q\phi \qquad (4.189)$$

as the total nonrelativistic energy of a particle. Correspondingly, the nonrelativistic Hamiltonian of the particle becomes

$$H_{\text{NR}} = \frac{\vec{\pi}^2}{2m} + q\phi. \qquad (4.190)$$

Now, if we use the procedure of changing the independent variable from time t to z, the variable along the straight optic axis of a system, starting with this Hamiltonian, we obtain

$$\begin{aligned} \mathcal{H}_{o,\text{NR}} &= -p_z = -\sqrt{2m(E - q\phi) - \vec{\pi}_\perp^2} - qA_z \\ &= -\sqrt{p_0^2 - \vec{\pi}_\perp^2 - 2mq\phi} - qA_z, \end{aligned} \qquad (4.191)$$

taking p_0 as the design momentum such that $E = \frac{p_0^2}{2m}$ is the total conserved energy of the beam particle in the system. Taking the nonrelativistic expression for the momentum $p_0 = mv_0$, we can write

$$\mathcal{H}_{o,\text{NR}} = -\sqrt{m^2 v_0^2 - \vec{\pi}_\perp^2 - 2mq\phi} - qA_z. \qquad (4.192)$$

We have found in (4.2) the relativistic classical beam optical Hamiltonian of a general electromagnetic optical element to be

$$\mathcal{H}_o = -\sqrt{p_0^2 - \vec{\pi}_\perp^2 - \frac{2Eq\phi}{c^2}} - qA_z, \qquad (4.193)$$

derived under the condition $E \gg q\phi$ as is usually the case in optical elements and which is anyway true for purely magnetic optical elements with $\phi = 0$. In (4.193), E is the total conserved energy of the particle given by $\sqrt{m^2c^4 + c^2 p_0^2} = \gamma mc^2$, with $p_0 = \gamma m v_0$. Thus, we can write

$$\mathcal{H}_o = -\sqrt{\gamma^2 m^2 v_0^2 - \pi_\perp^2 - 2\gamma mq\phi} - qA_z. \qquad (4.194)$$

Now, comparing (4.192) and (4.194), it is clear that the relativistic optical Hamiltonian can be obtained from the nonrelativistic optical Hamiltonian by the replacement

$$m \longrightarrow \gamma m. \qquad (4.195)$$

Thus, we find that one can extend the results based on the nonrelativistic optical Hamiltonian to get the corresponding relativistic results by replacing the rest mass m of the particle by γm (the so-called relativistic mass) under the condition $E \gg q\phi$. This is a practice often used in classical charged particle optics. Similarly, one can make nonrelativistic approximations of the results derived from the relativistic optical Hamiltonian by using the limit $\gamma \longrightarrow 1$ carefully.

5 Quantum Charged Particle Beam Optics
Scalar Theory for Spin-0 and Spinless Particles

5.1 GENERAL FORMALISM OF QUANTUM CHARGED PARTICLE BEAM OPTICS

In the classical theory of charged particle beam optics, our interest is to study the evolution of the state (position and momentum) of the beam particle along the optic axis of an optical system through which the beam is propagating. Such a study leads to an understanding of the performance of the concerned optical system based on classical mechanics. In the quantum theory of charged particle beam optics, our interest is to study the evolution of the state (wave function, or density operator) of the beam particle along the optic axis of an optical system through which the beam is propagating. This study should lead to an understanding of the performance of the concerned optical system based on quantum mechanics, and the understanding based on classical mechanics should emerge as an approximation. As we have seen already, the fundamental equations of quantum mechanics prescribe the time evolution of the state of a particle through a Hamiltonian operator governing the system. Exactly like the independent variable was changed from time to the coordinate along the optic axis of the concerned system in the classical Hamiltonian theory of charged particle beam optics, we have to rewrite the basic quantum mechanical equations as equations for the evolution of the quantum state of the particle along the coordinate, say z, of the optic axis of the concerned system, *i.e.*, we have to cast the concerned evolution equations in the form

$$i\hbar \frac{\partial \psi(\vec{r}_\perp, z)}{\partial z} = \widehat{\mathcal{H}}_o \psi(\vec{r}_\perp, z). \qquad (5.1)$$

Note that $i\hbar(\partial/\partial z) = -\widehat{p}_z$ corresponds to the quantum charged particle beam optical Hamiltonian operator $\widehat{\mathcal{H}}_o$.

In this chapter, we shall consider the scalar quantum theory of charged particle beam optics based on the relativistic Klein–Gordon equation and the nonrelativistic Schrödinger equation. The Klein–Gordon equation can be used for spin-0 particles, and all spinless relativistic particles, (*i.e.*, particles for which we can ignore the spin). The nonrelativistic Schrödinger equation can be used for any nonrelativistic particle for which we can ignore the spin. Since the nonrelativistic Schrödinger equation is an approximation of the Klein–Gordon equation, we shall consider first the

Klein–Gordon equation. This chapter is essentially based on the work of Jagannathan and Khan [82], Khan and Jagannathan [90], and Khan [91], on the nonrelativistic scalar theory of quantum charged particle beam optics based on the nonrelativistic Schrödinger equation, and the relativistic scalar theory of quantum charged particle beam optics based on the Klein–Gordon equation, following the earlier work by Jagannathan, Simon, Sudarshan, and Mukunda [79] and Jagannathan [80] on the spinor theory of quantum charged particle beam optics based on the Dirac equation. Spinor theory of quantum charged particle beam optics based on the Dirac–Pauli equation was developed later by Conte, Jagannathan, Khan, and Pusterla [26]. In the formulation of the relativistic quantum charged particle beam optics, we shall follow the philosophy of Hawkes and Kasper [72]: There is no need for a Lorentz-covariant formulation, and in practically all situations, the description in the laboratory frame is perfectly adequate.

5.2 RELATIVISTIC QUANTUM CHARGED PARTICLE BEAM OPTICS BASED ON THE KLEIN–GORDON EQUATION

5.2.1 GENERAL FORMALISM

We are concerned with a beam of identical particles of charge q and rest mass m propagating through a stationary electromagnetic field, constituting a charged particle optical system, such as a round magnetic lens in an electron microscope or a magnetic quadrupole in a particle accelerator. Let us ignore the spin of the particle and hence consider the Klein–Gordon equation as the appropriate equation for treating the quantum dynamics of the particle when its motion is relativistic. We consider the optical system to be, in practice, localized in a definite region of space so that the regions outside the system can be considered field-free or free space. We shall assume, throughout, the beam to be monoenergetic, paraxial or quasiparaxial, and forward moving along the optic axis of the system. All the beam particles are assumed to have the same positive energy $E = \sqrt{m^2c^4 + c^2 p_0^2}$, where p_0 is magnitude of the design momentum. We assume that all the particles of the beam enter the optical system from the free space outside the system with this constant energy. Let $\phi(\vec{r})$ and $\vec{A}(\vec{r})$ be the electric scalar potential and the magnetic vector potential of the time-independent electromagnetic field of the optical system. In passing through the system or in the process of scattering of the particle by the system, the total energy of the particle E will be conserved, and hence it is time-independent. After passing through the system, the particle will emerge outside in the free space with the same energy E it had when entering the system. In general, we shall consider the system to have a straight optic axis along the z-direction, except in the case of the bending magnet.

The scalar Klein–Gordon equation governing the quantum mechanics of the particle in an electromagnetic field is

$$\left(i\hbar\frac{\partial}{\partial t} - q\phi(\vec{r},t)\right)^2 \Psi(\vec{r},t) = \left[m^2c^4 + c^2\left(\vec{p} - q\vec{A}(\vec{r},t)\right)^2\right]\Psi(\vec{r},t), \qquad (5.2)$$

Scalar Theory: Spin-0 and Spinless Particles

as we have already seen (3.586). Since we are dealing with a time-independent system, we can take the wave function of the beam particle to be

$$\Psi(\vec{r},t) = e^{-iEt/\hbar}\psi(\vec{r}). \tag{5.3}$$

Then, the equation (5.2) becomes

$$e^{-iEt/\hbar}(E-q\phi)^2\psi(\vec{r}) = e^{-iEt/\hbar}\left(m^2c^4+c^2\widehat{\pi}^2\right)\psi(\vec{r}), \tag{5.4}$$

or

$$\left[(E-q\phi)^2 - m^2c^4 - c^2\widehat{\pi}_\perp^2 - c^2\widehat{\pi}_z^2\right]\psi(\vec{r}) = 0. \tag{5.5}$$

We can rewrite this equation as

$$c^2\widehat{\pi}_z^2\psi(\vec{r}) = \left[E^2 - m^2c^4 - q\phi(2E-q\phi) - c^2\widehat{\pi}_\perp^2\right]\psi(\vec{r}). \tag{5.6}$$

Dividing throughout by c^2, noting that $E^2 - m^2c^4 = c^2p_0^2$, and introducing the notation $q\phi(2E-q\phi) = c^2\tilde{p}^2$, we get

$$\widehat{\pi}_z^2\psi(\vec{r}) = \left(p_0^2 - \tilde{p}^2 - \widehat{\pi}_\perp^2\right)\psi(\vec{r}). \tag{5.7}$$

Getting the clue from the Feshbach–Villars formalism of the Klein–Gordon equation, let us introduce a two-component wave function

$$\begin{pmatrix} \psi_1 \\ \psi_2 \end{pmatrix} = \begin{pmatrix} \psi \\ \frac{\widehat{\pi}_z}{p_0}\psi \end{pmatrix}. \tag{5.8}$$

Then, we can write (5.7) equivalently as

$$\frac{\widehat{\pi}_z}{p_0}\begin{pmatrix} \psi_1 \\ \psi_2 \end{pmatrix} = \begin{pmatrix} 0 & 1 \\ \frac{1}{p_0^2}(p_0^2 - \tilde{p}^2 - \widehat{\pi}_\perp^2) & 0 \end{pmatrix}\begin{pmatrix} \psi_1 \\ \psi_2 \end{pmatrix}. \tag{5.9}$$

Now, let

$$\begin{pmatrix} \psi_+ \\ \psi_- \end{pmatrix} = M\begin{pmatrix} \psi_1 \\ \psi_2 \end{pmatrix} = \frac{1}{2}\begin{pmatrix} 1 & 1 \\ 1 & -1 \end{pmatrix}\begin{pmatrix} \psi_1 \\ \psi_2 \end{pmatrix}$$

$$= \frac{1}{2}\begin{pmatrix} \psi + \frac{\widehat{\pi}_z}{p_0}\psi \\ \psi - \frac{\widehat{\pi}_z}{p_0}\psi \end{pmatrix}. \tag{5.10}$$

Then,

$$\frac{\widehat{\pi}_z}{p_0}\begin{pmatrix} \psi_+ \\ \psi_- \end{pmatrix} = \frac{1}{p_0}\left(-i\hbar\frac{\partial}{\partial z} - qA_z\right)\begin{pmatrix} \psi_+ \\ \psi_- \end{pmatrix}$$

$$= M\begin{pmatrix} 0 & 1 \\ \frac{1}{p_0^2}(p_0^2-\tilde{p}^2-\widehat{\pi}_\perp^2) & 0 \end{pmatrix}M^{-1}\begin{pmatrix} \psi_+ \\ \psi_- \end{pmatrix}$$

$$= \begin{pmatrix} 1 - \frac{1}{2p_0^2}(\widehat{\pi}_\perp^2+\tilde{p}^2) & -\frac{1}{2p_0^2}(\widehat{\pi}_\perp^2+\tilde{p}^2) \\ \frac{1}{2p_0^2}(\widehat{\pi}_\perp^2+\tilde{p}^2) & -1+\frac{1}{2p_0^2}(\widehat{\pi}_\perp^2+\tilde{p}^2) \end{pmatrix}\begin{pmatrix} \psi_+ \\ \psi_- \end{pmatrix}. \tag{5.11}$$

Rearranging this equation, we get

$$i\hbar \frac{\partial}{\partial z}\begin{pmatrix} \psi_+ \\ \psi_- \end{pmatrix} = \widehat{H}\begin{pmatrix} \psi_+ \\ \psi_- \end{pmatrix},$$

$$\widehat{H} = -p_0\sigma_z + \widehat{\mathscr{E}} + \widehat{\mathscr{O}},$$

$$\widehat{\mathscr{E}} = -qA_z I + \frac{1}{2p_0}\left(\widehat{\pi}_\perp^2 + \tilde{p}^2\right)\sigma_z,$$

$$\widehat{\mathscr{O}} = \frac{i}{2p_0}\left(\widehat{\pi}_\perp^2 + \tilde{p}^2\right)\sigma_y. \tag{5.12}$$

Note that $\widehat{H}$ has been partitioned, apart from the leading term $-p_0\sigma_z$, into an even term $\widehat{\mathscr{E}}$ that does not couple ψ_+ and ψ_- and an odd term $\widehat{\mathscr{O}}$ that couples ψ_+ and ψ_-.

To understand (5.12) better, let us see what it implies in the case of propagation of the beam in free space. In free space, with $\phi = 0$ and $\vec{A} = (0,0,0)$, we have

$$i\hbar\frac{\partial}{\partial z}\begin{pmatrix}\psi_+ \\ \psi_-\end{pmatrix} = \widehat{H}\begin{pmatrix}\psi_+ \\ \psi_-\end{pmatrix}$$

$$\widehat{H} = -p_0\sigma_z + \widehat{\mathscr{E}} + \widehat{\mathscr{O}}$$

$$= -p_0\sigma_z + \frac{1}{2p_0}\widehat{p}_\perp^2 \sigma_z + \frac{i}{2p_0}\widehat{p}_\perp^2 \sigma_y$$

$$= \begin{pmatrix} -p_0 + \frac{\widehat{p}_\perp^2}{2p_0} & +\frac{\widehat{p}_\perp^2}{2p_0} \\ -\frac{\widehat{p}_\perp^2}{2p_0} & p_0 - \frac{\widehat{p}_\perp^2}{2p_0} \end{pmatrix}. \tag{5.13}$$

If we take

$$\psi(\vec{r}) = \phi_{\vec{p}_0}(\vec{r}) = \frac{1}{(2\pi\hbar)^{3/2}}e^{\frac{i}{\hbar}\vec{p}_0\cdot\vec{r}}, \tag{5.14}$$

a plane wave of momentum $\vec{p}_0 = (p_{0x}, p_{0y}, p_{0z})$, we have

$$\begin{pmatrix} \psi_+ \\ \psi_- \end{pmatrix} = \frac{1}{2}\begin{pmatrix} \phi_{\vec{p}_0} - \frac{i\hbar}{p_0}\frac{\partial \phi_{\vec{p}_0}}{\partial z} \\ \phi_{\vec{p}_0} + \frac{i\hbar}{p_0}\frac{\partial \phi_{\vec{p}_0}}{\partial z} \end{pmatrix} = \frac{1}{2}\begin{pmatrix} \left(1 + \frac{p_{0z}}{p_0}\right)\phi_{\vec{p}_0} \\ \left(1 - \frac{p_{0z}}{p_0}\right)\phi_{\vec{p}_0} \end{pmatrix}. \tag{5.15}$$

It can be easily checked that this $\begin{pmatrix}\psi_+ \\ \psi_-\end{pmatrix}$ satisfies (5.13). For a particle of a quasi-paraxial beam moving along and close to the $+z$-direction, associated with the plane wave $\phi_{\vec{p}_0}(\vec{r})$ with $p_{0z} > 0$ and $p_{0z} \approx p_0$, it is clear from the above equation that $\psi_+ \gg \psi_-$. By extending this observation, it is easy to see that for any wave packet of the form

$$\psi(\vec{r}) = \int d^3p_0\, C(\vec{p}_0)\phi_{\vec{p}_0}(\vec{r}), \quad \text{with} \int d^3p_0\, |C(\vec{p}_0)|^2 = 1,$$

$$|\vec{p}_0| = p_0,\ p_{0z} \approx p_0,\ p_{0z} > 0, \tag{5.16}$$

Scalar Theory: Spin-0 and Spinless Particles

representing a monochromatic quasiparaxial beam moving forward in the z-direction,

$$\begin{pmatrix} \psi_+(\vec{r}) \\ \psi_-(\vec{r}) \end{pmatrix} = \frac{1}{2} \begin{pmatrix} \int d^3 p_0 \, C(p_0) \left(1 + \frac{p_{0z}}{p_0}\right) \phi_{\vec{p}_0}(\vec{r}) \\ \int d^3 p_0 \, C(p_0) \left(1 - \frac{p_{0z}}{p_0}\right) \phi_{\vec{p}_0}(\vec{r}) \end{pmatrix},$$

$$\text{with } |\vec{p}_0| = p_0, \; p_{0z} \approx p_0, \; p_{0z} > 0, \qquad (5.17)$$

is such that $\psi_+(\vec{r}) \gg \psi_-(\vec{r})$. Thus, in general, in the representation of (5.12), ψ_+ is large compared to ψ_- for any monochromatic quasiparaxial beam passing along and close to the $+z$-direction through any system supporting beam propagation.

The purpose of casting (5.6) in the form of (5.12) is obvious when we compare the latter with the form of the Dirac equation in (3.711). Let us recall briefly. The Dirac equation for a spin-$\frac{1}{2}$ particle of charge q and mass m in an electromagnetic field is

$$i\hbar \frac{\partial}{\partial t} \begin{pmatrix} \Psi_u(\vec{r},t) \\ \Psi_\ell(\vec{r},t) \end{pmatrix} = \widehat{H} \begin{pmatrix} \Psi_u(\vec{r},t) \\ \Psi_\ell(\vec{r},t) \end{pmatrix}, \qquad (5.18)$$

where

$$\Psi_u = \begin{pmatrix} \Psi_1 \\ \Psi_2 \end{pmatrix}, \quad \Psi_\ell = \begin{pmatrix} \Psi_3 \\ \Psi_4 \end{pmatrix},$$

$$\widehat{H} = mc^2 \beta + \widehat{\mathscr{E}} + \widehat{\mathscr{O}},$$

$$\widehat{\mathscr{E}} = q\phi \begin{pmatrix} I & 0 \\ 0 & I \end{pmatrix}, \quad \widehat{\mathscr{O}} = c \begin{pmatrix} 0 & \vec{\sigma} \cdot \vec{\pi} \\ \vec{\sigma} \cdot \vec{\pi} & I \end{pmatrix}. \qquad (5.19)$$

For any positive energy Dirac Ψ in the nonrelativistic situation ($|c\vec{\pi}| \ll mc^2$), the upper components (Ψ_u) are larger compared to the lower components (Ψ_ℓ). The even operator ($\widehat{\mathscr{E}}$) does not couple the large and small components and the odd operator ($\widehat{\mathscr{O}}$) does couple them. Using mainly the algebraic property $\beta\widehat{\mathscr{O}} = -\widehat{\mathscr{O}}\beta$, the Foldy–Wouthuysen technique expands the Dirac Hamiltonian in a series with $1/mc^2$ as the expansion parameter. This leads to a good understanding of the nonrelativistic limit of the Dirac equation by showing how the Dirac Hamiltonian can be seen as consisting of a nonrelativistic part and a systematic series of relativistic correction terms. We have also seen that the Foldy–Wouthuysen transformation technique can be used to expand the Feshbach–Villars Hamiltonian of the Klein–Gordon equation in a series of nonrelativistic and relativistic correction terms.

The analogy between (5.12) and the Dirac equation is clear now. There is a correspondence as follows:

Forward propagation of the beam close to the z-direction	$\longrightarrow$	Positive energy Dirac particle
Paraxial beam ($|\vec{\pi}_\perp| \ll p_0$)	$\longrightarrow$	Nonrelativistic motion ($|\vec{\pi}| \ll mc$)
Deviation from paraxial condition (aberrating system)	$\longrightarrow$	Deviation from nonrelativistic situation (relativistic motion)

Also, it should be noted that in (5.12) σ_z plays the role of β in the Dirac Hamiltonian and $\widehat{\mathcal{O}}$ in (5.12) and σ_z obey the relation $\sigma_z\widehat{\mathcal{O}} = -\widehat{\mathcal{O}}\sigma_z$ analogous to the relation $\beta\widehat{\mathcal{O}} = -\widehat{\mathcal{O}}\beta$ obtained in the case of the Dirac Hamiltonian. Hence, by applying a Foldy–Wouthuysen-like technique to (5.12), it should be possible to analyse the beam propagation through an electromagnetic optical system in a systematic way up to any desired order of approximation starting with the paraxial approximation.

In the Foldy–Wouthuysen theory, a series of transformations are used to make the odd terms in the Hamiltonian as small as desired. We shall apply this technique to (5.12) to arrive at a representation in which the odd term will be as small as we desire. In this case, $1/p_0$ can be taken as the expansion parameter. The order of smallness of an odd term can be labeled by the lowest power of $1/p_0$ in it and higher order smallness will correspond to higher power. Thus, the odd term in (5.12) is of first order in $1/p_0$. Following the Foldy–Wouthuysen theory, let us define

$$\begin{pmatrix} \psi_+^{(1)} \\ \psi_-^{(1)} \end{pmatrix} = e^{i\widehat{S}_1}\begin{pmatrix} \psi_+ \\ \psi_- \end{pmatrix}, \quad \text{with } \widehat{S}_1 = \frac{i}{2p_0}\sigma_z\widehat{\mathcal{O}}. \tag{5.20}$$

The equation (5.12) is now transformed, as follows from (3.689), into

$$i\hbar\frac{\partial}{\partial z}\begin{pmatrix} \psi_+^{(1)} \\ \psi_-^{(1)} \end{pmatrix} = \left[e^{i\widehat{S}_1}\widehat{\mathbb{H}}e^{-i\widehat{S}_1} - i\hbar e^{i\widehat{S}_1}\frac{\partial}{\partial z}\left(e^{-i\widehat{S}_1}\right)\right]\begin{pmatrix} \psi_+^{(1)} \\ \psi_-^{(1)} \end{pmatrix}$$

$$= \widehat{\mathbb{H}}^{(1)}\begin{pmatrix} \psi_+^{(1)} \\ \psi_-^{(1)} \end{pmatrix}, \tag{5.21}$$

where $\widehat{\mathbb{H}}^{(1)}$ is to be calculated using (3.706), with $\widehat{H}$ replaced by $\widehat{\mathbb{H}}$, $\widehat{S}$ replaced by $\widehat{S}_1$, and t replaced by z. The result is

$$\widehat{\mathbb{H}}^{(1)} = -p_0\sigma_z + \widehat{\mathcal{E}}^{(1)} + \widehat{\mathcal{O}}^{(1)},$$

$$\widehat{\mathcal{E}}^{(1)} = -qA_zI + \left[\frac{1}{2p_0}\left(\widehat{\pi}_\perp^2 + \tilde{p}^2\right) + \frac{1}{8p_0^3}\left(\widehat{\pi}_\perp^2 + \tilde{p}^2\right)^2\right]\sigma_z$$

$$- \frac{1}{32p_0^4}\left\{\left[\left(\widehat{\pi}_\perp^2 + \tilde{p}^2\right), [\widehat{\pi}_\perp^2, qA_z]\right.\right.$$

$$\left.+ i\hbar q\left(\vec{\tilde{p}}_\perp \cdot \frac{\partial \vec{A}_\perp}{\partial z} + \frac{\partial \vec{A}_\perp}{\partial z}\cdot \vec{\tilde{p}}_\perp\right)\right]$$

$$\left.- \left[\widehat{\pi}_\perp^2, i\hbar\frac{\partial}{\partial z}(q^2A_\perp^2 + \tilde{p}^2)\right]\right\}I,$$

$$\widehat{\mathcal{O}}^{(1)} = -\frac{1}{2p_0^2}\left\{[\widehat{\pi}_\perp^2, qA_z] + i\hbar q\left(\vec{\tilde{p}}_\perp \cdot \frac{\partial \vec{A}_\perp}{\partial z} + \frac{\partial \vec{A}_\perp}{\partial z}\cdot \vec{\tilde{p}}_\perp\right)\right.$$

$$\left.+ i\hbar\frac{\partial}{\partial z}(q^2A_\perp^2 + \tilde{p}^2)\right\}i\sigma_y, \tag{5.22}$$

Scalar Theory: Spin-0 and Spinless Particles

with the odd term $\widehat{\mathscr{O}}^{(1)}$ of order $1/p_0^2$. By another transformation of the same type as in (5.20) with $\widehat{S}_2 = (i/2p_0)\sigma_z\widehat{\mathscr{O}}^{(1)}$, we can transform $\widehat{H}^{(1)}$ into a $\widehat{H}^{(2)}$ with an $\widehat{\mathscr{O}}^{(2)}$ of order $1/p_0^3$. Since $\widehat{\mathscr{E}}^{(1)}$, the even part of $\widehat{H}^{(1)}$, itself is a sufficiently good approximation for our purpose, we shall not continue the process of transformations further. Hence, we write

$$i\hbar\frac{\partial}{\partial z}\begin{pmatrix}\psi_+^{(1)}\\ \psi_-^{(1)}\end{pmatrix} = \widehat{H}^{(1)}\begin{pmatrix}\psi_+^{(1)}\\ \psi_-^{(1)}\end{pmatrix},$$

$$\widehat{H}^{(1)} = -p_0\sigma_z + \widehat{\mathscr{E}}^{(1)}, \tag{5.23}$$

dropping the odd term.

Let us now look at $\begin{pmatrix}\psi_+^{(1)}\\ \psi_-^{(1)}\end{pmatrix}$ corresponding to $\psi(\vec{r}) = \phi_{\vec{p}_0}(\vec{r})$, the plane wave, in free space. We get, using (5.15)

$$\begin{pmatrix}\psi_+^{(1)}\\ \psi_-^{(1)}\end{pmatrix} = e^{i\widehat{S}_1}\begin{pmatrix}\psi_+\\ \psi_-\end{pmatrix} = e^{-\frac{\vec{p}_\perp^2}{4p_0^2}\sigma_x}\begin{pmatrix}\psi_+\\ \psi_-\end{pmatrix}$$

$$\approx \frac{1}{2}\begin{pmatrix}1 & -\frac{\vec{p}_\perp^2}{4p_0^2}\\ -\frac{\vec{p}_\perp^2}{4p_0^2} & 1\end{pmatrix}\begin{pmatrix}\left(1+\frac{p_{0z}}{p_0}\right)\phi_{\vec{p}_0}\\ \left(1-\frac{p_{0z}}{p_0}\right)\phi_{\vec{p}_0}\end{pmatrix}$$

$$= \frac{1}{2}\begin{pmatrix}\left\{1+\frac{p_{0z}}{p_0} - \left(1-\frac{p_{0z}}{p_0}\right)\frac{p_{0\perp}^2}{4p_0^2}\right\}\phi_{\vec{p}_0}\\ \left\{1-\frac{p_{0z}}{p_0} - \left(1+\frac{p_{0z}}{p_0}\right)\frac{p_{0\perp}^2}{4p_0^2}\right\}\phi_{\vec{p}_0}\end{pmatrix}, \tag{5.24}$$

showing that $\psi_+^{(1)} \gg \psi_-^{(1)}$ for a quasiparaxial beam. The result easily extends to any wave packet of the form in (5.16). Thus, in general, we can take $\psi_+^{(1)} \gg \psi_-^{(1)}$ in (5.23) for the beam wave functions of interest to us. We can express this property that $\psi_+^{(1)} \gg \psi_-^{(1)}$, or essentially $\psi_-^{(1)} \approx 0$ compared to $\psi_+^{(1)}$, as

$$\sigma_z\begin{pmatrix}\psi_+^{(1)}\\ \psi_-^{(1)}\end{pmatrix} \approx I\begin{pmatrix}\psi_+^{(1)}\\ \psi_-^{(1)}\end{pmatrix}, \tag{5.25}$$

where I is the 2×2 identity matrix. Then, we can approximate (5.23) further to read

$$i\hbar \frac{\partial}{\partial z} \begin{pmatrix} \psi_+^{(1)} \\ \psi_-^{(1)} \end{pmatrix} = \widehat{H}^{(1)} \begin{pmatrix} \psi_+^{(1)} \\ \psi_-^{(1)} \end{pmatrix},$$

$$\widehat{H}^{(1)} \approx \left(-p_0 - qA_z + \frac{1}{2p_0} \left(\widehat{\pi}_\perp^2 + \tilde{p}^2 \right) + \frac{1}{8p_0^3} \left(\widehat{\pi}_\perp^2 + \tilde{p}^2 \right)^2 \right.$$

$$- \frac{1}{32p_0^4} \left\{ \left[\left(\widehat{\pi}_\perp^2 + \tilde{p}^2 \right), \left[\widehat{\pi}_\perp^2, qA_z \right] + i\hbar q \left(\vec{\tilde{p}}_\perp \cdot \frac{\partial \vec{A}_\perp}{\partial z} + \frac{\partial \vec{A}_\perp}{\partial z} \cdot \vec{\tilde{p}}_\perp \right) \right] \right.$$

$$\left. \left. - \left[\widehat{\pi}_\perp^2, i\hbar \frac{\partial}{\partial z} \left(q^2 A_\perp^2 + \tilde{p}^2 \right) \right] \right\} \right) I. \tag{5.26}$$

Now, we have two choices to proceed. If we want to study the z-evolution of the wave function of a system, we can specify it in the original representation as $\psi(\vec{r}_\perp, z)$, get it transformed to the representation

$$\begin{pmatrix} \psi_+^{(1)} \\ \psi_-^{(1)} \end{pmatrix} = e^{i\widehat{S}_1} M \begin{pmatrix} \psi \\ \frac{\widehat{\pi}_z}{p_0} \psi \end{pmatrix}, \tag{5.27}$$

and study its z-evolution according to (5.26) with $\widehat{H}^{(1)}$ as the optical Hamiltonian. In other words, we can model the wave function of the system directly in the representation (5.27). The Hermitian operators for the observables of interest, like $\vec{r}_\perp$ and $\vec{\tilde{p}}_\perp$, also need to be changed to the representation (5.27) as

$$\widehat{O}^{(1)} = e^{i\widehat{S}_1} M \widehat{O} I M^{-1} e^{-i\widehat{S}_1}, \tag{5.28}$$

where I is the 2×2 identity matrix.

The other choice, which we shall follow for the scalar theory, is to get the z-evolution equation for the original wave function $\psi(\vec{r}_\perp, z)$ by retracing the transformations and rewriting (5.26) in terms of $\begin{pmatrix} \psi_1 \\ \psi_2 \end{pmatrix}$ in which $\psi_1 = \psi$. In this case, we can use the usual Hermitian operators for the observables. It may look like that retracing the transformations will get us only back to the original evolution equation with the old Hamiltonian. This is not so. Of course, we will get back to the original representation for the wave function. But, due to the transformation and truncation of the resulting series, what we get will be the series of paraxial and nonparaxial approximations of the optical Hamiltonian in the original representation. With this understanding, let us proceed as follows. Substituting in (5.26),

$$\begin{pmatrix} \psi_1 \\ \psi_2 \end{pmatrix} = M^{-1} \begin{pmatrix} \psi_+ \\ \psi_- \end{pmatrix} = M^{-1} e^{-i\widehat{S}_1} \begin{pmatrix} \psi_+^{(1)} \\ \psi_-^{(1)} \end{pmatrix}, \tag{5.29}$$

Scalar Theory: Spin-0 and Spinless Particles 221

we get

$$i\hbar \frac{\partial}{\partial z} \begin{pmatrix} \psi_1 \\ \psi_2 \end{pmatrix} = i\hbar \frac{\partial}{\partial z} \left[M^{-1} e^{-i\widehat{S}_1} \begin{pmatrix} \psi_+^{(1)} \\ \psi_-^{(1)} \end{pmatrix} \right]$$

$$= M^{-1} i\hbar \frac{\partial}{\partial z} \left[e^{-i\widehat{S}_1} \begin{pmatrix} \psi_+^{(1)} \\ \psi_-^{(1)} \end{pmatrix} \right]$$

$$= M^{-1} \left[i\hbar \left(\frac{\partial}{\partial z} e^{-i\widehat{S}_1} \right) + e^{-i\widehat{S}_1} \widehat{H}^{(1)} \right] \begin{pmatrix} \psi_+^{(1)} \\ \psi_-^{(1)} \end{pmatrix}$$

$$= M^{-1} \left[i\hbar \left(\frac{\partial}{\partial z} e^{-i\widehat{S}_1} \right) + e^{-i\widehat{S}_1} \widehat{H}^{(1)} \right] e^{i\widehat{S}_1} M \begin{pmatrix} \psi_1 \\ \psi_2 \end{pmatrix}$$

$$= \left\{ M^{-1} \left[i\hbar \left(\frac{\partial}{\partial z} e^{-i\widehat{S}_1} \right) e^{i\widehat{S}_1} + e^{-i\widehat{S}_1} \widehat{H}^{(1)} e^{i\widehat{S}_1} \right] M \right\} \begin{pmatrix} \psi_1 \\ \psi_2 \end{pmatrix}$$

$$= \left\{ M^{-1} \left[-i\hbar e^{-i\widehat{S}_1} \left(\frac{\partial}{\partial z} e^{i\widehat{S}_1} \right) + e^{-i\widehat{S}_1} \widehat{H}^{(1)} e^{i\widehat{S}_1} \right] M \right\} \begin{pmatrix} \psi_1 \\ \psi_2 \end{pmatrix}$$

$$= \widehat{H}_o \begin{pmatrix} \psi_1 \\ \psi_2 \end{pmatrix}, \qquad (5.30)$$

where $\widehat{H}_o$ becomes

$$\widehat{H}_o \approx \left(-p_0 - qA_z + \frac{1}{2p_0} \left(\widehat{\pi}_\perp^2 + \tilde{p}^2 \right) \right) I - \frac{1}{4p_0^2} \left([\widehat{\pi}_\perp^2, qA_z] \right.$$

$$+ \left\{ i\hbar q \left(\vec{\tilde{p}}_\perp \cdot \frac{\partial \vec{A}_\perp}{\partial z} + \frac{\partial \vec{A}_\perp}{\partial z} \cdot \vec{\tilde{p}}_\perp \right) - i\hbar q^2 \frac{\partial A_\perp^2}{\partial z} \right\} \right) \sigma_z$$

$$+ \left(\frac{1}{8p_0^3} \left(\widehat{\pi}_\perp^2 + \tilde{p}^2 \right)^2 - \frac{1}{16p_0^4} \left\{ \left[(\widehat{\pi}_\perp^2 + \tilde{p}^2) \right), [\widehat{\pi}_\perp^2, qA_z] \right] \right.$$

$$\left. + i\hbar q \left(\vec{\tilde{p}}_\perp \cdot \frac{\partial \vec{A}_\perp}{\partial z} + \frac{\partial \vec{A}_\perp}{\partial z} \cdot \vec{\tilde{p}}_\perp \right) \right] - \left[\widehat{\pi}_\perp^2, i\hbar \frac{\partial}{\partial z} (q^2 A_\perp^2 + \tilde{p}^2) \right] \right\} \right) I. \qquad (5.31)$$

Since $\widehat{H}_o$ is diagonal, the equation (5.30) describes the z-evolution of ψ_1 and ψ_2 separately. We are interested in the z-evolution of $\psi = \psi_1$, and hence we write

$$i\hbar \frac{\partial \psi(\vec{r}_\perp, z)}{\partial z} = \widehat{H}_o \psi(\vec{r}_\perp, z),$$

$$\widehat{H}_o \approx -p_0 - qA_z + \frac{1}{2p_0} \left(\widehat{\pi}_\perp^2 + \tilde{p}^2 \right) - \frac{1}{4p_0^2} \left([\widehat{\pi}_\perp^2, qA_z] \right.$$

$$+ \left\{ i\hbar q \left(\vec{\tilde{p}}_\perp \cdot \frac{\partial \vec{A}_\perp}{\partial z} + \frac{\partial \vec{A}_\perp}{\partial z} \cdot \vec{\tilde{p}}_\perp \right) - i\hbar q^2 \frac{\partial A_\perp^2}{\partial z} \right\} \right)$$

$$+ \frac{1}{8p_0^3}\left(\widehat{\pi}_\perp^2 + \tilde{p}^2\right)^2 - \frac{1}{16p_0^4}\left\{\left[\left(\widehat{\pi}_\perp^2 + \tilde{p}^2\right), \left[\widehat{\pi}_\perp^2, qA_z\right]\right]\right.$$

$$\left. + i\hbar q\left(\vec{\tilde{p}}_\perp \cdot \frac{\partial \vec{A}_\perp}{\partial z} + \frac{\partial \vec{A}_\perp}{\partial z} \cdot \vec{\tilde{p}}_\perp\right)\right]$$

$$\left. - \left[\widehat{\pi}_\perp^2, i\hbar\frac{\partial}{\partial z}\left(q^2 A_\perp^2 + \tilde{p}^2\right)\right]\right\}. \quad (5.32)$$

We have derived the above z-evolution equation for $\psi(\vec{r}_\perp, z)$ starting from the Klein–Gordon equation which, as we have seen, does not have a proper probability interpretation. However, let us look at the case of free space in which $\phi = 0$ and $\vec{A} = (0,0,0)$. Then, the equation (5.32) reduces to

$$i\hbar\frac{\partial \psi(\vec{r}_\perp, z)}{\partial z} = \widehat{H}_o \psi(\vec{r}_\perp, z), \quad \widehat{H}_o = \left(-p_0 + \frac{\tilde{p}_\perp^2}{2p_0}\right), \quad (5.33)$$

which has a Hermitian Hamiltonian. It is analogous to the nonrelativistic two dimensional Schrödinger equation for a free particle in which z is taken as time, p_0 is taken as the mass of the particle, and the constant potential-like term $-p_0$ is ignorable. In this case, the z-evolution of $\psi(\vec{r}_\perp, z)$ is unitary, $\int\int dxdy\, |\psi(\vec{r}_\perp, z)|^2$ is conserved, and hence $|\psi(\vec{r}_\perp, z)|^2$ can be taken as the probability of finding the particle at the position (x,y) in the vertical xy-plane at the point z on the z-axis (optic axis). However, in general, it is clear that $\int\int dxdy\, |\psi(\vec{r}_\perp, z)|^2$ need not be conserved. Only $\int\int\int dxdydz\, |\psi(\vec{r}_\perp, z)|^2 = 1$ has to be conserved as it represents the probability of finding the particle somewhere in space. Hence, it is not surprising that the Hamiltonian in (5.32) is not Hermitian and the corresponding z-evolution is not unitary. In fact, one should expect a loss of intensity of the forward propagating beam along the optic axis, since there will, in general, be reflections at the boundaries of the system as we have seen in the case of scattering by potential well and potential barrier. Actually, we are dealing with the scattering states of the beam particle. Thus, it is natural that there are non-Hermitian terms in H_o (terms $\propto 1/p_0^2$). In the above expansion scheme, in general, among the terms $\propto$ (even powers of $1/p_0$), alternate terms will be non-Hermitian. Of course, the effect of these non-Hermitian terms can be expected to be quite small and negligible. Hence, we can approximate $\widehat{H}_o$, further, by Hermitianizing it, *i.e.*, writing it as the sum of its Hermitian part $\frac{1}{2}\left(\widehat{H}_o + \widehat{H}_o^\dagger\right)$ and the anti-Hermitian part $\frac{1}{2}\left(\widehat{H}_o - \widehat{H}_o^\dagger\right)$ and dropping the anti-Hermitian part. This leads to

$$i\hbar\frac{\partial \psi(\vec{r}_\perp, z)}{\partial z} = \widehat{\mathscr{H}}_o \psi(\vec{r}_\perp, z) \quad \text{or,} \quad i\hbar\frac{\partial}{\partial z}|\psi(z)\rangle = \widehat{\mathscr{H}}_o |\psi(z)\rangle,$$

$$\widehat{\mathscr{H}}_o = -p_0 - qA_z + \frac{1}{2p_0}\left(\widehat{\pi}_\perp^2 + \tilde{p}^2\right) + \frac{1}{8p_0^3}\left(\widehat{\pi}_\perp^2 + \tilde{p}^2\right)^2$$

$$- \frac{1}{16p_0^4}\left\{\left[\left(\widehat{\pi}_\perp^2 + \tilde{p}^2\right), \left[\widehat{\pi}_\perp^2, qA_z\right]\right]\right.$$

Scalar Theory: Spin-0 and Spinless Particles

$$+i\hbar q\left(\vec{\hat{p}}_\perp \cdot \frac{\partial \vec{A}_\perp}{\partial z} + \frac{\partial \vec{A}_\perp}{\partial z}\cdot \vec{\hat{p}}_\perp\right)\right] - \left[\hat{\pi}_\perp^2, i\hbar\frac{\partial}{\partial z}\left(q^2 A_\perp^2 + \hat{p}^2\right)\right]\right\},$$

with $p_0^2 = \frac{1}{c^2}\left(E^2 - m^2 c^4\right)$, $\hat{p}^2 = \frac{2qE\phi}{c^2}\left(1 - \frac{q\phi}{2E}\right)$, (5.34)

Thus, $\widehat{\mathcal{H}_o}$ is the basic Hamiltonian of the scalar theory of quantum charged particle beam optics applicable when the spin of the particle is zero or ignored. With the quantum charged particle beam optical Hamiltonian, or simply the quantum beam optical Hamiltonian, $\widehat{\mathcal{H}_o}$, being Hermitian we can take $\psi(\vec{r}_\perp, z)$ to be normalized over the xy-plane at any z, i.e., for any z, we shall take

$$\langle \psi(z)|\psi(z)\rangle = \iint dx\,dy\, |\psi(\vec{r}_\perp, z)|^2 = 1. \quad (5.35)$$

Hereafter, we shall write $\int d^2 r_\perp$ for $\iint dx\,dy$. It may be noted that we have obtained from the Kelin–Gordon a one-way propagation equation, along the $+z$-direction, for the wave function representing the charged particle beam using the Feshbach–Villars-like formalism and the Foldy–Wouthuysen-like transformation technique. Detailed and rigorous studies on one-way wave equation modeling in two-way wave propagation problems have been made by Fishman et al. (see Fishman, de Hoop, and van Stralen [51], Fishman [52], and references therein).

Having obtained the basic quantum beam optical Hamiltonian operator, we can proceed to get the relation between $\psi(\vec{r}_\perp, z_i)$, wave function in the plane vertical to the optic axis at the point z_i where the beam enters the system, and $\psi(\vec{r}_\perp, z)$ wave function in the plane vertical to the optical axis at any other point of interest (z). By formal integration of (5.34), using the analogous result (3.449) for the time-dependent Schrödinger equation, we can write

$$|\psi(z)\rangle = \widehat{U}(z, z_i)|\psi(z_i)\rangle,$$

$$\widehat{U}(z, z_i) = \mathsf{P}\left(e^{-\frac{i}{\hbar}\int_{z_i}^{z} dz\, \widehat{\mathcal{H}_o}(z)}\right)$$

$$= I - \frac{i}{\hbar}\int_{z_i}^{z} dz_1\, \widehat{\mathcal{H}_o}(z_1) + \left(-\frac{i}{\hbar}\right)^2 \int_{z_i}^{z} dz_2 \int_{z_i}^{z_2} dz_1\, \widehat{\mathcal{H}_o}(z_2)\widehat{\mathcal{H}_o}(z_1)$$

$$+ \left(-\frac{i}{\hbar}\right)^3 \int_{z_i}^{z} dz_3 \int_{z_i}^{z_3} dz_2 \int_{z_i}^{z_2} dz_1\, \widehat{\mathcal{H}_o}(z_3)\widehat{\mathcal{H}_o}(z_2)\widehat{\mathcal{H}_o}(z_1) + \cdots,$$

(5.36)

with I as the identity operator. The z-evolution operator $\widehat{U}(z, z_i)$ is seen to satisfy the relations

$$i\hbar\frac{\partial}{\partial z}\widehat{U}(z, z_i) = \widehat{\mathcal{H}_o}\widehat{U}(z, z_i),$$

$$i\hbar\frac{\partial}{\partial z}\widehat{U}^\dagger(z, z_i) = -\widehat{U}(z, z_i)^\dagger \widehat{\mathcal{H}_o},$$

$$\widehat{U}(z, z_i)^\dagger \widehat{U}(z, z_i) = I, \qquad \widehat{U}(z_i, z_i) = I. \quad (5.37)$$

We can write the z-evolution operator $\widehat{U}(z, z_i)$ in the Dyson form (3.454), *i.e.*, as an z-ordered exponential. For our purpose, the most convenient form of expression for $\widehat{U}(z, z_i)$ is the equivalent Magnus form

$$\widehat{U}(z, z_i) = e^{-\frac{i}{\hbar}\widehat{T}(z, z_i)},$$

$$\widehat{T}(z, z_i) = \int_{z_i}^{z} dz_1\, \widehat{\mathcal{H}_o}(z_1)$$

$$+ \frac{1}{2}\left(-\frac{i}{\hbar}\right) \int_{z_i}^{z} dz_2 \int_{z_i}^{z_2} dz_1 \left[\widehat{\mathcal{H}_o}(z_2), \widehat{\mathcal{H}_o}(z_1)\right]$$

$$+ \frac{1}{6}\left(-\frac{i}{\hbar}\right)^2 \int_{z_i}^{z} dz_3 \int_{z_i}^{z_3} dz_2 \int_{z_i}^{z_2} dz_1 \left\{\left[\left[\widehat{\mathcal{H}_o}(z_3), \widehat{\mathcal{H}_o}(z_2)\right],\right.\right.$$

$$\left.\left. \widehat{\mathcal{H}_o}(z_1)\right] + \left[\left[\widehat{\mathcal{H}_o}(z_1), \widehat{\mathcal{H}_o}(z_2)\right], \widehat{\mathcal{H}_o}(z_3)\right]\right\} + \cdots. \quad (5.38)$$

Note that $\widehat{T}(z, z_i)$ is Hermitian, when $\widehat{\mathcal{H}_o}(z)$ is Hermitian, such that $\widehat{U}(z, z_i)$ is unitary. Analogous to the semigroup property (3.167) of the time-evolution operator, we have the semigroup property for the z-evolution operator: for $z > z_n > z_{n-1} > \cdots > z_2 > z_1 > z_i$,

$$\widehat{U}(z, z_i) = \widehat{U}(z, z_n)\, \widehat{U}(z_n, z_{n-1}) \ldots \widehat{U}(z_2, z_1)\, \widehat{U}(z_1, z_i). \quad (5.39)$$

This follows from the observation that, for $z > z' > z_i$,

$$|\psi(z)\rangle = \widehat{U}(z, z')\,|\psi(z')\rangle, \quad |\psi(z')\rangle = \widehat{U}(z', z_i)\,|\psi(z_i)\rangle,$$

$$|\psi(z)\rangle = \widehat{U}(z, z_i)\,|\psi(z_i)\rangle, \quad (5.40)$$

The integral form of the scalar quantum charged particle beam optical Schrödinger equation (5.34) becomes

$$\psi(\vec{r}_\perp, z) = \int d^2 r_{\perp i}\, K(\vec{r}_\perp, z; \vec{r}_{\perp i}, z_i)\, \psi(\vec{r}_{\perp i}, z_i), \quad (5.41)$$

where the kernel of z-propagation, or the z-propagator, is given by

$$K(\vec{r}_\perp, z; \vec{r}_{\perp i}, z_i) = \langle \vec{r}_\perp | \widehat{U}(z, z_i) | \vec{r}_{\perp i} \rangle$$

$$= \int d^2 r'_\perp\, \delta^2(\vec{r}'_\perp - \vec{r}_\perp)\, \widehat{U}(z, z_i)\, \delta^2(\vec{r}'_\perp - \vec{r}_{\perp i}). \quad (5.42)$$

We have seen from Ehrenfest's theorem that quantum averages, or the expectation values, of observables behave like the corresponding classical observables. Hence, to study the quantum corrections to the classical transfer maps of the optical elements, we have to find the transfer maps for the corresponding quantum averages. When we want to relate the values of the quantum averages of the beam at two different points along the optic axis of the system, we can use the Heisenberg picture. For an observable O associated with the Hermitian quantum operator $\widehat{O}$, the average for the state $|\psi(z)\rangle$ at the xy-plane at the point z is given by

$$\langle O \rangle(z) = \langle \psi(z)|\widehat{O}|\psi(z)\rangle = \int d^2 r_\perp\, \psi^*(\vec{r}_\perp, z)\, \widehat{O} \psi(\vec{r}_\perp, z), \quad (5.43)$$

Scalar Theory: Spin-0 and Spinless Particles

with $\langle \psi(z) | \psi(z) \rangle = 1$. Sometimes we may denote $\langle O \rangle(z)$ by $\langle \widehat{O} \rangle(z)$ also. Now, from (5.36), we have

$$\begin{aligned}\langle O \rangle(z) &= \langle \psi(z) | \widehat{O} | \psi(z) \rangle \\ &= \left\langle \psi(z_i) | \widehat{U}^\dagger(z,z_i) \, \widehat{O} \widehat{U}(z,z_i) | \psi(z_i) \right\rangle \\ &= \left\langle \widehat{U}^\dagger(z,z_i) \, \widehat{O} \widehat{U}(z,z_i) \right\rangle(z_i), \end{aligned} \quad (5.44)$$

leading to the required transfer map giving the expectation values of the observables in the plane at z in terms of their values in the plane at z_i.

We are considering only the single particle theory and assuming that all the particles of the beam are in the same quantum state, pure or mixed. If a particle of the monoenergetic quasiparaxial beam entering the optical system, from the free space outside it, is associated with a wave function as we have assumed above, then a representation for the wave function is

$$\Psi(\vec{r},t) = e^{-iEt/\hbar} \psi(\vec{r}_\perp, z), \quad E = \sqrt{m^2 c^4 + c^2 p_0^2},$$

$$\psi(\vec{r}_\perp, z) = \frac{1}{(2\pi\hbar)^{3/2}} \int\int dp_x dp_y \, C(p_x, p_y) e^{\frac{i}{\hbar}(\vec{p}_\perp \cdot \vec{r}_\perp + p_z z)},$$

$$p_\perp^2 + p_z^2 = p_0^2, \quad |\vec{p}_\perp| \ll p_z \approx p_0,$$

$$\int\int dp_x dp_y \, |C(p_x, p_y)|^2 = 1. \quad (5.45)$$

When the particle is not in a pure state representable by a wave function as above, it has to be represented by a mixed state density operator

$$\widehat{\rho}(z,t) = \sum_j p_j e^{-iEt/\hbar} |\psi_j(z)\rangle \langle \psi_j(z)| e^{iEt/\hbar} = \sum_j p_j |\psi_j(z)\rangle \langle \psi_j(z)|,$$

$$\sum_j p_j = 1, \quad \langle \psi_j(z) | \psi_j(z) \rangle = 1, \quad (5.46)$$

where for each j

$$\langle \vec{r}_\perp | \psi_j(z) \rangle = \psi_j(\vec{r}_\perp, z) = \frac{1}{(2\pi\hbar)^{3/2}} \int\int dp_x dp_y \, C_j(p_x, p_y) e^{\frac{i}{\hbar}(\vec{p}_\perp \cdot \vec{r}_\perp + p_z z)},$$

$$p_\perp^2 + p_z^2 = p_0^2, \quad |\vec{p}_\perp| \ll p_z \approx p_0,$$

$$\int\int dp_x dp_y \, |C_j(p_x, p_y)|^2 = 1. \quad (5.47)$$

Note that $\widehat{\rho}(z,t)$ is independent of time when the beam is monoenergetic which, of course, is an ideal situation never realized in practice. However, we shall assume the beam to be monoenergetic. In the position representation, the matrix elements of $\widehat{\rho}$ in the xy-plane at z are given by

$$\langle \vec{r}_\perp | \widehat{\rho}(z) | \vec{r}'_\perp \rangle = \sum_j p_j \psi_j(\vec{r}_\perp, z) \psi_j^*(\vec{r}'_\perp, z). \quad (5.48)$$

Note that at any z we have

$$\text{Tr}(\widehat{\rho}(z)) = \sum_j p_j \text{Tr}\left(|\psi_j(z)\rangle\langle\psi_j(z)|\right)$$
$$= \sum_j p_j \langle\psi_j(z)|\psi_j(z)\rangle = \sum_j p_j = 1, \quad (5.49)$$

as required. If we are representing the beam particle by such a density matrix, then the average of an observable in the xy-plane at z will be given by

$$\langle O \rangle(z) = \text{Tr}\left(\widehat{\rho}(z)\widehat{O}\right). \quad (5.50)$$

It follows from (5.36) that the z-evolution of the density matrix will be given by

$$\widehat{\rho}(z) = \widehat{U}(z,z_i)\widehat{\rho}(z_i)\widehat{U}^\dagger(z,z_i). \quad (5.51)$$

This leads to the transfer map

$$\langle O \rangle(z) = \text{Tr}\left(\widehat{U}(z,z_i)\widehat{\rho}(z_i)\widehat{U}^\dagger(z,z_i)\widehat{O}\right)$$
$$= \text{Tr}\left(\widehat{\rho}(z_i)\widehat{U}^\dagger(z,z_i)\widehat{O}\widehat{U}(z,z_i)\right). \quad (5.52)$$

Note that, as follows from (5.44) and (5.52), this transfer map depends essentially on $\widehat{U}(z,z_i)$.

The z-dependent operator

$$\widehat{O}_H(z) = \widehat{U}^\dagger(z,z_i)\widehat{O}\widehat{U}(z,z_i) \quad (5.53)$$

represents the observable O in the beam optical Heisenberg picture in which the initial state of the particle $|\psi(z_i)\rangle$, or $\widehat{\rho}(z_i)$, remains constant and the observables evolve along the optic axis of the system. From (5.44) we have

$$\langle\widehat{O}\rangle(z) = \langle\psi(z_i)|\widehat{O}_H|\psi(z_i)\rangle = \langle\widehat{O}_H\rangle(z_i) \quad (5.54)$$

From (5.37), it follows that

$$i\hbar\frac{d\widehat{O}_H}{dz} = i\hbar\left\{\left(\frac{\partial}{\partial z}\widehat{U}^\dagger(z,z_i)\right)\widehat{O}\widehat{U}(z,z_i) + \widehat{U}^\dagger(z,z_i)\widehat{O}\left(\frac{\partial}{\partial z}\widehat{U}(z,z_i)\right)\right\}$$
$$+ i\hbar\left(\widehat{U}^\dagger(z,z_i)\frac{\partial\widehat{O}}{\partial z}\widehat{U}(z,z_i)\right)$$
$$= \left\{-\widehat{U}^\dagger(z,z_i)\widehat{\mathscr{H}}_o\widehat{O}\widehat{U}(z,z_i) + \widehat{U}^\dagger(z,z_i)\widehat{O}\widehat{\mathscr{H}}_o\widehat{U}(z,z_i)\right\}$$
$$+ i\hbar\left(\widehat{U}^\dagger(z,z_i)\frac{\partial\widehat{O}}{\partial z}\widehat{U}(z,z_i)\right)$$
$$= \left\{-\widehat{U}^\dagger(z,z_i)\widehat{\mathscr{H}}_o\widehat{U}(z,z_i)\widehat{U}^\dagger(z,z_i)\widehat{O}\widehat{U}(z,z_i)\right.$$

Scalar Theory: Spin-0 and Spinless Particles

$$+\widehat{U}^\dagger(z,z_i)\,\widehat{O}\widehat{\mathcal{H}_o}\widehat{U}(z,z_i)\,\widehat{U}^\dagger(z,z_i)\,\widehat{U}(z,z_i)\bigg\}$$

$$+i\hbar\left(\widehat{U}^\dagger(z,z_i)\frac{\partial\widehat{O}}{\partial z}\widehat{U}(z,z_i)\right)$$

$$=\left[-\widehat{\mathcal{H}}_{o,\mathrm{H}},\widehat{O}_\mathrm{H}\right]+i\hbar\left(\frac{\partial\widehat{O}}{\partial z}\right)_\mathrm{H}. \tag{5.55}$$

or

$$\frac{d\widehat{O}_\mathrm{H}}{dz}=\frac{1}{i\hbar}\left[-\widehat{\mathcal{H}}_{o,\mathrm{H}},\widehat{O}_\mathrm{H}\right]+\left(\frac{\partial\widehat{O}}{\partial z}\right)_\mathrm{H}. \tag{5.56}$$

This equation is the quantum beam optical equation of motion for observables and in the classical limit, with the correspondence $\left[\widehat{A},\widehat{B}\right]/i\hbar \longrightarrow \{A,B\}$, it becomes the classical beam optical equation of motion (2.96). Note that a z-dependent $\widehat{\mathcal{H}_o}$ need not commute with $\widehat{U}(z,z_i)$, and hence $\widehat{\mathcal{H}}_{o,\mathrm{H}} \neq \widehat{\mathcal{H}_o}$, unless $\widehat{\mathcal{H}_o}$ is z-independent. The equation (5.53) is the integral version of (5.56). Since

$$\left[-\widehat{\mathcal{H}}_{o,\mathrm{H}},\widehat{O}_\mathrm{H}\right]=\left[-\widehat{\mathcal{H}_o},\widehat{O}\right]_\mathrm{H} \tag{5.57}$$

we can also write

$$\frac{d\widehat{O}_\mathrm{H}}{dz}=\frac{1}{i\hbar}\left[-\widehat{\mathcal{H}_o},\widehat{O}\right]_\mathrm{H}+\left(\frac{\partial\widehat{O}}{\partial z}\right)_\mathrm{H}. \tag{5.58}$$

Correspondingly, from (5.54), we get the quantum beam optical equation of motion for the average $\left\langle\widehat{O}\right\rangle(z)$ as

$$\frac{d\left\langle\widehat{O}\right\rangle(z)}{dz}=\frac{d}{dz}\left\langle\psi(z_i)\left|\widehat{O}_\mathrm{H}\right|\psi(z_i)\right\rangle=\left\langle\psi(z_i)\left|\frac{d\widehat{O}_\mathrm{H}}{dz}\right|\psi(z_i)\right\rangle$$

$$=\left\langle\psi(z_i)\left|\frac{1}{i\hbar}\left[-\widehat{\mathcal{H}_o},\widehat{O}\right]_\mathrm{H}+\left(\frac{\partial\widehat{O}}{\partial z}\right)_\mathrm{H}\right|\psi(z_i)\right\rangle$$

$$=\frac{1}{i\hbar}\left\langle\left[-\widehat{\mathcal{H}_o},\widehat{O}\right]\right\rangle(z)+\left\langle\frac{\partial\widehat{O}}{\partial z}\right\rangle(z). \tag{5.59}$$

When $\widehat{O}$ does not have an explicit z-dependence, the equations of motion become

$$\frac{d\widehat{O}_\mathrm{H}}{dz}=\frac{1}{i\hbar}\left[-\widehat{\mathcal{H}_o},\widehat{O}\right]_\mathrm{H}, \tag{5.60}$$

and

$$\frac{d\left\langle\widehat{O}\right\rangle(z)}{dz}=\frac{1}{i\hbar}\left\langle\left[-\widehat{\mathcal{H}_o},\widehat{O}\right]\right\rangle(z). \tag{5.61}$$

Particularly, for $\vec{r}_\perp$ and $\vec{p}_\perp/p_0$, in which we shall be primarily interested, we have

$$\frac{d\vec{r}_{\perp,\mathrm{H}}}{dz} = \frac{1}{i\hbar}\left[-\widehat{\mathscr{H}}_o, \vec{r}_\perp\right]_{\mathrm{H}} = \frac{i}{\hbar}\left[\widehat{\mathscr{H}}_o, \vec{r}_\perp\right]_{\mathrm{H}},$$

$$\frac{1}{p_0}\frac{d\vec{p}_{\perp,\mathrm{H}}}{dz} = \frac{1}{i\hbar p_0}\left[-\widehat{\mathscr{H}}_o, \vec{p}_\perp\right]_{\mathrm{H}} = \frac{i}{\hbar p_0}\left[\widehat{\mathscr{H}}_o, \vec{p}_\perp\right]_{\mathrm{H}}, \quad (5.62)$$

and

$$\frac{d\langle\vec{r}_\perp\rangle(z)}{dz} = \frac{1}{i\hbar}\left\langle\left[-\widehat{\mathscr{H}}_o, \vec{r}_\perp\right]\right\rangle(z) = \frac{i}{\hbar}\left\langle\left[\widehat{\mathscr{H}}_o, \vec{r}_\perp\right]\right\rangle(z),$$

$$\frac{1}{p_0}\frac{d\langle\vec{p}_\perp\rangle(z)}{dz} = \frac{1}{i\hbar p_0}\left\langle\left[-\widehat{\mathscr{H}}_o, \vec{p}_\perp\right]\right\rangle(z) = \frac{i}{\hbar p_0}\left\langle\left[\widehat{\mathscr{H}}_o, \vec{p}_\perp\right]\right\rangle(z). \quad (5.63)$$

To find the transfer maps for $\langle\vec{r}_\perp\rangle$ and $\langle\vec{p}_\perp\rangle$, we can use, depending on convenience, either (5.63) or

$$\begin{aligned}
\langle\vec{r}_\perp\rangle(z) &= \langle\vec{r}_{\perp,\mathrm{H}}\rangle(z_i) \\
&= \left\langle\psi(z_i)\left|\widehat{U}^\dagger(z,z_i)\vec{r}_\perp\widehat{U}(z,z_i)\right|\psi(z_i)\right\rangle \\
&= \left\langle\psi(z_i)\left|e^{\frac{i}{\hbar}\widehat{T}(z,z_i)}\vec{r}_\perp e^{-\frac{i}{\hbar}\widehat{T}(z,z_i)}\right|\psi(z_i)\right\rangle \\
\frac{1}{p_0}\langle\vec{p}_\perp\rangle(z) &= \frac{1}{p_0}\langle\vec{p}_{\perp,\mathrm{H}}\rangle(z_i) \\
&= \frac{1}{p_0}\left\langle\psi(z_i)\left|\widehat{U}^\dagger(z,z_i)\vec{p}_\perp\widehat{U}(z,z_i)\right|\psi(z_i)\right\rangle \\
&= \frac{1}{p_0}\left\langle\psi(z_i)\left|e^{\frac{i}{\hbar}\widehat{T}(z,z_i)}\vec{p}_\perp e^{-\frac{i}{\hbar}\widehat{T}(z,z_i)}\right|\psi(z_i)\right\rangle. \quad (5.64)
\end{aligned}$$

Let us now recall the identities (3.142 and 3.143): with $:\widehat{A}:\widehat{B} = \left[\widehat{A},\widehat{B}\right]$,

$$e^{\widehat{A}}\widehat{B}e^{-\widehat{A}} = \left(I + :\widehat{A}: + \frac{1}{2!}:\widehat{A}:^2 + \frac{1}{3!}:\widehat{A}:^3 + \cdots\right)\widehat{B} = e^{:\widehat{A}:}\widehat{B}, \quad (5.65)$$

and

$$e^{:a\widehat{A}:} = e^{a:\widehat{A}:}. \quad (5.66)$$

Using these identities in (5.64), with $\widehat{A} = i\widehat{T}/\hbar$ and $\widehat{B} = \vec{r}_\perp$, and $\vec{p}_\perp$, we have

$$\begin{aligned}
\langle\vec{r}_\perp\rangle(z) &= \langle\psi(z_i)|e^{\frac{i}{\hbar}:\widehat{T}(z,z_i):}\vec{r}_\perp|\psi(z_i)\rangle, \\
&= \left\langle\psi(z_i)\left|\left\{\left(I + \frac{i}{\hbar}:\widehat{T}(z,z_i): + \frac{1}{2!}\left(\frac{i}{\hbar}\right)^2:\widehat{T}(z,z_i):^2\right.\right.\right.\right. \\
&\qquad\left.\left.\left.\left. + \frac{1}{3!}\left(\frac{i}{\hbar}\right)^3:\widehat{T}(z,z_i):^3 + \cdots\right)\vec{r}_\perp\right\}\right|\psi(z_i)\right\rangle
\end{aligned}$$

Scalar Theory: Spin-0 and Spinless Particles

$$\frac{1}{p_0}\left\langle \vec{\hat{p}}_\perp \right\rangle(z) = \frac{1}{p_0}\left\langle \psi(z_i) \left| \left\{ \left(I + \frac{i}{\hbar} : \widehat{T}(z,z_i): + \frac{1}{2!}\left(\frac{i}{\hbar}\right)^2 : \widehat{T}(z,z_i):^2 \right.\right.\right.\right.$$
$$\left.\left.\left.\left. + \frac{1}{3!}\left(\frac{i}{\hbar}\right)^3 : \widehat{T}(z,z_i):^3 + \cdots \right) \vec{\hat{p}}_\perp \right\} \right| \psi(z_i) \right\rangle. \quad (5.67)$$

In general, for any observable O, with the quantum operator $\widehat{O}$, we will have

$$\left\langle \widehat{O} \right\rangle(z) = \left\langle \psi(z_i) \left| e^{\frac{i}{\hbar}:\widehat{T}(z,z_i):} \widehat{O} \right| \psi(z_i) \right\rangle \quad (5.68)$$

Note that

$$e^{\frac{i}{\hbar}:\widehat{T}(z,z_i):}\widehat{O} = e^{\frac{i}{\hbar}\widehat{T}(z,z_i)}\widehat{O}e^{-\frac{i}{\hbar}\widehat{T}(z,z_i)} = \widehat{U}^\dagger(z,z_i)\widehat{O}\widehat{U}(z,z_i) = \widehat{O}_{\mathrm{H}} \quad (5.69)$$

The beam optical Heisenberg equation of motion for any $\widehat{O}$ with no explicit z-dependence is

$$\frac{d\widehat{O}_{\mathrm{H}}}{dz} = \frac{1}{i\hbar}\left[-\widehat{\mathscr{H}_o}, \widehat{O}_{\mathrm{H}}\right]. \quad (5.70)$$

Directly integrating this equation, *á la* Magnus (see the Appendix at the end of Chapter 3), with $\widehat{O}_{\mathrm{H}}(z_i) = \widehat{O}$, we get an expression for $\widehat{O}_{\mathrm{H}}(z)$ equivalent to the one given in (5.69):

$$\widehat{O}_{\mathrm{H}}(z) = e^{\widehat{\mathsf{T}}(z,z_i)}\widehat{O}, \quad (5.71)$$

where

$$\widehat{\mathsf{T}}(z,z_i) = \frac{1}{i\hbar}\int_{z_i}^{z}dz : -\widehat{\mathscr{H}_o}:$$
$$+ \left(\frac{1}{i\hbar}\right)^2 \frac{1}{2}\int_{z_i}^{z}dz_2 \int_{z_i}^{z_2}dz_1 \left[:-\widehat{\mathscr{H}_o}:(z_2), :-\widehat{\mathscr{H}_o}:(z_1)\right]$$
$$+ \frac{1}{6}\left(\frac{1}{i\hbar}\right)^3 \int_{z_i}^{z}dz_3 \int_{z_i}^{z_3}dz_2 \int_{z_i}^{z_2}dz_1$$
$$\left\{\left[\left[:-\widehat{\mathscr{H}_o}:(z_3), :-\widehat{\mathscr{H}_o}:(z_2)\right], :-\widehat{\mathscr{H}_o}:(z_1)\right]\right.$$
$$\left.+ \left[\left[:-\widehat{\mathscr{H}_o}:(z_1), :-\widehat{\mathscr{H}_o}:(z_2)\right], :-\widehat{\mathscr{H}_o}:(z_3)\right]\right\}$$
$$+ \cdots, \quad (5.72)$$

where

$$\left[:-\widehat{\mathscr{H}_o}:(z_1), :-\widehat{\mathscr{H}_o}:(z_2)\right]$$
$$= :-\widehat{\mathscr{H}_o}:(z_1):-\widehat{\mathscr{H}_o}:(z_2) - :-\widehat{\mathscr{H}_o}:(z_2):-\widehat{\mathscr{H}_o}:(z_1)$$
$$= :\left[-\widehat{\mathscr{H}_o}(z_1), -\widehat{\mathscr{H}_o}(z_2)\right]:, \quad (5.73)$$

as follows from (3.32). In the classical limit

$$\frac{1}{i\hbar} : -\widehat{\mathcal{H}_o} : \widehat{O} \longrightarrow : -\mathcal{H}_o : O, \qquad (5.74)$$

the equation (5.72) is seen to become the classical equation

$$O(z) = e^{T(z,z_i)} O(z_i), \qquad (5.75)$$

with $T(z, z_i)$ given by (4.97). This shows that in the classical limit, the quantum equation (5.68), which we shall be using throughout in the quantum beam optics for getting the transfer maps, becomes the classical equation (5.75), which is the basis for the Lie transfer operator method for getting the transfer maps in classical beam optics.

5.2.2 FREE PROPAGATION: DIFFRACTION

Let us now consider the propagation of a quasiparaxial beam through free space, or between two optical elements, around the z-axis. The corresponding quantum beam optical Hamiltonian is

$$\widehat{\mathcal{H}_o} \approx -p_0 + \frac{1}{2p_0}\widehat{p}_\perp^2 + \frac{1}{8p_0^3}\widehat{p}_\perp^4, \qquad (5.76)$$

obtained from (5.34) by putting $\phi = 0$ and $\vec{A} = (0,0,0)$, and consequently taking $\widehat{\pi}_\perp^2 = \widehat{p}_\perp^2$ and $\tilde{p}^2 = 0$. For an ideal paraxial beam, we can take

$$\widehat{\mathcal{H}_o} = -p_0 + \frac{1}{2p_0}\widehat{p}_\perp^2, \qquad (5.77)$$

retaining only terms up to second order in $\vec{p}_\perp$ in view of the paraxiality condition $\left|\vec{p}_\perp\right| \ll p_0$ assumed to be valid for all particles of the beam. Actually, the free propagation can be treated exactly. The expression for the quantum beam optical Hamiltonian $\widehat{\mathcal{H}_o}$ in (5.76) is an approximation for the exact result $-\sqrt{p_0^2 - \widehat{p}_\perp^2}$, corresponding to $-p_z$, which will be obtained in the infinite series form if we continue the Foldy–Wouthuysen-like transformation process up to all orders starting with (5.13) corresponding to free propagation.

First, let us look at free propagation of a paraxial beam in the Schrödinger picture. With the corresponding optical Hamiltonian given by (5.77), the Schrödinger equation for free propagation of a paraxial beam along the z-direction becomes

$$i\hbar \frac{\partial}{\partial z}|\psi(z)\rangle = \widehat{\mathcal{H}_o}|\psi(z)\rangle = \left(-p_0 + \frac{\widehat{p}_\perp^2}{2p_0}\right)|\psi(z)\rangle. \qquad (5.78)$$

On integration, we have

$$|\psi(z)\rangle = e^{-\frac{i\Delta z}{\hbar}\left(-p_0 + \frac{\widehat{p}_\perp^2}{2p_0}\right)}|\psi(z_i)\rangle, \quad \Delta z = z - z_i. \qquad (5.79)$$

Scalar Theory: Spin-0 and Spinless Particles

In position representation, we can write

$$\psi(\vec{r}_\perp, z) = \int d^2 r_{\perp i} \, K(\vec{r}_\perp, z; \vec{r}_{\perp i}, z_i) \, \psi(\vec{r}_{\perp i}, z_i),$$

$$K(\vec{r}_\perp, z; \vec{r}_{\perp i}, z_i) = \left\langle \vec{r}_\perp \left| e^{-\frac{i\Delta z}{\hbar}\left(-p_0 + \frac{\hat{p}_\perp^2}{2p_0}\right)} \right| \vec{r}_{\perp i} \right\rangle. \tag{5.80}$$

The Schrödinger equation for free propagation of a paraxial beam (5.78), with z replaced by t, is the same as the nonrelativistic Schrödinger equation of a free particle of mass p_0 moving in the xy-plane, in a constant potential $-p_0$. We can ignore the constant potential. We have already calculated in (3.313) the propagator for a free particle of mass m. It is straightforward to use (3.313) to derive the desired result:

$$K(\vec{r}_\perp, z; \vec{r}_{\perp i}, z_i) = \left(\frac{p_0}{2\pi i \hbar \Delta z}\right) e^{\left\{\frac{i}{\hbar} p_0 \Delta z + \frac{i p_0}{2\hbar \Delta z} |\vec{r}_\perp - \vec{r}_{\perp i}|^2\right\}}. \tag{5.81}$$

Now, the equation (5.80) becomes

$$\psi(\vec{r}_\perp, z) = \left(\frac{p_0}{2\pi i \hbar \Delta z}\right) e^{\frac{i}{\hbar} p_0 \Delta z} \int d^2 r_{\perp i} \, e^{\frac{i p_0}{2\hbar \Delta z} |\vec{r}_\perp - \vec{r}_{\perp i}|^2} \psi(\vec{r}_{\perp i}, z_i). \tag{5.82}$$

Substituting $2\pi\hbar/p_0 = \lambda_0$, the de Broglie wavelength of the particle, we get

$$\psi(x,y,z) = \left(\frac{1}{i\lambda_0(z-z_i)}\right) e^{\frac{i 2\pi}{\lambda_0}(z-z_i)}$$

$$\times \int\int dx_i dy_i \, e^{\frac{i\pi}{\lambda_0(z-z_i)}\left[(x-x_i)^2 + (y-y_i)^2\right]} \psi(x_i, y_i, z_i), \tag{5.83}$$

which is the well-known Fresnel diffraction formula. Here, the xy-plane at z_i is the plane of the diffracting object and the plane at z is the observation plane. The wave function on the exit side of the diffracting object is $\psi(\vec{r}_\perp, z_i)$, and $|\psi(\vec{r}_\perp, z)|^2$ gives the intensity distribution of the diffraction pattern (probability distribution for the particle) at the observation plane. It is clear that the paraxial approximation used in deriving (5.77), dropping terms of order higher than second in $\vec{\hat{p}}_\perp$, essentially corresponds to the traditional approximation used in deriving the Fresnel diffraction formula from the general Kirchchoff's result. Equations (5.34), (5.36), (5.41), and (5.42), represent, in operator form, the general theory of charged particle diffraction in the presence of electromagnetic fields (for more details of the wave theory of electron diffraction, see Hawkes and Kasper [72]).

Next, let us work out the transfer maps for the expectation values of $\vec{r}_\perp$ and $\vec{\hat{p}}_\perp$. From (5.36) and (5.38),

$$|\psi(z)\rangle = \hat{U}(z, z_i) |\psi(z_i)\rangle,$$

$$\hat{U}(z, z_i) = e^{\left\{-\frac{i}{\hbar} \int_{z_i}^{z} dz \, \widehat{\mathcal{H}_o}\right\}} = e^{-\frac{i}{\hbar} \widehat{T}(z, z_i)},$$

$$\widehat{T}(z, z_i) = \Delta z \widehat{\mathcal{H}_o} = \Delta z \left(-p_0 + \frac{1}{2p_0} \hat{p}_\perp^2\right), \qquad \Delta z = (z - z_i). \tag{5.84}$$

Note that $\widehat{U}(z,z_i)$ is an ordinary exponential since $\widehat{\mathcal{H}_o}$ is z-independent. Now, from (5.67), with

$$\begin{aligned}
:\widehat{T}(z,z_i):\vec{r}_\perp &= \left[\widehat{T}(z,z_i),\vec{r}_\perp\right] = \left[\Delta z\left(-p_0 + \frac{1}{2p_0}\widehat{p}_\perp^2\right),\vec{r}_\perp\right] \\
&= \left[\Delta z\left(\frac{1}{2p_0}\widehat{p}_\perp^2\right),\vec{r}_\perp\right] = -i\hbar\Delta z\frac{\widehat{p}_\perp}{p_0}, \\
:\widehat{T}(z,z_i):\vec{\widehat{p}}_\perp &= \left[\widehat{T}(z,z_i),\vec{\widehat{p}}_\perp\right] = \left[\Delta z\left(-p_0 + \frac{1}{2p_0}\widehat{p}_\perp^2\right),\vec{\widehat{p}}_\perp\right] \\
&= \left[\Delta z\left(\frac{1}{2p_0}\widehat{p}_\perp^2\right),\vec{\widehat{p}}_\perp\right] = 0, \\
:\widehat{T}(z,z_i):^2 \vec{r}_\perp &= \left[\widehat{T}(z,z_i),\left[\widehat{T}(z,z_i),\vec{r}_\perp\right]\right] \\
&= -i\hbar\frac{\Delta z}{p_0}\left[\widehat{T}(z,z_i),\vec{\widehat{p}}_\perp\right] = 0, \quad (5.85)
\end{aligned}$$

we get

$$\begin{aligned}
\langle\vec{r}_\perp\rangle(z) &= \left\langle\psi(z_i)\left|\left(\vec{r}_\perp + \Delta z\frac{\widehat{p}_\perp}{p_0}\right)\right|\psi(z_i)\right\rangle, \\
&= \langle\vec{r}_\perp\rangle(z_i) + \frac{\Delta z}{p_0}\langle\vec{\widehat{p}}_\perp\rangle(z_i), \\
\langle\vec{\widehat{p}}_\perp\rangle(z) &= \langle\psi(z_i)|\vec{\widehat{p}}_\perp|\psi(z_i)\rangle = \langle\vec{\widehat{p}}_\perp\rangle(z_i). \quad (5.86)
\end{aligned}$$

Thus, the resulting transfer map,

$$\begin{pmatrix} \langle\vec{r}_\perp\rangle(z) \\ \frac{1}{p_0}\langle\vec{\widehat{p}}_\perp\rangle(z) \end{pmatrix} = \begin{pmatrix} 1 & \Delta z \\ 0 & 1 \end{pmatrix} \begin{pmatrix} \langle\vec{r}_\perp\rangle(z_i) \\ \frac{1}{p_0}\langle\vec{\widehat{p}}_\perp\rangle(z_i) \end{pmatrix}, \quad (5.87)$$

is exactly the same as the transfer map (4.16) for the classical $\vec{r}_\perp$ and $\vec{p}_\perp$ in accordance with Ehrenfest's theorem.

5.2.3 AXIALLY SYMMETRIC MAGNETIC LENS: ELECTRON OPTICAL IMAGING

5.2.3.1 Paraxial Approximation: Point-to-Point Imaging

Let us now consider a round magnetic lens. Let us recall that it comprises an axially symmetric magnetic field, and there is no electric field, i.e., $\phi = 0$. Taking the z-axis as the optic axis, the vector potential can be taken, in general, as

$$\vec{A} = \left(-\frac{1}{2}y\Pi(\vec{r}_\perp,z), \frac{1}{2}x\Pi(\vec{r}_\perp,z), 0\right), \quad (5.88)$$

Scalar Theory: Spin-0 and Spinless Particles

with

$$\Pi(\vec{r}_\perp, z) = \sum_{n=0}^{\infty} \frac{1}{n!(n+1)!} \left(-\frac{r_\perp^2}{4}\right)^n B^{(2n)}(z)$$
$$= B(z) - \frac{1}{8} r_\perp^2 B''(z) + \frac{1}{192} r_\perp^4 B''''(z) - \cdots, \quad (5.89)$$

where $r_\perp^2 = x^2 + y^2$, $B^{(0)}(z) = B(z)$, $B'(z) = dB(z)/dz$, $B''(z) = d^2B(z)/dz^2$, $B'''(z) = d^3B(z)/dz^3$, $B''''(z) = d^4B(z)/dz^4, \ldots$, and $B^{(2n)}(z) = d^{2n}B(z)/dz^{2n}$. The corresponding magnetic field is given by

$$\vec{B}_\perp = -\frac{1}{2}\left(B'(z) - \frac{1}{8}B'''(z)r_\perp^2 + \cdots\right)\vec{r}_\perp,$$
$$B_z = B(z) - \frac{1}{4}B''(z)r_\perp^2 + \frac{1}{64}B''''(z)r_\perp^4 - \cdots. \quad (5.90)$$

The practical boundaries of the lens, say z_ℓ and z_r, are determined by where $B(z)$ becomes negligible, *i.e.*, $B(z < z_\ell) \approx 0$ and $B(z > z_r) \approx 0$.

First, we are concerned with the propagation of a monoenergetic paraxial beam with design momentum p_0 such that $|\vec{p}| = p_0 \approx p_z$ for all its constituent particles. For a paraxial beam propagating through the round magnetic lens, along the optic axis, the vector potential $\vec{A}$ can be taken as

$$\vec{A} \approx \vec{A}_0 = \left(-\frac{1}{2}B(z)y, \frac{1}{2}B(z)x, 0\right), \quad (5.91)$$

corresponding to the field

$$\vec{B} \approx \vec{B}_0 = \left(-\frac{1}{2}B'(z)x, -\frac{1}{2}B'(z)y, B(z)\right). \quad (5.92)$$

In the paraxial approximation, only the lowest order terms in $\vec{r}_\perp$ are considered to contribute to the effective field felt by the particles moving close to the optic axis. Further, it entails approximating the quantum beam optical Hamiltonian by dropping from it the terms of order higher than second in $\left|\vec{p}_\perp\right|/p_0$ in view of the condition $|\vec{p}_\perp|/p_0 \ll 1$. Thus, the quantum beam optical Hamiltonian of the round magnetic lens derived from (5.34) in the paraxial approximation is

$$\widehat{\mathcal{H}_o} \approx -p_0 + \frac{1}{2p_0}\hat{p}_\perp^2 + \frac{q^2 B(z)^2}{8p_0}r_\perp^2 - \frac{qB(z)}{2p_0}\hat{L}_z$$
$$+ \frac{\hbar^2 q^2}{16p_0^4} B(z)B'(z)\left(\vec{r}_\perp \cdot \vec{\hat{p}}_\perp + \vec{\hat{p}}_\perp \cdot \vec{r}_\perp\right), \quad (5.93)$$

where

$$B(z) \begin{cases} \neq 0 & \text{in the lens region } (z_\ell \leq z \leq z_r) \\ = 0 & \text{outside the lens region } (z < z_\ell, z > z_r), \end{cases} \quad (5.94)$$

and $\widehat{L}_z = (x\widehat{p}_y - y\widehat{p}_x)$ is the z-component of the orbital angular momentum operator. In the above $\widehat{\mathscr{H}_o}$, the last term is proportional to $\hbar^2$ and is very small compared to the other terms. We will call the terms proportional to powers of $\hbar$ as $\hbar$-dependent since they contain $\hbar$ explicitly such that their quantum averages, being proportional to powers of $\hbar$, would vanish in the classical limit $\hbar \longrightarrow 0$. Note that the components of $\vec{\widehat{p}} = -i\hbar\vec{\nabla}$ also contain $\hbar$, but we are able to identify their quantum averages with the components of the classical momentum vector $\vec{p}$. For the present, we shall omit the last $\hbar$-dependent term from the above $\widehat{\mathscr{H}_o}$ and take the optical Hamiltonian of the round magnetic lens as

$$\widehat{\mathscr{H}_o} = -p_0 + \frac{1}{2p_0}\widehat{p}_\perp^2 + \frac{q^2 B(z)^2}{8p_0}r_\perp^2 - \frac{qB(z)}{2p_0}\widehat{L}_z. \tag{5.95}$$

Note that the replacement of the quantum operators $\left(\vec{r}_\perp, \vec{\widehat{p}}_\perp\right)$ by the classical variables $(\vec{r}_\perp, \vec{p}_\perp)$ takes $\widehat{\mathscr{H}_o}$ exactly to the classical $\mathscr{H}_o$ in (4.24). As a result, without the $\hbar$-dependent term, the paraxial $\widehat{\mathscr{H}_o}$ leads exactly to the same performance of the lens as in the classical paraxial theory, except for the replacement of the classical $(\vec{r}_\perp, \vec{p}_\perp)$ by their quantum averages. Later we shall see the effect of the $\hbar$-dependent nonclassical term at the paraxial level which, of course, will be very small. Introducing the notation

$$\alpha(z) = \frac{qB(z)}{2p_0}, \tag{5.96}$$

we shall write

$$\widehat{\mathscr{H}_o} = \widehat{\mathscr{H}}_{o,L} + \widehat{\mathscr{H}}_{o,R},$$
$$\widehat{\mathscr{H}}_{o,L} = -p_0 + \frac{1}{2p_0}\widehat{p}_\perp^2 + \frac{1}{2}p_0\alpha^2(z)r_\perp^2,$$
$$\widehat{\mathscr{H}}_{o,R} = -\alpha(z)\widehat{L}_z. \tag{5.97}$$

Let us consider the propagation of a paraxial beam from the xy-plane at $z_i < z_\ell$ to the xy-plane at $z > z_r$. We have from (5.36) and (5.38)

$$|\psi(z)\rangle = \widehat{U}_p(z,z_i)|\psi(z_i)\rangle,$$
$$\widehat{U}_p(z,z_i) = \mathrm{P}\left(e^{-\frac{i}{\hbar}\int_{z_i}^z dz\, \widehat{\mathscr{H}_o}}\right) = e^{-\frac{i}{\hbar}\widehat{T}(z,z_i)}, \tag{5.98}$$

where it has to be noted that $\widehat{\mathscr{H}_o}$ is z-dependent. Now, using (5.38), we have

$$\widehat{T}(z,z_i) = \int_{z_i}^z dz_1 \left(\widehat{\mathscr{H}}_{o,L}(z_1) + \widehat{\mathscr{H}}_{o,R}(z_1)\right) +$$
$$\frac{1}{2}\left(-\frac{i}{\hbar}\right)\int_{z_i}^z dz_2 \int_{z_i}^{z_2} dz_1 \left[\left(\widehat{\mathscr{H}}_{o,L}(z_2) + \widehat{\mathscr{H}}_{o,R}(z_2)\right),\right.$$
$$\left.\left(\widehat{\mathscr{H}}_{o,L}(z_1) + \widehat{\mathscr{H}}_{o,R}(z_1)\right)\right] +$$

Scalar Theory: Spin-0 and Spinless Particles

$$\frac{1}{6}\left(-\frac{i}{\hbar}\right)^2 \int_{z_i}^{z} dz_3 \int_{z_i}^{z_3} dz_2 \int_{z_i}^{z_2} dz_1 \left\{ \left[\left[\left(\widehat{\mathcal{H}}_{o,L}(z_3) + \widehat{\mathcal{H}}_{o,R}(z_3)\right),\right.\right.\right.$$
$$\left.\left(\widehat{\mathcal{H}}_{o,L}(z_2) + \widehat{\mathcal{H}}_{o,R}(z_2)\right)\right], \left(\widehat{\mathcal{H}}_{o,L}(z_1) + \widehat{\mathcal{H}}_{o,R}(z_1)\right)\right] +$$
$$\left[\left[\left(\widehat{\mathcal{H}}_{o,L}(z_1) + \widehat{\mathcal{H}}_{o,R}(z_1)\right), \left(\widehat{\mathcal{H}}_{o,L}(z_2) + \widehat{\mathcal{H}}_{o,R}(z_2)\right)\right],\right.$$
$$\left.\left.\left(\widehat{\mathcal{H}}_{o,L}(z_3) + \widehat{\mathcal{H}}_{o,R}(z_3)\right)\right]\right\} + \cdots. \tag{5.99}$$

Note that

$$\begin{aligned}
\left[p_0, \widehat{L}_z\right] &= 0, \quad \text{since } p_0 = \text{constant}, \\
\left[r_\perp^2, \widehat{L}_z\right] &= [x^2 + y^2, (x\widehat{p}_y - y\widehat{p}_x)] = [y^2, x\widehat{p}_y] - [x^2, y\widehat{p}_x] \\
&= 2i\hbar(xy - yx) = 0, \\
\left[\widehat{p}_\perp^2, \widehat{L}_z\right] &= [\widehat{p}_x^2 + \widehat{p}_y^2, (x\widehat{p}_y - y\widehat{p}_x)] = [\widehat{p}_x^2, x\widehat{p}_y] - [\widehat{p}_y^2, y\widehat{p}_x] \\
&= 2i\hbar(\widehat{p}_x \widehat{p}_y - \widehat{p}_y \widehat{p}_x) = 0.
\end{aligned} \tag{5.100}$$

This makes

$$\begin{aligned}
\left[\widehat{\mathcal{H}}_{o,L}(z'), \widehat{\mathcal{H}}_{o,R}(z'')\right] &= \left[-p_0 + \frac{1}{2p_0}\widehat{p}_\perp^2 + \frac{1}{2}p_0\alpha^2(z')r_\perp^2, \alpha(z'')\widehat{L}_z\right] \\
&= 0, \quad \text{for all } z', z'', \\
\left[\widehat{\mathcal{H}}_{o,R}(z'), \widehat{\mathcal{H}}_{o,R}(z'')\right] &= \left[\alpha(z')\widehat{L}_z, \alpha(z'')\widehat{L}_z\right] \\
&= 0, \quad \text{for all } z', z'', \\
\left[\widehat{\mathcal{H}}_{o,L}(z'), \widehat{\mathcal{H}}_{o,L}(z'')\right] &= \left[-p_0 + \frac{1}{2p_0}\widehat{p}_\perp^2 + \frac{1}{2}p_0\alpha^2(z')r_\perp^2,\right. \\
&\quad \left. -p_0 + \frac{1}{2p_0}\widehat{p}_\perp^2 + \frac{1}{2}p_0\alpha^2(z'')r_\perp^2\right] \\
&= \frac{i\hbar}{2}\left(\alpha^2(z') - \alpha^2(z'')\right)\left(\vec{r}_\perp \cdot \vec{\widehat{p}}_\perp + \vec{\widehat{p}}_\perp \cdot \vec{r}_\perp\right) \\
&\neq 0, \text{ in general.}
\end{aligned} \tag{5.101}$$

As a result, we get

$$\widehat{T}(z, z_i) = \widehat{T}_R(z, z_i) + \widehat{T}_L(z, z_i), \tag{5.102}$$

with

$$\begin{aligned}
\widehat{T}_R(z, z_i) &= \int_{z_i}^{z} dz\, \widehat{\mathcal{H}}_{o,L}(z) = -\theta(z, z_i)\widehat{L}_z, \\
\theta(z, z_i) &= \int_{z_i}^{z} dz\, \alpha(z) = \frac{q}{2p_0}\int_{z_i}^{z} dz\, B(z),
\end{aligned} \tag{5.103}$$

and

$$\widehat{T}_L(z,z_i) = \int_{z_i}^{z} dz\, \widehat{\mathcal{H}}_{o,L}(z) + \frac{1}{2}\left(-\frac{i}{\hbar}\right)\int_{z_i}^{z} dz_2 \int_{z_i}^{z_2} dz_1 \left[\widehat{\mathcal{H}}_{o,L}(z_2), \widehat{\mathcal{H}}_{o,L}(z_1)\right] +$$
$$\frac{1}{6}\left(-\frac{i}{\hbar}\right)^2 \int_{z_i}^{z} dz_3 \int_{z_i}^{z_3} dz_2 \int_{z_i}^{z_2} dz_1 \left\{\left[\left[\left(\widehat{\mathcal{H}}_{o,L}(z_3)\right),\left(\widehat{\mathcal{H}}_{o,L}(z_2)\right)\right],\right.\right.$$
$$\left.\left.\left(\widehat{\mathcal{H}}_{o,L}(z_1)\right)\right] + \left[\left[\left(\widehat{\mathcal{H}}_{o,L}(z_1)\right),\left(\widehat{\mathcal{H}}_{o,L}(z_2)\right)\right],\left(\widehat{\mathcal{H}}_{o,L}(z_3)\right)\right]\right\} + \cdots.$$
(5.104)

Since
$$\left[\widehat{T}_R(z,z_i), \widehat{T}_L(z,z_i)\right] = 0, \tag{5.105}$$

we have
$$\widehat{U}_p(z,z_i) = e^{-\frac{i}{\hbar}\widehat{T}(z,z_i)} = e^{-\frac{i}{\hbar}\left(\widehat{T}_R(z,z_i)+\widehat{T}_L(z,z_i)\right)}$$
$$= e^{-\frac{i}{\hbar}\widehat{T}_R(z,z_i)} e^{-\frac{i}{\hbar}\widehat{T}_L(z,z_i)} = e^{-\frac{i}{\hbar}\widehat{T}_L(z,z_i)} e^{-\frac{i}{\hbar}\widehat{T}_R(z,z_i)}$$
$$= \widehat{U}_R(z,z_i)\widehat{U}_L(z,z_i) = \widehat{U}_L(z,z_i)\widehat{U}_R(z,z_i). \tag{5.106}$$

Thus, in the Schrödinger picture, we have

$$\psi(\vec{r}_\perp, z) = \int d^2 r_{\perp,i}\, K\left(\vec{r}_\perp, z; \vec{r}_{\perp,i}, z_i\right) \psi\left(\vec{r}_{\perp,i}, z_i\right)$$
$$= \int d^2 r_{\perp,i} \left\langle \vec{r}_\perp, z \left| \widehat{U}_p(z,z_i) \right| \vec{r}_{\perp,i}, z_i \right\rangle \psi\left(\vec{r}_{\perp,i}, z_i\right)$$
$$= \int d^2 r_{\perp,i} \int d^2 r'_\perp \left\langle \vec{r}_\perp \left| \widehat{U}_R(z,z_i) \right| \vec{r}'_\perp \right\rangle \left\langle \vec{r}'_\perp \left| \widehat{U}_L(z,z_i) \right| \vec{r}_{\perp,i} \right\rangle \psi\left(\vec{r}_{\perp,i}, z_i\right).$$
(5.107)

Let us first compute the required matrix element of the rotation operator $\widehat{U}_R$ through an angle θ around the z-axis. To this end, let us choose the cylindrical coordinate system: $x = \rho \cos\vartheta, y = \rho \sin\vartheta, z = z$. Then,

$$e^{\frac{i}{\hbar}\theta \widehat{L}_z} \psi(x,y,z) = e^{\frac{i}{\hbar}\theta(x\widehat{p}_y - y\widehat{p}_x)} \psi(x,y,z)$$
$$= e^{\theta(\partial/\partial\vartheta)} \psi(\rho,\vartheta,z) = \psi(\rho, \vartheta+\theta, z)$$
$$= \psi(x(\theta), y(\theta), z),$$
$$\text{with } \begin{pmatrix} x(\theta) \\ y(\theta) \end{pmatrix} = \begin{pmatrix} \cos\theta & -\sin\theta \\ \sin\theta & \cos\theta \end{pmatrix} \begin{pmatrix} x \\ y \end{pmatrix}. \tag{5.108}$$

Using this result, we get

$$\left\langle \vec{r}'_\perp \left| \widehat{U}_R(z,z_i) \right| \vec{r}''_\perp \right\rangle = \left\langle \vec{r}'_\perp \left| e^{\frac{i}{\hbar}\theta(z,z_i)\widehat{L}_z} \right| \vec{r}''_\perp \right\rangle$$
$$= \int d^2 r_\perp\, \delta^2\left(\vec{r}'_\perp - \vec{r}_\perp\right) e^{\frac{i}{\hbar}\theta(z,z_i)\widehat{L}_z} \delta^2\left(\vec{r}_\perp - \vec{r}''_\perp\right)$$

Scalar Theory: Spin-0 and Spinless Particles

$$= \int d^2 r_\perp \, \delta^2 \left(\vec{r}_\perp - \vec{r}'_\perp \right) \delta^2 \left(\vec{r}_\perp \left(\theta \left(z, z_i \right) \right) - \vec{r}''_\perp \right)$$

$$= \delta^2 \left(\vec{r}'_\perp \left(\theta \left(z, z_i \right) \right) - \vec{r}''_\perp \right), \tag{5.109}$$

where $\vec{r}'_\perp \left(\theta \left(z, z_i \right) \right)$ is $\vec{r}_\perp \left(\theta \left(z, z_i \right) \right)$ with (x, y) replaced by (x', y') (see (5.108)). Thus, substituting the required matrix element,

$$\left\langle \vec{r}_\perp \left| \widehat{U}_R \left(z, z_i \right) \right| \vec{r}'_\perp \right\rangle = \delta^2 \left(\vec{r}_\perp \left(\theta \left(z, z_i \right) \right) - \vec{r}'_\perp \right), \tag{5.110}$$

in (5.107), we get

$$\psi \left(\vec{r}_\perp, z \right) = \int d^2 r_{\perp,i} \int d^2 r'_\perp \, \delta^2 \left(\vec{r}_\perp \left(\theta \left(z, z_i \right) \right) - \vec{r}'_\perp \right)$$
$$\times \left\langle \vec{r}'_\perp \left| \widehat{U}_L \left(z, z_i \right) \right| \vec{r}_{\perp,i} \right\rangle \psi \left(\vec{r}_{\perp,i}, z_i \right)$$
$$= \int d^2 r_{\perp,i} \left\langle \vec{r}_\perp \left(\theta \left(z, z_i \right) \right) \left| \widehat{U}_L \left(z, z_i \right) \right| \vec{r}_{\perp,i} \right\rangle \psi \left(\vec{r}_{\perp,i}, z_i \right)$$
$$= \int d^2 r_{\perp,i} \, K_L \left(\vec{r}_\perp \left(\theta \left(z, z_i \right) \right), z; \vec{r}_{\perp,i}, z_i \right) \psi \left(\vec{r}_{\perp,i}, z_i \right). \tag{5.111}$$

The exact expression for the propagator $K_L \left(\vec{r}_\perp \left(\theta \left(z, z_i \right) \right), z; \vec{r}_{\perp,i}, z_i \right)$ can be written down. Note that $\widehat{\mathscr{H}}_{o,L}$, except for the additive constant $-p_0$, is exactly like the Hamiltonian of a two-dimensional isotropic harmonic oscillator with time-dependent frequency. Such a connection between optics and harmonic oscillator is of course well known (see, *e.g.*, Agarwal and Simon [2] in the context of light optics and Dattoli, Renieri, and Torre [31] in the context of charged particle optics). For such quantum systems with time-dependent Hamiltonians quadratic in $\vec{r}$ and $\vec{p}$, the exact propagators have been found using the Feynman path integral method (see, *e.g.*, Khandekar and Lawande [121], and Kleinert [122]). We shall closely follow Wolf [190] who has given a method, using techniques of canonical transformations and Lie algebra, to get the propagator for any quantum system with a time-dependent Hamiltonian quadratic in $\vec{r}$ and $\vec{p}$ (see the Appendix at the end of this chapter for an outline of the method applied in our case). The resulting expression for K_L is

$$K_L \left(\vec{r}_\perp \left(\theta \left(z, z_i \right) \right), z; \vec{r}_{\perp,i}, z_i \right)$$
$$= \frac{p_0}{i 2\pi \hbar h \left(z, z_i \right)} \exp \left\{ \frac{i}{\hbar} p_0 \Delta z \right.$$
$$\left. + \frac{i p_0}{2 \hbar h \left(z, z_i \right)} \left[g \left(z, z_i \right) r_{\perp,i}^2 - 2 \vec{r}_{\perp,i} \cdot \vec{r}_\perp \left(\theta \left(z, z_i \right) \right) + h' \left(z, z_i \right) r_\perp^2 \right] \right\},$$
$$\text{if } h \left(z, z_i \right) \neq 0, \tag{5.112}$$

and

$$K_L\left(\vec{r}_\perp\left(\theta(z,z_i),z;\vec{r}_{\perp,i},z_i\right)\right)$$
$$= \frac{1}{g(z,z_i)}\exp\left(\frac{i}{\hbar}p_0\Delta z + \frac{ip_0 g'(z,z_i)}{2\hbar g(z,z_i)}r_\perp^2\right)\delta^2\left(\vec{r}_{\perp,i} - \frac{\vec{r}_\perp(\theta(z,z_i))}{g(z,z_i)}\right),$$
$$\text{if } h(z,z_i) = 0, \qquad (5.113)$$

where $g(z,z_i)$, $h(z,z_i)$, $g'(z,z_i)$, and $h'(z,z_i)$ are the elements of the classical transfer matrix (4.53) for $\left(\vec{r}_\perp(z_i),\vec{P}_\perp(z_i)/p_0\right) \longrightarrow \left(\vec{r}_\perp(z),\vec{P}_\perp(z)/p_0\right)$, the two linearly independent solutions of the classical paraxial equation (4.51), satisfying the initial conditions

$$g(z_i,z_i) = h'(z_i,z_i) = 1, \quad h(z_i,z_i) = g'(z_i,z_i) = 0, \qquad (5.114)$$

and the relation

$$g(z,z_i)h'(z,z_i) - h(z,z_i)g'(z,z_i) = 1, \qquad \text{for any } z \geq z_i. \qquad (5.115)$$

Note that when $z = z_i$, $K_L = \delta^2\left(\vec{r}_{\perp,i} - \vec{r}_\perp\right)$ as required. Now, from (5.111)–(5.113), it follows that

$$\psi(\vec{r}_\perp,z) = \frac{1}{i\lambda_0 h(z,z_i)}\exp\left(\frac{i2\pi}{\lambda_0}\left[(z-z_i) + \frac{h'(z,z_i)}{2h(z,z_i)}r_\perp^2\right]\right)$$
$$\times \int d^2r_{\perp,i}\exp\left\{\frac{i\pi}{\lambda_0 h(z,z_i)}\left[g(z,z_i)r_{\perp,i}^2 - 2\vec{r}_{\perp,i}\cdot\vec{r}_\perp(\theta(z,z_i))\right]\right\},$$
$$\text{if } h(z,z_i) \neq 0,$$
$$\psi(\vec{r}_\perp,z) = \frac{1}{g(z,z_i)}\exp\left(\frac{i2\pi}{\lambda_0}\left[(z-z_i) + \frac{g'(z,z_i)}{2g(z,z_i)}r_\perp^2\right]\right)$$
$$\times \psi\left(\frac{\vec{r}_\perp(\theta(z,z_i))}{g(z,z_i)},z_i\right),$$
$$\text{if } h(z,z_i) = 0. \qquad (5.116)$$

This equation represents the well-known general law of propagation of the paraxial wave function in the case of a round magnetic lens (see Glaser [59, 62]), Glaser and Schiske [60, 61], and Hawkes and Kasper [72]). The phase factor $\exp\{i2\pi(z-z_i)/\lambda_0\}$ would not be there if we drop the constant term $-p_0$ in the optical Hamiltonian. Usually, a rotated coordinate frame is introduced, as we have done in the classical treatment, to avoid the rotation factor $\theta(z,z_i)$ in the final z-plane. Equation (5.116) is the basis for the development of Fourier transform methods in the electron optical imaging techniques (for details, see Hawkes and Kasper [72]).

Let us take $z_i = z_{obj} < z_\ell$, where the object plane is, and $z = z_{img} > z_r$, where the image of the object is formed. As is clear from (5.116), if $h(z_{img},z_{obj}) = 0$, then we have, with $M = g(z_{img},z_{obj})$ and $\theta_L = \theta(z_\ell,z_r) = \theta(z_{img},z_{obj})$,

Scalar Theory: Spin-0 and Spinless Particles

$$\psi\left(\vec{r}_{\perp,img}, z_{img}\right) = \frac{1}{M} \exp\left(\frac{i2\pi}{\lambda_0}\left[(z_{img} - z_{obj}) + \frac{g'(z_{img}, z_{obj})}{2M} r_{\perp,img}^2\right]\right)$$

$$\times \psi\left(\frac{1}{M}\vec{r}_{\perp,img}(\theta_L), z_{obj}\right),$$

$$\left|\psi\left(\vec{r}_{\perp,img}, z_{img}\right)\right|^2 = \frac{1}{M^2}\left|\psi\left(\frac{1}{M}\vec{r}_{\perp,img}(\theta_L), z_{obj}\right)\right|^2. \quad (5.117)$$

This equation shows that at the image plane at $z = z_{img}$, where $h(z_{img}, z_{obj}) = 0$, the intensity distribution at the object plane is reproduced exactly, point-to-point, with the magnification $M = g(z_{img}, z_{obj})$ and a rotation through the angle

$$\theta_L = \theta(z_{obj}, z_{img}) = \int_{z_{obj}}^{z_{img}} dz\, \alpha(z)$$

$$= \frac{q}{2p_0}\int_{z_{obj}}^{z_{img}} dz\, B(z) = \frac{q}{2p_0}\int_{z_\ell}^{z_r} dz\, B(z). \quad (5.118)$$

Note that $B(z) = 0$ outside the lens region ($z < z_\ell, z > z_r$). This image rotation in a single-stage electron optical imaging using a round magnetic lens is the effect of Larmor rotation of any charged particle in a magnetic field. Let us observe that, as should be, the total intensity is conserved:

$$\int d^2 r_{\perp,img} \left|\psi(\vec{r}_{\perp,img}, z_{img})\right|^2 = \frac{1}{M^2}\int d^2 r_{\perp,img} \left|\psi\left(\frac{1}{M}\vec{r}_{\perp,img}(\theta_L), z_{obj}\right)\right|^2$$

$$= \frac{1}{M^2}\int d^2 r_{\perp,img}(\theta_L) \left|\psi\left(\frac{1}{M}\vec{r}_{\perp,img}(\theta_L), z_{obj}\right)\right|^2$$

$$= \int \frac{1}{M^2} d^2 r_{\perp,obj} \left|\psi\left(\frac{1}{M}\vec{r}_{\perp,obj}, z_{obj}\right)\right|^2$$

$$= \int d^2 r_{\perp,obj} \left|\psi(\vec{r}_{\perp,obj}, z_{obj})\right|^2. \quad (5.119)$$

Let us now understand the Gaussian, point-to-point, or stigmatic imaging of an object by the round magnetic lens through the transfer maps for $\langle \vec{r}_\perp \rangle$ and $\langle \vec{p}_\perp \rangle / p_0$. To this end, we shall use the z-evolution equations in (5.63) for $\langle \vec{r}_\perp \rangle$ and $\langle \vec{p}_\perp \rangle / p_0$. We get

$$\frac{d}{dz}\langle x \rangle(z) = \frac{i}{\hbar}\left\langle \left[\widehat{\mathcal{H}_o}, x\right]\right\rangle(z)$$

$$= \left\langle \frac{i}{\hbar}\left[-p_0 + \frac{1}{2p_0}\widehat{p}_\perp^2 + \frac{1}{2}p_0 \alpha^2(z) r_\perp^2 - \alpha(z)\widehat{L}_z, x\right]\right\rangle(z)$$

$$= \frac{1}{p_0}\langle \widehat{p}_x \rangle(z) + \alpha(z)\langle y \rangle(z),$$

$$\frac{d}{dz}\langle y \rangle(z) = \frac{i}{\hbar}\left\langle \left[\widehat{\mathcal{H}_o}, y\right]\right\rangle(z)$$

$$= \left\langle \frac{i}{\hbar} \left[-p_0 + \frac{1}{2p_0}\widehat{p}_\perp^2 + \frac{1}{2}p_0\alpha^2(z)r_\perp^2 - \alpha(z)\widehat{L}_z \, , \, y \right] \right\rangle (z)$$

$$= \frac{1}{p_0} \langle \widehat{p}_y \rangle (z) - \alpha(z)\langle x \rangle (z),$$

$$\frac{1}{p_0}\frac{d}{dz}\langle \widehat{p}_x \rangle (z) = \frac{i}{\hbar p_0}\left\langle \left[\widehat{\mathscr{H}}_o \, , \, \widehat{p}_x \right] \right\rangle (z)$$

$$= \left\langle \frac{i}{\hbar p_0} \left[-p_0 + \frac{1}{2p_0}\widehat{p}_\perp^2 + \frac{1}{2}p_0\alpha^2(z)r_\perp^2 - \alpha(z)\widehat{L}_z \, , \, \widehat{p}_x \right] \right\rangle (z)$$

$$= -\alpha^2(z)\langle x \rangle (z) + \frac{1}{p_0}\alpha(z)\langle \widehat{p}_y \rangle (z)$$

$$\frac{1}{p_0}\frac{d}{dz}\langle \widehat{p}_y \rangle (z) = \frac{i}{\hbar p_0}\left\langle \left[\widehat{\mathscr{H}}_o \, , \, \widehat{p}_y \right] \right\rangle (z)$$

$$= \left\langle \frac{i}{\hbar p_0} \left[-p_0 + \frac{1}{2p_0}\widehat{p}_\perp^2 + \frac{1}{2}p_0\alpha^2(z)r_\perp^2 - \alpha(z)\widehat{L}_z \, , \, \widehat{p}_y \right] \right\rangle (z)$$

$$= -\alpha^2(z)\langle y \rangle (z) - \frac{1}{p_0}\alpha(z)\langle \widehat{p}_x \rangle (z). \tag{5.120}$$

We can write these equations of motion as

$$\frac{d}{dz}\begin{pmatrix} \langle x \rangle (z) \\ \langle y \rangle (z) \\ \frac{1}{p_0}\langle \widehat{p}_x \rangle (z) \\ \frac{1}{p_0}\langle \widehat{p}_y \rangle (z) \end{pmatrix} = \begin{pmatrix} 0 & \alpha(z) & 1 & 0 \\ -\alpha(z) & 0 & 0 & 1 \\ -\alpha^2(z) & 0 & 0 & \alpha(z) \\ 0 & -\alpha^2(z) & -\alpha(z) & 0 \end{pmatrix} \begin{pmatrix} \langle x \rangle (z) \\ \langle y \rangle (z) \\ \frac{1}{p_0}\langle \widehat{p}_x \rangle (z) \\ \frac{1}{p_0}\langle \widehat{p}_y \rangle (z) \end{pmatrix}. \tag{5.121}$$

Let us write this as

$$\frac{d}{dz}\begin{pmatrix} \langle x \rangle (z) \\ \langle y \rangle (z) \\ \frac{1}{p_0}\langle \widehat{p}_x \rangle (z) \\ \frac{1}{p_0}\langle \widehat{p}_y \rangle (z) \end{pmatrix} = [\mu(z) \otimes I + I \otimes \rho(z)] \begin{pmatrix} \langle x \rangle (z) \\ \langle y \rangle (z) \\ \frac{1}{p_0}\langle \widehat{p}_x \rangle (z) \\ \frac{1}{p_0}\langle \widehat{p}_y \rangle (z) \end{pmatrix}, \tag{5.122}$$

with

$$\mu(z) = \begin{pmatrix} 0 & 1 \\ -\alpha^2(z) & 0 \end{pmatrix}, \tag{5.123}$$

and

$$\rho(z) = \alpha(z)\begin{pmatrix} 0 & 1 \\ -1 & 0 \end{pmatrix}. \tag{5.124}$$

It may be noted that these equations are exactly the same as in the classical theory, except for the replacement of the classical $\vec{r}_\perp$ and $\vec{p}_\perp$ by their corresponding quantum averages $\langle \vec{r}_\perp \rangle (z)$ and $\langle \vec{\widehat{p}}_\perp \rangle (z)$. So, to make the present chapter on quantum theory self-contained, we shall be essentially repeating the same equations and statements in the chapter on classical theory.

Scalar Theory: Spin-0 and Spinless Particles

Let us now integrate (5.122). Noting that $\mu(z) \otimes I$ and $I \otimes \rho(z)$ commute with each other, let

$$\begin{pmatrix} \langle x \rangle(z) \\ \langle y \rangle(z) \\ \frac{1}{p_0}\langle \widehat{p}_x \rangle(z) \\ \frac{1}{p_0}\langle \widehat{p}_y \rangle(z) \end{pmatrix} = \mathscr{M}(z,z_i)\mathscr{R}(z,z_i) \begin{pmatrix} \langle x \rangle(z_i) \\ \langle y \rangle(z_i) \\ \frac{1}{p_0}\langle \widehat{p}_x \rangle(z_i) \\ \frac{1}{p_0}\langle \widehat{p}_y \rangle(z_i) \end{pmatrix}$$

for any $z \geq z_i$, (5.125)

where

$$\frac{d\mathscr{M}(z,z_i)}{dz} = (\mu(z) \otimes I)\mathscr{M}(z,z_i), \qquad \mathscr{M}(z_i,z_i) = I, \quad (5.126)$$

$$\frac{d\mathscr{R}(z,z_i)}{dz} = (I \times \rho(z))\mathscr{R}(z,z_i), \qquad \mathscr{R}(z_i,z_i) = I. \quad (5.127)$$

It is clear that we can write

$$\mathscr{M}(z,z_i) = M(z,z_i) \otimes I, \qquad \frac{d\mathscr{M}(z,z_i)}{dz} = \frac{dM(z,z_i)}{dz} \otimes I, \quad (5.128)$$

with

$$\frac{dM(z,z_i)}{dz} = \mu(z)M(z,z_i), \quad (5.129)$$

and

$$\mathscr{R}((z,z_i)) = I \otimes R(z,z_i), \qquad \frac{d\mathscr{R}(z,z_i)}{dz} = I \otimes \frac{dR(z,z_i)}{dz}, \quad (5.130)$$

with

$$\frac{dR(z,z_i)}{dz} = \rho(z)R(z,z_i). \quad (5.131)$$

Note that $\mathscr{M}(z,z_i)$ and $\mathscr{R}(z,z_i)$ commute with each other, $d\mathscr{M}(z,z_i)/dz$ commutes with $\mathscr{R}((z,z_i))$, and $d\mathscr{R}(z,z_i)/dz$ commutes with $\mathscr{M}((z,z_i))$. Then,

$$\frac{d}{dz}\begin{pmatrix} \langle x \rangle(z) \\ \langle y \rangle(z) \\ \frac{1}{p_0}\langle \widehat{p}_x \rangle(z) \\ \frac{1}{p_0}\langle \widehat{p}_y \rangle(z) \end{pmatrix}$$

$$= \left(\frac{d\mathscr{M}(z,z_i)}{dz}\mathscr{R}(z,z_i) + \mathscr{M}(z,z_i)\frac{d\mathscr{R}(z,z_i)}{dz} \right) \begin{pmatrix} \langle x \rangle(z_i) \\ \langle y \rangle(z_i) \\ \frac{1}{p_0}\langle \widehat{p}_x \rangle(z_i) \\ \frac{1}{p_0}\langle \widehat{p}_y \rangle(z_i) \end{pmatrix}$$

$$= [\mu(z) \otimes I + I \otimes \rho(z)] \mathscr{M}(z,z_i) \mathscr{R}(z,z_i) \begin{pmatrix} \langle x \rangle (z_i) \\ \langle y \rangle (z_i) \\ \frac{1}{p_0} \langle \hat{p}_x \rangle (z_i) \\ \frac{1}{p_0} \langle \hat{p}_y \rangle (z_i) \end{pmatrix}$$

$$= [\mu(z) \otimes I + I \otimes \rho(z)] \begin{pmatrix} \langle x \rangle (z) \\ \langle y \rangle (z) \\ \frac{1}{p_0} \langle \hat{p}_x \rangle (z) \\ \frac{1}{p_0} \langle \hat{p}_y \rangle (z) \end{pmatrix}, \quad (5.132)$$

as required by (5.122).

The equation for $\mathscr{R}(z,z_i)$, (5.127), can be readily integrated by ordinary exponentiation, since $[\rho(z'), \rho(z'')] = 0$ for any z' and z''. Thus, we have

$$\begin{aligned} \mathscr{R}(z,z_i) &= e^{\int_{z_i}^{z} dz\, \rho(z)} \\ &= \exp\left\{ I \otimes \left[\theta(z_i,z) \begin{pmatrix} 0 & 1 \\ -1 & 0 \end{pmatrix} \right] \right\} \\ &= I \otimes \begin{pmatrix} \cos\theta(z_i,z) & \sin\theta(z_i,z) \\ -\sin\theta(z_i,z) & \cos\theta(z_i,z) \end{pmatrix} \\ &= I \otimes R(z,z_i), \end{aligned} \quad (5.133)$$

with

$$\theta(z_i,z) = \int_{z_i}^{z} dz\, \alpha(z) = \frac{q}{2p_0} \int_{z_i}^{z} dz\, B(z). \quad (5.134)$$

Let us now introduce an XYz-coordinate system with its X and Y axes rotating along the z-axis at the rate $d\theta(z)/dz = \alpha(z) = qB(z)/2p_0$, such that we can write

$$\begin{pmatrix} x(z) \\ y(z) \end{pmatrix} = R(z,z_i) \begin{pmatrix} X(z) \\ Y(z) \end{pmatrix}, \quad (5.135)$$

or

$$\begin{aligned} \begin{pmatrix} X(z) \\ Y(z) \end{pmatrix} &= R^{-1}(z,z_i) \begin{pmatrix} x(z) \\ y(z) \end{pmatrix} \\ &= \begin{pmatrix} \cos\theta(z_i,z) & -\sin\theta(z_i,z) \\ \sin\theta(z_i,z) & \cos\theta(z_i,z) \end{pmatrix} \begin{pmatrix} x(z) \\ y(z) \end{pmatrix}. \end{aligned} \quad (5.136)$$

Then, we have

$$\begin{pmatrix} p_x(z) \\ p_y(z) \end{pmatrix} = R(z,z_i) \begin{pmatrix} P_X(z) \\ P_Y(z) \end{pmatrix}, \quad (5.137)$$

or

$$\begin{pmatrix} P_X(z) \\ P_Y(z) \end{pmatrix} = \begin{pmatrix} \cos\theta(z_i,z) & -\sin\theta(z_i,z) \\ \sin\theta(z_i,z) & \cos\theta(z_i,z) \end{pmatrix} \begin{pmatrix} p_x(z) \\ p_y(z) \end{pmatrix}. \quad (5.138)$$

where P_X and P_Y are the components of momentum in the rotating coordinate system. Note that in the vertical plane at $z = z_i$, where the particle enters the system,

Scalar Theory: Spin-0 and Spinless Particles

XYz coordinate system coincides with the xyz coordinate system. Thus, we can write (5.125) as, for any $z \geq z_i$,

$$\mathscr{R}(z,z_i) \begin{pmatrix} \langle X \rangle(z) \\ \langle Y \rangle(z) \\ \frac{1}{p_0}\langle \widehat{P}_X \rangle(z) \\ \frac{1}{p_0}\langle \widehat{P}_Y \rangle(z) \end{pmatrix} = \mathscr{M}(z,z_i)\mathscr{R}(z,z_i) \begin{pmatrix} \langle X \rangle(z_i) \\ \langle Y \rangle(z_i) \\ \frac{1}{p_0}\langle \widehat{P}_X \rangle(z_i) \\ \frac{1}{p_0}\langle \widehat{P}_Y \rangle(z_i) \end{pmatrix}$$

$$= \mathscr{R}(z,z_i)\mathscr{M}(z,z_i) \begin{pmatrix} \langle X \rangle(z_i) \\ \langle Y \rangle(z_i) \\ \frac{1}{p_0}\langle \widehat{P}_X \rangle(z_i) \\ \frac{1}{p_0}\langle \widehat{P}_Y \rangle(z_i) \end{pmatrix}.$$

(5.139)

In other words, the quantum averages of position and momentum of the particle with reference to the rotating coordinate system evolve along the optic axis according to

$$\begin{pmatrix} \langle X \rangle(z) \\ \langle Y \rangle(z) \\ \frac{1}{p_0}\langle \widehat{P}_X \rangle(z) \\ \frac{1}{p_0}\langle \widehat{P}_Y \rangle(z) \end{pmatrix} = \mathscr{M}(z,z_i) \begin{pmatrix} \langle X \rangle(z_i) \\ \langle Y \rangle(z_i) \\ \frac{1}{p_0}\langle \widehat{P}_X \rangle(z_i) \\ \frac{1}{p_0}\langle \widehat{P}_Y \rangle(z_i) \end{pmatrix}, \quad \text{for any } z \geq z_i. \quad (5.140)$$

From (5.123) and (5.126), the corresponding equations of motion follow:

$$\frac{d}{dz}\begin{pmatrix} \langle \vec{R}_\perp \rangle(z) \\ \frac{1}{p_0}\langle \vec{\widehat{P}}_\perp \rangle(z) \end{pmatrix} = \mu(z) \begin{pmatrix} \langle \vec{R}_\perp \rangle(z) \\ \frac{1}{p_0}\langle \vec{\widehat{P}}_\perp \rangle(z) \end{pmatrix}$$

$$= \begin{pmatrix} 0 & 1 \\ -\alpha^2(z) & 0 \end{pmatrix} \begin{pmatrix} \langle \vec{R}_\perp \rangle(z) \\ \frac{1}{p_0}\langle \vec{\widehat{P}}_\perp \rangle(z) \end{pmatrix}, \quad (5.141)$$

with

$$\langle \vec{R}_\perp \rangle(z) = \begin{pmatrix} \langle X \rangle(z) \\ \langle Y \rangle(z) \end{pmatrix}, \quad \langle \vec{\widehat{P}}_\perp \rangle(z) = \begin{pmatrix} \langle \widehat{P}_X \rangle(z) \\ \langle \widehat{P}_Y \rangle(z) \end{pmatrix}. \quad (5.142)$$

From this, it follows that

$$\frac{d^2}{dz^2}\begin{pmatrix} \langle \vec{R}_\perp \rangle(z) \\ \frac{1}{p_0}\langle \vec{\widehat{P}}_\perp \rangle(z) \end{pmatrix} = \begin{pmatrix} -\alpha^2(z) & 0 \\ -2\alpha(z)\alpha'(z) & -\alpha^2(z) \end{pmatrix} \begin{pmatrix} \langle \vec{R}_\perp \rangle(z) \\ \frac{1}{p_0}\langle \vec{\widehat{P}}_\perp \rangle(z) \end{pmatrix},$$

(5.143)

or,

$$\left\langle \vec{R}_\perp \right\rangle''(z) + \alpha^2(z)\left\langle \vec{R}_\perp \right\rangle(z) = 0,$$
$$\frac{1}{p_0}\left\langle \vec{P}_\perp \right\rangle''(z) + 2\alpha(z)\alpha'(z)\left\langle \vec{R}_\perp \right\rangle(z) + \frac{1}{p_0}\alpha^2(z)\left\langle \vec{P}_\perp \right\rangle(z) = 0. \quad (5.144)$$

First of the equations (5.144) is the quantum paraxial equation of motion in the rotating coordinate system. The second equation is not independent of (5.144), since it is just the consequence of the relation $\vec{P}_\perp(z)/p_0 = d\vec{R}_\perp(z)/dz$.

Equation (5.141) cannot be integrated by ordinary exponentiation since $\mu(z')$ and $\mu(z'')$ do not commute when $z' \neq z''$ if $\alpha^2(z') \neq \alpha^2(z'')$. So, to integrate (5.141), we have to follow the same method used to integrate the time-dependent Schrödinger equation (3.439) as in (3.440) with $\widehat{U}(t,t_0)$ given by the series expression in (3.449). Thus, replacing t by z, and $-i\widehat{H}(t)/\hbar$ by $\mu(z)$, we get

$$\begin{pmatrix} \left\langle \vec{R}_\perp \right\rangle(z) \\ \frac{1}{p_0}\left\langle \vec{P}_\perp \right\rangle(z) \end{pmatrix} = \begin{pmatrix} g(z,z_i) & h(z,z_i) \\ g'(z,z_i) & h'(z,z_i) \end{pmatrix} \begin{pmatrix} \left\langle \vec{R}_\perp \right\rangle(z_i) \\ \frac{1}{p_0}\left\langle \vec{P}_\perp \right\rangle(z_i) \end{pmatrix}$$
$$= M(z,z_i) \begin{pmatrix} \left\langle \vec{R}_\perp \right\rangle(z_i) \\ \frac{1}{p_0}\left\langle \vec{P}_\perp \right\rangle(z_i) \end{pmatrix}, \quad (5.145)$$

where

$$M(z,z_i) = I + \int_{z_i}^z dz_1 \mu(z_1) + \int_{z_i}^z dz_2 \int_{z_i}^{z_2} dz_1 \mu(z_2)\mu(z_1)$$
$$+ \int_{z_i}^z dz_3 \int_{z_i}^{z_3} dz_2 \int_{z_i}^{z_2} dz_1 \mu(z_3)\mu(z_2)\mu(z_1)$$
$$+ \cdots. \quad (5.146)$$

It can be directly verified that

$$\frac{dM(z,z_i)}{dz} = \mu(z)M(z,z_i), \quad (5.147)$$

and hence

$$\frac{d}{dz}\begin{pmatrix} \left\langle \vec{R}_\perp \right\rangle(z) \\ \frac{1}{p_0}\left\langle \vec{P}_\perp \right\rangle(z) \end{pmatrix} = \frac{dM(z,z_i)}{dz}\begin{pmatrix} \left\langle \vec{R}_\perp \right\rangle(z_i) \\ \frac{1}{p_0}\left\langle \vec{P}_\perp \right\rangle(z_i) \end{pmatrix} = \mu(z)\begin{pmatrix} \left\langle \vec{R}_\perp \right\rangle(z_i) \\ \frac{1}{p_0}\left\langle \vec{P}_\perp \right\rangle(z_i) \end{pmatrix}, \quad (5.148)$$

Scalar Theory: Spin-0 and Spinless Particles

as required in (5.141). Explicit expressions for the elements of M can be written down from (5.146) as follows:

$$g(z,z_i) = 1 - \int_{z_i}^{z} dz_2 \int_{z_i}^{z_2} dz_1\, \alpha^2(z_1)$$
$$+ \int_{z_i}^{z} dz_4 \int_{z_i}^{z_4} dz_3\, \alpha^2(z_3) \int_{z_i}^{z_3} dz_2 \int_{z_i}^{z_2} dz_1\, \alpha^2(z_1) - \cdots,$$

$$h(z,z_i) = (z-z_i) - \int_{z_i}^{z} dz_2 \int_{z_i}^{z_2} dz_1\, \alpha^2(z_1)(z_1 - z_i)$$
$$+ \int_{z_i}^{z} dz_4 \int_{z_i}^{z_4} dz_3\, \alpha^2(z_3) \int_{z_i}^{z_3} dz_2 \int_{z_i}^{z_2} dz_1\, \alpha^2(z_1)(z_1 - z_i) - \cdots,$$

$$g'(z,z_i) = - \int_{z_i}^{z} dz_1\, \alpha^2(z_1) + \int_{z_i}^{z} dz_3\, \alpha^2(z_3) \int_{z_i}^{z_3} dz_2 \int_{z_i}^{z_2} dz_1\, \alpha^2(z_1) - \cdots,$$

$$h'(z,z_i) = 1 - \int_{z_i}^{z} dz_1\, \alpha^2(z_1)(z_1 - z_i)$$
$$+ \int_{z_i}^{z} dz_3\, \alpha^2(z_3) \int_{z_i}^{z_3} dz_2 \int_{z_i}^{z_2} dz_1\, \alpha^2(z_1)(z_1 - z_i) - \cdots. \quad (5.149)$$

Note that

$$\frac{dg(z,z_i)}{dz} = g'(z,z_i), \qquad \frac{dh(z,z_i)}{dz} = h'(z,z_i), \quad (5.150)$$

consistent with the relation $d\langle \vec{R}_\perp \rangle(z)/dz = \langle \vec{P}_\perp \rangle(z)/p_0$. From the differential equation (5.147) for M, we know that we can write

$$M(z,z_i) = \lim_{N \to \infty} \lim_{\Delta z \to 0} \left\{ e^{\Delta z \mu(z_i + N\Delta z)} e^{\Delta z \mu(z_i + (N-1)\Delta z)} e^{\Delta z \mu(z_i + (N-2)\Delta z)} \cdots \right.$$
$$\left. \cdots e^{\Delta z \mu(z_i + 2\Delta z)} e^{\Delta z \mu(z_i + \Delta z)} \right\}, \quad (5.151)$$

with $N\Delta z = (z - z_i)$. Since the trace of $\mu(z)$ is zero, each of the factors in the above product expression, of the type $e^{\Delta z \mu(z_i + j\Delta z)}$, has unit determinant. Thus, the matrix M has unit determinant, i.e.,

$$g(z,z_i) h'(z,z_i) - h(z,z_i) g'(z,z_i) = 1, \quad (5.152)$$

for any (z,z_i). Note that the solution for the transfer matrix for the quantum averages of $\left(\vec{r}_\perp, \vec{p}_\perp/p_0 \right)$ is exactly same as the transfer matrix for the corresponding classical variables, since the quantum paraxial equation of motion (5.144) is same as the classical paraxial equation of motion (4.51), except for the replacement of classical $(\vec{r}_\perp, \vec{p}_\perp/p_0)$ by their quantum averages.

In a simplified picture of electron microscope, let us consider the monoenergetic paraxial beam comprising of electrons scattered elastically from the specimen (the object to be imaged) being illuminated. The beam transmitted by the specimen carries the information about the structure of the specimen. This beam going through the magnetic lens gets magnified and is recorded at the image plane. Let us take the positions of the object plane and the image plane in the z-axis, the optic axis, as z_{obj} and z_{img} respectively. As already mentioned, we shall take the practical boundaries of the lens to be at z_ℓ and z_r. The magnetic field of the lens is practically confined to the region between z_ℓ and z_r and $z_{obj} < z_\ell < z_r < z_{img}$. The region between z_{obj} and z_ℓ and the region between z_r and z_{img} are free spaces. Thus, the electron beam transmitted by the specimen (object) travels first through the free space between z_{obj} and z_ℓ, then the lens between z_ℓ and z_r, and finally, the free space between z_r and z_{img} where it is recorded (image). Hence, the z-evolution matrix, or the transfer matrix, $M(z_{img}, z_{obj})$ can be written as

$$M(z_{img}, z_{obj}) = M_D(z_{img}, z_r) M_L(z_r, z_\ell) M_D(z_\ell, z_{obj}), \qquad (5.153)$$

where subscript D indicates the drift region and L indicates the lens region. Since $\alpha^2(z) = 0$ in the free space regions, or drift regions, we get from (5.149)

$$M_D(z_{img}, z_r) = \begin{pmatrix} 1 & (z_{img} - z_r) \\ 0 & 1 \end{pmatrix}, \quad M_D(z_\ell, z_{obj}) = \begin{pmatrix} 1 & (z_\ell - z_{obj}) \\ 0 & 1 \end{pmatrix}. \qquad (5.154)$$

For the lens region, we have from (5.149)

$$M_L(z_r, z_\ell) = \begin{pmatrix} g(z_r, z_\ell) & h(z_r, z_\ell) \\ g'(z_r, z_\ell) & h'(z_r, z_\ell) \end{pmatrix} = \begin{pmatrix} g_L & h_L \\ g'_L & h'_L \end{pmatrix}. \qquad (5.155)$$

Thus, we get

$$M(z_{img}, z_{obj}) = \begin{pmatrix} 1 & (z_{img} - z_r) \\ 0 & 1 \end{pmatrix} \begin{pmatrix} g_L & h_L \\ g'_L & h'_L \end{pmatrix} \begin{pmatrix} 1 & (z_\ell - z_{obj}) \\ 0 & 1 \end{pmatrix}$$

$$= \begin{pmatrix} g_L + g'_L(z_{img} - z_r) & \begin{aligned}[t] & [g_L(z_\ell - z_{obj}) + h_L \\ & + g'_L(z_{img} - z_r)(z_\ell - z_{obj}) \\ & + h'_L(z_{img} - z_r)] \end{aligned} \\ g'_L & g'_L(z_\ell - z_{obj}) + h'_L \end{pmatrix}$$

$$= \begin{pmatrix} g(z_{img}, z_{obj}) & h(z_{img}, z_{obj}) \\ g'(z_{img}, z_{obj}) & h'(z_{img}, z_{obj}) \end{pmatrix} \qquad (5.156)$$

Note that $M_D(z_\ell, z_{obj})$ and $M_D(z_{img}, z_r)$ are of unit determinant. From the general theory, we know that $M_L(z_r, z_\ell)$ should also be of unit determinant, i.e.,

$$g_L h'_L - h_L g'_L = 1. \qquad (5.157)$$

Since z_{img} is the position of the image plane $h(z_{img}, z_{obj})$, the 12-element of the matrix $M(z_{img}, z_{obj})$ should vanish such that

$$\langle \vec{R}_\perp \rangle (z_{img}) \propto \langle \vec{R}_\perp \rangle (z_{obj}). \qquad (5.158)$$

This means that we should have

$$g_L(z_\ell - z_{obj}) + h_L + g'_L(z_{img} - z_r)(z_\ell - z_{obj}) + h'_L(z_{img} - z_r) = 0. \quad (5.159)$$

Now, let

$$u = (z_\ell - z_{obj}) + \frac{h'_L - 1}{g'_L}, \qquad v = (z_{img-r}) + \frac{g_L - 1}{g'_L}, \quad (5.160)$$

and

$$f = -\frac{1}{g'_L}. \quad (5.161)$$

We are to interpret u as the object distance, v as the image distance, and f as the focal length of the lens. Using straightforward algebra, and the relation (5.157), one can verify that the familiar lens equation,

$$\frac{1}{u} + \frac{1}{v} = \frac{1}{f}, \quad (5.162)$$

implies (5.159). From (5.160) it is seen that the principal planes from which u and v are to be measured in the case of a thick lens are situated at

$$z_{P_{obj}} = z_\ell + \frac{h'_L - 1}{g'_L} = z_\ell + f(1 - h'_L),$$

$$z_{P_{img}} = z_r - \frac{g_L - 1}{g'_L} = z_r - f(1 - g_L), \quad (5.163)$$

such that

$$u = z_{P_{obj}} - z_{obj}, \qquad v = z_{img} - z_{P_{img}}. \quad (5.164)$$

The explicit expression for the focal length f follows from (5.161) and (5.149):

$$\frac{1}{f} = \int_{z_\ell}^{z_r} dz\, \alpha^2(z) - \int_{z_\ell}^{z_r} dz_2\, \alpha^2(z_2) \int_{z_\ell}^{z_2} dz_1 \int_{z_\ell}^{z_1} dz\, \alpha^2(z) + \cdots$$

$$= \frac{q^2}{4p_0^2} \int_{z_\ell}^{z_r} dz\, B^2(z) - \frac{q^4}{16p_0^4} \int_{z_\ell}^{z_r} dz_2\, B^2(z_2) \int_{z_\ell}^{z_2} dz_1 \int_{z_\ell}^{z_1} dz\, B^2(z) + \cdots.$$

$$(5.165)$$

To understand the behavior of the lens, let us consider the idealized model in which $B(z) = B$ is a constant in the lens region and zero outside. Then,

$$\frac{1}{f} = \frac{qB}{2p_0} \sin\left(\frac{qBw}{2p_0}\right), \quad (5.166)$$

where $w = (z_r - z_\ell)$ is the width, or thickness, of the lens. This shows that the focal length is always positive to start with and is then periodic with respect to the variation of the field strength. Thus, the round magnetic lens is convergent up to a certain strength of the field. Round magnetic lenses commonly used in electron microscopy are always convergent. Note that $\int_{z_\ell}^{z_r} dz\, \alpha^2(z)$ has the dimension of reciprocal length. The round magnetic lens is considered weak when

$$\int_{z_\ell}^{z_r} dz\, \alpha^2(z) \ll \frac{1}{w}. \tag{5.167}$$

For such a weak lens, the expression for the focal length (4.74) can be approximated by

$$\frac{1}{f} \approx \int_{z_\ell}^{z_r} dz\, \alpha^2(z) = \frac{q^2}{4p_0^2} \int_{z_\ell}^{z_r} dz\, B^2(z) = \frac{q^2}{4p_0^2} \int_{-\infty}^{\infty} dz\, B^2(z), \tag{5.168}$$

the Busch formula (Busch [17]). A weak lens is said to be thin, since in this case $1/f \ll 1/w$, or $f \gg w$. Thus, we now understand how quantum mechanics leads to the classical Busch formula for an axially symmetric thin magnetic lens.

The paraxial z-evolution matrix from the object plane to the image plane (5.156) becomes, taking into account (5.159),

$$\begin{aligned}
M(z_{img}, z_{obj}) &= \begin{pmatrix} g_L - \frac{z_{img}-z_r}{f} & 0 \\ -\frac{1}{f} & h'_L - \frac{z_\ell - z_{obj}}{f} \end{pmatrix} \\
&= \begin{pmatrix} g_L - \frac{v+f(g_L-1)}{f} & 0 \\ -\frac{1}{f} & h'_L - \frac{u+f(h'_L-1)}{f} \end{pmatrix} \\
&= \begin{pmatrix} 1-\frac{v}{f} & 0 \\ -\frac{1}{f} & 1-\frac{u}{f} \end{pmatrix} = \begin{pmatrix} -\frac{v}{u} & 0 \\ -\frac{1}{f} & -\frac{u}{v} \end{pmatrix} \\
&= \begin{pmatrix} M & 0 \\ -\frac{1}{f} & \frac{1}{M} \end{pmatrix} \\
&= \begin{pmatrix} g(z_{img}, z_{obj}) & 0 \\ g'(z_{img}, z_{obj}) & 1/g(z_{img}, z_{obj}) \end{pmatrix}
\end{aligned} \tag{5.169}$$

where $g(z_{img}, z_{obj}) = M$ is the magnification, as we have identified (see (5.117)) while understanding imaging in the Schrödinger picture. Further, we find that $g'(z_{img}, z_{obj}) = -1/f$, and hence we can write, from (5.117),

$$\psi\left(\vec{r}_{\perp,img}, z_{img}\right) = \frac{1}{M} \exp\left(\frac{i2\pi}{\lambda_0}\left[(z_{img} - z_{obj}) - \frac{1}{2fM} r_{\perp,img}^2\right]\right)$$
$$\times \psi\left(\frac{1}{M}\vec{r}_{\perp,img}(\theta_L), z_{obj}\right). \tag{5.170}$$

In our notation, both u and v are positive and hence $M = -v/u$ is negative, showing the inverted nature of the image, as should be in the case of imaging by a convergent lens.

For a thin lens with $f \gg w$, we can take $w \approx 0$, and the two principal planes collapse into a single plane at the center of the lens, *i.e.*, we can take $z_P = (z_\ell + z_r)/2$.

Scalar Theory: Spin-0 and Spinless Particles

Then, from (5.149) it is clear that for the thin lens, we can make the approximation

$$M_{TL}(z_r, z_\ell) \approx \begin{pmatrix} 1 & 0 \\ -\frac{1}{f} & 1 \end{pmatrix}, \qquad (5.171)$$

where the subscript *TL* stands for thin lens. Thus, corresponding to (5.156), we have for the thin lens

$$M(z_{img}, z_{obj}) \approx \begin{pmatrix} 1 & (z_{img} - z_P) \\ 0 & 1 \end{pmatrix} \begin{pmatrix} 1 & 0 \\ -\frac{1}{f} & 1 \end{pmatrix} \begin{pmatrix} 1 & (z_P - z_{obj}) \\ 0 & 1 \end{pmatrix}$$

$$= \begin{pmatrix} 1 - \frac{1}{f}(z_{img} - z_P) & \begin{array}{c} [(z_P - z_{obj}) \\ -\frac{1}{f}(z_{img} - z_P)(z_P - z_{obj}) \\ + (z_{img} - z_P)] \end{array} \\ -\frac{1}{f} & 1 - \frac{1}{f}(z_P - z_{obj}) \end{pmatrix}. \qquad (5.172)$$

Since $(z_P - z_{obj}) = u$ and $(z_{img} - z_P) = v$, and $1/u + 1/v = 1/f$, we get

$$M(z_{img}, z_{obj}) = \begin{pmatrix} 1 - \frac{v}{f} & u - \frac{uv}{f} + v \\ -\frac{1}{f} & 1 - \frac{u}{f} \end{pmatrix}$$

$$= \begin{pmatrix} -\frac{v}{u} & 0 \\ -\frac{1}{f} & -\frac{u}{v} \end{pmatrix} = \begin{pmatrix} M & 0 \\ -\frac{1}{f} & \frac{1}{M} \end{pmatrix}, \qquad (5.173)$$

where $M = -v/u$ is the magnification. Note that the effect of passage through a thin lens of negligible width with the transfer matrix $\begin{pmatrix} 1 & 0 \\ -1/f & 1 \end{pmatrix}$ on the wave function will be to have

$$\psi(\vec{r}_{\perp,z_r}, z_r) = e^{-\frac{i\pi}{\lambda_0 f} r^2_{\perp,zr}} \psi(\vec{r}_{\perp,z_r}(\theta_L), z_\ell), \qquad (5.174)$$

i.e., multiplication of the wave function by the phase factor $\exp\left(-\frac{i\pi}{\lambda_0 f} r^2_\perp\right)$, as is well known (see Hawkes and Kasper [72] for more details of electron optical imaging).

So far, we have been describing the behavior of the lens in the rotated XYz-coordinate system. Now, if we return to the original xyz-coordinate system, we would have, as seen from (5.125), (5.128), and (5.130),

$$\begin{pmatrix} \langle \vec{r}_\perp \rangle (z_{img}) \\ \frac{1}{p_0} \langle \vec{p}_\perp \rangle (z_{img}) \end{pmatrix} = M(z_{img}, z_{obj}) \otimes R(z_{img}, z_{obj}) \begin{pmatrix} \langle \vec{r}_\perp \rangle (z_{obj}) \\ \frac{1}{p_0} \langle \vec{p}_\perp \rangle (z_{obj}) \end{pmatrix}, \qquad (5.175)$$

with $\langle \vec{r}_\perp \rangle = \begin{pmatrix} \langle x \rangle \\ \langle y \rangle \end{pmatrix}$ and $\langle \vec{p}_\perp \rangle = \begin{pmatrix} \langle \hat{p}_x \rangle \\ \langle \hat{p}_y \rangle \end{pmatrix}$. Since the rotation of the image is the effect of only the magnetic field of the lens

$$R(z_{img}, z_{obj}) = R(z_\ell, z_r) = \begin{pmatrix} \cos\theta_L & \sin\theta_L \\ -\sin\theta_L & \cos\theta_L \end{pmatrix}, \qquad (5.176)$$

where
$$\theta_L = \theta(z_\ell, z_r) = \frac{q}{2p_0} \int_{z_\ell}^{z_r} dz\, B(z). \tag{5.177}$$

From (5.172) and (5.175), it is clear that apart from rotation and drifts through field-free regions in the front and back of the lens, the effect of a thin convergent lens is essentially described by the transfer matrix $\begin{pmatrix} 1 & 0 \\ -1/f & 1 \end{pmatrix}$. This can also be seen simply as follows. If a particle of a paraxial beam enters a thin lens in the transverse xy-plane with $(\langle x \rangle, \langle y \rangle) = (\langle x \rangle, 0)$ and $(\langle \widehat{p}_x \rangle, \langle \widehat{p}_y \rangle) = (0,0)$, $i.e.$, in the xz-plane and parallel to the optic (z) axis, then we have

$$\begin{pmatrix} 1 & f \\ 0 & 1 \end{pmatrix} \begin{pmatrix} 1 & 0 \\ -\frac{1}{f} & 1 \end{pmatrix} \begin{pmatrix} \langle x \rangle \\ 0 \end{pmatrix} = \begin{pmatrix} 0 \\ -\frac{\langle x \rangle}{f} \end{pmatrix}. \tag{5.178}$$

This means that all the particles of the paraxial beam hitting the lens parallel to the optic axis in the xz-plane meet the optic axis at a distance f from the lens independent of the $\langle x \rangle$ with which it enters the lens. This implies that the lens is a convergent lens with the focal length f. If the lens is a thin divergent lens of focal length $-f$, then correspondingly we would have

$$\begin{pmatrix} 1 & -f \\ 0 & 1 \end{pmatrix} \begin{pmatrix} 1 & 0 \\ \frac{1}{f} & 1 \end{pmatrix} \begin{pmatrix} \langle x \rangle \\ 0 \end{pmatrix} = \begin{pmatrix} 0 \\ \frac{\langle x \rangle}{f} \end{pmatrix}, \tag{5.179}$$

implying that the particles appear virtually to meet behind the lens at a distance f. Thus, the effect of a thin divergent lens is essentially described by the transfer matrix $\begin{pmatrix} 1 & 0 \\ 1/f & 1 \end{pmatrix}$. Note that, in general, if the transfer matrix of a lens system is $\begin{pmatrix} g & h \\ g' & h' \end{pmatrix}$, then it will be focusing or defocusing depending on whether g' is negative or positive, respectively. This is so because if g' is negative then $\langle \widehat{p}_x \rangle / p_0$ decreases proportional to the $\langle x \rangle$ of the beam particle driving it towards the optic axis, and if g' is positive then $\langle \widehat{p}_x \rangle / p_0$ increases proportional to the $\langle x \rangle$ of the beam particle driving it away from the optic axis.

5.2.3.2 Going Beyond the Paraxial Approximation: Aberrations

Paraxial beam is an idealization. Let us now turn to the realistic case of quasiparaxial beam that would lead to aberrations due to the necessity to include terms of order higher than quadratic in $\left(\vec{r}_\perp, \vec{\widehat{p}}_\perp \right)$ in the quantum beam optical Hamiltonian $\widehat{\mathcal{H}_o}$. We have to treat the nonparaxial terms in $\widehat{\mathcal{H}_o}$, which will be anyway small compared to the paraxial terms, as perturbations and use the well-known techniques of time-dependent perturbation theory of quantum mechanics utilizing the interaction picture approach. In the classical limit, the formalism of quantum theory of paraxial and aberrating charged particle optical systems presented here tends to the similar approach to the classical theory of charged particle optics, including accelerator

Scalar Theory: Spin-0 and Spinless Particles

optics, based on Lie transfer operator methods developed extensively by Dragt et al. (see Dragt [34], Dragt and Forest [35], Dragt et al. [36], Forest, Berz, and Irwin [54], Rangarajan, Dragt, and Neri [156], Forest and Hirata [55], Forest [56], Radlička [154], and references therein; see also Berz [11], Mondragon and Wolf [135], Wolf [191], Rangarajan and Sachidanand [157], Lakshminarayanan, Sridhar, and Jagannathan [124], and Wolski [192]).

When the beam deviates from the ideal paraxial condition, as is always the case in practice, going beyond the paraxial approximation, the next approximation entails retaining in $\widehat{\mathcal{H}_o}$ terms of order up to fourth in $\left(\vec{r}_\perp, \vec{p}_\perp\right)$. To this end, we substitute in $\widehat{\mathcal{H}_o}$ obtained in (5.34)

$$\phi(\vec{r}) = 0,$$
$$\vec{A} = \left(-\frac{1}{2}y\Pi(\vec{r}_\perp, z), \frac{1}{2}x\Pi(\vec{r}_\perp, z), 0\right),$$
$$\text{with } \Pi(\vec{r}_\perp, z) = B(z) - \frac{1}{8}r_\perp^2 B''(z), \tag{5.180}$$

expand, and approximate as desired. Then, we get, with $\alpha(z) = qB(z)/2p_0$,

$$\widehat{\mathcal{H}_o} = \widehat{\mathcal{H}_{o,p}} + \widehat{\mathcal{H}_o'}$$
$$\widehat{\mathcal{H}_{o,p}} = \frac{1}{2p_0}\widehat{p}_\perp^2 + \frac{1}{2}p_0\alpha^2(z)r_\perp^2 - \alpha(z)\widehat{L}_z$$
$$\widehat{\mathcal{H}_o'} = \frac{1}{8p_0^3}\widehat{p}_\perp^4 - \frac{\alpha(z)}{2p_0^2}\widehat{p}_\perp^2\widehat{L}_z - \frac{\alpha^2(z)}{8p_0}\left(\vec{r}_\perp \cdot \vec{p}_\perp + \vec{p}_\perp \cdot \vec{r}_\perp\right)^2$$
$$+ \frac{3\alpha^2(z)}{8p_0}\left(r_\perp^2 \widehat{p}_\perp^2 + \widehat{p}_\perp^2 r_\perp^2\right) + \frac{1}{8}\left(\alpha''(z) - 4\alpha^3(z)\right)\widehat{L}_z r_\perp^2$$
$$+ \frac{p_0}{8}\left(\alpha^4(z) - \alpha(z)\alpha''(z)\right)r_\perp^4, \tag{5.181}$$

in which we have dropped unimportant constant terms, giving rise only to multiplicative phase factors, and retained only the Hermitian part without the $\hbar$-dependent Hermitian terms. In $\widehat{\mathcal{H}_o}$, $\widehat{\mathcal{H}_{o,p}}$ is the paraxial term, same as in (5.97) except for the additive constant term $-p_0$, and $\widehat{\mathcal{H}_o'}$ represents the lowest order nonparaxial, or perturbation, term. Note that $\widehat{\mathcal{H}_{o,p}}$ is a homogeneous quadratic polynomial in $\left(\vec{r}_\perp, \vec{p}_\perp\right)$ and $\widehat{\mathcal{H}_o'}$ is a homogeneous fourth-degree polynomial in $\left(\vec{r}_\perp, \vec{p}_\perp\right)$. Since the system is rotationally symmetric about the z-axis, the optical Hamiltonian does not contain any odd-degree polynomial in $\left(\vec{r}_\perp, \vec{p}_\perp\right)$.

Now the z-evolution equation for the system is

$$i\hbar\frac{\partial |\psi(z)\rangle}{\partial z} = \widehat{\mathcal{H}_o}|\psi(z)\rangle. \tag{5.182}$$

Following the time-dependent perturbation theory, let

$$|\psi_i(z)\rangle = \widehat{U}_p^\dagger(z, z_i)|\psi(z)\rangle, \quad \widehat{U}_p(z, z_i) = \mathsf{P}\left(e^{-\frac{i}{\hbar}\int_{z_i}^z dz\, \widehat{\mathcal{H}_{o,p}}}\right). \tag{5.183}$$

Note that
$$|\psi_{\mathrm{i}}(z_i)\rangle = \widehat{U}_p^\dagger(z_i,z_i)|\psi(z_i)\rangle = |\psi(z_i)\rangle. \quad (5.184)$$

Then, we have
$$\begin{aligned}
i\hbar\frac{\partial}{\partial z}|\psi_{\mathrm{i}}(z)\rangle &= i\hbar\frac{\partial}{\partial z}\widehat{U}_p^\dagger(z,z_i)|\psi(z)\rangle + \widehat{U}_p^\dagger(z,z_i)i\hbar\frac{\partial}{\partial z}|\psi(z)\rangle \\
&= -\widehat{U}_p^\dagger(z,z_i)\widehat{\mathscr{H}}_{o,p}|\psi(z)\rangle + \widehat{U}_p^\dagger(z,z_i)\widehat{\mathscr{H}}_o|\psi\rangle \\
&= -\widehat{U}_p^\dagger(z,z_i)\widehat{\mathscr{H}}_{o,p}\widehat{U}_p(z,z_i)|\psi_{\mathrm{i}}(z)\rangle + \widehat{U}_p^\dagger(z,z_i)\widehat{\mathscr{H}}_o\widehat{U}_p(z,z_i)|\psi_{\mathrm{i}}(z)\rangle \\
&= \widehat{U}_p^\dagger(z,z_i)\left(\widehat{\mathscr{H}}_o - \widehat{\mathscr{H}}_{o,p}\right)\widehat{U}_p(z,z_i)|\psi_{\mathrm{i}}(z)\rangle \\
&= \widehat{U}_p^\dagger(z,z_i)\widehat{\mathscr{H}}_o'\widehat{U}_p(z,z_i)|\psi_{\mathrm{i}}(z)\rangle. \quad (5.185)
\end{aligned}$$

Defining
$$\widehat{\mathscr{H}}_{o,\mathrm{i}}' = \widehat{U}_p^\dagger(z,z_i)\widehat{\mathscr{H}}_o'\widehat{U}_p(z,z_i), \quad (5.186)$$

we have
$$i\hbar\frac{\partial}{\partial z}|\psi_{\mathrm{i}}(z)\rangle = \widehat{\mathscr{H}}_{o,\mathrm{i}}'|\psi_{\mathrm{i}}(z)\rangle, \quad (5.187)$$

where the subscript i denotes the interaction picture. Integrating (5.187), we get
$$\begin{aligned}
|\psi_{\mathrm{i}}(z)\rangle &= \widehat{U}_{\mathrm{i}}(z,z_i)|\psi_{\mathrm{i}}(z_i)\rangle = \widehat{U}_{\mathrm{i}}(z,z_i)|\psi(z_i)\rangle, \\
\widehat{U}_{\mathrm{i}}(z,z_i) &\approx e^{-\frac{i}{\hbar}\widehat{T}_{\mathrm{i}}(z,z_i)}, \\
\widehat{T}_{\mathrm{i}}(z,z_i) &= \int_{z_i}^z dz\,\widehat{\mathscr{H}}_{o,\mathrm{i}}' = \int_{z_i}^z dz\,\widehat{U}_p^\dagger(z,z_i)\widehat{\mathscr{H}}_o'\widehat{U}_p(z,z_i), \quad (5.188)
\end{aligned}$$

where we have approximated $\widehat{U}_{\mathrm{i}}(z,z_i)$ by an ordinary integral disregarding all the commutator terms in (5.38), since they lead to polynomials of degree higher than four in $\left(\vec{r}_\perp,\vec{p}_\perp\right)$.

From the paraxial theory, we know
$$\begin{pmatrix} \langle x\rangle(z) \\ \langle y\rangle(z) \\ \frac{1}{p_0}\langle\widehat{p}_x\rangle(z) \\ \frac{1}{p_0}\langle\widehat{p}_y\rangle(z) \end{pmatrix} = M(z,z_i)\otimes R(\theta(z,z_i))\begin{pmatrix} \langle x\rangle(z_i) \\ \langle y\rangle(z_i) \\ \frac{1}{p_0}\langle\widehat{p}_x\rangle(z_i) \\ \frac{1}{p_0}\langle\widehat{p}_y\rangle(z_i) \end{pmatrix}$$
$$M(z,z_i) = \begin{pmatrix} g(z,z_i) & h(z,z_i) \\ g'(z,z_i) & h'(z,z_i) \end{pmatrix}$$
$$R(\theta(z,z_i)) = \begin{pmatrix} \cos\theta(z,z_i) & \sin\theta(z,z_i) \\ -\sin\theta(z,z_i) & \cos\theta(z,z_i) \end{pmatrix}. \quad (5.189)$$

Scalar Theory: Spin-0 and Spinless Particles

Since

$$\begin{pmatrix} \langle x \rangle (z) \\ \langle y \rangle (z) \\ \frac{1}{p_0} \langle \widehat{p}_x \rangle (z) \\ \frac{1}{p_0} \langle \widehat{p}_y \rangle (z) \end{pmatrix} = \begin{pmatrix} \langle \psi(z) | x | \psi(z) \rangle \\ \langle \psi(z) | y | \psi(z) \rangle \\ \frac{1}{p_0} \langle \psi(z) | \widehat{p}_x | \psi(z) \rangle \\ \frac{1}{p_0} \langle \psi(z) | \widehat{p}_y | \psi(z) \rangle \end{pmatrix}$$

$$= \begin{pmatrix} \langle \psi(z_i) | \widehat{U}_p^\dagger (z,z_i) x \widehat{U}_p (z,z_i) | \psi(z_i) \rangle \\ \langle \psi(z_i) | \widehat{U}_p^\dagger (z,z_i) y \widehat{U}_p (z,z_i) | \psi(z_i) \rangle \\ \frac{1}{p_0} \langle \psi(z_i) | \widehat{U}_p^\dagger (z,z_i) \widehat{p}_x \widehat{U}_p (z,z_i) | \psi(z_i) \rangle \\ \frac{1}{p_0} \langle \psi(z_i) | \widehat{U}_p^\dagger (z,z_i) \widehat{p}_y \widehat{U}_p (z,z_i) | \psi(z_i) \rangle \end{pmatrix},$$

$$\begin{pmatrix} \langle x \rangle (z_i) \\ \langle y \rangle (z_i) \\ \frac{1}{p_0} \langle \widehat{p}_x \rangle (z_i) \\ \frac{1}{p_0} \langle \widehat{p}_y \rangle (z_i) \end{pmatrix} = \begin{pmatrix} \langle \psi(z_i) | x | \psi(z_i) \rangle \\ \langle \psi(z_i) | y | \psi(z_i) \rangle \\ \frac{1}{p_0} \langle \psi(z_i) | \widehat{p}_x | \psi(z_i) \rangle \\ \frac{1}{p_0} \langle \psi(z_i) | \widehat{p}_y | \psi(z_i) \rangle \end{pmatrix}, \quad (5.190)$$

we have

$$\begin{pmatrix} \widehat{U}_p^\dagger (z,z_i) x \widehat{U}_p (z,z_i) \\ \widehat{U}_p^\dagger (z,z_i) y \widehat{U}_p (z,z_i) \\ \frac{1}{p_0} \widehat{U}_p^\dagger (z,z_i) \widehat{p}_x \widehat{U}_p (z,z_i) \\ \frac{1}{p_0} \widehat{U}_p^\dagger (z,z_i) \widehat{p}_y \widehat{U}_p (z,z_i) \end{pmatrix} = M(z,z_i) \otimes R(\theta(z,z_i)) \begin{pmatrix} x \\ y \\ \frac{1}{p_0} \widehat{p}_x \\ \frac{1}{p_0} \widehat{p}_y \end{pmatrix}.$$

(5.191)

Using this result, we can compute $\widehat{\mathcal{H}}'_{o,\mathbf{i}}$, $\widehat{T}_{\mathbf{i}}(z,z_i)$, and $\widehat{U}_{\mathbf{i}}(z,z_i)$. Note that for any $\widehat{O}$
$\widehat{U}_p^\dagger (z,z_i) \widehat{O}^n \widehat{U}_p (z,z_i) = \left(\widehat{U}_p^\dagger (z,z_i) \widehat{O} \widehat{U}_p (z,z_i) \right)^n$. After a considerable, but straightforward algebra, we obtain

$$\widehat{U}_{\mathbf{i}}(z,z_i) = e^{-\frac{i}{\hbar} \widehat{T}_{\mathbf{i}}(z,z_i)}$$

$$= \exp\left\{ -\frac{i}{\hbar} \left[\frac{1}{4p_0^3} C(z,z_i) \widehat{p}_\perp^4 + \frac{1}{4p_0^2} K(z,z_i) \left\{ \widehat{p}_\perp^2, \vec{\widehat{p}}_\perp \cdot \vec{r}_\perp + \vec{r}_\perp \cdot \vec{\widehat{p}}_\perp \right\} \right.\right.$$

$$+ \frac{1}{p_0^2} k(z,z_i) \widehat{p}_\perp^2 \widehat{L}_z + \frac{1}{4p_0} A(z,z_i) \left(\vec{\widehat{p}}_\perp \cdot \vec{r}_\perp + \vec{r}_\perp \cdot \vec{\widehat{p}}_\perp \right)^2$$

$$+ \frac{1}{2p_0} a(z,z_i) \left(\vec{\widehat{p}}_\perp \cdot \vec{r}_\perp + \vec{r}_\perp \cdot \vec{\widehat{p}}_\perp \right) \widehat{L}_z + \frac{1}{4p_0} F(z,z_i) \left(\widehat{p}_\perp^2 r_\perp^2 + r_\perp^2 \widehat{p}_\perp^2 \right)$$

$$+ \frac{1}{4} D(z,z_i) \left\{ \vec{\widehat{p}}_\perp \cdot \vec{r}_\perp + \vec{r}_\perp \cdot \vec{\widehat{p}}_\perp , r_\perp^2 \right\} + d(z,z_i) r_\perp^2 \widehat{L}_z$$

$$\left.\left. + \frac{p_0}{4} E(z,z_i) r_\perp^4 \right] \right\}, \quad (5.192)$$

where $\left\{\widehat{A},\widehat{B}\right\} = \widehat{A}\widehat{B} + \widehat{B}\widehat{A}$ and

$$C(z,z_i) = \frac{1}{2}\int_{z_i}^{z} dz \left\{ \left(\alpha^4 - \alpha\alpha''\right) h^4 + 2\alpha^2 h^2 h'^2 + h'^4 \right\},$$

$$K(z,z_i) = \frac{1}{2}\int_{z_i}^{z} dz \left\{ \left(\alpha^4 - \alpha\alpha''\right) gh^3 + \alpha^2 (gh)' hh' + g'h'^3 \right\},$$

$$k(z,z_i) = \int_{z_i}^{z} dz \left\{ \left(\frac{1}{8}\alpha'' - \frac{1}{2}\alpha^3\right) h^2 - \frac{1}{2}\alpha h'^2 \right\},$$

$$A(z,z_i) = \frac{1}{2}\int_{z_i}^{z} dz \left\{ \left(\alpha^4 - \alpha\alpha''\right) g^2 h^2 + 2\alpha^2 gg'hh' + g'^2 h'^2 - \alpha^2 \right\},$$

$$a(z,z_i) = \int_{z_i}^{z} dz \left\{ \left(\frac{1}{4}\alpha'' - \alpha^3\right) gh - \alpha g'h' \right\},$$

$$F(z,z_i) = \frac{1}{2}\int_{z_i}^{z} dz \left\{ \left(\alpha^4 - \alpha\alpha''\right) g^2 h^2 \right.$$
$$\left. + \alpha^2 \left(g^2 h'^2 + g'^2 h^2\right) + g'^2 h'^2 + 2\alpha^2 \right\},$$

$$D(z,z_i) = \frac{1}{2}\int_{z_i}^{z} dz \left\{ \left(\alpha^4 - \alpha\alpha''\right) g^3 h + \alpha^2 gg'(gh)' + g'^3 h' \right\},$$

$$d(z,z_i) = \int_{z_i}^{z} dz \left\{ \left(\frac{1}{8}\alpha'' - \frac{1}{2}\alpha^3\right) g^2 - \frac{1}{2}\alpha g'^2 \right\},$$

$$E(z,z_i) = \frac{1}{2}\int_{z_i}^{z} dz \left\{ \left(\alpha^4 - \alpha\alpha''\right) g^4 + 2\alpha g^2 g'^2 + g'^4 \right\}, \quad (5.193)$$

with $g = g(z,z_i)$, $h = h(z,z_i)$, $g' = g'(z,z_i)$, and $h' = h'(z,z_i)$.

From (5.183) and (5.188), we have

$$|\psi(z)\rangle = \widehat{U}_p(z,z_i)\widehat{U}_{\mathrm{i}}(z,z_i)|\psi(z_i)\rangle, \quad (5.194)$$

representing the generalization of the paraxial propagation law (5.116) including the lowest order aberrations. Now, the transfer map becomes

$$\langle \vec{r}_\perp \rangle (z) = \left\langle \widehat{U}_{\mathrm{i}}^\dagger U_p^\dagger \vec{r}_\perp \widehat{U}_p \widehat{U}_{\mathrm{i}} \right\rangle (z_i),$$
$$\left\langle \vec{\widehat{p}}_\perp \right\rangle (z) = \left\langle \widehat{U}_{\mathrm{i}}^\dagger U_p^\dagger \vec{\widehat{p}}_\perp \widehat{U}_p \widehat{U}_{\mathrm{i}} \right\rangle (z_i), \quad (5.195)$$

with $\widehat{U}_p = \widehat{U}_p(z,z_i)$ and $\widehat{U}_{\mathrm{i}} = \widehat{U}_{\mathrm{i}}(z,z_i)$. Explicitly writing,

$$\begin{pmatrix} \langle x \rangle (z) \\ \langle y \rangle (z) \\ \frac{1}{p_0}\langle \widehat{p}_x \rangle (z) \\ \frac{1}{p_0}\langle \widehat{p}_y \rangle (z) \end{pmatrix} = (M(z,z_i) \otimes R(\theta(z,z_i))) \begin{pmatrix} \left\langle \widehat{U}_{\mathrm{i}}^\dagger x \widehat{U}_{\mathrm{i}} \right\rangle (z_i) \\ \left\langle \widehat{U}_{\mathrm{i}}^\dagger y \widehat{U}_{\mathrm{i}} \right\rangle (z_i) \\ \frac{1}{p_0}\left\langle \widehat{U}_{\mathrm{i}}^\dagger \widehat{p}_x \widehat{U}_{\mathrm{i}} \right\rangle (z_i) \\ \frac{1}{p_0}\left\langle \widehat{U}_{\mathrm{i}}^\dagger \widehat{p}_y \widehat{U}_{\mathrm{i}} \right\rangle (z_i) \end{pmatrix}. \quad (5.196)$$

Scalar Theory: Spin-0 and Spinless Particles

We shall write, with $\widehat{T_i} = \widehat{T_i}(z, z_i)$,

$$\begin{pmatrix} \langle \widehat{U}_i^\dagger x \widehat{U}_i \rangle (z_i) \\ \langle \widehat{U}_i^\dagger y \widehat{U}_i \rangle (z_i) \\ \frac{1}{p_0} \langle \widehat{U}_i^\dagger \widehat{p}_x \widehat{U}_i \rangle (z_i) \\ \frac{1}{p_0} \langle \widehat{U}_i^\dagger \widehat{p}_y \widehat{U}_i \rangle (z_i) \end{pmatrix} = \begin{pmatrix} \langle e^{\frac{i}{\hbar}\widehat{T}_i} x e^{-\frac{i}{\hbar}\widehat{T}_i} \rangle (z_i) \\ \langle e^{\frac{i}{\hbar}\widehat{T}_i} y e^{-\frac{i}{\hbar}\widehat{T}_i} \rangle (z_i) \\ \frac{1}{p_0} \langle e^{\frac{i}{\hbar}\widehat{T}_i} \widehat{p}_x e^{-\frac{i}{\hbar}\widehat{T}_i} \rangle (z_i) \\ \frac{1}{p_0} \langle e^{\frac{i}{\hbar}\widehat{T}_i} \widehat{p}_y e^{-\frac{i}{\hbar}\widehat{T}_i} \rangle (z_i) \end{pmatrix} = \begin{pmatrix} \langle e^{\frac{i}{\hbar}:\widehat{T}_i:} x \rangle (z_i) \\ \langle e^{\frac{i}{\hbar}:\widehat{T}_i:} y \rangle (z_i) \\ \frac{1}{p_0} \langle e^{\frac{i}{\hbar}:\widehat{T}_i:} \widehat{p}_x \rangle (z_i) \\ \frac{1}{p_0} \langle e^{\frac{i}{\hbar}:\widehat{T}_i:} \widehat{p}_y \rangle (z_i) \end{pmatrix}. \tag{5.197}$$

Now, let us make the approximation

$$e^{\frac{i}{\hbar}:\widehat{T}_i:} \approx I + \frac{i}{\hbar} : \widehat{T}_i :, \tag{5.198}$$

such that we retain only the single commutator terms since the remaining multiple commutator terms lead to polynomials of degree ≥ 5 in $(\vec{r}_\perp, \vec{p}_\perp)$, which are to be ignored consistent with the approximation we are considering. Then, we have

$$\begin{pmatrix} \langle x \rangle (z) \\ \langle y \rangle (z) \\ \frac{1}{p_0} \langle \widehat{p}_x \rangle (z) \\ \frac{1}{p_0} \langle \widehat{p}_y \rangle (z) \end{pmatrix} \approx (M(z, z_i) \otimes R(\theta(z, z_i)))$$

$$\times \left(\begin{pmatrix} \langle x \rangle (z_i) \\ \langle y \rangle (z_i) \\ \frac{1}{p_0} \langle \widehat{p}_x \rangle (z_i) \\ \frac{1}{p_0} \langle \widehat{p}_y \rangle (z_i) \end{pmatrix} + \begin{pmatrix} \langle \frac{i}{\hbar} [\widehat{T}_i, x] \rangle (z_i) \\ \langle \frac{i}{\hbar} [\widehat{T}_i, y] \rangle (z_i) \\ \frac{1}{p_0} \langle \frac{i}{\hbar} [\widehat{T}_i, \widehat{p}_x] \rangle (z_i) \\ \frac{1}{p_0} \langle \frac{i}{\hbar} [\widehat{T}_i, \widehat{p}_y] \rangle (z_i) \end{pmatrix} \right)$$

$$= (M(z, z_i) \otimes R(\theta(z, z_i)))$$

$$\times \left(\begin{pmatrix} \langle x \rangle (z_i) \\ \langle y \rangle (z_i) \\ \frac{1}{p_0} \langle \widehat{p}_x \rangle (z_i) \\ \frac{1}{p_0} \langle \widehat{p}_y \rangle (z_i) \end{pmatrix} + \begin{pmatrix} (\delta x)(z_i) \\ (\delta y)(z_i) \\ \frac{1}{p_0}(\delta p_x)(z_i) \\ \frac{1}{p_0}(\delta p_y)(z_i) \end{pmatrix} \right)$$

$$= \left(\begin{pmatrix} \langle x \rangle_p (z) \\ \langle y \rangle_p (z) \\ \frac{1}{p_0} \langle \widehat{p}_x \rangle_p (z) \\ \frac{1}{p_0} \langle \widehat{p}_y \rangle_p (z) \end{pmatrix} + \begin{pmatrix} (\Delta x)(z) \\ (\Delta y)(z) \\ \frac{1}{p_0}(\Delta p_x)(z) \\ \frac{1}{p_0}(\Delta p_y)(z) \end{pmatrix} \right), \quad (5.199)$$

where $(\Delta x)(z)$, $(\Delta y)(z)$, $\frac{1}{p_0}(\Delta p_x)(z)$, and $\frac{1}{p_0}(\Delta p_y)(z)$ are third-order aberrations, deviations from the paraxial results involving the averages of third-order polynomials in $\left(\vec{r}_\perp, \vec{\widehat{p}}_\perp \right)$.

Obviously, the plane at which the influence of aberrations is to be known is the image plane at $z = z_{img}$:

$$\begin{pmatrix} \langle x \rangle (z_{img}) \\ \langle y \rangle (z_{img}) \\ \frac{1}{p_0} \langle \widehat{p}_x \rangle (z_{img}) \\ \frac{1}{p_0} \langle \widehat{p}_y \rangle (z_{img}) \end{pmatrix} = \left(\begin{pmatrix} M & 0 \\ -\frac{1}{f} & \frac{1}{M} \end{pmatrix} \otimes R(\theta_L) \right) \times$$

$$\left(\begin{pmatrix} \langle x \rangle (z_{obj}) \\ \langle y \rangle (z_{obj}) \\ \frac{1}{p_0} \langle \widehat{p}_x \rangle (z_{obj}) \\ \frac{1}{p_0} \langle \widehat{p}_y \rangle (z_{obj}) \end{pmatrix} + \begin{pmatrix} (\delta x)(z_{obj}) \\ (\delta y)(z_{obj}) \\ \frac{1}{p_0}(\delta p_x)(z_{obj}) \\ \frac{1}{p_0}(\delta p_y)(z_{obj}) \end{pmatrix} \right)$$

$$= \begin{pmatrix} \langle x \rangle_p (z_{img}) \\ \langle y \rangle_p (z_{img}) \\ \frac{1}{p_0} \langle \widehat{p}_x \rangle_p (z_{img}) \\ \frac{1}{p_0} \langle \widehat{p}_y \rangle_p (z_{img}) \end{pmatrix}$$

$$+ \left(\begin{pmatrix} M & 0 \\ -\frac{1}{f} & \frac{1}{M} \end{pmatrix} \otimes R(\theta_L) \right) \times \begin{pmatrix} (\delta x)(z_{obj}) \\ (\delta y)(z_{obj}) \\ \frac{1}{p_0}(\delta p_x)(z_{obj}) \\ \frac{1}{p_0}(\delta p_y)(z_{obj}) \end{pmatrix},$$

$$(5.200)$$

with

$$(\delta x)(z_{obj}) = \frac{C_s}{p_0^3} \langle \widehat{p}_x \widehat{p}_\perp^2 \rangle (z_{obj})$$

$$+ \frac{K}{2p_0^2} \langle \{ \widehat{p}_x, (\vec{\widehat{p}}_\perp \cdot \vec{r}_\perp + \vec{r}_\perp \cdot \vec{\widehat{p}}_\perp) \} + \{ x, \widehat{p}_\perp^2 \} \rangle (z_{obj})$$

$$+ \frac{k}{p_0^2} \langle \{ \widehat{p}_x, \widehat{L}_z \} - \frac{1}{2} \{ y, \widehat{p}_\perp^2 \} \rangle (z_{obj})$$

Scalar Theory: Spin-0 and Spinless Particles   257

$$+ \frac{A}{2p_0} \left\langle \left\{ x, \left(\vec{\hat{p}}_\perp \cdot \vec{r}_\perp + \vec{r}_\perp \cdot \vec{\hat{p}}_\perp \right) \right\} \right\rangle (z_{obj})$$

$$+ \frac{a}{2p_0} \left\langle \left\{ x, \hat{L}_z \right\} - \left\{ y, \left(\vec{\hat{p}}_\perp \cdot \vec{r}_\perp + \vec{r}_\perp \cdot \vec{\hat{p}}_\perp \right) \right\} \right\rangle (z_{obj})$$

$$+ \frac{F}{2p_0} \left\langle \left\{ \hat{p}_x, r_\perp^2 \right\} \right\rangle (z_{obj}) + D \left\langle x r_\perp^2 \right\rangle (z_{obj}) - d \left\langle y r_\perp^2 \right\rangle (z_{obj}), \quad (5.201)$$

$$(\delta y)(z_{obj}) = \frac{C_s}{p_0^3} \left\langle \hat{p}_y \hat{p}_\perp^2 \right\rangle (z_{obj})$$

$$+ \frac{K}{2p_0^2} \left\langle \left\{ \hat{p}_y, \left(\vec{\hat{p}}_\perp \cdot \vec{r}_\perp + \vec{r}_\perp \cdot \vec{\hat{p}}_\perp \right) \right\} + \left\{ y, \hat{p}_\perp^2 \right\} \right\rangle (z_{obj})$$

$$+ \frac{k}{p_0^2} \left\langle \left\{ \hat{p}_y, \hat{L}_z \right\} + \frac{1}{2} \left\{ x, \hat{p}_\perp^2 \right\} \right\rangle (z_{obj})$$

$$+ \frac{A}{2p_0} \left\langle \left\{ y, \left(\vec{\hat{p}}_\perp \cdot \vec{r}_\perp + \vec{r}_\perp \cdot \vec{\hat{p}}_\perp \right) \right\} \right\rangle (z_{obj})$$

$$+ \frac{a}{2p_0} \left\langle \left\{ y, \hat{L}_z \right\} - \left\{ x, \left(\vec{\hat{p}}_\perp \cdot \vec{r}_\perp + \vec{r}_\perp \cdot \vec{\hat{p}}_\perp \right) \right\} \right\rangle (z_{obj})$$

$$+ \frac{F}{2p_0} \left\langle \left\{ \hat{p}_y, r_\perp^2 \right\} \right\rangle (z_{obj}) + D \left\langle y r_\perp^2 \right\rangle (z_{obj}) + d \left\langle x r_\perp^2 \right\rangle (z_{obj}), \quad (5.202)$$

$$(\delta p_x)(z_{obj}) = - \frac{K}{p_0^2} \left\langle \hat{p}_x \hat{p}_\perp^2 \right\rangle (z_{obj})$$

$$- \frac{k}{p_0^2} \left\langle \hat{p}_y \hat{p}_\perp^2 \right\rangle (z_{obj})$$

$$- \frac{A}{2p_0} \left\langle \left\{ \hat{p}_x, \left(\vec{\hat{p}}_\perp \cdot \vec{r}_\perp + \vec{r}_\perp \cdot \vec{\hat{p}}_\perp \right) \right\} \right\rangle (z_{obj})$$

$$- \frac{a}{2p_0} \left\langle \left\{ \hat{p}_x, \hat{L}_z \right\} + \left\{ \hat{p}_y, \left(\vec{\hat{p}}_\perp \cdot \vec{r}_\perp + \vec{r}_\perp \cdot \vec{\hat{p}}_\perp \right) \right\} \right\rangle (z_{obj})$$

$$- \frac{F}{2p_0} \left\langle \left\{ x, \hat{p}_\perp^2 \right\} \right\rangle (z_{obj})$$

$$- \frac{D}{2} \left\langle \left\{ \hat{p}_x, r_\perp^2 \right\} + \left\{ x, \left(\vec{\hat{p}}_\perp \cdot \vec{r}_\perp + \vec{r}_\perp \cdot \vec{\hat{p}}_\perp \right) \right\} \right\rangle (z_{obj})$$

$$- d \left\langle \left\{ x, \hat{L}_z \right\} + \frac{1}{2} \left\{ \hat{p}_y, r_\perp^2 \right\} \right\rangle (z_{obj}) - E p_0 \left\langle x r_\perp^2 \right\rangle (z_{obj}), \quad (5.203)$$

$$(\delta p_y)(z_{obj}) = - \frac{K}{p_0^2} \left\langle \hat{p}_y \hat{p}_\perp^2 \right\rangle (z_{obj})$$

$$+ \frac{k}{p_0^2} \left\langle \hat{p}_x \hat{p}_\perp^2 \right\rangle (z_{obj})$$

$$- \frac{A}{2p_0} \left\langle \left\{ \hat{p}_y, \left(\vec{\hat{p}}_\perp \cdot \vec{r}_\perp + \vec{r}_\perp \cdot \vec{\hat{p}}_\perp \right) \right\} \right\rangle (z_{obj})$$

$$-\frac{a}{2p_0}\left\langle\left\{\widehat{p}_y,\widehat{L}_z\right\}+\left\{\widehat{p}_x,\left(\vec{\widehat{p}}_\perp\cdot\vec{r}_\perp+\vec{r}_\perp\cdot\vec{\widehat{p}}_\perp\right)\right\}\right\rangle(z_{obj})$$
$$-\frac{F}{2p_0}\left\langle\left\{y,\widehat{p}_\perp^2\right\}\right\rangle(z_{obj})$$
$$-\frac{D}{2}\left\langle\left\{\widehat{p}_y,r_\perp^2\right\}+\left\{y,\left(\vec{\widehat{p}}_\perp\cdot\vec{r}_\perp+\vec{r}_\perp\cdot\vec{\widehat{p}}_\perp\right)\right\}\right\rangle(z_{obj})$$
$$-d\left\langle\left\{y,\widehat{L}_z\right\}-\frac{1}{2}\left\{\widehat{p}_x,r_\perp^2\right\}\right\rangle(z_{obj})-Ep_0\langle yr_\perp^2\rangle(z_{obj}), \quad (5.204)$$

with

$$\begin{aligned}
&C_s = C(z_{img},z_{obj}), \quad K = K(z_{img},z_{obj}), \quad k = k(z_{img},z_{obj}),\\
&A = A(z_{img},z_{obj}), \quad a = a(z_{img},z_{obj}), \quad F = F(z_{img},z_{obj}),\\
&D = D(z_{img},z_{obj}), \quad d = d(z_{img},z_{obj}), \quad E = E(z_{img},z_{obj}).
\end{aligned} \quad (5.205)$$

With reference to the aberrations of position, (5.201) and (5.202), the constants C_s, K, k, A, a, F, D, and d, respectively, are known as the aberration coefficients corresponding to spherical aberration, coma, anisotropic coma, astigmatism, anisotropic astigmatism, curvature of field, distortion, and anisotropic distortion. The gradient aberrations, (5.203) and (5.204), do not affect the single-stage image, but should be taken into account as the input to the next stage when the lens forms a part of a complex imaging system. It should be noted that the expressions for the various aberration coefficients obtained above are exactly same as the classical expressions for them. For a detailed picture of the effects of these aberrations on the quality of the image and the classical methods of computation of these aberrations, see Hawkes and Kasper [70]. Ximen [194] has given a treatment of the classical theory of aberrations using position, momentum, and Hamilton's equations of motion.

Introducing the notations

$$u = x+iy, \quad v = \frac{1}{p_0}(\widehat{p}_x+i\widehat{p}_y), \quad (5.206)$$

the transfer map including the third-order aberrations, (5.200–5.204) can be written in a compact matrix form (see Hawkes and Kasper [70] for details) as follows:

$$\begin{pmatrix}\langle u\rangle(z_{img})\\ \langle v\rangle(z_{img})\end{pmatrix} = e^{-i\theta_L}\begin{pmatrix}M & 0\\ -\frac{1}{f} & \frac{1}{M}\end{pmatrix}$$
$$\times\begin{pmatrix}1 & 0 & C_s & 2K & 2k & F\\ 0 & 1 & ik-K & ia-2A & -a & id-D\\ & & D+id & 2A+ia & -a & K+ik\\ & & -E & -2D & 2d & -F\end{pmatrix}$$

Scalar Theory: Spin-0 and Spinless Particles

$$\times \begin{pmatrix} \langle u \rangle (z_{obj}) \\ \langle v \rangle (z_{obj}) \\ \langle vv^\dagger v \rangle (z_{obj}) \\ \frac{1}{4} \langle \{v, u^\dagger v + v^\dagger u\} \rangle (z_{obj}) \\ \frac{1}{4i} \langle \{v, u^\dagger v - v^\dagger u\} \rangle (z_{obj}) \\ \frac{1}{2} \langle \{v, u^\dagger u\} \rangle (z_{obj}) \\ \langle uu^\dagger u \rangle (z_{obj}) \\ \frac{1}{4} \langle \{u, v^\dagger u + u^\dagger v\} \rangle (z_{obj}) \\ \frac{1}{4i} \langle \{u, v^\dagger u - u^\dagger v\} \rangle (z_{obj}) \\ \frac{1}{2} \langle \{u, v^\dagger v\} \rangle (z_{obj}) \end{pmatrix}. \quad (5.207)$$

The wave function at the image plane is

$$\begin{aligned} \psi(\vec{r}_{\perp,img}, z_{img}) &= \int d^2 r_{\perp,obj} \int d^2 r_\perp \langle \vec{r}_{\perp,img} | \widehat{U}_p(z_{img}, z_{obj}) | \vec{r}_\perp \rangle \\ &\quad \times \langle \vec{r}_\perp | \widehat{U}_i(z_{img}, z_{obj}) | \vec{r}_{\perp,obj} \rangle \psi(\vec{r}_{\perp,obj}, z_{obj}) \\ &\sim \frac{1}{M} \exp\left(-\frac{i\pi r^2_{\perp,img}}{\lambda_0 f}\right) \\ &\quad \times \int d^2 r_{\perp,obj} \int d^2 r_\perp \delta^2\left(\vec{r}_\perp - \frac{1}{M}\vec{r}_{\perp,img}(\theta_L)\right) \\ &\quad \times \langle \vec{r}_\perp | \widehat{U}_i(z_{img}, z_{obj}) | \vec{r}_{\perp,obj} \rangle \psi(\vec{r}_{\perp,obj}, z_{obj}) \\ &= \frac{1}{M} \exp\left(-\frac{i\pi r^2_{\perp,img}}{\lambda_0 f}\right) \\ &\quad \times \int d^2 r_{\perp,obj} \langle \vec{r}_{\perp,img}(\theta_L)/M | \widehat{U}_i(z_{img}, z_{obj}) | \vec{r}_{\perp,obj} \rangle \\ &\quad \times \psi(\vec{r}_{\perp,obj}, z_{obj}). \end{aligned} \quad (5.208)$$

When there are no aberrations

$$\langle \vec{r}_{\perp,img}(\theta_L)/M | \widehat{U}_i(z_{img}, z_{obj}) | \vec{r}_{\perp,obj} \rangle = \delta^2\left(\vec{r}_{\perp,obj} - \frac{1}{M}\vec{r}_{\perp,img}(\theta_L)\right), \quad (5.209)$$

and hence one obtains the point-to-point imaging as seen earlier. It is clear from (5.208) that, when aberrations are present, the intensity distribution in the image plane will represent only a blurred and distorted version of the image. Note that in $\widehat{U}_i(z_{img}, z_{obj})$ the most dominant aberration term is the spherical aberration term that is independent of the position of the object point, as seen from (5.201) and (5.202) (for details on the practical aspects of electron optical imaging, see Hawkes and Kasper [70, 71, 72]).

5.2.3.3 Quantum Corrections to the Classical Results

Before closing the discussion on the quantum mechanics of round magnetic lens, comments on the effects of $\hbar$-dependent Hermitian and anti-Hermitian terms, which have been dropped in deriving the final quantum charged particle beam optical Hamiltonian for the axially symmetric magnetic lens (5.181) are in order.

Let us first consider, for example, the $\hbar$-dependent anti-Hermitian term:

$$\widehat{\mathbb{A}} = \frac{i\hbar}{2p_0^2}\left(\alpha(z)\alpha'(z)r_\perp^2 - p_0\alpha'(z)\widehat{L}_z\right), \qquad (5.210)$$

that has got dropped in (5.181) in the process of making the quantum beam optical Hamiltonian Hermitian. This term is a paraxial term, since it is quadratic in $\left(\vec{r}_\perp,\vec{\widehat{p}}_\perp\right)$. If we retain any such anti-Hermitian term in the paraxial Hamiltonian, the resultant z-evolution operator will be nonunitary and will have the form

$$\begin{aligned}\widehat{\mathrm{T}}_p(z,z_i) &= \exp\left\{-\frac{i}{\hbar}\left(\widehat{T}(z,z_i) + \widehat{t}_\mathrm{h}(z,z_i) + \widehat{t}_\mathrm{a}(z,z_i)\right)\right\} \\ &\approx \exp\left(-\frac{i}{\hbar}\widehat{T}(z,z_i)\right)\exp\left(-\frac{i}{\hbar}\widehat{t}_\mathrm{h}(z,z_i)\right)\exp\left(-\frac{i}{\hbar}\widehat{t}_\mathrm{a}(z,z_i)\right),\end{aligned} \qquad (5.211)$$

where $\widehat{t}_\mathrm{h}(z,z_i)$ and $\widehat{t}_\mathrm{a}(z,z_i)$ are, respectively, the Hermitian and anti-Hermitian correction terms to the main part $\widehat{T}(z,z_i)$. It may be noted that any term of the type $i\left[\widehat{A},\widehat{B}\right]/\hbar$ is Hermitian when both $\widehat{A}$ and $\widehat{B}$ are Hermitian or anti-Hermitian, and such a term is anti-Hermitian if one of the two operators $\left(\widehat{A},\widehat{B}\right)$ is Hermitian and the other is anti-Hermitian. We can use $\widehat{\mathrm{T}}_p(z,z_i)$ to calculate the transfer maps as

$$\begin{aligned}\langle\vec{r}_\perp\rangle(z_i) &\longrightarrow \langle\vec{r}_\perp\rangle(z) \\ &= \frac{\langle\psi(z)|\vec{r}_\perp|\psi(z)\rangle}{\langle\psi(z)|\psi(z)\rangle} \\ &= \frac{\left\langle\psi(z_i)\left|\widehat{\mathrm{T}}_p^\dagger(z,z_i)\vec{r}_\perp\widehat{\mathrm{T}}_p(z,z_i)\right|\psi(z_i)\right\rangle}{\left\langle\psi(z_i)\left|\widehat{\mathrm{T}}_p^\dagger(z,z_i)\widehat{\mathrm{T}}_p(z,z_i)\right|\psi(z_i)\right\rangle}, \\[6pt] \left\langle\vec{\widehat{p}}_\perp\right\rangle(z_i) &\longrightarrow \left\langle\vec{\widehat{p}}_\perp\right\rangle(z) \\ &= \frac{\left\langle\psi(z)\left|\vec{\widehat{p}}_\perp\right|\psi(z)\right\rangle}{\langle\psi(z)|\psi(z)\rangle} \\ &= \frac{\left\langle\psi(z_i)\left|\widehat{\mathrm{T}}_p^\dagger(z,z_i)\vec{\widehat{p}}_\perp\widehat{\mathrm{T}}_p(z,z_i)\right|\psi(z_i)\right\rangle}{\left\langle\psi(z_i)\left|\widehat{\mathrm{T}}_p^\dagger(z,z_i)\widehat{\mathrm{T}}_p(z,z_i)\right|\psi(z_i)\right\rangle}.\end{aligned} \qquad (5.212)$$

Scalar Theory: Spin-0 and Spinless Particles

Note that since the z-evolution of $|\psi(z)\rangle$ is not unitary, for calculating the average values, we have to take the normalized state at the plane at any z as $|\psi(z)\rangle / \sqrt{\langle \psi(z)|\psi(z)\rangle}$, where $|\psi(z)\rangle = \widehat{T}_p(z, z_i) |\psi(z_i)\rangle$. It is seen that the Hermitian correction term modifies the paraxial map while the anti-Hermitian term leads to an overall real scaling factor $\sim 1/\langle \psi(z_{obj})|\exp(-2i\widehat{t}_a/\hbar)|\psi(z_{obj})\rangle$, affecting the image magnification as a consequence of nonconservation of intensity, and contributes to aberrations since the terms like $\exp(-i\widehat{t}_a/\hbar)^\dagger \vec{r}_\perp \exp(-i\widehat{t}_a/\hbar)$ and $\exp(-i\widehat{t}_a/\hbar)^\dagger \vec{p}_\perp \exp(-i\widehat{t}_a/\hbar)$ lead on expansion, respectively, to Hermitian terms of the form $\vec{r}_\perp$ + nonlinear terms in $(\vec{r}_\perp, \vec{p}_\perp)$ and $\vec{p}_\perp$ + nonlinear terms in $(\vec{r}_\perp, \vec{p}_\perp)$ only. In this present case, the anti-Hermitian term $\widehat{A}$ in (5.210) does not lead to any Hermitian correction term (note that $[\widehat{A}(z'), \widehat{A}(z'')] = 0$ for any z' and z''), and its contribution to the functioning of the lens is only through the anti-Hermitian correction term affecting the conservation of intensity and adding to the aberrations.

Let us now turn to the $\hbar$-dependent Hermitian terms

$$\widehat{\mathbb{H}}_p = \frac{\hbar^2}{4p_0^2} \alpha(z)\alpha'(z) \left(\vec{r}_\perp \cdot \vec{p}_\perp + \vec{p}_\perp \cdot \vec{r}_\perp\right)$$

$$\widehat{\mathbb{H}}' = \frac{\hbar^2}{32p_0^3} \Big\{ \alpha'''(z)\widehat{L}_z \left(\vec{r}_\perp \cdot \vec{p}_\perp + \vec{p}_\perp \cdot \vec{r}_\perp\right)$$

$$- p_0 \left(\alpha'(z)\alpha''(z) + \alpha(z)\alpha'''(z)\right) \left\{ r_\perp^2, \left(\vec{r}_\perp \cdot \vec{p}_\perp + \vec{p}_\perp \cdot \vec{r}_\perp\right) \right\} \Big\}. \quad (5.213)$$

which have also been dropped in (5.181). Taking into account the influence of these terms is straightforward. Note that $\widehat{\mathbb{H}}_p$ is a paraxial term and should be added to $\widehat{\mathscr{H}}_{o,L}$ while studying the functioning of the lens in the paraxial approximation. The rotation does not get affected by the addition of the term $\widehat{\mathbb{H}}_p$. In the rotated coordinate system, the equation of motion (5.141) gets replaced by

$$\frac{d}{dz} \begin{pmatrix} \langle \vec{R}_\perp \rangle(z) \\ \frac{1}{p_0} \langle \vec{P}_\perp \rangle(z) \end{pmatrix} = \bar{\mu}(z) \begin{pmatrix} \langle \vec{R}_\perp \rangle(z) \\ \frac{1}{p_0} \langle \vec{P}_\perp \rangle(z) \end{pmatrix}$$

$$= \begin{pmatrix} \frac{\hbar^2}{2p_0^2} \alpha(z)\alpha'(z) & 1 \\ -\alpha^2(z) & -\frac{\hbar^2}{2p_0^2} \alpha(z)\alpha'(z) \end{pmatrix}$$

$$\times \begin{pmatrix} \langle \vec{R}_\perp \rangle(z) \\ \frac{1}{p_0} \langle \vec{P}_\perp \rangle(z) \end{pmatrix}, \quad (5.214)$$

This equation can be integrated to give the transfer map

$$\begin{pmatrix} \langle \vec{R}_\perp \rangle (z) \\ \frac{1}{p_0} \langle \vec{P}_\perp \rangle (z) \end{pmatrix} = \bar{M}(z, z_i) \begin{pmatrix} \langle \vec{R}_\perp \rangle (z_i) \\ \frac{1}{p_0} \langle \vec{P}_\perp \rangle (z_i) \end{pmatrix}$$

$$= \begin{pmatrix} \bar{g}(z, z_i) & \bar{h}(z, z_i) \\ \bar{g}'(z, z_i) & \bar{h}'(z, z_i) \end{pmatrix} \begin{pmatrix} \langle \vec{R}_\perp \rangle (z_i) \\ \frac{1}{p_0} \langle \vec{P}_\perp \rangle (z_i) \end{pmatrix}, \quad (5.215)$$

where the matrix elements of $\bar{M}(z, z_i)$ can be obtained from (5.146) by replacing $\mu(z)$ by $\bar{\mu}(z)$. The resulting expression for $\bar{g}'(z, z_i)$ shows that the focal length gets an additive contribution $\sim \hbar^2$ vanishing in the classical limit. The other Hermitian term mentioned above, $\widehat{\mathbb{H}}'$, has to be added to the perturbation term $\mathscr{H}'_o$ in the quantum beam optical Hamiltonian (5.181). The corresponding computation of the transfer maps, using the z-dependent perturbation theory in terms of the interaction picture, leads to the modification of aberration coefficients. For example, the modified spherical aberration coefficient turns out to be

$$\bar{C}_s = \frac{1}{2} \int_{z_{obj}}^{z_{img}} dz \left\{ \left[(\alpha^4 - \alpha \alpha'') \right. \right.$$
$$+ \frac{\hbar^4}{2 p_0^4} \left(\alpha^4 \alpha'^2 + \alpha \alpha'^2 \alpha'' + \alpha^2 \alpha' \alpha''' \right) \bigg] \bar{h}^4$$
$$+ \left[\frac{\hbar^2}{p_0^2} (2\alpha \alpha' - \alpha' \alpha'' - \alpha \alpha''') - \frac{\hbar^4}{p_0^4} \alpha^4 \alpha'^2 \right] \bar{h}^3 \bar{h}'$$
$$+ \left[2\alpha^2 + \frac{3\hbar^4}{2p_0^4} \alpha^2 \alpha'^2 \right] \bar{h}^2 \bar{h}'^2$$
$$+ \left[\frac{\hbar^2}{p_0^2} (2\alpha \alpha' - \alpha' \alpha'' - \alpha \alpha''') \right] \bar{h} \bar{h}'^3 + \bar{h}'^4 \bigg\}, \quad (5.216)$$

where $\bar{h} = \bar{h}(z, z_i)$ and $\bar{h}' = \bar{h}'(z, z_i)$ in (5.215). Again, the tiny $\hbar$-dependent contributions to C_s, and similarly to other aberrations, vanish in the classical limit. The positivity of the spherical aberration coefficient C_s is a celebrated theorem of Scherzer [167], and the quantum corrections do not affect this result. This is why the classical theory of charged particle optics is working so well!

5.2.4 NORMAL MAGNETIC QUADRUPOLE

Let us now look at the quantum mechanics of a normal magnetic quadrupole lens. Let a quasiparaxial beam of particles of charge q and rest mass m be propagating through a normal magnetic quadrupole lens. We shall take the optic axis of the lens to be along the z-direction and the practical boundaries of the lens to be at z_ℓ and z_r with $z_\ell < z_r$. The field of the normal magnetic quadrupole is

$$\vec{B}(\vec{r}) = (-Q_n y, -Q_n x, 0), \quad (5.217)$$

Scalar Theory: Spin-0 and Spinless Particles

where

$$Q_n = \begin{cases} \text{constant in the lens region} & (z_\ell < z < z_r) \\ 0 \text{ outside the lens region} & (z < z_\ell, z > z_r) \end{cases} \quad (5.218)$$

There is no electric field and hence $\phi(\vec{r}) = 0$. The vector potential of the field can be taken as

$$\vec{A}(\vec{r}) = \left(0, 0, \frac{1}{2} Q_n \left(x^2 - y^2\right)\right). \quad (5.219)$$

Now, from (5.34), the quantum beam optical Hamiltonian operator is seen to be, keeping terms up to fourth power in $\left(\vec{r}_\perp, \vec{\hat{p}}_\perp\right)$,

$$\widehat{\mathcal{H}_o} = \widehat{\mathcal{H}_{o,p}} + \widehat{\mathcal{H}_o'} + \widehat{\mathbb{H}_p} + \widehat{\mathbb{A}}$$

$$\widehat{\mathcal{H}_{o,p}} = -p_0 + \frac{\hat{p}_\perp^2}{2p_0} - \frac{1}{2} p_0 K_n \left(x^2 - y^2\right), \quad \widehat{\mathcal{H}_o'} = \frac{1}{8p_0^3} \hat{p}_\perp^4,$$

$$\widehat{\mathbb{H}}_p = \frac{\hbar^2 K_n}{4 p_0^3} \left(\hat{p}_x^2 - \hat{p}_y^2\right), \quad \widehat{\mathbb{A}} = \frac{i \hbar K_n}{2 p_0} \left(x \hat{p}_x - y \hat{p}_y\right), \quad (5.220)$$

where $K_n = q Q_n / p_0$.

Let us consider the propagation of a paraxial beam from the transverse xy-plane at $z = z_i < z_\ell$ to the transverse plane at $z > z_r$. The z-evolution of the beam is governed by the paraxial quantum beam optical Hamiltonian

$$\widehat{\mathcal{H}_{o,p}} = -p_0 + \frac{\hat{p}_\perp^2}{2p_0} - \frac{1}{2} p_0 K_n \left(x^2 - y^2\right), \quad (5.221)$$

obtained from $\widehat{\mathcal{H}_o}$ by dropping the nonparaxial perturbation term $\widehat{\mathcal{H}_o'}$, $\hbar$-dependent paraxial Hermitian term $\widehat{\mathbb{H}}_p$, and the $\hbar$-dependent anti-Hermitian term $\widehat{\mathbb{A}}$. The beam propagates in free space from z_i to z_ℓ, passes through the lens from z_ℓ to z_r, and propagates through free space again from z_r to z. Following (5.44), the transfer map for the quantum averages of transverse coordinates and momenta of a beam particle is given by

$$\begin{pmatrix} \langle x \rangle(z) \\ \frac{1}{p_0} \langle \hat{p}_x \rangle(z) \\ \langle y \rangle(z) \\ \frac{1}{p_0} \langle \hat{p}_y \rangle(z) \end{pmatrix} = \begin{pmatrix} \left\langle \hat{U}^\dagger(z,z_i) x \hat{U}(z,z_i) \right\rangle(z_i) \\ \frac{1}{p_0} \left\langle \hat{U}^\dagger(z,z_i) \hat{p}_x \hat{U}(z,z_i) \right\rangle(z_i) \\ \left\langle \hat{U}^\dagger(z,z_i) y \hat{U}(z,z_i) \right\rangle(z_i) \\ \frac{1}{p_0} \left\langle \hat{U}^\dagger(z,z_i) \hat{p}_y \hat{U}(z,z_i) \right\rangle(z_i) \end{pmatrix} \quad (5.222)$$

where $\hat{U}(z, z_i)$ is the z-evolution operator corresponding to the Hamiltonian $\widehat{\mathcal{H}_{o,p}}$. Using the semigroup property of the z-evolution operator (5.39), we can write

$$\hat{U}(z, z_i) = \hat{U}(z, z_r) \hat{U}(z_r, z_\ell) \hat{U}(z_\ell, z_i)$$
$$= \hat{U}_D(z, z_r) \hat{U}_L(z_r, z_\ell) \hat{U}_D(z_\ell, z_i), \quad (5.223)$$

where the subscripts D and L stand for drift and propagation through the lens, respectively, and the intervals (z_i, z_ℓ), (z_ℓ, z_r), and (z_r, z) correspond to free space, lens, and free space, respectively. Since within each of these regions $\widehat{\mathcal{H}}_{o,p}$ is independent of z, we get

$$\widehat{U}_D(z, z_r) = \exp\left\{-\frac{i}{\hbar}(z - z_r)\left[-p_0 + \frac{\widehat{p}_\perp^2}{2p_0}\right]\right\},$$

$$\widehat{U}_L(z_r, z_\ell) = \exp\left\{-\frac{i}{\hbar}(z_r - z_\ell)\left[-p_0 + \frac{\widehat{p}_\perp^2}{2p_0} - \frac{1}{2}p_0 K_n(x^2 - y^2)\right]\right\},$$

$$\widehat{U}_D(z_\ell, z_i) = \exp\left\{-\frac{i}{\hbar}(z_\ell - z_i)\left[-p_0 + \frac{\widehat{p}_\perp^2}{2p_0}\right]\right\}. \tag{5.224}$$

Substituting the above expression for $\widehat{U}(z, z_i)$ in (5.222), we can write

$$\langle x \rangle(z) = \left\langle e^{\frac{i}{\hbar}(z-z_r):\frac{\widehat{p}_\perp^2}{2p_0}:} e^{\frac{i}{\hbar}(z_r-z_\ell):\left(\frac{\widehat{p}_\perp^2}{2p_0} - \frac{1}{2}p_0 K_n(x^2-y^2)\right):} e^{\frac{i}{\hbar}(z_\ell-z_i):\frac{\widehat{p}_\perp^2}{2p_0}:} x \right\rangle(z_i),$$

$$\frac{1}{p_0}\langle \widehat{p}_x \rangle(z) = \frac{1}{p_0}\left\langle e^{\frac{i}{\hbar}(z-z_r):\frac{\widehat{p}_\perp^2}{2p_0}:} e^{\frac{i}{\hbar}(z_r-z_\ell):\left(\frac{\widehat{p}_\perp^2}{2p_0} - \frac{1}{2}p_0 K_n(x^2-y^2)\right):} e^{\frac{i}{\hbar}(z_\ell-z_i):\frac{\widehat{p}_\perp^2}{2p_0}:} \widehat{p}_x \right\rangle(z_i),$$

$$\langle y \rangle(z) = \left\langle e^{\frac{i}{\hbar}(z-z_r):\frac{\widehat{p}_\perp^2}{2p_0}:} e^{\frac{i}{\hbar}(z_r-z_\ell):\left(\frac{\widehat{p}_\perp^2}{2p_0} - \frac{1}{2}p_0 K_n(x^2-y^2)\right):} e^{\frac{i}{\hbar}(z_\ell-z_i):\frac{\widehat{p}_\perp^2}{2p_0}:} y \right\rangle(z_i),$$

$$\frac{1}{p_0}\langle \widehat{p}_y \rangle(z) = \frac{1}{p_0}\left\langle e^{\frac{i}{\hbar}(z-z_r):\frac{\widehat{p}_\perp^2}{2p_0}:} e^{\frac{i}{\hbar}(z_r-z_\ell):\left(\frac{\widehat{p}_\perp^2}{2p_0} - \frac{1}{2}p_0 K_n(x^2-y^2)\right):} e^{\frac{i}{\hbar}(z_\ell-z_i):\frac{\widehat{p}_\perp^2}{2p_0}:} \widehat{p}_y \right\rangle(z_i).$$

$$\tag{5.225}$$

From the results

$$\frac{i}{\hbar}:\frac{\widehat{p}_\perp^2}{2p_0}:\begin{pmatrix} x \\ \frac{\widehat{p}_x}{p_0} \\ y \\ \frac{\widehat{p}_y}{p_0} \end{pmatrix} = \begin{pmatrix} 0 & 1 & 0 & 0 \\ 0 & 0 & 0 & 0 \\ 0 & 0 & 0 & 1 \\ 0 & 0 & 0 & 0 \end{pmatrix}\begin{pmatrix} x \\ \frac{\widehat{p}_x}{p_0} \\ y \\ \frac{\widehat{p}_y}{p_0} \end{pmatrix}, \tag{5.226}$$

and

$$\frac{i}{\hbar}:\left(\frac{\widehat{p}_\perp^2}{2p_0} - \frac{1}{2}p_0 K_n(x^2 - y^2)\right):\begin{pmatrix} x \\ \frac{\widehat{p}_x}{p_0} \\ y \\ \frac{\widehat{p}_y}{p_0} \end{pmatrix} = \begin{pmatrix} 0 & 1 & 0 & 0 \\ K_n & 0 & 0 & 0 \\ 0 & 0 & 0 & 1 \\ 0 & 0 & -K_n & 0 \end{pmatrix}\begin{pmatrix} x \\ \frac{\widehat{p}_x}{p_0} \\ y \\ \frac{\widehat{p}_y}{p_0} \end{pmatrix},$$

$$\tag{5.227}$$

Scalar Theory: Spin-0 and Spinless Particles

we get

$$\begin{pmatrix} \langle x \rangle(z) \\ \frac{1}{p_0}\langle \hat{p}_x \rangle(z) \\ \langle y \rangle(z) \\ \frac{1}{p_0}\langle \hat{p}_y \rangle(z) \end{pmatrix} = \mathcal{T}_D(z,z_r)\,\mathcal{T}_L(z_r,z_\ell)\,\mathcal{T}_D(z_\ell,z_i) \begin{pmatrix} \langle x \rangle(z_i) \\ \frac{1}{p_0}\langle \hat{p}_x \rangle(z_i) \\ \langle y \rangle(z_i) \\ \frac{1}{p_0}\langle \hat{p}_y \rangle(z_i) \end{pmatrix},$$

$$\mathcal{T}_D(z,z_r) = \begin{pmatrix} 1 & (z-z_r) & 0 & 0 \\ 0 & 1 & 0 & 0 \\ 0 & 0 & 1 & (z-z_r) \\ 0 & 0 & 0 & 1 \end{pmatrix}$$

$$\mathcal{T}_L(z_r,z_\ell) = \left(\begin{pmatrix} \cosh(w\sqrt{K_n}) & \frac{1}{\sqrt{K_n}}\sinh(w\sqrt{K_n}) \\ \sqrt{K_n}\sinh(w\sqrt{K_n}) & \cosh(w\sqrt{K_n}) \end{pmatrix} \oplus \right.$$
$$\left. \begin{pmatrix} \cos(w\sqrt{K_n}) & \frac{1}{\sqrt{K_n}}\sin(w\sqrt{K_n}) \\ -\sqrt{K_n}\sin(w\sqrt{K_n}) & \cos(w\sqrt{K_n}) \end{pmatrix} \right)$$

$$\mathcal{T}_D(z_\ell,z_i) = \begin{pmatrix} 1 & (z_\ell-z_i) & 0 & 0 \\ 0 & 1 & 0 & 0 \\ 0 & 0 & 1 & (z_\ell-z_i) \\ 0 & 0 & 0 & 1 \end{pmatrix}$$

$$\text{with } w = (z_r - z_\ell). \tag{5.228}$$

This transfer map for the quantum averages $(\langle x \rangle(z), \langle \hat{p}_x \rangle(z)/p_0, \langle y \rangle(z), \langle \hat{p}_y \rangle(z)/p_0)$ is seen to be exactly the same as the transfer map (4.142) for the classical variables $(x(z), p_x(z)/p_0, y(z), p_y(z)/p_0)$ in the paraxial approximation. Thus, the properties of the normal magnetic quadrupole lens we have derived earlier, in the context of the classical transfer map (4.142), are valid in the present context of the quantum transfer map (5.228). Let us recall. When $K_n > 0$, the lens is divergent in the xz-plane and convergent in the yz-plane. In other words, the normal magnetic quadrupole lens produces a line focus. When $K_n < 0$, the lens is convergent in the xz-plane and divergent in the yz-plane. It is seen that K_n has the dimensions of length^{-2}. In the weak field case, when $w\sqrt{|K_n|} \ll 1$, the lens can be considered as a thin lens with the focal lengths

$$\frac{1}{f^{(x)}} = -wK_n, \qquad \frac{1}{f^{(y)}} = wK_n. \tag{5.229}$$

A lens of focal length $\pm f$ can be considered effectively to be two lenses of focal length $\pm 2f$, respectively, joined together without any gap between them. A doublet of focusing and defocusing lenses of focal lengths f and $-f$ with a distance d between them would have the effective focal length $f^2/d > 0$ and hence would behave as a focusing lens. A triplet of quadrupoles can be arranged in such a way to get net focusing effect in both the transverse planes. Series of such quadrupole triplets are the main design elements of long beam transport lines or circular accelerators to provide a periodic focusing structure called a FODO-channel. FODO stands for F(ocusing)O(nonfocusing)D(efocusing)O(nonfocusing), where O can be a drift space or a bending magnet.

Now, if we retain the $\hbar$-dependent Hermitian term $\widehat{\mathbb{H}}_p$, we have the paraxial quantum beam optical Hamiltonian as

$$\widehat{\mathscr{H}}_{o,p} = -p_0 + \frac{1}{2p_0}\left(1 + \frac{\hbar^2 K_n}{2p_0^2}\right)\widehat{p}_x^2 + \frac{1}{2p_0}\left(1 - \frac{\hbar^2 K_n}{2p_0^2}\right)\widehat{p}_y^2 - \frac{1}{2}p_0 K_n \left(x^2 - y^2\right). \tag{5.230}$$

This modifies the equation (5.227) as

$$\frac{i}{\hbar} : \left(\frac{1}{2p_0}\left(1 + \frac{\hbar^2 K_n}{2p_0^2}\right)\widehat{p}_x^2 + \frac{1}{2p_0}\left(1 - \frac{\hbar^2 K_n}{2p_0^2}\right)\widehat{p}_y^2 \right.$$

$$\left. - \frac{1}{2}p_0 K_n \left(x^2 - y^2\right)\right) : \begin{pmatrix} x \\ \frac{\widehat{p}_x}{p_0} \\ y \\ \frac{\widehat{p}_y}{p_0} \end{pmatrix}$$

$$= \begin{pmatrix} 0 & 1 + \frac{\hbar^2 K_n}{2p_0^2} & 0 & 0 \\ K_n & 0 & 0 & 0 \\ 0 & 0 & 0 & 1 - \frac{\hbar^2 K_n}{2p_0^2} \\ 0 & 0 & -K_n & 0 \end{pmatrix} \begin{pmatrix} x \\ \frac{\widehat{p}_x}{p_0} \\ y \\ \frac{\widehat{p}_y}{p_0} \end{pmatrix}, \tag{5.231}$$

leading to the transfer map

$$\begin{pmatrix} \langle x \rangle (z) \\ \frac{1}{p_0}\langle \widehat{p}_x \rangle (z) \\ \langle y \rangle (z) \\ \frac{1}{p_0}\langle \widehat{p}_y \rangle (z) \end{pmatrix} = \mathscr{T}_D(z, z_r)\, \mathscr{T}_L(z_r, z_\ell)\, \mathscr{T}_D(z_\ell, z_i) \begin{pmatrix} \langle x \rangle (z_i) \\ \frac{1}{p_0}\langle \widehat{p}_x \rangle (z_i) \\ \langle y \rangle (z_i) \\ \frac{1}{p_0}\langle \widehat{p}_y \rangle (z_i) \end{pmatrix},$$

$$\mathscr{T}_D(z, z_r) = \begin{pmatrix} 1 & (z - z_r) & 0 & 0 \\ 0 & 1 & 0 & 0 \\ 0 & 0 & 1 & (z - z_r) \\ 0 & 0 & 0 & 1 \end{pmatrix}$$

$$\mathscr{T}_L(z_r, z_\ell) = \left(\begin{pmatrix} \cosh\left(w\sqrt{K_n 1^+}\right) & \sqrt{\frac{1^+}{K_n}}\sinh\left(w\sqrt{K_n 1^+}\right) \\ \sqrt{\frac{K_n}{1^+}}\sinh\left(w\sqrt{K_n 1^+}\right) & \cosh\left(w\sqrt{K_n 1^+}\right) \end{pmatrix} \right.$$

$$\left. \oplus \begin{pmatrix} \cos\left(w\sqrt{K_n 1^-}\right) & \sqrt{\frac{1^-}{K_n}}\sin\left(w\sqrt{K_n 1^-}\right) \\ -\sqrt{\frac{K_n}{1^-}}\sin\left(w\sqrt{K_n 1^-}\right) & \cos\left(w\sqrt{K_n 1^-}\right) \end{pmatrix}\right),$$

with $1^+ = 1 + \frac{\hbar^2 K_n}{2p_0^2}$, $1^- = 1 - \frac{\hbar^2 K_n}{2p_0^2}$,

$$\mathscr{T}_D(z_\ell, z_i) = \begin{pmatrix} 1 & (z_\ell - z_i) & 0 & 0 \\ 0 & 1 & 0 & 0 \\ 0 & 0 & 1 & (z_\ell - z_i) \\ 0 & 0 & 0 & 1 \end{pmatrix}. \tag{5.232}$$

Scalar Theory: Spin-0 and Spinless Particles

For a thin lens, with $w\sqrt{|K_n|} \ll 1$ and $w \approx 0$, the above transfer map reduces to

$$\begin{pmatrix} \langle x \rangle (z) \\ \frac{1}{p_0} \langle \widehat{p}_x \rangle (z) \\ \langle y \rangle (z) \\ \frac{1}{p_0} \langle \widehat{p}_y \rangle (z) \end{pmatrix} = \mathcal{T}_D(z, z_r) \, \mathcal{T}_L(z_r, z_\ell) \, \mathcal{T}_D(z_\ell, z_i) \begin{pmatrix} \langle x \rangle (z_i) \\ \frac{1}{p_0} \langle \widehat{p}_x \rangle (z_i) \\ \langle y \rangle (z_i) \\ \frac{1}{p_0} \langle \widehat{p}_y \rangle (z_i) \end{pmatrix},$$

$$\mathcal{T}_D(z, z_r) = \begin{pmatrix} 1 & (z - z_r) & 0 & 0 \\ 0 & 1 & 0 & 0 \\ 0 & 0 & 1 & (z - z_r) \\ 0 & 0 & 0 & 1 \end{pmatrix},$$

$$\mathcal{T}_L(z_r, z_\ell) \approx \left(\begin{pmatrix} 1 + \frac{1}{2} w^2 K_n 1^+ & w 1^+ + \frac{1}{3!} w^3 K_n (1^+)^2 \\ w K_n + w^3 K_n^2 1^+ & 1 + \frac{1}{2} w^2 K_n 1^+ \end{pmatrix} \right.$$
$$\left. \oplus \begin{pmatrix} 1 - \frac{1}{2} w^2 K_n 1^- & w 1^- - \frac{1}{3!} w^3 K_n (1^-)^2 \\ -w K_n + w^3 K_n^2 1^- & 1 - \frac{1}{2} w^2 K_n 1^- \end{pmatrix} \right),$$

$$\approx \begin{pmatrix} 1 & 0 & 0 & 0 \\ w K_n & 1 & 0 & 0 \\ 0 & 0 & 1 & 0 \\ 0 & 0 & -w K_n & 1 \end{pmatrix},$$

$$\mathcal{T}_D(z_\ell, z_i) = \begin{pmatrix} 1 & (z_\ell - z_i) & 0 & 0 \\ 0 & 1 & 0 & 0 \\ 0 & 0 & 1 & (z_\ell - z_i) \\ 0 & 0 & 0 & 1 \end{pmatrix}, \qquad (5.233)$$

showing that the two focal lengths of the thin normal magnetic quadrupole lens remain the same,

$$\frac{1}{f^{(x)}} = -w K_n, \qquad \frac{1}{f^{(y)}} = w K_n, \qquad (5.234)$$

without any appreciable change due to the $\hbar$-dependent Hermitian term in the optical Hamiltonian. As we have already mentioned, the $\hbar$-dependent anti-Hermitian term in the optical Hamiltonian also does not affect the lens properties in any appreciable way. This is why accelerator optics works so well without any need to consider quantum mechanics in its design and operation.

5.2.5 SKEW MAGNETIC QUADRUPOLE

Quantum mechanics of the skew magnetic quadrupole can be analysed in a similar way. As seen in the classical theory, the skew magnetic quadrupole will be equivalent to the normal magnetic quadrupole rotated by an angle $\frac{\pi}{4}$ about the optic axis. A skew magnetic quadrupole is associated with the magnetic field

$$\vec{B}(\vec{r}) = (-Q_s x, Q_s y, 0), \qquad (5.235)$$

corresponding to the vector potential

$$\vec{A}(\vec{r}) = (0, 0, -Q_s xy). \qquad (5.236)$$

Let z_ℓ and $z_r > z_\ell$ be the boundaries of the quadrupole. Then, Q_s is a nonzero constant for $z_\ell \leq z \leq z_r$ and $Q_s = 0$ outside the quadrupole ($z < z_\ell$, $z > z_r$). For a paraxial beam propagating through this skew magnetic quadrupole along its optic axis, z-axis, the quantum beam optical Hamiltonian will be

$$\widehat{\mathcal{H}}_o = \begin{cases} -p_0 + \frac{\widehat{p}_\perp^2}{2p_0}, & \text{for } z < z_\ell, z > z_r, \\ -p_0 + \frac{\widehat{p}_\perp^2}{2p_0} + p_0 K_s xy, & \text{for } z_\ell \leq z \leq z_r, \end{cases} \quad (5.237)$$

where $K_s = qQ_s/p_0$. Now, if we make the transformation

$$\begin{pmatrix} x \\ \widehat{p}_x \\ y \\ \widehat{p}_y \end{pmatrix} = \frac{1}{\sqrt{2}} \begin{pmatrix} 1 & 0 & 1 & 0 \\ 0 & 1 & 0 & 1 \\ -1 & 0 & 1 & 0 \\ 0 & -1 & 0 & 1 \end{pmatrix} \begin{pmatrix} x' \\ \widehat{p}'_x \\ y' \\ \widehat{p}'_y \end{pmatrix} = \mathbb{R} \begin{pmatrix} x' \\ \widehat{p}'_x \\ y' \\ \widehat{p}'_y \end{pmatrix}, \quad (5.238)$$

the Hamiltonian is transformed to

$$\widehat{\mathcal{H}}'_o = \begin{cases} -p_0 + \frac{\widehat{p}'^2_\perp}{2p_0}, & \text{for } z < z_\ell, z > z_r, \\ -p_0 + \frac{\widehat{p}'^2_\perp}{2p_0} - \frac{1}{2} p_0 K_s \left(x'^2 - y'^2 \right), & \text{for } z_\ell \leq z \leq z_r, \end{cases} \quad (5.239)$$

which is same as for the propagation of the paraxial beam through a normal magnetic quadrupole (5.220) in the new (x', y') coordinate system, with K_n replaced by K_s. The above transformation corresponds to a clockwise rotation of the (x, y) coordinate axes by $\frac{\pi}{4}$ about the optic axis. Hence, the transfer map for the skew magnetic quadrupole can be obtained from the transfer map for the normal magnetic quadrupole as

$$\mathcal{T}_{sq} = \mathbb{R} \mathcal{T}_{nq} \mathbb{R}^{-1}, \quad (5.240)$$

where the subscripts sq and nq stand for the skew magnetic quadrupole and the normal magnetic quadrupole, respectively. Using the same arguments as in the classical theory of the skew magnetic quadrupole, we find that we can write

$$\begin{aligned} \mathcal{T}_{sq}(z, z_i) &= \mathcal{T}_D(z, z_r) \mathcal{T}_{sq,L}(z_r, z_\ell) \mathcal{T}_D(z_\ell, z_i) \\ &= \mathcal{T}_D(z, z_r) \mathbb{R} \mathcal{T}_{nq,L}(z_r, z_\ell) \mathbb{R}^{-1} \mathcal{T}_D(z_\ell, z_i), \end{aligned} \quad (5.241)$$

where $\mathbb{R}$ is as in (5.238), $\mathcal{T}_D(z, z_r)$ and $\mathcal{T}_D(z_\ell, z_i)$ are the same as in (5.228), and $\mathcal{T}_{nq,L}(z_r, z_\ell)$ is the same as $\mathcal{T}_L(z_r, z_\ell)$ in (5.228), with K_n replaced by K_s. One can get $\mathcal{T}_{sq,L}(z_r, z_\ell)$ directly as follows. Observe that

$$\frac{i}{\hbar} : \left(\frac{\widehat{p}_\perp^2}{2p_0} + p_0 K_s xy \right) : \begin{pmatrix} x \\ \frac{\widehat{p}_x}{p_0} \\ y \\ \frac{\widehat{p}_y}{p_0} \end{pmatrix} = \begin{pmatrix} 0 & 1 & 0 & 0 \\ 0 & 0 & -K_s & 0 \\ 0 & 0 & 0 & 1 \\ -K_s & 0 & 0 & 0 \end{pmatrix} \begin{pmatrix} x \\ \frac{\widehat{p}_x}{p_0} \\ y \\ \frac{\widehat{p}_y}{p_0} \end{pmatrix}. \quad (5.242)$$

Scalar Theory: Spin-0 and Spinless Particles

Then, one gets

$$\begin{pmatrix} \langle x \rangle (z_r) \\ \frac{\langle \hat{p}_x \rangle (z_r)}{p_0} \\ \langle y \rangle (z_r) \\ \frac{\langle \hat{p}_y \rangle (z_r)}{p_0} \end{pmatrix} = \mathcal{T}_{sq,L}(z_r, z_\ell) \begin{pmatrix} \langle x \rangle (z_\ell) \\ \frac{\langle \hat{p}_x \rangle (z_\ell)}{p_0} \\ \langle y \rangle (z_\ell) \\ \frac{\langle \hat{p}_y \rangle (z_\ell)}{p_0} \end{pmatrix}$$

$$= \exp\left[w \begin{pmatrix} 0 & 1 & 0 & 0 \\ 0 & 0 & -K_s & 0 \\ 0 & 0 & 0 & 1 \\ -K_s & 0 & 0 & 0 \end{pmatrix} \right] \begin{pmatrix} \langle x \rangle (z_\ell) \\ \frac{\langle \hat{p}_x \rangle (z_\ell)}{p_0} \\ \langle y \rangle (z_\ell) \\ \frac{\langle \hat{p}_y \rangle (z_\ell)}{p_0} \end{pmatrix}$$

$$= \frac{1}{2} \begin{pmatrix} C^+ & \frac{1}{\sqrt{K_s}}S^+ & C^- & \frac{1}{\sqrt{K_s}}S^- \\ -\sqrt{K_s}S^- & C^+ & -\sqrt{K_s}S^+ & C^- \\ C^- & \frac{1}{\sqrt{K_s}}S^- & C^+ & \frac{1}{\sqrt{K_s}}S^+ \\ -\sqrt{K_s}S^+ & C^- & -\sqrt{K_s}S^- & C^+ \end{pmatrix}$$

$$\times \begin{pmatrix} \langle x \rangle (z_\ell) \\ \frac{\langle \hat{p}_x \rangle (z_\ell)}{p_0} \\ \langle y \rangle (z_\ell) \\ \frac{\langle \hat{p}_y \rangle (z_\ell)}{p_0} \end{pmatrix},$$

with $w(z_r - z_\ell)$, (5.243)

where

$$C^\pm = \cos(w\sqrt{K_s}) \pm \cosh(w\sqrt{K_s}),$$
$$S^\pm = \sin(w\sqrt{K_s}) \pm \sinh(w\sqrt{K_s}). \quad (5.244)$$

It can be verified that $\mathcal{T}_{sq,L} = \mathbb{R}\mathcal{T}_{nq,L}(z_r, z_\ell) \mathbb{R}^{-1}$.

5.2.6 AXIALLY SYMMETRIC ELECTROSTATIC LENS

An axially symmetric electrostatic lens, or a round electrostatic lens, with the axis along the z-direction consists of the electric field corresponding to the potential

$$\phi(\vec{r}_\perp, z) = \sum_{n=0}^{\infty} \frac{(-1)^n}{(n!)^2 4^n} \phi^{(2n)}(z) r_\perp^{2n}$$

$$= \phi(z) - \frac{1}{4}\phi''(z) r_\perp^2 + \frac{1}{64}\phi''''(z) r_\perp^4 - \cdots, \quad (5.245)$$

inside the lens region ($z_\ell < z < z_r$). Outside the lens, i.e., ($z < z_\ell$, $z > z_r$), $\phi(z) = 0$. And, there is no magnetic field, i.e., $\vec{A}(\vec{r}) = (0,0,0)$.

To get the paraxial quantum beam optical Hamiltonian of the round electrostatic lens, we have to start with (5.34) and make the paraxial approximation by dropping terms of order higher than quadratic in $\left(\vec{r}_\perp, \vec{\hat{p}}_\perp\right)$. We also drop the $\hbar$-dependent terms. Then, we get

$$\widehat{\mathcal{H}_o} = -\left(1 - \frac{qE}{c^2 p_0^2}\phi(z) - \frac{q^2 E^2}{2c^4 p_0^4}\phi^2(z)\right) p_0 + \left(1 + \frac{qE}{c^2 p_0^2}\phi(z)\right)\frac{\hat{p}_\perp^2}{2p_0}$$
$$- \frac{qE}{4c^2 p_0}\left(1 + \frac{qE}{c^2 p_0^2}\phi(z)\right)\phi''(z) r_\perp^2. \quad (5.246)$$

This quantum beam optical Hamiltonian of the round electrostatic lens is seen to be very similar to the quantum beam optical Hamiltonian of the round magnetic lens (5.95), except for the absence of the rotation term and the presence of an z-dependent coefficient for the drift term $\hat{p}_\perp^2/2p_0$. The first constant term, though z-dependent, can be ignored as before since it has vanishing commutators with $\left(\vec{r}_\perp, \vec{\hat{p}}_\perp\right)$. The above paraxial quantum beam optical Hamiltonian, being quadratic in $\vec{r}_\perp$ and $\vec{\hat{p}}_\perp$, leads to a linear transfer map for the quantum averages of $\left(\vec{r}_\perp, \vec{\hat{p}}_\perp\right)$, and it is straightforward to calculate the map following the same procedure used in the case of round magnetic lens. When the beam is not paraxial, there will be aberrations and we have to go beyond the paraxial approximation. We shall not pursue the quantum mechanics of this lens further. As already mentioned, electrostatic lenses are used in the extraction, preparation, and initial acceleration of electron and ion beams in a variety of applications.

5.2.7 ELECTROSTATIC QUADRUPOLE LENS

The electric field of an ideal electrostatic quadrupole lens with the optic axis along the z-direction corresponds to the potential

$$\phi(\vec{r}_\perp, z) = \begin{cases} \frac{1}{2}Q_e\left(x^2 - y^2\right) & \text{in the lens region } (z_\ell < z < z_r), \\ 0 & \text{outside the lens } (z < z_\ell, z > z_r), \end{cases} \quad (5.247)$$

where Q_e is a constant and $z_r - z_\ell = w$ is the width of the lens. Starting with (5.34), using (5.247) and $\vec{A} = (0, 0, 0)$ since there is no magnetic field, and making the paraxial approximation by dropping terms of order higher than quadratic in $\left(\vec{r}_\perp, \vec{\hat{p}}_\perp\right)$, we get

$$\widehat{\mathcal{H}_o} \approx -p_0 + \frac{\hat{p}_\perp^2}{2p_0} + \frac{1}{2}p_0 K_e\left(x^2 - y^2\right), \quad \text{with } K_e = \frac{qEQ_e}{c^2 p_0^2}, \quad (5.248)$$

as the paraxial quantum beam optical Hamiltonian operator for the electrostatic quadrupole lens. Actually, no $\hbar$-dependent terms appear up to this approximation.

Simply by comparing the quantum beam optical Hamiltonian of the normal magnetic quadrupole lens (5.220) with the quantum beam optical Hamiltonian of the electrostatic quadrupole lens (5.248), it is readily seen that the electrostatic quadrupole lens is convergent in the xz-plane and divergent in the yz-plane when

Scalar Theory: Spin-0 and Spinless Particles

$K_e > 0$. When $K_e < 0$, this lens is divergent in the xz-plane and convergent in the yz-plane. At the paraxial level, there is no difference between the classical and quantum theories of the electrostatic quadrupole lens, except for the replacement of the classical variables by their quantum averages. Let us recall from the earlier discussion of the classical theory of electrostatic quadrupole lens. K_e has the dimension of length^{-2}. In the weak field case, when $w^2 \ll 1/|K_e|$, the lens can be considered as thin and has focal lengths

$$\frac{1}{f^{(x)}} = wK_e, \qquad \frac{1}{f^{(y)}} = -wK_e. \tag{5.249}$$

Deviations from the ideal behavior result from nonparaxial conditions, and to treat these deviations, we have to go beyond the paraxial approximation. We shall not pursue the quantum mechanics of this lens further here.

5.2.8 BENDING MAGNET

We have already discussed the classical mechanics of bending the charged particle beam by a dipole magnet. A constant magnetic field in the vertical direction produced by a dipole magnet bends the beam along a circular arc in the horizontal plane. The circular arc of radius of curvature ρ, the design trajectory of the particle, is the optic axis of the bending magnet. It is natural to use the arclength, say S, measured along the optic axis from some reference point as the independent coordinate, instead of z. Let the reference particle moving along the design trajectory carry an orthonormal XY-coordinate frame with it. The X-axis is taken to be perpendicular to the tangent to the design orbit and in the same horizontal plane as the trajectory, and the Y-axis is taken to be in the vertical direction perpendicular to both the X-axis and the trajectory. The curved S-axis is along the design trajectory and perpendicular to both the X and Y axes at any point on the design trajectory. The instantaneous position of the reference particle in the design trajectory at an arclength S from the reference point corresponds to $X = 0$ and $Y = 0$. Let any particle of the beam have coordinates (x,y,z) with respect to a fixed right-handed Cartesian coordinate frame, with its origin at the reference point on the design trajectory from which the arclength S is measured. Then, the two sets of coordinates of any particle of the beam, (X,Y,S) and (x,y,z), will be related as follows:

$$x = (\rho + X)\cos\left(\frac{S}{\rho}\right) - \rho, \quad z = (\rho + X)\sin\left(\frac{S}{\rho}\right), \quad y = Y. \tag{5.250}$$

To understand the quantum mechanics of bending of the beam particle by the dipole magnet, we have to change the corresponding Klein–Gordon equation to the curved (X,Y,S)-coordinate system and find the quantum beam optical Hamiltonian governing the S-evolution of the beam variables. To this end, we proceed as follows.

The free particle Klein–Gordon equation is

$$\left(\nabla^2 - \frac{1}{c^2}\frac{\partial^2}{\partial t^2}\right)\Psi(\vec{r},t) = \left(\frac{mc}{\hbar}\right)^2 \Psi(\vec{r},t). \tag{5.251}$$

Multiplying throughout by $c^2\hbar^2$, we can write this equation as

$$-\hbar^2 \frac{\partial^2 \Psi(\vec{r},t)}{\partial t^2} = \left(-c^2\hbar^2 \nabla^2 + m^2 c^4\right) \Psi(\vec{r},t). \tag{5.252}$$

In the curved (X,Y,S) coordinate frame, the line element ds, distance between two infinitesimally close points (X,Y,S) and $(X+dX, Y+dY, S+dS)$, is given by

$$ds^2 = dx^2 + dy^2 + dz^2 = dX^2 + dY^2 + \zeta^2 dS^2, \tag{5.253}$$

where $\zeta = 1 + \kappa X$ and $\kappa = 1/\rho$ is the curvature of the design orbit. Correspondingly, we know (see *e.g.*, Arfken, Weber, and Harris [3]) that the Laplacian ∇^2 in terms of (X,Y,S) coordinates is given by

$$\begin{aligned}\nabla^2 &= \frac{1}{\zeta} \frac{\partial}{\partial X}\left(\zeta \frac{\partial}{\partial X}\right) + \frac{\partial^2}{\partial Y^2} + \frac{1}{\zeta^2} \frac{\partial^2}{\partial S^2} \\ &= \frac{\partial^2}{\partial X^2} + \frac{\partial^2}{\partial Y^2} + \frac{1}{\zeta^2} \frac{\partial^2}{\partial S^2} + \frac{\kappa}{\zeta} \frac{\partial}{\partial X}.\end{aligned} \tag{5.254}$$

If we substitute this expression for ∇^2 in (5.252) and change $\Psi(\vec{r},t)$ to $\Psi\left(\vec{R}_\perp, S, t\right)$, we get

$$-\hbar^2 \frac{\partial^2 \Psi\left(\vec{R}_\perp, S, t\right)}{\partial t^2} = \left\{-c^2\hbar^2 \left[\frac{\partial^2}{\partial X^2} + \frac{\partial^2}{\partial Y^2} + \frac{1}{\zeta^2} \frac{\partial^2}{\partial S^2}\right.\right.$$
$$\left.\left. + \frac{\kappa}{\zeta} \frac{\partial}{\partial X}\right] + m^2 c^4\right\} \Psi\left(\vec{R}_\perp, S, t\right). \tag{5.255}$$

where $\vec{R}_\perp = (X,Y)$. While on the left-hand side $-\hbar^2 \left(\partial^2/\partial t^2\right)$ is Hermitian, the right-hand side is not Hermitian because of the last term in the expression for ∇^2 in (5.254). So, let us replace this term by the Hermitian term

$$\frac{1}{2}\left[\left(\frac{\kappa}{\zeta}\frac{\partial}{\partial X}\right) + \left(\frac{\kappa}{\zeta}\frac{\partial}{\partial X}\right)^\dagger\right] = \frac{1}{2}\left(\frac{\kappa}{\zeta}\frac{\partial}{\partial X} - \frac{\partial}{\partial X}\frac{\kappa}{\zeta}\right)$$
$$= \frac{1}{2}\left[\frac{\kappa}{\zeta}, \frac{\partial}{\partial X}\right] = \frac{\kappa^2}{2\zeta^2}, \tag{5.256}$$

where we have to remember that ζ is a function of X. Thus, we shall take the free particle Klein–Gordon equation in the curved (X,Y,S) coordinate system to be

$$-\hbar^2 \frac{\partial^2 \Psi\left(\vec{R}_\perp, S, t\right)}{\partial t^2} = \left\{-c^2\hbar^2 \left[\frac{\partial^2}{\partial X^2} + \frac{\partial^2}{\partial Y^2} + \frac{1}{\zeta^2} \frac{\partial^2}{\partial S^2}\right.\right.$$
$$\left.\left. + \frac{\kappa^2}{2\zeta^2}\right] + m^2 c^4\right\} \Psi\left(\vec{R}_\perp, S, t\right). \tag{5.257}$$

Scalar Theory: Spin-0 and Spinless Particles

Let us now rewrite this equation as

$$\left(i\hbar\frac{\partial}{\partial t}\right)^2 \Psi\left(\vec{R}_\perp, S, t\right)$$
$$= \left\{c^2\left[\left(-i\hbar\frac{\partial}{\partial X}\right)^2 + \left(-i\hbar\frac{\partial}{\partial Y}\right)^2 + \frac{1}{\zeta^2}\left(-i\hbar\frac{\partial}{\partial S}\right)^2\right]\right.$$
$$\left. -\frac{c^2\hbar^2\kappa^2}{2\zeta^2} + m^2c^4\right\}\Psi\left(\vec{R}_\perp, S, t\right). \quad (5.258)$$

This suggests that in the (X, Y, S) coordinate system, we can take the Klein–Gordon equation for a charged particle in an electromagnetic field as

$$\left(i\hbar\frac{\partial}{\partial t} - q\phi\right)^2 \Psi(\vec{R}_\perp, S, t)$$
$$= \left\{c^2\left[\left(-i\hbar\frac{\partial}{\partial X} - qA_X\right)^2 + \left(-i\hbar\frac{\partial}{\partial Y} - qA_Y\right)^2\right.\right.$$
$$\left.\left. +\frac{1}{\zeta^2}\left(-i\hbar\frac{\partial}{\partial S} - q\zeta A_S\right)^2\right] - \frac{c^2\hbar^2\kappa^2}{2\zeta^2} + m^2c^4\right\}\Psi\left(\vec{R}_\perp, S, t\right),$$
$$(5.259)$$

where (A_X, A_Y, A_S) are the magnetic vector potentials and ϕ is the electric scalar potential, and using the principle of minimal coupling, we have made the replacements

$$-i\hbar\frac{\partial}{\partial X} \longrightarrow -i\hbar\frac{\partial}{\partial X} - qA_X,$$
$$-i\hbar\frac{\partial}{\partial Y} \longrightarrow -i\hbar\frac{\partial}{\partial Y} - qA_Y,$$
$$\frac{1}{\zeta}\left(-i\hbar\frac{\partial}{\partial S}\right) \longrightarrow \frac{1}{\zeta}\left(-i\hbar\frac{\partial}{\partial S}\right) - qA_S$$
$$= \frac{1}{\zeta}\left(-i\hbar\frac{\partial}{\partial S} - q\zeta A_S\right). \quad (5.260)$$

Let us now consider the beam to be monoenergetic and a beam particle to be associated with the wave function

$$\Psi\left(\vec{R}_\perp, S, t\right) = e^{-iEt/\hbar}\psi\left(\vec{R}_\perp, S\right), \quad \text{with } E = \sqrt{m^2c^4 + c^2p_0^2}, \quad (5.261)$$

where p_0 is the design momentum with which the particle enters the bending magnet from the free space outside. The time-independent Klein–Gordon equation becomes

$$(E-q\phi)^2 \psi\left(\vec{R}_\perp, S\right)$$
$$= \left\{ c^2 \left[\left(-i\hbar \frac{\partial}{\partial X} - qA_X\right)^2 + \left(-i\hbar \frac{\partial}{\partial Y} - qA_Y\right)^2 \right. \right.$$
$$\left. \left. + \left(\frac{-i\hbar}{\zeta} \frac{\partial}{\partial S} - qA_S\right)^2 \right] - \frac{c^2 \hbar^2 \kappa^2}{2\zeta^2} + m^2 c^4 \right\} \psi\left(\vec{R}_\perp, S\right),$$
(5.262)

Let
$$\left(-i\hbar \frac{\partial}{\partial X} - qA_X\right) = \widehat{P}_X - qA_X = \widehat{\pi}_X,$$
$$\left(-i\hbar \frac{\partial}{\partial Y} - qA_Y\right) = \widehat{P}_Y - qA_Y = \widehat{\pi}_Y,$$
$$\left(\frac{-i\hbar}{\zeta} \frac{\partial}{\partial S} - qA_S\right) = \widehat{P}_S - qA_S = \widehat{\pi}_S, \quad (5.263)$$

Then, by dividing throughout by c^2 and rearranging the terms in (5.262), we can rewrite the time-independent Klein–Gordon equation as

$$\widehat{\pi}_S^2 \psi\left(\vec{R}_\perp, S\right) = \left(p_0^2 - \widehat{\pi}_X^2 - \widehat{\pi}_Y^2 - \mathsf{p}^2\right) \psi\left(\vec{R}_\perp, S\right), \quad (5.264)$$

where
$$\mathsf{p}^2 = \frac{2qE\phi}{c^2}\left(1 - \frac{q\phi}{2E}\right) - \frac{\hbar^2 \kappa^2}{2\zeta^2}. \quad (5.265)$$

For the dipole magnet, we have to take

$$\phi = 0, \quad A_X = 0, \quad A_Y = 0, \quad A_S = -B_0\left(X - \frac{\kappa X^2}{2\zeta}\right), \quad (5.266)$$

so that the magnetic field is

$$B_X = 0, \quad B_Y = B_0, \quad B_S = 0, \quad (5.267)$$

as we have already seen in the classical theory of bending magnet. Then, we have

$$\widehat{\pi}_S^2 \psi\left(\vec{R}_\perp, S\right) = \left(p_0^2 - \widehat{P}_\perp^2 + \widetilde{\mathsf{p}}^2\right) \psi\left(\vec{R}_\perp, S\right), \quad (5.268)$$

with
$$\widetilde{\mathsf{p}}^2 = \frac{\hbar^2 \kappa^2}{2\zeta^2} \quad (5.269)$$

This equation can be equivalently written as

$$\frac{\widehat{\pi}_S}{p_0}\begin{pmatrix} \psi \\ \frac{\widehat{\pi}_S}{p_0}\psi \end{pmatrix} = \begin{pmatrix} 0 & 1 \\ \frac{1}{p_0^2}\left(p_0^2 - \widehat{P}_\perp^2 + \widetilde{\mathsf{p}}^2\right) & 0 \end{pmatrix} \begin{pmatrix} \psi \\ \frac{\widehat{\pi}_S}{p_0}\psi \end{pmatrix}. \quad (5.270)$$

Scalar Theory: Spin-0 and Spinless Particles

Let us introduce the two-component wave function

$$\begin{pmatrix} \psi_+ \\ \psi_- \end{pmatrix} = M \begin{pmatrix} \psi \\ \frac{\widehat{\pi}_S}{p_0}\psi \end{pmatrix} = \frac{1}{2}\begin{pmatrix} 1 & 1 \\ 1 & -1 \end{pmatrix}\begin{pmatrix} \psi \\ \frac{\widehat{\pi}_S}{p_0}\psi \end{pmatrix}$$

$$= \frac{1}{2}\begin{pmatrix} \psi + \frac{\widehat{\pi}_S}{p_0}\psi \\ \psi - \frac{\widehat{\pi}_S}{p_0}\psi \end{pmatrix}. \tag{5.271}$$

Now,

$$\frac{\widehat{\pi}_S}{p_0}\begin{pmatrix} \psi_+ \\ \psi_- \end{pmatrix} = \frac{1}{p_0}\left(\frac{-i\hbar}{\zeta}\frac{\partial}{\partial S} - qA_S\right)\begin{pmatrix} \psi_+ \\ \psi_- \end{pmatrix}$$

$$= M\begin{pmatrix} 0 & 1 \\ \frac{1}{p_0^2}\left(p_0^2 - \widehat{P}_\perp^2 + \widetilde{p}^2\right) & 0 \end{pmatrix}M^{-1}\begin{pmatrix} \psi_+ \\ \psi_- \end{pmatrix}$$

$$= \begin{pmatrix} 1 - \frac{1}{2p_0^2}\left(\widehat{P}_\perp^2 - \widetilde{p}^2\right) & -\frac{1}{2p_0^2}\left(\widehat{P}_\perp^2 - \widetilde{p}^2\right) \\ \frac{1}{2p_0^2}\left(\widehat{P}_\perp^2 - \widetilde{p}^2\right) & -1 + \frac{1}{2p_0^2}\left(\widehat{P}_\perp^2 - \widetilde{p}^2\right) \end{pmatrix}\begin{pmatrix} \psi_+ \\ \psi_- \end{pmatrix}. \tag{5.272}$$

Rearranging this equation, we get

$$i\hbar\frac{\partial}{\partial S}\begin{pmatrix} \psi_+ \\ \psi_- \end{pmatrix} = \widehat{H}\begin{pmatrix} \psi_+ \\ \psi_- \end{pmatrix},$$

$$\widehat{H} = -p_0\sigma_z + \widehat{\mathcal{E}} + \widehat{\mathcal{O}},$$

$$\widehat{\mathcal{E}} = -q\zeta A_S I + \left(\frac{\zeta}{2p_0}\left(\widehat{P}_\perp^2 - \widetilde{p}^2\right) - p_0\kappa X\right)\sigma_z,$$

$$\widehat{\mathcal{O}} = \frac{i\zeta}{2p_0}\left(\widehat{P}_\perp^2 - \widetilde{p}^2\right)\sigma_y. \tag{5.273}$$

Let us now apply a Foldy–Wouthuysen-like transformation

$$\begin{pmatrix} \psi_+^{(1)} \\ \psi_-^{(1)} \end{pmatrix} = e^{i\widehat{S}_1}\begin{pmatrix} \psi_+ \\ \psi_- \end{pmatrix}, \quad \text{with } \widehat{S}_1 = \frac{i}{2p_0}\sigma_z\widehat{\mathcal{O}}. \tag{5.274}$$

The result is

$$i\hbar\frac{\partial}{\partial S}\begin{pmatrix} \psi_+^{(1)} \\ \psi_-^{(1)} \end{pmatrix} = \left[e^{i\widehat{S}_1}\widehat{H}e^{-i\widehat{S}_1} - i\hbar e^{i\widehat{S}_1}\frac{\partial}{\partial S}\left(e^{-i\widehat{S}_1}\right)\right]\begin{pmatrix} \psi_+^{(1)} \\ \psi_-^{(1)} \end{pmatrix}$$

$$= e^{i\widehat{S}_1}\widehat{H}e^{-i\widehat{S}_1}\begin{pmatrix} \psi_+^{(1)} \\ \psi_-^{(1)} \end{pmatrix} = H^{(1)}\begin{pmatrix} \psi_+^{(1)} \\ \psi_-^{(1)} \end{pmatrix}, \tag{5.275}$$

since $\widehat{S}_1$ is independent of the coordinate S. Calculating $H^{(1)}$ we have, up to the paraxial approximation required for our purpose,

$$H^{(1)} \approx -q\zeta A_S I + \left[\frac{1}{2p_0}\widehat{P}_\perp^2 - \zeta\left(p_0 + \widetilde{p}^2\right)\right]\sigma_z. \tag{5.276}$$

As we have done previously for deriving the general quantum beam optical Hamiltonian (5.34) for any system with a straight optic axis, we can retrace the above transformation and identify the upper component of the resulting two-component wave function with the wave function of the particle moving forward along the curved path. Then, we get the required S-evolution equation for the wave function as

$$i\hbar \frac{\partial}{\partial S} \psi\left(\vec{R}_\perp, S\right) = \left[-\left(p_0 + \frac{\hbar^2 \kappa^2}{4p_0}\right) + \frac{1}{2p_0}\widehat{P}_\perp^2 + \frac{1}{2}qB_0\kappa X^2 \right.$$
$$\left. + \left(qB_0 - \kappa p_0 - \frac{\hbar^2 \kappa^3}{4p_0}\right)X\right]\psi\left(\vec{R}_\perp, S\right), \quad (5.277)$$

where we have taken $\kappa X \ll 1$, reasonably, and hence $1/\zeta \approx (1 - \kappa X)$. With the curvature of the design trajectory matched to the dipole magnetic field as

$$B_0 \rho = \frac{p_0}{q}, \quad \text{or,} \quad (qB_0 - \kappa p_0) = 0, \quad (5.278)$$

we have

$$i\hbar \frac{\partial}{\partial S}\psi\left(\vec{R}_\perp, S\right) = \widetilde{\mathcal{H}}_{o,d}\psi\left(\vec{R}_\perp, S\right),$$
$$\widetilde{\mathcal{H}}_{o,d} = \left(-p_0 + \frac{1}{2p_0}\widehat{P}_\perp^2 + \frac{1}{2}p_0\kappa^2 X^2 - \frac{\hbar^2\kappa^2}{4p_0}(1 + \kappa X)\right)\psi\left(\vec{R}_\perp, S\right). \quad (5.279)$$

Note that $\widetilde{\mathcal{H}}_{o,d}$ reproduces the classical beam optical Hamiltonian of the dipole magnet $\mathcal{H}_{o,d}$ in (4.185) when $\vec{P}_\perp$ and $\vec{R}_\perp$ are taken as the corresponding classical variables, and the $\hbar$-dependent term is dropped.

The quantum beam optical Hamiltonian for the dipole, $\widetilde{\mathcal{H}}_{o,d}$, does not depend on S. Thus, the quantum transfer map for the dipole becomes, with $\Delta S = S - S_i$,

$$\begin{pmatrix} \langle X \rangle(S) \\ \frac{1}{p_0}\langle \widehat{P}_X \rangle(S) \\ \langle Y \rangle(S) \\ \frac{1}{p_0}\langle \widehat{P}_Y \rangle(S) \end{pmatrix} = \begin{pmatrix} \left\langle e^{\frac{i}{\hbar}\Delta S : \widetilde{\mathcal{H}}_{o,d}:} X \right\rangle(S_i) \\ \frac{1}{p_0}\left\langle e^{\frac{i}{\hbar}\Delta S : \widetilde{\mathcal{H}}_{o,d}:} \widehat{P}_X \right\rangle(S_i) \\ \left\langle e^{\frac{i}{\hbar}\Delta S : \widetilde{\mathcal{H}}_{o,d}:} Y \right\rangle(S_i) \\ \frac{1}{p_0}\left\langle e^{\frac{i}{\hbar}\Delta S : \widetilde{\mathcal{H}}_{o,d}:} \widehat{P}_Y \right\rangle(S_i) \end{pmatrix}. \quad (5.280)$$

With

$$\frac{i}{\hbar} : \widetilde{\mathcal{H}}_{o,d} : X = \frac{i}{\hbar}\left[\frac{1}{2p_0}\widehat{P}_\perp^2 + \frac{1}{2}p_0\kappa^2 X^2 - \frac{\hbar^2\kappa^3}{4p_0}X, X\right] = \frac{\widehat{P}_X}{p_0},$$

$$\frac{i}{\hbar} : \widetilde{\mathcal{H}}_{o,d} : \frac{\widehat{P}_X}{p_0} = \frac{i}{\hbar}\left[\frac{1}{2p_0}\widehat{P}_\perp^2 + \frac{1}{2}p_0\kappa^2 X^2 - \frac{\hbar^2\kappa^3}{4p_0}X, \frac{\widehat{P}_X}{p_0}\right] = -\kappa^2 X + \frac{\hbar^2\kappa^3}{4p_0^2},$$

Scalar Theory: Spin-0 and Spinless Particles

$$\frac{i}{\hbar}:\widetilde{\mathscr{H}}_{o,d}:Y = \frac{i}{\hbar}\left[\frac{1}{2p_0}\widehat{P}_\perp^2 + \frac{1}{2}p_0\kappa^2 X^2 - \frac{\hbar^2\kappa^3}{4p_0}X, Y\right] = \frac{\widehat{P}_Y}{p_0},$$

$$\frac{i}{\hbar}:\widetilde{\mathscr{H}}_{o,d}:\frac{\widehat{P}_Y}{p_0} = \frac{i}{\hbar}\left[\frac{1}{2p_0}\widehat{P}_\perp^2 + \frac{1}{2}p_0\kappa^2 X^2 - \frac{\hbar^2\kappa^3}{4p_0}X, \frac{\widehat{P}_Y}{p_0}\right] = 0, \quad (5.281)$$

we get

$$\begin{pmatrix} \langle X\rangle(S) \\ \frac{1}{p_0}\langle \widehat{P}_X\rangle(S) \\ \langle Y\rangle(S) \\ \frac{1}{p_0}\langle \widehat{P}_Y\rangle(S) \end{pmatrix} = \left(\begin{pmatrix} \cos(\kappa\Delta S) & \frac{1}{\kappa}\sin(\kappa\Delta S) \\ -\kappa\sin(\kappa\Delta S) & \cos(\kappa\Delta S) \end{pmatrix}\right.$$

$$\left.\oplus \begin{pmatrix} 1 & \Delta S \\ 0 & 1 \end{pmatrix}\right)\begin{pmatrix} \langle X\rangle(S_i) \\ \frac{1}{p_0}\langle \widehat{P}_X\rangle(S_i) \\ \langle Y\rangle(S_i) \\ \frac{1}{p_0}\langle \widehat{P}_Y\rangle(S_i) \end{pmatrix}$$

$$+ \begin{pmatrix} -\frac{\hbar^2\kappa}{4p_0^2}(\cos(\kappa\Delta S)-1) \\ -\frac{\hbar^2\kappa^2}{4p_0^2}\sin(\kappa\Delta S) \\ 0 \\ 0 \end{pmatrix}. \quad (5.282)$$

This shows that a particle entering the dipole magnet along the design trajectory, i.e., with ($\langle X\rangle(S_i) = 0$, $\langle \widehat{P}_X\rangle(S_i) = 0$, $\langle Y\rangle(S_i) = 0$, $\langle \widehat{P}_Y\rangle(S_i) = 0$), will follow the curved design trajectory, except for some tiny quantum kicks in the X coordinate ($\sim \lambda_0^2/\rho$) and the X-gradient ($\sim \lambda_0^2/\rho^2$), where λ_0 is the de Broglie wavelength! This again proves the remarkable effectiveness of classical mechanics in the design and operation of accelerator optics.

5.3 EFFECT OF QUANTUM UNCERTAINTIES ON ABERRATIONS IN ELECTRON MICROSCOPY AND NONLINEARITIES IN ACCELERATOR OPTICS

Now we have to emphasize an important aspect of phase space transfer maps, as revealed by the quantum theory in contrast to the classical theory. We have identified the quantum averages $\langle \vec{r}_\perp\rangle(z)$ and $\langle \vec{p}_\perp\rangle(z)/p_0$ as the classical ray coordinates corresponding to the position and the slope of the ray intersecting the xy-plane at z. As we have already noted, for the round magnetic lens, central to electron microscopy, the expressions we have derived from quantum theory for the various aberration coefficients are the same as their respective classical expressions, of course, under the approximations considered. However, the quantum expressions in (5.201–5.204) would correspond exactly to the classical expressions for aberrations of position and gradient, provided we can replace $\langle \hat{p}_x \hat{p}_\perp^2 \rangle$, $\left\langle \left\{ \hat{p}_x, \left(\vec{\hat{p}}_\perp \cdot \vec{r}_\perp + \vec{r}_\perp \cdot \vec{\hat{p}}_\perp \right) \right\} \right\rangle$, $\langle \{x, \hat{p}_\perp^2\} \rangle$, etc., respectively, by $\langle \hat{p}_x \rangle \left(\langle \hat{p}_x \rangle^2 + \langle \hat{p}_y \rangle^2 \right)$, $4\left(\langle x \rangle \langle \hat{p}_x \rangle^2 + \langle y \rangle \langle \hat{p}_x \rangle \langle \hat{p}_y \rangle \right)$, $2\langle x \rangle \left(\langle \hat{p}_x \rangle^2 + \langle \hat{p}_y \rangle^2 \right)$, etc. But, that cannot be done. In quantum mechanics, in general, for any observable O, $\langle \psi | f(\hat{O}) | \psi \rangle \neq f\left(\langle \psi | \hat{O} | \psi \rangle \right)$ unless $|\psi\rangle$ is an eigenstate of O. And, for any two observables O_1 and O_2, $\langle \psi | f(\hat{O}_1, \hat{O}_2) | \psi \rangle \neq f\left(\langle \psi | \hat{O}_1 | \psi \rangle, \langle \psi | \hat{O}_2 | \psi \rangle \right)$ unless $|\psi\rangle$ is a simultaneous eigenstate of both $\hat{O}_1$ and $\hat{O}_2$. It is thus clear that for the wave packets involved in electron optical imaging the replacement as mentioned above is not allowed. As an illustration, consider the term $\sim \langle \{\vec{r}_\perp, \hat{p}_\perp^2\} \rangle (z_{obj})$, one of the terms contributing to coma (see (5.201) and (5.202)) which, being linear in position, is the dominant aberration next to the spherical aberration. The corresponding classical term, $((dx/dz)^2 + (dy/dz)^2)\vec{r}_\perp$ at z_{obj}, vanishes obviously for an object point on the axis. But, for a quantum wave packet with $\langle \vec{r}_\perp \rangle (z_{obj}) = (0,0)$ the value of $\langle \{\vec{r}_\perp, \hat{p}_\perp^2\} \rangle (z_{obj})$ need not be zero since it is not linear in $\langle \vec{r}_\perp \rangle (z_{obj})$. This can be seen more explicitly as follows. Let

$$\widehat{\delta x} = x - \langle x \rangle, \qquad \widehat{\delta y} = y - \langle y \rangle,$$
$$\widehat{\delta p_x} = \hat{p}_x - \langle \hat{p}_x \rangle, \qquad \widehat{\delta p_y} = \hat{p}_y - \langle \hat{p}_y \rangle. \qquad (5.283)$$

Then,

$$\langle \{\vec{r}_\perp, \hat{p}_\perp^2\} \rangle (z_{obj}) = \left\langle \left\{ \langle \vec{r}_\perp \rangle + \widehat{\delta \vec{r}}_\perp, \left(\langle \hat{p}_x \rangle + \widehat{\delta p_x} \right)^2 \right. \right.$$
$$\left. \left. + \left(\langle \hat{p}_y \rangle + \widehat{\delta p_y} \right)^2 \right\} \right\rangle (z_{obj})$$
$$= \left\langle \left\{ \langle \vec{r}_\perp \rangle + \widehat{\delta \vec{r}}_\perp, \langle \hat{p}_x \rangle^2 + \left(\widehat{\delta p_x}\right)^2 + \langle \hat{p}_y \rangle^2 + \left(\widehat{\delta p_y}\right)^2 \right. \right.$$
$$\left. \left. + 2\left(\langle \hat{p}_x \rangle \widehat{\delta p_x} + \langle \hat{p}_y \rangle \widehat{\delta p_y} \right) \right\} \right\rangle (z_{obj})$$
$$= 2\langle \vec{r}_\perp \rangle (z_{obj}) \langle \vec{\hat{p}}_\perp \rangle (z_{obj})^2$$
$$+ 2\langle \vec{r}_\perp \rangle (z_{obj}) \left\langle \left(\widehat{\delta p_x}\right)^2 + \left(\widehat{\delta p_y}\right)^2 \right\rangle (z_{obj})$$
$$+ \left\langle \left\{ \widehat{\delta \vec{r}}_\perp, \left(\widehat{\delta p_x}\right)^2 + \left(\widehat{\delta p_y}\right)^2 \right\} \right\rangle (z_{obj})$$
$$+ 2\left\langle \left\{ \widehat{\delta \vec{r}}_\perp, \widehat{\delta p_x} \right\} \right\rangle (z_{obj}) \langle \hat{p}_x \rangle (z_{obj})$$
$$+ 2\left\langle \left\{ \widehat{\delta \vec{r}}_\perp, \widehat{\delta p_y} \right\} \right\rangle (z_{obj}) \langle \hat{p}_y \rangle (z_{obj}), \qquad (5.284)$$

showing clearly that this coma term is not necessarily zero for an object point on the axis, *i.e.*, when $\langle \vec{r}_\perp \rangle = (0,0)$. The above equation also shows how this coma term for

Scalar Theory: Spin-0 and Spinless Particles

off-axis points ($\langle \vec{r}_\perp \rangle \neq (0,0)$) depends also on the higher order central moments of the wave packet besides the position ($\langle \vec{r}_\perp \rangle (z_{obj})$) and the slope ($\langle \vec{\hat{p}}_\perp \rangle (z_{obj}) / p_0$) of the corresponding classical ray. When an aperture is introduced in the path of the beam to limit the transverse momentum spread, one will be introducing uncertainties in position coordinates

$$\Delta x = \sqrt{\left\langle \widehat{\delta x}^2 \right\rangle}, \qquad \Delta y = \sqrt{\left\langle \widehat{\delta y}^2 \right\rangle}, \qquad (5.285)$$

and hence the corresponding momentum uncertainties

$$\Delta p_x = \sqrt{\left\langle \widehat{\delta p_x}^2 \right\rangle}, \qquad \Delta p_y = \sqrt{\left\langle \widehat{\delta p_y}^2 \right\rangle}, \qquad (5.286)$$

in accordance with the Heisenberg uncertainty principle, and this would influence the aberrations. However, in practice, such tiny quantum effects might be masked by classical uncertainties caused by instrumental imperfections.

In the context of accelerator optics, let us consider a sextupole magnet normally used for the control of chromaticity arising due to the beam being not monoenergetic. Sextupoles are used in electron microscopy for the correction of spherical aberration. A sextupole magnet, with its straight optic axis along the z-direction, is associated with the magnetic field

$$\vec{B} = \left(Q_{sx} xy, \frac{1}{2} Q_{sx} \left(x^2 - y^2\right), 0 \right), \qquad (5.287)$$

corresponding to the vector potential

$$\vec{A} = \left(0, 0, -\frac{1}{6} Q_{sx} \left(x^3 - 3xy^2\right) \right), \qquad (5.288)$$

where Q_{sx} is a constant in the sextupole region $z_\ell \leq z \leq z_r$ and zero outside. For a quasiparaxial beam propagating in the $+z$-direction through the sextupole, the quantum beam optical Hamiltonian is seen to be, from (5.34),

$$\widehat{\mathcal{H}_o} = \begin{cases} -p_0 + \frac{\hat{p}_\perp^2}{2p_0} + \frac{1}{6} p_0 K_{sx} \left(x^3 - 3xy^2\right), & z_\ell \leq z \leq z_r, \\ -p_0 + \frac{\hat{p}_\perp^2}{2p_0}, & z < z_\ell, z > z_r, \end{cases} \qquad (5.289)$$

with $K_{sx} = qQ_{sx}/p_0$. Note that in deriving the above $\widehat{\mathcal{H}_o}$, we have to keep terms of order up to third power in (x,y), even for a paraxial beam, since the paraxial approximation is meaningless in this case; in the paraxial approximation, the sextupole will disappear. Let us assume the sextupole to be thin such that $w = z_r - z_\ell \approx 0$. Then, for the propagation of the beam through the sextupole, from z_ℓ to z_r, the phase space transfer map is seen to be

$$\begin{pmatrix} \langle x \rangle (z_r) \\ \frac{1}{p_0} \langle \widehat{p}_x \rangle (z_r) \\ \langle y \rangle (z_r) \\ \frac{1}{p_0} \langle \widehat{p}_y \rangle (z_r) \end{pmatrix} = \begin{pmatrix} \left\langle e^{\frac{i}{\hbar} w: \left(-p_0 + \frac{\widehat{p}_\perp^2}{2p_0} + \frac{1}{6} p_0 K_{sx}(x^3 - 3xy^2)\right):} x \right\rangle (z_\ell) \\ \frac{1}{p_0} \left\langle e^{\frac{i}{\hbar} w: \left(-p_0 + \frac{\widehat{p}_\perp^2}{2p_0} + \frac{1}{6} p_0 K_{sx}(x^3 - 3xy^2)\right):} \widehat{p}_x \right\rangle (z_\ell) \\ \left\langle e^{\frac{i}{\hbar} w: \left(-p_0 + \frac{\widehat{p}_\perp^2}{2p_0} + \frac{1}{6} p_0 K_{sx}(x^3 - 3xy^2)\right):} y \right\rangle (z_\ell) \\ \frac{1}{p_0} \left\langle e^{\frac{i}{\hbar} w: \left(-p_0 + \frac{\widehat{p}_\perp^2}{2p_0} + \frac{1}{6} p_0 K_{sx}(x^3 - 3xy^2)\right):} \widehat{p}_y \right\rangle (z_\ell) \end{pmatrix},$$

$$\approx \begin{pmatrix} \langle x \rangle (z_\ell) + w \frac{\langle p_x \rangle (z_\ell)}{p_0} \\ \frac{1}{p_0} \langle \widehat{p}_x \rangle (z_\ell) - \frac{1}{2} w K_{sx} \left(\langle x^2 \rangle (z_\ell) - \langle y^2 \rangle (z_\ell)\right) \\ \langle y \rangle (z_\ell) + w \frac{\langle p_y \rangle (z_\ell)}{p_0} \\ \frac{1}{p_0} \langle \widehat{p}_y \rangle (z_\ell) + w K_{sx} \langle xy \rangle (z_\ell) \end{pmatrix}.$$

(5.290)

Let us look at the nonlinear maps for the momentum components

$$\langle \widehat{p}_x \rangle (z_r) \approx \langle \widehat{p}_x \rangle (z_\ell) - \frac{1}{2} w p_0 K_{sx} \left(\langle x^2 \rangle (z_\ell) - \langle y^2 \rangle (z_\ell)\right),$$
$$\langle \widehat{p}_y \rangle (z_r) \approx \langle \widehat{p}_y \rangle (z_\ell) + w p_0 K_{sx} \langle xy \rangle (z_\ell). \quad (5.291)$$

It is clear that in the classical theory, one would have the corresponding maps as

$$p_x(z_r) \approx p_x(z_\ell) - \frac{1}{2} w p_0 K_{sx} \left(x(z_\ell)^2 - y(z_\ell)^2\right),$$
$$p_y(z_r) \approx p_y(z_\ell) + w p_0 K_{sx} x(z_\ell) y(z_\ell). \quad (5.292)$$

Substituting $x = \langle x \rangle + \widehat{\delta x}$ and $y = \langle x \rangle + \widehat{\delta y}$, the quantum maps (5.291) become

$$\langle \widehat{p}_x \rangle (z_r) \approx \langle \widehat{p}_x \rangle (z_\ell) - \frac{1}{2} w p_0 K_{sx} \left[\left(\langle x \rangle^2 (z_\ell) + \left\langle \widehat{\delta x}^2 \right\rangle (z_\ell)\right) \right.$$
$$\left. - \left(\langle y \rangle^2 (z_\ell) + \left\langle \widehat{\delta y}^2 \right\rangle (z_\ell)\right) \right],$$

$$\langle \widehat{p}_y \rangle (z_r) \approx \langle \widehat{p}_y \rangle (z_\ell) + w p_0 K_{sx} \left[\langle x \rangle (z_\ell) \langle y \rangle (z_\ell) \right.$$
$$\left. + \left\langle \left(\widehat{\delta x}\right) \left(\widehat{\delta y}\right) \right\rangle (z_\ell) \right]. \quad (5.293)$$

This shows that, generally, the leading quantum effects on the nonlinear accelerator optics can be expected to be due to the uncertainties in the position coordinates and the momentum components of the particles of the beam entering the optical elements. Such quantum effects involve $\hbar$ only through the uncertainties and not explicitly. This has already been pointed out by Heifets and Yan [73], who have shown that the quantum effects due to such uncertainties affect substantially the classical results of tracking for trajectories close to the separatrix, and hence the quantum maps can be useful in quick findings of the nonlinear resonances. Accelerator beams are complicated nonlinear, stochastic, many-particle systems. For details on the nonlinear optics of accelerator beams, see, *e.g.*, the references on accelerator physics mentioned earlier (Berz, Makino, and Wan [12], Conte and MacKay [27], Lee [127], Reiser [158], Seryi [168], Weidemann [187], Wolski [192], Chao, Mess, Tigner, and Zimmermann [20]), and also Mais [131], Todesco [184], and references therein.

5.4 NONRELATIVISTIC QUANTUM CHARGED PARTICLE BEAM OPTICS: SPIN-0 AND SPINLESS PARTICLES

The nonrelativistic Schrödinger equation for a particle of charge q and mass m moving in the time-independent electromagnetic field of an optical system is

$$i\hbar \frac{\partial \Psi(\vec{r},t)}{\partial t} = \left[\frac{1}{2m} \left(\vec{\hat{p}} - q\vec{A}(\vec{r}) \right)^2 + q\phi(\vec{r}) \right] \Psi(\vec{r},t), \quad (5.294)$$

where $\vec{A}(\vec{r})$ and $\phi(\vec{r})$ are the magnetic vector potential and the electric scalar potential. Let us consider a particle of a nonrelativistic monoenergetic quasiparaxial beam with total energy $E = p_0^2/2m$, where p_0 is the design momentum with which it enters the optical system from the free space outside and $|\vec{p}_{0\perp}| \ll p_0$. We assume that the total energy E is conserved when the beam propagates through any optical system we are considering. Choosing the wave function of the particle as

$$\Psi(\vec{r},t) = e^{-iEt/\hbar} \psi(\vec{r}), \quad (5.295)$$

we get the time-independent Schrödinger equation obeyed by $\psi(\vec{r}_\perp, z)$:

$$\left(\frac{\hat{\pi}^2}{2m} + q\phi \right) \psi(\vec{r}_\perp, z) = E \psi(\vec{r}_\perp, z), \quad (5.296)$$

where $\vec{\hat{\pi}} = \vec{\hat{p}} - q\vec{A}$. Multiplying the equation on both sides by $2m$, taking

$$2mE = p_0^2, \qquad 2mq\phi = \tilde{p}^2, \quad (5.297)$$

and rearranging the terms, we have

$$\hat{\pi}_z^2 \psi = \left(p_0^2 - \tilde{p}^2 - \hat{\pi}_\perp^2 \right) \psi. \quad (5.298)$$

Note that this equation is identical, except for the expressions of the symbols p_0^2 and $\tilde{p}^2$, to the time-independent Klein–Gordon equation (5.7) which is the starting point for our derivation of the relativistic quantum beam optical Hamiltonian (5.34). Now, starting with (5.298), we can follow the same procedure, going through the Feshbach–Villars-like formalism and the Foldy–Wouthuysen-like transformations, and finally arrive at the nonrelativistic quantum beam optical Schrödinger equation,

$$i\hbar\frac{\partial \psi(\vec{r}_\perp,z)}{\partial z} = \widehat{\mathcal{H}}_{o,\text{NR}} \psi(\vec{r}_\perp,z),$$

$$\widehat{\mathcal{H}}_{o,\text{NR}} = -p_0 - qA_z + \frac{1}{2p_0}\left(\widehat{\pi}_\perp^2 + \tilde{p}^2\right) + \frac{1}{8p_0^3}\left(\widehat{\pi}_\perp^2 + \tilde{p}^2\right)^2$$

$$- \frac{1}{16p_0^4}\Bigg\{\left[\left(\widehat{\pi}_\perp^2 + \tilde{p}^2\right), \left[\widehat{\pi}_\perp^2, qA_z\right]\right.$$

$$+ i\hbar q\left(\vec{\hat{p}}_\perp \cdot \frac{\partial \vec{A}_\perp}{\partial z} + \frac{\partial \vec{A}_\perp}{\partial z}\cdot\vec{\hat{p}}_\perp\right)\Bigg]$$

$$- \left[\widehat{\pi}_\perp^2, i\hbar\frac{\partial}{\partial z}\left(q^2 A_\perp^2 + \tilde{p}^2\right)\right]\Bigg\},$$

$$\text{with } p_0^2 = 2m\text{E}, \quad \tilde{p}^2 = 2mq\phi, \qquad (5.299)$$

which is identical to the relativistic quantum beam optical equation (5.34), except for the expressions of p_0^2 and $\tilde{p}^2$. This shows that one can approximate the expressions and formulae derived from the relativistic quantum beam optical equation (5.34) using the nonrelativistic approximation

$$p_0^2 = \frac{1}{c^2}\left(E^2 - m^2c^4\right) = \frac{1}{c^2}\left(E + mc^2\right)\left(E - mc^2\right) \approx \frac{1}{c^2}2mc^2\text{E} = 2m\text{E},$$

$$\tilde{p}^2 = \frac{2qE\phi}{c^2}\left(1 - \frac{q\phi}{2E}\right) \approx \frac{2qE\phi}{c^2} \approx \frac{2qmc^2\phi}{c^2} = 2mq\phi, \qquad (5.300)$$

as is appropriate for the charged particle beam optical systems for which $E = mc^2 + \text{E}$, $\text{E} \ll mc^2$ and $q\phi \ll mc^2$. Conversely, since the nonrelativistic and the relativistic quantum beam optical equations of motion, (5.299) and (5.34) are identical, except for the expressions of the symbols E, p_0^2, and $\tilde{p}^2$, it is possible to convert the expressions and formulae derived from the nonrelativistic equation to relativistic expressions and formulae by the replacement

$$p_0 = \sqrt{2m\text{E}} = mv_0$$

$$\longrightarrow p_0 = \frac{1}{c}\sqrt{E^2 - m^2c^4}$$

$$= \frac{1}{c}\sqrt{m^2c^4\left(\frac{1}{1-\frac{v_0^2}{c^2}} - 1\right)} = \frac{mv_0}{\sqrt{1-\frac{v_0^2}{c^2}}}, \qquad (5.301)$$

or,

$$m \longrightarrow \frac{m}{\sqrt{1-\frac{v_0^2}{c^2}}}, \qquad (5.302)$$

i.e., replacement of the rest mass by the so-called relativistic mass, as has been the common practice in electron optics.

5.5 APPENDIX: PROPAGATOR FOR A SYSTEM WITH TIME-DEPENDENT QUADRATIC HAMILTONIAN

Let it be required to find the propagator for a one-dimensional system obeying the Schrödinger equation

$$i\hbar \frac{\partial |\Psi(t)\rangle}{\partial t} = \widehat{H}(t)|\Psi(t)\rangle, \tag{5.303}$$

where the time-dependent Hamiltonian has the form

$$\widehat{H}(t) = A(t)\widehat{p}_x^2 + B(t)(x\widehat{p}_x + \widehat{p}_x x) + C(t)x^2. \tag{5.304}$$

with $A(t)$, $B(t)$, and $C(t)$ being real functions of t.

Using the Magnus formula (3.768), we can write

$$|\Psi(t)\rangle = \widehat{U}(t,t')|\Psi(t')\rangle \tag{5.305}$$

with

$$\widehat{U}(t,t') = \exp\left\{ -\frac{i}{\hbar} \int_{t'}^{t} dt_1\, \widehat{H}(t_1) \right.$$
$$+ \frac{1}{2}\left(-\frac{i}{\hbar}\right)^2 \int_{t'}^{t} dt_2 \int_{t'}^{t_2} dt_1 \left[\widehat{H}(t_2), \widehat{H}(t_1)\right]$$
$$+ \frac{1}{6}\left(-\frac{i}{\hbar}\right)^3 \int_{t'}^{t} dt_3 \int_{t'}^{t_3} dt_2 \int_{t'}^{t_2} dt_1 \left(\left[\left[\widehat{H}(t_3),\widehat{H}(t_2)\right],\widehat{H}(t_1)\right]\right.$$
$$\left.\left. + \left[\left[\widehat{H}(t_1),\widehat{H}(t_2)\right],\widehat{H}(t_3)\right]\right) + \cdots \right\}. \tag{5.306}$$

The required propagator is given by

$$K(x,t;x',t') = \left\langle x \left| \widehat{U}(t,t') \right| x' \right\rangle, \tag{5.307}$$

such that

$$\Psi(x,t) = \int dx\, K(x,t;x',t')\, \Psi(x',t'). \tag{5.308}$$

Note that the operators $\left(\widehat{p}_x^2, (x\widehat{p}_x + \widehat{p}_x x), x^2\right)$ are closed under commutation leading to the Lie algebra

$$[\widehat{p}_x^2, (x\widehat{p}_x + \widehat{p}_x x)] = -4i\hbar \widehat{p}_x^2,$$
$$[\widehat{p}_x^2, x^2] = -2i\hbar (x\widehat{p}_x + \widehat{p}_x x),$$
$$[(x\widehat{p}_x + \widehat{p}_x x), x^2] = -4i\hbar x^2. \tag{5.309}$$

Then, it is clear that we can write

$$\widehat{U}(t,t') = \exp\left\{ -\frac{i}{\hbar}\left[a(t,t')\,\widehat{p}_x^2 + b(t,t')\,(x\widehat{p}_x + \widehat{p}_x x) + c(t,t')\,x^2\right]\right\}, \tag{5.310}$$

where $a(t,t')$, $b(t,t')$, and $c(t,t')$ are infinite series expressions in terms of $A(t)$, $B(t)$, and $C(t)$. To obtain the precise form of the equation (5.310) we proceed as follows.

Substituting $|\Psi(t)\rangle = \widehat{U}(t,t')|\Psi(t')\rangle$ in (5.303), it is seen that

$$i\hbar\frac{\partial \widehat{U}(t,t')}{\partial t} = \widehat{H}(t)\widehat{U}(t,t'), \qquad \widehat{U}(t',t') = I, \tag{5.311}$$

where I is the identity operator. Following Wolf [190], let us write

$$a = \frac{\varphi\beta}{2\sin\varphi}, \quad b = \frac{\varphi(\alpha-\delta)}{4\sin\varphi}, \quad c = -\frac{\varphi\gamma}{2\sin\varphi}, \quad \cos\varphi = \frac{1}{2}(\alpha+\delta). \tag{5.312}$$

Then, we have

$$\widehat{U}(t,t') = \exp\left\{-\frac{i}{\hbar}\left(\frac{\varphi(t,t')}{2\sin\varphi(t,t')}\right)\left[\beta(t,t')\,\widehat{p}_x^2\right.\right.$$
$$\left.\left. +\frac{1}{2}\left(\alpha(t,t')-\delta(t,t')\right)(x\widehat{p}_x+\widehat{p}_x x) - \gamma(t,t')\,x^2\right]\right\}.$$

(5.313)

Substituting this expression for $\widehat{U}(t,t')$ in (5.311), with $H(t)$ given by (5.304), shows that $\alpha(t,t')$, $\beta(t,t')$, $\gamma(t,t')$, and $\delta(t,t')$ satisfy the following equations:

$$A\ddot{\alpha} - \dot{A}\dot{\alpha} + \left[4A(AC - B^2) + 2\dot{A}B - 2A\dot{B}\right]\alpha = 0,$$
$$\alpha(t',t') = 1, \qquad \dot{\alpha}(t',t') = 2B(t'), \tag{5.314}$$
$$\alpha\ddot{\beta} - \beta\ddot{\alpha} - 2\dot{A} = 0, \quad \beta(t',t') = 0, \quad \dot{\beta}(t',t') = 2A(t'), \tag{5.315}$$
$$\dot{\alpha} - 2\alpha B = 2A\gamma, \tag{5.316}$$
$$\alpha\delta - \beta\gamma = 1, \tag{5.317}$$

where $\dot{\alpha} = d\alpha/dt$, $\ddot{\alpha} = d^2\alpha/dt^2$, etc. It is thus clear that, given $A(t)$, $B(t)$, and $C(t)$ in $\widehat{H}(t)$, the above equations provide $\alpha(t,t')$, $\beta(t,t')$, $\gamma(t,t')$, and $\delta(t,t')$, and $\varphi(t,t')$ is given by $2\cos\varphi = \alpha + \delta$. Hence, $\widehat{U}(t,t')$ is now completely determined through (5.313).

To find the required propagator, observe that

$$\begin{pmatrix} \widehat{U}^\dagger(t,t')x\widehat{U}(t,t') \\ \widehat{U}^\dagger(t,t')\widehat{p}_x\widehat{U}(t,t') \end{pmatrix} = \begin{pmatrix} \alpha(t,t') & \beta(t,t') \\ \gamma(t,t') & \delta(t,t') \end{pmatrix} \begin{pmatrix} x \\ \widehat{p}_x \end{pmatrix}. \tag{5.318}$$

Note that $\widehat{U}(t,t')$ generates a real linear canonical transformation of the conjugate pair $(x, \widehat{p}_x)$, i.e., if

$$\widehat{X} = \alpha x + \beta \widehat{p}_x, \quad \widehat{P} = \gamma x + \delta \widehat{p}_x, \tag{5.319}$$

then

$$\left[\widehat{X}, \widehat{P}\right] = (\alpha\delta - \beta\gamma)[x, \widehat{p}_x] = i\hbar. \tag{5.320}$$

Scalar Theory: Spin-0 and Spinless Particles

We can rewrite (5.318) as

$$x\widehat{U}(t,t')|\psi\rangle = \widehat{U}(t,t')\left(\alpha(t,t')x + \beta(t,t')\widehat{p}_x\right)|\psi\rangle$$
$$\widehat{p}_x\widehat{U}(t,t')|\psi\rangle = \widehat{U}(t,t')\left(\gamma(t,t')x + \delta(t,t')\widehat{p}_x\right)|\psi\rangle, \quad (5.321)$$

for any $|\psi\rangle$. Writing out this equation explicitly in terms of the matrix elements, it is possible to solve for $\langle x|\widehat{U}(t,t')|x'\rangle = K(x,t;x',t')$, up to a multiplicative constant phase factor (see Wolf [189, 190] for details of the solution). The result is

$$K(x,t;x',t') = \frac{1}{\sqrt{2\pi i\hbar\beta(t,t')}} e^{\frac{i}{2\hbar\beta(t,t')}\left[\alpha(t,t')x'^2 - 2xx' + \delta(t,t')x^2\right]},$$

$$\text{if } \beta(t,t') \neq 0, \quad (5.322)$$

and

$$K(x,t;x',t') = \frac{e^{\frac{i\gamma(t,t')}{2\hbar\alpha(t,t')}}}{\sqrt{\alpha(t,t')}} \delta\left(x' - \frac{x}{\alpha(t,t')}\right),$$

$$\text{if } \beta(t,t') = 0. \quad (5.323)$$

In the two-dimensional case, if the Hamiltonian is of the form

$$\widehat{H}(t) = A(t)\widehat{p}_\perp^2 + B(t)\left(\vec{r}_\perp \cdot \vec{\widehat{p}}_\perp + \vec{\widehat{p}}_\perp \cdot \vec{r}_\perp\right) + c(t)r_\perp^2, \quad (5.324)$$

the variables x and y are independent and separable, and the above results can be extended in a straightforward manner with the replacements $x^2 \longrightarrow r_\perp^2$, $\widehat{p}_x^2 \longrightarrow \widehat{p}_\perp^2$, and $(x\widehat{p}_x + \widehat{p}_x x) \longrightarrow \left(\vec{r}_\perp \cdot \vec{\widehat{p}}_\perp + \vec{\widehat{p}}_\perp \cdot \vec{r}_\perp\right)$. Then, we can write $K(\vec{r}_\perp,t;\vec{r}'_\perp,t') = K(x,t;x',t')K(y,t;y',t')$.

In the case of the round magnetic lens taking (t,t') as (z,z_i), one has in (5.97)

$$A(z) = \frac{1}{2p_0}, \quad B(z) = 0, \quad C(z) = \frac{1}{2}p_0\alpha^2(z). \quad (5.325)$$

Then, writing

$$\alpha(z,z_i) = g(z,z_i), \quad \beta(z,z_i) = \frac{1}{p_0}h(z,z_i), \quad (5.326)$$

we have from (5.314)–(5.317)

$$g''(z,z_i) + \alpha^2(z)g(z,z_i) = 0, \quad g(z_i,z_i) = 1, \quad g'(z_i,z_i) = 0,$$
$$h''(z,z_i) + \alpha^2(z)h(z,z_i) = 0, \quad h(z_i,z_i) = 0, \quad h'(z_i,z_i) = 1,$$
$$\gamma(z,z_i) = p_0 g'(z,z_i),$$
$$\delta(z,z_i) = \frac{1}{g(z,z_i)}\left(1 + h(z,z_i)g'(z,z_i)\right),$$
$$\alpha(z,z_i)\delta(z,z_i) - \beta(z,z_i)\gamma(z,z_i) = 1, \quad (5.327)$$

where $g' = dg/dz$, $g'' = d^2g/dz^2$, etc. showing that $g(z,z_i)$ and $h(z,z_i)$ are two linearly independent solutions of the paraxial equation (4.51) satisfying the initial conditions (5.114) and the relation (5.115).

6 Quantum Charged Particle Beam Optics
Spinor Theory for Spin-$\frac{1}{2}$ Particles

6.1 RELATIVISTIC QUANTUM CHARGED PARTICLE BEAM OPTICS BASED ON THE DIRAC–PAULI EQUATION

6.1.1 GENERAL FORMALISM

For electrons, or any spin-$\frac{1}{2}$ particle, the proper equation to be the basis for the theory of beam optics is the Dirac equation. Sudarshan, Simon, and Mukunda have shown that there is a simple rule by which paraxial scalar wave optics based on the Helmholtz equation can be generalized consistently to paraxial Maxwell wave optics, *i.e.*, to Fourier optics for the Maxwell field (Sudarshan, Simon, and Mukunda [175], Mukunda, Simon, and Sudarshan [137, 138], Simon, Sudarshan, and Mukunda [170, 171]). Searching for a similar passage from the scalar electron optics based on the nonrelativistic Schrödinger equation to the spinor electron optics based on the Dirac equation, a formalism of quantum electron beam optics was initiated with application to round magnetic lens by Jagannathan, Simon, Sudarshan, and Mukunda [79]. A more comprehensive formalism of quantum charged particle beam optics, with applications to several other electromagnetic optical systems, was developed later by Jagannathan [80], Khan and Jagannathan [90], Jagannathan and Khan [82], Conte, Jagannathan, Khan and Pusterla [26], and Khan [91]. This chapter on the spinor theory of quantum charged particle beam optics follows mostly the work by Jagannathan, Simon, Sudarshan, and Mukunda [79], Jagannathan [80], Khan and Jagannathan [90], Jagannathan and Khan [82], Conte, Jagannathan, Khan, and Pusterla [26], and Khan [91] (see also Jagannathan [83, 84, 85], Khan [92, 93, 95, 99]).

There were some preliminary studies by Rubinowicz [161, 162, 163, 164], Durand [38], Phan-Van-Loc [144, 145, 146, 147, 148, 149] on the use of the Dirac equation in electron optics. In 1986, Ferwerda, Hoenders, and Slump [48, 49] concluded that the use of the scalar Klein–Gordon equation in electron microscopy could be vindicated because a scalar approximation of the Dirac spinor theory would be justifiable under the conditions obtaining in electron microscopy (see also Lubk [129]). However, with the technological developments in electron microscopy, like the Low Energy Electron Microscopy (LEEM) (see Bauer [7]) and Spin Polarized Low Energy Electron Microscopy (SPLEEM) (see Rougemaille and

Schmid [160]), the need for a proper quantum theory of electron beam optics based on the Dirac equation is imminent.

As we have already seen, the Dirac equation is the relativistic equation linear in the first derivative with respect to time, possessing proper probability and current densities obeying a continuity equation, and naturally incorporating an intrinsic angular momentum, or spin, $\hbar/2$ for the particle described by it. Being linear in all the first derivatives with respect to the space and time coordinates (x, y, z, t), it is, in a way, already in a quantum beam optical form. Let us recall. The Dirac equation for a free particle of mass m is

$$i\hbar \frac{\partial |\Psi(t)\rangle}{\partial t} = \widehat{H}_{\text{D,f}} |\Psi(t)\rangle,$$
$$\widehat{H}_{\text{D,f}} = mc^2 \beta + c\vec{\alpha} \cdot \vec{p},$$
$$\beta = \begin{pmatrix} \mathsf{I} & \mathsf{O} \\ \mathsf{O} & -\mathsf{I} \end{pmatrix}, \quad \vec{\alpha} = \begin{pmatrix} \mathsf{O} & \vec{\sigma} \\ \vec{\sigma} & \mathsf{O} \end{pmatrix}, \quad (6.1)$$

where the state vector $|\Psi(t)\rangle$ is the four-component Dirac spinor,

$$\mathsf{I} = \begin{pmatrix} 1 & 0 \\ 0 & 1 \end{pmatrix}, \quad \mathsf{O} = \begin{pmatrix} 0 & 0 \\ 0 & 0 \end{pmatrix}, \quad (6.2)$$

and

$$\sigma_x = \begin{pmatrix} 0 & 1 \\ 1 & 0 \end{pmatrix}, \quad \sigma_y = \begin{pmatrix} 0 & -i \\ i & 0 \end{pmatrix}, \quad \sigma_z = \begin{pmatrix} 1 & 0 \\ 0 & -1 \end{pmatrix}, \quad (6.3)$$

are the Pauli matrices. The matrices α_x, α_y, α_z, and β are known as the Dirac matrices. For a particle of charge q, mass m, and anomalous magnetic moment μ_a, in a time-independent electromagnetic field $\left(\vec{E}((\vec{r})), \vec{B}(\vec{r})\right)$, with the magnetic vector potential $\vec{A}(\vec{r})$ and the electric scalar potential $\phi(\vec{r})$, the Dirac–Pauli equation is

$$i\hbar \left(\frac{\partial}{\partial t} - q\phi) \right) |\Psi(t)\rangle = \widehat{H}_{\text{DP}} |\Psi(t)\rangle,$$
$$\widehat{H}_{\text{DP}} = mc^2 \beta + c\vec{\alpha} \cdot \left(\vec{p} - q\vec{A}\right) - \mu_a \beta \vec{\Sigma} \cdot \vec{B}, \quad (6.4)$$

obtained from the free particle equation (6.1) by the principle of electromagnetic minimal coupling and adding the Pauli term to take into account the anomalous magnetic moment. In general, for any spin-$\frac{1}{2}$ particle, with charge q and mass m, we can write the total spin magnetic moment as $\vec{\mu} = (gq/2m)\vec{S} = (gq\hbar/4m)\vec{\sigma} = \mu \vec{\sigma}$. Writing $g = 2(1+a)$, we have $\mu = [(1+a)q\hbar/2m]$ with $q\hbar/2m$ as the Dirac magnetic moment corresponding to $g = 2$ and $\mu_a = aq\hbar/2m$ as the anomalous magnetic moment corresponding to the magnetic anomaly $a = (g/2) - 1$.

First, we shall get the general formalism to study the propagation of a quasiparaxial Dirac particle beam through electromagnetic optical systems with straight optic axis along the z-direction. To this end, we proceed as follows. Let us take

$$\Psi(\vec{r}, t) = e^{-iEt/\hbar} \psi(\vec{r}_\perp, z), \quad (6.5)$$

Spinor Theory for Spin-1/2 Particles

as the four-component wave function of a particle of the monoenergetic quasiparaxial beam moving forward along the optic axis in which $\underline{\psi}(\vec{r}_\perp, z)$ is the time-independent four-component wave function. With $\vec{p}_0 = p_0 \vec{k}$ as the design momentum, and $|\vec{p}_{0,\perp}| \ll p_0$, we shall take the energy of the particle to be positive given by $E = +\sqrt{m^2 c^4 + c^2 p_0^2}$. Substituting the expression for $\Psi(\vec{r}, t)$ in (6.4) we get

$$(E - q\phi)\underline{\psi}(\vec{r}_\perp, z) = \left[mc^2 \beta + c\vec{\alpha} \cdot \left(\vec{p} - q\vec{A} \right) - \mu_a \beta \vec{\Sigma} \cdot \vec{B} \right] \underline{\psi}(\vec{r}_\perp, z). \quad (6.6)$$

Multiplying this equation throughout from left by α_z/c and rearranging the terms, we can write

$$i\hbar \frac{\partial |\underline{\psi}(z)\rangle}{\partial z} = \widehat{\mathscr{H}} |\underline{\psi}(z)\rangle,$$

$$\widehat{\mathscr{H}} = -p_0 \beta \chi \alpha_z - qA_z \mathbb{I} + (q/c)\phi \alpha_z + \alpha_z \vec{\alpha}_\perp \cdot \vec{\pi}_\perp + \mu_a \beta \alpha_z \vec{\Sigma} \cdot \vec{B},$$

$$\text{with } \chi = \begin{pmatrix} \xi \mathbb{I} & 0 \\ 0 & -\xi^{-1} \mathbb{I} \end{pmatrix}, \quad \xi = \sqrt{\frac{E + mc^2}{E - mc^2}}, \quad (6.7)$$

where $\mathbb{I}$ is the 4×4 identity matrix. It is seen that $\widehat{\mathscr{H}}$ is not Hermitian. This is related to the fact that, as we have already noted in the case of the scalar theory, the probability of finding the particle in the vertical plane at the point z, $\int d^2 r_\perp \underline{\psi}^\dagger(\vec{r}_\perp, z) \underline{\psi}(\vec{r}_\perp, z)$, need not be conserved along the z-axis.

Noting that, with

$$\mathrm{M} = \frac{1}{\sqrt{2}} (\mathbb{I} + \chi \alpha_z), \quad \mathrm{M}^{-1} = \frac{1}{\sqrt{2}} (\mathbb{I} - \chi \alpha_z), \quad (6.8)$$

one has

$$\mathrm{M} \beta \chi \alpha_z \mathrm{M}^{-1} = \beta, \quad (6.9)$$

we define

$$\left| \underline{\psi}'(z) \right\rangle = \mathrm{M} \left| \underline{\psi}(z) \right\rangle. \quad (6.10)$$

Then,

$$i\hbar \frac{\partial}{\partial z} \left| \underline{\psi}'(z) \right\rangle = \widehat{\mathscr{H}}' \left| \underline{\psi}'(z) \right\rangle,$$

$$\widehat{\mathscr{H}}' = \mathrm{M} \widehat{\mathscr{H}} \mathrm{M}^{-1} = -p_0 \beta + \widehat{\mathscr{E}} + \widehat{\mathscr{O}},$$

$$\widehat{\mathscr{E}} = -qA_z \mathbb{I} + \frac{q\phi}{2c} \left(\xi + \xi^{-1} \right) \beta$$

$$\quad - \frac{\mu_a}{2c} \left[\left(\xi + \xi^{-1} \right) \vec{B}_\perp \cdot \vec{\Sigma}_\perp + \left(\xi - \xi^{-1} \right) \beta B_z \Sigma_z \right],$$

$$\widehat{\mathscr{O}} = \chi \left\{ \vec{\alpha}_\perp \cdot \vec{\pi}_\perp + \frac{\mu_a}{2c} \left[\left(\xi - \xi^{-1} \right) \left(\sigma_y \otimes (B_x \sigma_y - B_y \sigma_x) \right) \right. \right.$$

$$\quad \left. \left. + \left(\xi + \xi^{-1} \right) B_z (\sigma_x \otimes \mathbb{I}) \right] \right\}. \quad (6.11)$$

To see the effect of the transformation (6.10), let us apply it to a free time-independent paraxial positive-energy Dirac plane wave propagating in the $+z$-direction:

$$\begin{pmatrix} \psi_1'(\vec{r}_\perp, z) \\ \psi_2'(\vec{r}_\perp, z) \\ \psi_3'(\vec{r}_\perp, z) \\ \psi_4'(\vec{r}_\perp, z) \end{pmatrix} = \frac{1}{\sqrt{2}} \begin{pmatrix} 1 & 0 & \xi & 0 \\ 0 & 1 & 0 & -\xi \\ -\xi^{-1} & 0 & 1 & 0 \\ 0 & \xi^{-1} & 0 & 1 \end{pmatrix} \times$$

$$\frac{1}{4}\sqrt{\frac{cp_0\xi}{\pi^3\hbar^3 E}} \begin{pmatrix} a_+ \\ a_- \\ \frac{1}{p_0\xi}(a_+ p_z + a_- p_-) \\ \frac{1}{p_0\xi}(a_+ p_+ - a_- p_z) \end{pmatrix} e^{\frac{i}{\hbar}(\vec{p}_\perp \cdot \vec{r}_\perp + p_z z)},$$

with $|\vec{p}| = p_0$, $p_+ = p_x + ip_y$, $p_- = p_x - ip_y$,

$$|\vec{p}_\perp| \ll p_z \approx p_0, \quad |a_+|^2 + |a_-|^2 = 1, \qquad (6.12)$$

where we have used the positive-energy solutions of the free-particle Dirac equation (see (3.637), (3.640), and (3.641)). The result is

$$\begin{pmatrix} \psi_1'(\vec{r}_\perp, z) \\ \psi_2'(\vec{r}_\perp, z) \\ \psi_3'(\vec{r}_\perp, z) \\ \psi_4'(\vec{r}_\perp, z) \end{pmatrix} = \frac{1}{4}\sqrt{\frac{cp_0\xi}{2\pi^3\hbar^3 E}} \begin{pmatrix} \frac{1}{p_0}[a_+(p_0+p_z) + a_- p_-] \\ \frac{1}{p_0}[a_-(p_0+p_z) - a_+ p_+] \\ -\frac{1}{p_0\xi}[a_+(p_0-p_z) - a_- p_-] \\ \frac{1}{p_0\xi}[a_-(p_0-p_z) + a_+ p_+] \end{pmatrix} e^{\frac{i}{\hbar}(\vec{p}_\perp \cdot \vec{r}_\perp + p_z z)}.$$

(6.13)

It is seen that the upper pair of components of $|\psi'(z)\rangle$ are very large compared to the lower pair of components since $p_z \approx p_0$ for any particle of the paraxial beam. The wave packet representing any particle of a positive-energy monoenergetic paraxial beam will be of the form

$$\underline{\Psi}(\vec{r}, t) = e^{-iEt/\hbar} \underline{\psi}(\vec{r}_\perp, z),$$

$$\underline{\psi}(\vec{r}_\perp, z) = \int_{|\vec{p}|=p_0} d^2 p_\perp \, \underline{u}(\vec{p}) \, e^{\frac{i}{\hbar}(\vec{p}_\perp \cdot \vec{r}_\perp + p_z z)},$$

$$\underline{u}(\vec{p}) = \frac{1}{4}\sqrt{\frac{cp_0\xi}{\pi^3\hbar^3 E}} \begin{pmatrix} a_+(\vec{p}) \\ a_-(\vec{p}) \\ \frac{1}{p_0\xi}(a_+(\vec{p}) p_z + a_-(\vec{p}) p_-) \\ \frac{1}{p_0\xi}(a_+(\vec{p}) p_+ - a_-(\vec{p}) p_z) \end{pmatrix},$$

with $\int_{|\vec{p}|=p_0} d^2 p_\perp \left(|a_+(\vec{p})|^2 + |a_-(\vec{p})|^2 \right) = 1,$

$$|\vec{p}_\perp| \ll p_z \approx p_0, \quad E = \sqrt{c^2 p_0^2 + m^2 c^4}. \qquad (6.14)$$

Spinor Theory for Spin-1/2 Particles

From (6.13), it is clear that for any such paraxial wave packet, $\left|\psi'(z)\right\rangle = \mathrm{M}\left|\underline{\psi}(z)\right\rangle$ will have its upper pair of components large compared to the lower pair of components. Thus, in the paraxial situation, the even operator $\widehat{\mathcal{E}}$ of $\widehat{\mathcal{H}}'$ in (6.11) does not couple the large upper components and small lower components of $\left|\psi'(z)\right\rangle$ while the odd operator $\widehat{\mathcal{O}}$ of $\widehat{\mathcal{H}}'$ does couple them. This is exactly the same as in the nonrelativistic situation obtained in the standard Dirac theory with respect to time evolution. This and the striking resemblance of (6.11) to the standard Dirac equation (3.711) make us turn to the Foldy–Wouthuysen transformation technique to analyse (6.11) further.

Let us recall that the Foldy–Wouthuysen technique is useful in analysing the Dirac equation systematically as a sum of nonrelativistic part and a series of relativistic correction terms. It is essentially based on the fact that β commutes with any even operator that does not couple the upper and lower pairs of components of the Dirac spinor and anticommutes with any odd operator that couples the upper and lower pairs of components of the Dirac spinor. So, applying a Foldy–Wouthuysen-like transformation to (6.11) should help us analyse it as a sum of the paraxial part and a series of nonparaxial, or aberration, correction terms. To this end, we make the first transformation

$$\left|\underline{\psi}^{(1)}\right\rangle = e^{i\widehat{S}_1}\left|\underline{\psi}'\right\rangle, \qquad \widehat{S}_1 = \frac{i}{2p_0}\beta\widehat{\mathcal{O}}. \tag{6.15}$$

The resulting equation for $\left|\underline{\psi}^{(1)}\right\rangle$ is

$$i\hbar\frac{\partial}{\partial z}\left|\underline{\psi}^{(1)}\right\rangle = \widehat{\mathcal{H}}^{(1)}\left|\underline{\psi}^{(1)}\right\rangle,$$

$$\widehat{\mathcal{H}}^{(1)} = e^{-\frac{1}{2p_0}\beta\widehat{\mathcal{O}}}\widehat{\mathcal{H}}'e^{\frac{1}{2p_0}\beta\widehat{\mathcal{O}}} - i\hbar e^{-\frac{1}{2p_0}\beta\widehat{\mathcal{O}}}\frac{\partial}{\partial z}\left(e^{\frac{1}{2p_0}\beta\widehat{\mathcal{O}}}\right)$$

$$= -p_0\beta + \widehat{\mathcal{E}}^{(1)} + \widehat{\mathcal{O}}^{(1)},$$

$$\widehat{\mathcal{E}}^{(1)} \approx \widehat{\mathcal{E}} - \frac{1}{2p_0}\beta\widehat{\mathcal{O}}^2 - \frac{1}{8p_0^2}\left[\widehat{\mathcal{O}},\left(\left[\widehat{\mathcal{O}},\widehat{\mathcal{E}}\right] + i\hbar\frac{\partial\widehat{\mathcal{O}}}{\partial z}\right)\right] - \frac{1}{8p_0^3}\beta\widehat{\mathcal{O}}^4,$$

$$\widehat{\mathcal{O}}^{(1)} \approx -\frac{1}{2p_0}\beta\left(\left[\widehat{\mathcal{O}},\widehat{\mathcal{E}}\right] + i\hbar\frac{\partial\widehat{\mathcal{O}}}{\partial z}\right) - \frac{1}{3p_0^2}\widehat{\mathcal{O}}^3. \tag{6.16}$$

Before proceeding further, let us find out the nature of $\left|\underline{\psi}^{(1)}\right\rangle$ by looking at the free particle case. From (6.13), for the free particle positive-energy plane wave spinor, we get

$$\underline{\psi}^{(1)}(\vec{r}_\perp, z) = e^{-\frac{1}{2p_0}\beta\chi\vec{\alpha}_\perp\cdot\vec{\hat{p}}_\perp} \underline{\psi}'(\vec{r}_\perp, z)$$

$$\approx \frac{1}{4}\sqrt{\frac{cp_0\xi}{2\pi^3\hbar^3 E}} \left(1 - \frac{1}{2p_0}\beta\chi\vec{\alpha}_\perp\cdot\vec{\hat{p}}_\perp\right)$$

$$\times \begin{pmatrix} \frac{1}{p_0}[a_+(p_0+p_z)+a_-p_-] \\ \frac{1}{p_0}[a_-(p_0+p_z)-a_+p_+] \\ -\frac{1}{p_0\xi}[a_+(p_0-p_z)-a_-p_-] \\ \frac{1}{p_0\xi}[a_-(p_0-p_z)+a_+p_+] \end{pmatrix} e^{\frac{i}{\hbar}(\vec{p}_\perp\cdot\vec{r}_\perp + p_z z)}$$

$$= \frac{1}{4}\sqrt{\frac{cp_0\xi}{2\pi^3\hbar^3 E}} \begin{pmatrix} 1 & 0 & 0 & -\frac{\xi p_-}{2p_0} \\ 0 & 1 & -\frac{\xi p_+}{2p_0} & 0 \\ 0 & -\frac{\xi p_-}{2p_0} & 1 & 0 \\ -\frac{\xi p_+}{2p_0} & 0 & 0 & 1 \end{pmatrix}$$

$$\times \begin{pmatrix} \frac{1}{p_0}[a_+(p_0+p_z)+a_-p_-] \\ \frac{1}{p_0}[a_-(p_0+p_z)-a_+p_+] \\ -\frac{1}{p_0\xi}[a_+(p_0-p_z)-a_-p_-] \\ \frac{1}{p_0\xi}[a_-(p_0-p_z)+a_+p_+] \end{pmatrix} e^{\frac{i}{\hbar}(\vec{p}_\perp\cdot\vec{r}_\perp + p_z z)}$$

$$= \frac{1}{4}\sqrt{\frac{cp_0\xi}{2\pi^3\hbar^3 E}}$$

$$\times \begin{pmatrix} a_+\left(1+\frac{p_z}{p_0}-\frac{p_\perp^2}{2p_0^2}\right) + \frac{1}{2}a_-\left[\frac{p_-}{p_0}\left(1+\frac{p_z}{p_0}\right)\right] \\ a_-\left(1+\frac{p_z}{p_0}+\frac{p_\perp^2}{2p_0^2}\right) - \frac{1}{2}a_+\left[\frac{p_+}{p_0}\left(1+\frac{p_z}{p_0}\right)\right] \\ -\frac{1}{\xi}\left\{a_+\left(1-\frac{p_z}{p_0}-\frac{p_\perp^2}{2p_0^2}\right) + \frac{1}{2}a_-\left[\frac{p_-}{p_0}\left(1-\frac{p_z}{p_0}\right)\right]\right\} \\ \frac{1}{\xi}\left\{a_-\left(1-\frac{p_z}{p_0}-\frac{p_\perp^2}{2p_0^2}\right) + \frac{1}{2}a_+\left[\frac{p_+}{p_0}\left(1-\frac{p_z}{p_0}\right)\right]\right\} \end{pmatrix}$$

$$\times e^{\frac{i}{\hbar}(\vec{p}_\perp\cdot\vec{r}_\perp + p_z z)}, \tag{6.17}$$

showing clearly that the transformation (6.15) preserves the largeness of the upper pair of components of the Dirac spinor compared to its lower pair of components.

It is seen that the effect of the transformation (6.15) is to eliminate from the odd part of $\widehat{\mathcal{H}}'$, the terms of zeroth order in $1/p_0$. The odd part of $\widehat{\mathcal{H}}^{(1)}$, $\widehat{\mathcal{O}}^{(1)}$, contains only terms of first and higher orders in $1/p_0$. By a series of successive transformations with the same recipe (6.15), one can eliminate the odd parts up to any desired order in $1/p_0$. This process will lead to the required quantum beam optical Hamiltonian necessary for finding the transfer maps for observables of interest. In the following section, we shall consider some examples of this general formalism.

Spinor Theory for Spin-1/2 Particles

6.1.1.1 Free Propagation: Diffraction

Let us now consider the propagation of a monoenergetic beam of Dirac particles of mass m through free space along the $+z$-direction. Let us start with the equation (6.11) in the field-free case. Substituting $\phi = 0$ and $\vec{A} = (0,0,0)$ in (6.11), we have

$$i\hbar \frac{\partial}{\partial z}\left|\psi'(z)\right\rangle = \widehat{\mathscr{H}'}\left|\psi'(z)\right\rangle, \qquad \widehat{\mathscr{H}'} = -p_0\beta + \chi\vec{\alpha}_\perp \cdot \vec{\widehat{p}}_\perp. \qquad (6.18)$$

It is seen that $\widehat{\mathscr{H}'}$ is not Hermitian since there is no conservation of probability along the z-axis. Note that

$$\left(\widehat{\mathscr{H}'}\right)^2 = \begin{pmatrix} -p_0 \mathbb{I} & \xi\vec{\sigma}_\perp \cdot \vec{\widehat{p}}_\perp \\ -\xi^{-1}\vec{\sigma}_\perp \cdot \vec{\widehat{p}}_\perp & p_0 \mathbb{I} \end{pmatrix}^2 = (p_0^2 - \widehat{p}_\perp^2)\,\mathbb{I}. \qquad (6.19)$$

indicating that $\widehat{\mathscr{H}'}$ can be identified with the classical beam optical Hamiltonian, $-\sqrt{p_0^2 - p_\perp^2}$, for free propagation of a monoenergetic beam with the square root taken in the Dirac way. Although in the present case it may look like as if one can take such a square root using only the 2×2 Pauli matrices, it is necessary to use the 4×4 Dirac matrices to take into account the two-component spin and the propagations in the forward and backward directions along the z-axis considered separately. The spinor wave function

$$\underline{\psi}'(\vec{r}_\perp, z) = \begin{pmatrix} a_1 \\ a_2 \\ \frac{a_2 p_+}{\xi(p_0 + p_z)} \\ \frac{a_1 p_-}{\xi(p_0 + p_z)} \end{pmatrix} e^{\frac{i}{\hbar}(p_z z + \vec{p}_\perp \cdot \vec{r}_\perp)}, \qquad p_0 = \sqrt{p_z^2 + p_\perp^2}, \qquad (6.20)$$

with arbitrary coefficients a_1 and a_2 including normalization constants is a general solution of (6.18) corresponding to a particle moving in the $+z$-direction with $p_z = \sqrt{p_0^2 - p_\perp^2}$. It is seen that this spinor wave function has large upper pair of components compared to the lower pair of components when the motion is paraxial, i.e., $|\vec{p}_\perp| \ll p_z \approx p_0$. This is analogous to the nonrelativistic positive-energy solutions of the free-particle Dirac equation having large upper pair of components compared to the lower pair of components.

In the same way as the free-particle Dirac Hamiltonian can be diagonalized by a Foldy–Wouthuysen transformation, $\widehat{\mathscr{H}'}$ can be diagonalized exactly by a Foldy–Wouthuysen-like transformation. Define

$$\left|\underline{\psi}^{(1)}(z)\right\rangle = e^{-\theta\beta\chi\vec{\alpha}_\perp \cdot \vec{\widehat{p}}_\perp}\left|\underline{\psi}'(z)\right\rangle, \quad \text{with } \tanh\left(2\left|\vec{\widehat{p}}_\perp\right|\theta\right) = \frac{\left|\vec{\widehat{p}}_\perp\right|}{p_0}. \qquad (6.21)$$

Then, we have

$$i\hbar \frac{\partial\left|\underline{\psi}^{(1)}(z)\right\rangle}{\partial z} = \widehat{\mathscr{H}}^{(1)}\left|\underline{\psi}^{(1)}(z)\right\rangle$$

$$\widehat{\mathscr{H}}^{(1)} = e^{-\theta\beta\chi\vec{\alpha}_\perp \cdot \vec{\widehat{p}}_\perp}\widehat{\mathscr{H}'}e^{\theta\beta\chi\vec{\alpha}_\perp \cdot \vec{\widehat{p}}_\perp}. \qquad (6.22)$$

Let us write

$$e^{\theta \beta \chi \vec{\alpha}_\perp \cdot \vec{p}_\perp} = \exp\left\{\left(\frac{\beta \chi \vec{\alpha}_\perp \cdot \vec{p}_\perp}{\left|\vec{p}_\perp\right|}\right)\left|\vec{p}_\perp\right|\theta\right\}, \quad \text{with} \quad \left(\frac{\beta \chi \vec{\alpha}_\perp \cdot \vec{p}_\perp}{\left|\vec{p}_\perp\right|}\right)^2 = \mathbb{I}. \quad (6.23)$$

Then, we have

$$e^{\theta \beta \chi \vec{\alpha}_\perp \cdot \vec{p}_\perp} = \cosh\left(\left|\vec{p}_\perp\right|\theta\right)\mathbb{I} + \sinh\left(\left|\vec{p}_\perp\right|\theta\right)\left(\frac{\beta \chi \vec{\alpha}_\perp \cdot \vec{p}_\perp}{\left|\vec{p}_\perp\right|}\right) \quad (6.24)$$

Thus,

$$\begin{aligned}
\widehat{\mathcal{H}}^{(1)} &= \left(\cosh\left(\left|\vec{p}_\perp\right|\theta\right)\mathbb{I} - \sinh\left(\left|\vec{p}_\perp\right|\theta\right)\left(\frac{\beta \chi \vec{\alpha}_\perp \cdot \vec{p}_\perp}{\left|\vec{p}_\perp\right|}\right)\right) \\
&\quad \times \left(-p_0 \beta + \chi \vec{\alpha}_\perp \cdot \vec{p}_\perp\right) \\
&\quad \times \left(\cosh\left(\left|\vec{p}_\perp\right|\theta\right)\mathbb{I} + \sinh\left(\left|\vec{p}_\perp\right|\theta\right)\left(\frac{\beta \chi \vec{\alpha}_\perp \cdot \vec{p}_\perp}{\left|\vec{p}_\perp\right|}\right)\right) \\
&= \left(\cosh\left(\left|\vec{p}_\perp\right|\theta\right)\mathbb{I} - \sinh\left(\left|\vec{p}_\perp\right|\theta\right)\left(\frac{\beta \chi \vec{\alpha}_\perp \cdot \vec{p}_\perp}{\left|\vec{p}_\perp\right|}\right)\right) \\
&\quad \beta\left(-p_0 + \left|\vec{p}_\perp\right| \times \left(\frac{\beta \chi \vec{\alpha}_\perp \cdot \vec{p}_\perp}{\left|\vec{p}_\perp\right|}\right)\right) \\
&\quad \times \left(\cosh\left(\left|\vec{p}_\perp\right|\theta\right)\mathbb{I} + \sinh\left(\left|\vec{p}_\perp\right|\theta\right)\left(\frac{\beta \chi \vec{\alpha}_\perp \cdot \vec{p}_\perp}{\left|\vec{p}_\perp\right|}\right)\right) \\
&= \beta\left(-p_0 + \left|\vec{p}_\perp\right|\left(\frac{\beta \chi \vec{\alpha}_\perp \cdot \vec{p}_\perp}{\left|\vec{p}_\perp\right|}\right)\right) \\
&\quad \times \left(\cosh\left(\left|\vec{p}_\perp\right|\theta\right)\mathbb{I} + \sinh\left(\left|\vec{p}_\perp\right|\theta\right)\left(\frac{\beta \chi \vec{\alpha}_\perp \cdot \vec{p}_\perp}{\left|\vec{p}_\perp\right|}\right)\right)^2 \\
&= \beta\left(-p_0 + \left|\vec{p}_\perp\right|\left(\frac{\beta \chi \vec{\alpha}_\perp \cdot \vec{p}_\perp}{\left|\vec{p}_\perp\right|}\right)\right) e^{2\theta \beta \chi \vec{\alpha}_\perp \cdot \vec{p}_\perp} \\
&= \beta\left(-p_0 + \left|\vec{p}_\perp\right|\left(\frac{\beta \chi \vec{\alpha}_\perp \cdot \vec{p}_\perp}{\left|\vec{p}_\perp\right|}\right)\right)
\end{aligned}$$

Spinor Theory for Spin-1/2 Particles

$$\times \left(\cosh\left(2\left|\vec{p}_\perp\right|\theta\right) \mathbb{I} + \sinh\left(2\left|\vec{p}_\perp\right|\theta\right) \left(\frac{\beta\chi\vec{\alpha}_\perp \cdot \vec{p}_\perp}{\left|\vec{p}_\perp\right|} \right) \right)$$

$$= -\beta\left(p_0 \cosh\left(2\left|\vec{p}_\perp\right|\theta\right) - \left|\vec{p}_\perp\right| \sinh\left(2\left|\vec{p}_\perp\right|\theta\right) \right)$$

$$+ \left(\frac{\beta\chi\vec{\alpha}_\perp \cdot \vec{p}_\perp}{\left|\vec{p}_\perp\right|} \right) \left(\left|\vec{p}_\perp\right| \cosh\left(2\left|\vec{p}_\perp\right|\theta\right) - p_0 \sinh\left(2\left|\vec{p}_\perp\right|\theta\right) \right)$$

$$= -\left(\sqrt{p_0^2 - \vec{p}_\perp^2}\right)\beta, \quad \text{with} \quad \tanh\left(2\left|\vec{p}_\perp\right|\theta\right) = \frac{\left|\vec{p}_\perp\right|}{p_0}. \tag{6.25}$$

This shows that in the $\psi^{(1)}$-representation the quantum beam optical Dirac Hamiltonian for free propagation takes the classical form for each component of the spinor wave function. In this representation, the four-component Dirac spinor representing the beam moving in the forward z-direction has the lower pair of components as zero and the four-component spinor representing the beam moving in the backward z-direction has the upper pair of components as zero.

We are interested in the propagation of a monoenergetic quasiparaxial beam for which $|\vec{p}_\perp| \ll p_z \approx p_0$, where p_0 is the design momentum. In this case,

$$\tanh\left(2\left|\vec{p}_\perp\right|\theta\right) = \frac{\left|\vec{p}_\perp\right|}{p_0} \ll 1, \tag{6.26}$$

and therefore, we have

$$\tanh\left(2\left|\vec{p}_\perp\right|\theta\right) \approx 2\left|\vec{p}_\perp\right|\theta = \frac{\left|\vec{p}_\perp\right|}{p_0}, \quad \text{or,} \quad \theta \approx \frac{1}{2p_0}. \tag{6.27}$$

Let us make the first Foldy–Wouthuysen-like transformation, as in (6.16), with the result

$$\left|\psi^{(1)}(z)\right\rangle = e^{-\frac{1}{2p_0}\beta\chi\vec{\alpha}_\perp \cdot \vec{p}_\perp} \left|\psi'(z)\right\rangle,$$

$$i\hbar\frac{\partial}{\partial z}\left|\psi^{(1)}(z)\right\rangle = \widehat{\mathscr{H}}^{(1)}\left|\psi^{(1)}(z)\right\rangle$$

$$\widehat{\mathscr{H}}^{(1)} = e^{-\frac{1}{2p_0}\beta\chi\vec{\alpha}_\perp \cdot \vec{p}_\perp} \widehat{\mathscr{H}} e^{\frac{1}{2p_0}\beta\chi\vec{\alpha}_\perp \cdot \vec{p}_\perp}$$

$$= e^{-\frac{1}{2p_0}\beta\chi\vec{\alpha}_\perp \cdot \vec{p}_\perp} \left(-p_0\beta + \chi\vec{\alpha}_\perp \cdot \vec{p}_\perp \right) e^{\frac{1}{2p_0}\beta\chi\vec{\alpha}_\perp \cdot \vec{p}_\perp}$$

$$\approx \left(-p_0 + \frac{1}{2p_0}\vec{p}_\perp^2 - \frac{1}{8p_0^3}\vec{p}_\perp^4 \right)\beta, \tag{6.28}$$

where we have omitted the odd term of second order in $1/p_0$. The expression on the right-hand side gives the first three terms of the series expansion of $-\left(\sqrt{p_0^2 - \widehat{p}_\perp^2}\right)\beta$. For an ideal paraxial situation, we have

$$\widehat{\mathscr{H}}^{(1)} \approx \left(-p_0 + \frac{1}{2p_0}\widehat{p}_\perp^2\right)\beta. \tag{6.29}$$

Since the lower pair of components of $\left|\underline{\psi}^{(1)}\right\rangle$ are too small, in fact almost zero, compared to the upper pair of components, we can replace β in $\widehat{\mathscr{H}}^{(1)}$ by $\mathbb{I}$. Hence, we can take

$$\widehat{\mathscr{H}}^{(1)} \approx \left(-p_0 + \frac{1}{2p_0}\widehat{p}_\perp^2\right). \tag{6.30}$$

Let us now retrace the transformations to get back to the original Dirac representation. We get

$$\left|\underline{\psi}(z)\right\rangle = M^{-1} e^{\frac{1}{2p_0}\beta\chi\vec{\alpha}_\perp\cdot\vec{\widehat{p}}_\perp} \left|\underline{\psi}^{(1)}(z)\right\rangle,$$

$$i\hbar\frac{\partial}{\partial z}\left|\underline{\psi}(z)\right\rangle = \widehat{\mathscr{H}}_o\left|\underline{\psi}(z)\right\rangle,$$

$$\widehat{\mathscr{H}}_o = M^{-1} e^{\frac{1}{2p_0}\beta\chi\vec{\alpha}_\perp\cdot\vec{\widehat{p}}_\perp} \widehat{\mathscr{H}}^{(1)} e^{-\frac{1}{2p_0}\beta\chi\vec{\alpha}_\perp\cdot\vec{\widehat{p}}_\perp} M$$

$$= M^{-1} e^{\frac{1}{2p_0}\beta\chi\vec{\alpha}_\perp\cdot\vec{\widehat{p}}_\perp} \left(-p_0 + \frac{1}{2p_0}\widehat{p}_\perp^2\right) e^{-\frac{1}{2p_0}\beta\chi\vec{\alpha}_\perp\cdot\vec{\widehat{p}}_\perp} M$$

$$= \left(-p_0 + \frac{1}{2p_0}\widehat{p}_\perp^2\right). \tag{6.31}$$

Thus, we arrive at the quantum beam optical form of the free particle Dirac equation

$$i\hbar\frac{\partial \underline{\psi}(\vec{r}_\perp, z)}{\partial z} \approx \left(-p_0 + \frac{1}{2p_0}\widehat{p}_\perp^2\right)\underline{\psi}(\vec{r}_\perp, z), \tag{6.32}$$

which is same as in the scalar relativistic Klein–Gordon or nonrelativistic Schrödinger theory, except for the wave function having four components.

Let us look at the transfer maps for transverse position coordinates and momentum components. Recall that the position operator $\vec{r}$ of the nonrelativistic Schrödinger theory has the zitterbewegung motion in the Dirac theory. In the Foldy–Wouthuysen representation of the Dirac theory, $\vec{r}$ behaves as a normal position operator and has been interpreted as the mean position operator. In the Dirac representation, the Newton–Wigner position operator behaves as a normal position operator, and it transforms to $\vec{r}$, the mean position operator, in the Foldy–Wouthuysen representation. Now, let us observe that in the $\underline{\psi}^{(1)}$-representation $\vec{r}_\perp$ behaves as the normal transverse position operator. To this end, let us define the Heisenberg picture operator corresponding to $\vec{r}_\perp$ as

$$\vec{r}(z)_\perp = e^{\frac{i}{\hbar}\widehat{\mathscr{H}}^{(1)}z}\vec{r}_\perp e^{-\frac{i}{\hbar}\widehat{\mathscr{H}}^{(1)}z}. \tag{6.33}$$

Then, it follows that

$$\begin{aligned}\frac{d\vec{r}(z)_\perp}{dz} &= \frac{i}{\hbar} e^{\frac{i}{\hbar}\widehat{\mathscr{H}}^{(1)}z}\left[\widehat{\mathscr{H}}^{(1)}, \vec{r}_\perp\right] e^{-\frac{i}{\hbar}\widehat{\mathscr{H}}^{(1)}z} \\ &= \frac{i}{\hbar} e^{\frac{i}{\hbar}\widehat{\mathscr{H}}^{(1)}z}\left[\left(-p_0 + \frac{1}{2p_0}\vec{\hat{p}}_\perp^2\right), \vec{r}_\perp\right] e^{-\frac{i}{\hbar}\widehat{\mathscr{H}}^{(1)}z} \\ &= \frac{1}{p_0} e^{\frac{i}{\hbar}\widehat{\mathscr{H}}^{(1)}z}\vec{\hat{p}}_\perp e^{-\frac{i}{\hbar}\widehat{\mathscr{H}}^{(1)}z},\end{aligned} \quad (6.34)$$

as should be. This also shows that the transverse momentum operator is $\vec{\hat{p}}_\perp$. So, we can take $\vec{r}_\perp$ and $\vec{\hat{p}}_\perp$ to be the mean transverse position operator and transverse momentum operator, respectively, in the $\underline{\psi}^{(1)}$-presentation. The corresponding mean transverse position operator in the quantum beam optical Dirac representation, or $\underline{\psi}$-representation, would be

$$\begin{aligned}\vec{\tilde{r}}_\perp &= \mathrm{M}^{-1} e^{\frac{1}{2p_0}\beta\chi\vec{\alpha}_\perp\cdot\vec{\hat{p}}_\perp} \vec{r}_\perp e^{-\frac{1}{2p_0}\beta\chi\vec{\alpha}_\perp\cdot\vec{\hat{p}}_\perp} \mathrm{M} \\ &\approx \mathrm{M}^{-1}\left(\vec{r}_\perp + \frac{1}{2p_0}\left[\beta\chi\vec{\alpha}_\perp\cdot\vec{\hat{p}}_\perp, \vec{r}_\perp\right]\right)\mathrm{M} \\ &= \vec{r}_\perp - \frac{i\hbar}{2p_0}\beta\chi\vec{\alpha}_\perp.\end{aligned} \quad (6.35)$$

Since $\vec{\tilde{r}}_\perp$ is non-Hermitian, we can take its Hermitian part

$$\begin{aligned}\vec{\bar{r}}_\perp &= \frac{1}{2}\left(\vec{\tilde{r}}_\perp + \vec{\tilde{r}}_\perp^\dagger\right) \\ &= \vec{r}_\perp - \frac{i\hbar}{4p_0}\left(\xi - \xi^{-1}\right)\beta\vec{\alpha}_\perp \\ &= \vec{r}_\perp - \frac{imc\hbar}{2p_0^2}\beta\vec{\alpha}_\perp = \vec{r}_\perp - \frac{i\lambda_0^2}{4\pi\lambda_c}\beta\vec{\alpha}_\perp,\end{aligned} \quad (6.36)$$

as the mean transverse position operator, where λ_0 is the de Broglie wavelength and λ_c is the Compton wavelength. Noting that in $\vec{\bar{r}}_\perp$ the constant term added to $\vec{r}_\perp$ is tiny, we can as well drop it and take the mean transverse position operator in the quantum beam optical Dirac representation to be $\vec{r}_\perp$ itself. The transverse momentum operator remains the same, $\vec{\hat{p}}_\perp$, in the quantum beam optical Dirac representation, since it commutes with $\mathrm{M}^{-1}e^{\frac{1}{2p_0}\beta\chi\vec{\alpha}_\perp\cdot\vec{\hat{p}}_\perp}$ and hence is unchanged under the transformation from the $\underline{\psi}^{(1)}$-representation.

Since the quantum beam optical paraxial Dirac Hamiltonian $\widehat{\mathscr{H}_o}$ is exactly same as the quantum beam optical paraxial Hamiltonian of the scalar theory (5.77), the transfer map is same as in (5.87),

$$\begin{pmatrix}\langle\vec{r}_\perp\rangle(z) \\ \frac{1}{p_0}\langle\vec{\hat{p}}_\perp\rangle(z)\end{pmatrix} = \begin{pmatrix}1 & \Delta z \\ 0 & 1\end{pmatrix}\begin{pmatrix}\langle\vec{r}_\perp\rangle(z_i) \\ \frac{1}{p_0}\langle\vec{\hat{p}}_\perp\rangle(z_i)\end{pmatrix}, \quad (6.37)$$

where the beam travels from the vertical plane at z_i to the vertical plane at z. The difference now is that

$$\langle \vec{r}_\perp \rangle (z) = \left\langle \underline{\psi}(z) \left| \vec{r}_\perp \right| \underline{\psi}(z) \right\rangle = \int d^2 r_\perp \, \underline{\psi}^\dagger(\vec{r}_\perp, z) \vec{r}_\perp \underline{\psi}(\vec{r}_\perp, z)$$

$$= \sum_{j=1}^{4} \int d^2 r_\perp \, \underline{\psi}_j^*(\vec{r}_\perp, z) \vec{r}_\perp \underline{\psi}_j(\vec{r}_\perp, z),$$

$$\langle \vec{p}_\perp \rangle (z) = \left\langle \underline{\psi}(z) \left| \vec{p}_\perp \right| \underline{\psi}(z) \right\rangle = \int d^2 r_\perp \, \underline{\psi}^\dagger(\vec{r}_\perp, z) \vec{p}_\perp \underline{\psi}(\vec{r}_\perp, z)$$

$$= \sum_{j=1}^{4} \int d^2 r_\perp \, \underline{\psi}_j^*(\vec{r}_\perp, z) \vec{p}_\perp \underline{\psi}_j(\vec{r}_\perp, z). \tag{6.38}$$

Since the quantum beam optical free-particle Hamiltonian $\widehat{\mathscr{H}_o}$ is Hermitian, we have taken $\left| \underline{\psi}(z) \right\rangle$ to be normalized, i.e.,

$$\left\langle \underline{\psi}(z) \middle| \underline{\psi}(z) \right\rangle = \int d^2 r_\perp \, \underline{\psi}^\dagger(\vec{r}_\perp, z) \, \underline{\psi}(\vec{r}_\perp, z)$$

$$= \sum_{j=1}^{4} \int d^2 r_\perp \, \underline{\psi}_j^*(\vec{r}_\perp, z) \, \underline{\psi}_j(\vec{r}_\perp, z) = 1. \tag{6.39}$$

If it is desired to keep any non-Hermitian terms in a quantum beam optical Dirac Hamiltonian, the four-component state vector $\left| \underline{\psi}(z) \right\rangle$ will have to be normalized at each z since the normalization will not be conserved. Then, the average value of any operator $\widehat{O}$ corresponding to an observable $\widehat{O}$, which may be a 4×4 matrix with operator entries like, say, $\vec{\alpha} \cdot \vec{p}$, can be defined by

$$\langle \widehat{O} \rangle (z) = \frac{\left\langle \underline{\psi}(z) \left| \widehat{O} \right| \underline{\psi}(z) \right\rangle}{\left\langle \underline{\psi}(z) \middle| \underline{\psi}(z) \right\rangle}$$

$$= \frac{\int d^2 r_\perp \, \underline{\psi}^\dagger(\vec{r}_\perp, z) \widehat{O} \underline{\psi}(\vec{r}_\perp, z)}{\int d^2 r_\perp \, \underline{\psi}^\dagger(\vec{r}_\perp, z) \underline{\psi}(\vec{r}_\perp, z)}$$

$$= \frac{\sum_{j,k=1}^{4} \int d^2 r_\perp \, \psi_j^*(\vec{r}_\perp, z) \widehat{O}_{jk} \psi_k(\vec{r}_\perp, z)}{\sum_{j=1}^{4} \int d^2 r_\perp \, \psi_j^*(\vec{r}_\perp, z) \psi_j(\vec{r}_\perp, z)}. \tag{6.40}$$

The quantum beam optical Dirac Hamiltonian for free propagation is same as in the scalar theory, except that it now acts on a four-component wave function. Integrating (6.32), we have

$$\left| \underline{\psi}(z) \right\rangle = e^{-\frac{i \Delta z}{\hbar} \left(-p_0 + \frac{\hat{p}_\perp^2}{2p_0} \right) \mathbb{I}} \left| \underline{\psi}(z_i) \right\rangle, \qquad \Delta z = z - z_i. \tag{6.41}$$

Spinor Theory for Spin-1/2 Particles

In position representation,

$$\underline{\psi}(\vec{r}_\perp, z) = \int d^2 r_{\perp i} \left[K(\vec{r}_\perp, z; \vec{r}_{\perp i}, z_i) \mathbb{I} \right] \underline{\psi}(\vec{r}_{\perp i}, z_i),$$

$$K(\vec{r}_\perp, z; \vec{r}_{\perp i}, z_i) = \left\langle \vec{r}_\perp \left| e^{-\frac{i\Delta z}{\hbar}\left(-p_0 + \frac{\hat{p}_\perp^2}{2p_0}\right)} \right| \vec{r}_{\perp i} \right\rangle. \quad (6.42)$$

The expression for $K(\vec{r}_\perp, z; \vec{r}_{\perp i}, z_i)$ is already known to us from the scalar theory:

$$K(\vec{r}_\perp, z; \vec{r}_{\perp i}, z_i) = \left(\frac{p_0}{2\pi i \hbar \Delta z}\right) e^{\left\{\frac{i}{\hbar} p_0 \Delta z + \frac{i p_0}{2\hbar \Delta z} |\vec{r}_\perp - \vec{r}_{\perp i}|^2\right\}}. \quad (6.43)$$

Using this result, we have

$$\underline{\psi}(\vec{r}_\perp, z) = \left(\frac{p_0}{2\pi i \hbar \Delta z}\right) e^{\frac{i}{\hbar} p_0 \Delta z} \int d^2 r_{\perp i} \left(e^{\frac{i p_0}{2\hbar \Delta z} |\vec{r}_\perp - \vec{r}_{\perp i}|^2} \mathbb{I} \right) \underline{\psi}(\vec{r}_{\perp i}, z_i). \quad (6.44)$$

With $2\pi\hbar/p_0 = \lambda_0$, the de Broglie wavelength, we get

$$\psi_j(x, y, z) = \left(\frac{1}{i\lambda_0 (z - z_i)}\right) e^{\frac{i 2\pi}{\lambda_0}(z - z_i)}$$
$$\times \iint dx_i dy_i \, e^{\frac{i\pi}{\lambda_0(z-z_i)}\left[(x-x_i)^2 + (y-y_i)^2\right]} \psi_j(x_i, y_i, z_i),$$
$$j = 1, 2, 3, 4, \quad (6.45)$$

which is the generalization of the Fresnel diffraction formula for a paraxial beam of free Dirac particles. Thus, it is obvious that the diffraction pattern due to a paraxial beam of free Dirac particles will be the superposition of the patterns due to the four individual components of the spinor $\left|\underline{\psi}(z)\right\rangle$ representing the beam and the intensity distribution of the diffraction pattern at the xy-plane at z will be given by

$$I(x, y, z) \propto \sum_{j=1}^{4} \left| \iint dx_i dy_i \, e^{\frac{i\pi}{\lambda_0(z-z_i)}\left[(x-x_i)^2 + (y-y_i)^2\right]} \psi_j(x_i, y_i, z_i) \right|^2 \quad (6.46)$$

where the plane of the diffracting object is at z_i and the plane at z is the observation plane. When the presence of a field makes the quantum beam optical Hamiltonian $\widehat{\mathcal{H}_o}$ acquire a matrix component, the propagator $K(\vec{r}_\perp, z; \vec{r}_{\perp i}, z_i)$ would have nontrivial matrix structure, instead of being just $\propto \mathbb{I}$, leading to interference between the four diffracted components of $\left|\underline{\psi}(z)\right\rangle$.

When the monoenergetic beam is not ideally paraxial to allow the relevant approximations to be made, one can directly use the free z-evolution equation,

$$i\hbar \frac{\partial}{\partial z} \left|\underline{\psi}(z)\right\rangle = -\left\{ p_0 \beta \chi \alpha_z + i(\Sigma_x \hat{p}_y - \Sigma_y \hat{p}_x) \right\} \left|\underline{\psi}(z)\right\rangle, \quad (6.47)$$

obtained by setting $\phi = 0$ and $\vec{A} = (0,0,0)$ in (6.7). On integration, we get

$$\left|\underline{\psi}(z)\right\rangle = \exp\left\{\frac{i}{\hbar}\Delta z\left(p_0\beta\chi\alpha_z + i(\Sigma_x\widehat{p}_y - \Sigma_y\widehat{p}_x)\right)\right\}\left|\underline{\psi}(z_i)\right\rangle, \qquad (6.48)$$

the general law of propagation of the free Dirac wave function in the $+z$-direction, showing the subtle way in which the spinor components are mixed up. For some detailed studies on the optics of general free Dirac waves, particularly diffraction, see Rubinowicz ([161, 162, 163, 164], Durand [38], and Phan-Van-Loc [144, 145, 146, 147, 148, 149], Boxem, Partoens, and Verbeeck [15], and references therein). A path integral approach to the optics of Dirac particles has been developed by Liñares [128].

6.1.1.2 Axially Symmetric Magnetic Lens

We shall study the propagation of a quasiparaxial charged Dirac particle beam through a round magnetic lens without taking into account the anomalous magnetic moment of the particle. We shall take the z-axis to be the optic axis of the lens. Since there is no electric field, $\phi = 0$. The vector potential can be taken, in general, as

$$\vec{A} = \left(-\frac{1}{2}y\Pi(\vec{r}_\perp, z), \frac{1}{2}x\Pi(\vec{r}_\perp, z), 0\right), \qquad (6.49)$$

with

$$\Pi(\vec{r}_\perp, z) = B(z) - \frac{1}{8}r_\perp^2 B''(z) + \frac{1}{192}r_\perp^4 B''''(z) - \cdots . \qquad (6.50)$$

The corresponding magnetic field is given by

$$\vec{B}_\perp = -\frac{1}{2}\left(B'(z) - \frac{1}{8}B'''(z)r_\perp^2 + \cdots\right)\vec{r}_\perp,$$

$$B_z = B(z) - \frac{1}{4}B''(z)r_\perp^2 + \frac{1}{64}B''''(z)r_\perp^4 - \cdots . \qquad (6.51)$$

The practical boundaries of the lens, say z_ℓ and z_r, are determined by where $B(z)$ becomes negligible, *i.e.*, $B(z < z_\ell) \approx 0$ and $B(z > z_r) \approx 0$.

As we have already seen, the z-evolution equation for the time-independent spinor wave function in the $\left|\psi'\right\rangle$-representation (6.11) reads

$$i\hbar\frac{\partial}{\partial z}\left|\underline{\psi}'(z)\right\rangle = \widehat{\mathscr{H}}\left|\underline{\psi}'(z)\right\rangle,$$

$$\widehat{\mathscr{H}} = -p_0\beta + \widehat{\mathscr{E}} + \widehat{\mathscr{O}},$$

$$\widehat{\mathscr{E}} = -qA_z\mathbb{I},$$

$$\widehat{\mathscr{O}} = \chi\vec{\alpha}_\perp \cdot \widehat{\vec{\pi}}_\perp = \chi\vec{\alpha}_\perp \cdot \left(\widehat{\vec{p}}_\perp - q\vec{A}_\perp\right). \qquad (6.52)$$

Let us recall that the $|\underline{\psi}'\rangle$ is related to the Dirac $|\underline{\psi}\rangle$ by

$$|\underline{\psi}'\rangle = \mathrm{M}|\underline{\psi}(z)\rangle,$$

$$\mathrm{M} = \mathrm{M} = \frac{1}{\sqrt{2}}(\mathbb{I} + \chi\alpha_z),$$

$$\chi = \begin{pmatrix} \xi\mathbb{I} & 0 \\ 0 & -\xi^{-1}\mathbb{I} \end{pmatrix}, \quad \text{with } \xi = \sqrt{\frac{E+mc^2}{E-mc^2}}, \quad (6.53)$$

where E is the conserved total energy $\sqrt{m^2c^4 + c^2p_0^2}$ of a particle of the beam with the design momentum p_0. To obtain the desired quantum beam optical representation of the z-evolution equation, we have to perform a series of successive Foldy–Wouthuysen-like transformations outlined earlier. The first transformation is

$$|\underline{\psi}^{(1)}\rangle = e^{i\widehat{S}_1}|\underline{\psi}'\rangle,$$

$$\widehat{S}_1 = \frac{i}{2p_0}\beta\widehat{\mathcal{O}} = \frac{i}{2p_0}\beta\chi\vec{\alpha}_\perp \cdot \vec{\pi}_\perp. \quad (6.54)$$

The resulting equation for $|\underline{\psi}^{(1)}\rangle$ is

$$i\hbar\frac{\partial}{\partial z}|\underline{\psi}^{(1)}\rangle = \widehat{\mathcal{H}}^{(1)}|\underline{\psi}^{(1)}\rangle,$$

$$\widehat{\mathcal{H}}^{(1)} = e^{-\frac{1}{2p_0}\beta\widehat{\mathcal{O}}}\widehat{\mathcal{H}}' e^{\frac{1}{2p_0}\beta\widehat{\mathcal{O}}} - i\hbar e^{-\frac{1}{2p_0}\beta\widehat{\mathcal{O}}}\frac{\partial}{\partial z}\left(e^{\frac{1}{2p_0}\beta\widehat{\mathcal{O}}}\right)$$

$$= -p_0\beta + \widehat{\mathcal{E}}^{(1)} + \widehat{\mathcal{O}}^{(1)},$$

$$\widehat{\mathcal{E}}^{(1)} \approx \widehat{\mathcal{E}} - \frac{1}{2p_0}\beta\widehat{\mathcal{O}}^2 - \frac{1}{8p_0^2}\left[\widehat{\mathcal{O}},\left([\widehat{\mathcal{O}},\widehat{\mathcal{E}}] + i\hbar\frac{\partial\widehat{\mathcal{O}}}{\partial z}\right)\right] - \frac{1}{8p_0^3}\beta\widehat{\mathcal{O}}^4,$$

$$\widehat{\mathcal{O}}^{(1)} \approx -\frac{1}{2p_0}\beta\left([\widehat{\mathcal{O}},\widehat{\mathcal{E}}] + i\hbar\frac{\partial\widehat{\mathcal{O}}}{\partial z}\right) - \frac{1}{3p_0^2}\widehat{\mathcal{O}}^3, \quad (6.55)$$

in which we have to substitute $\widehat{\mathcal{E}} = -qA_z\mathbb{I}$ and $\widehat{\mathcal{O}} = \chi\vec{\alpha}_\perp \cdot \vec{\pi}_\perp$ and calculate the expressions for $\widehat{\mathcal{E}}^{(1)}$ and $\widehat{\mathcal{O}}^{(1)}$. The second transformation will have the same prescription

$$|\underline{\psi}^{(2)}\rangle = e^{i\widehat{S}_2}|\underline{\psi}^{(1)}\rangle, \quad \widehat{S}_2 = \frac{i}{2p_0}\beta\widehat{\mathcal{O}}^{(1)}, \quad (6.56)$$

leading to the equation

$$i\hbar \frac{\partial}{\partial z}\left|\psi^{(2)}\right\rangle = \widehat{\mathscr{H}}^{(2)}\left|\psi^{(2)}\right\rangle,$$

$$\widehat{\mathscr{H}}^{(2)} = e^{-\frac{1}{2p_0}\beta\widehat{\mathscr{O}}^{(1)}} \widehat{\mathscr{H}}^{(1)} e^{\frac{1}{2p_0}\beta\widehat{\mathscr{O}}^{(1)}} - i\hbar e^{-\frac{1}{2p_0}\beta\widehat{\mathscr{O}}^{(1)}} \frac{\partial}{\partial z}\left(e^{\frac{1}{2p_0}\beta\widehat{\mathscr{O}}^{(1)}}\right)$$

$$= -p_0\beta + \widehat{\mathscr{E}}^{(2)} + \widehat{\mathscr{O}}^{(2)},$$

$$\widehat{\mathscr{E}}^{(2)} \approx \widehat{\mathscr{E}}^{(1)} - \frac{1}{2p_0}\beta\left(\widehat{\mathscr{O}}^{(1)}\right)^2$$

$$- \frac{1}{8p_0^2}\left[\widehat{\mathscr{O}}^{(1)}, \left(\left[\widehat{\mathscr{O}}^{(1)}, \widehat{\mathscr{E}}^{(1)}\right] + i\hbar\frac{\partial \widehat{\mathscr{O}}^{(1)}}{\partial z}\right)\right]$$

$$- \frac{1}{8p_0^3}\beta\left(\widehat{\mathscr{O}}^{(1)}\right)^4,$$

$$\widehat{\mathscr{O}}^{(2)} \approx -\frac{1}{2p_0}\beta\left(\left[\widehat{\mathscr{O}}^{(1)}, \widehat{\mathscr{E}}^{(1)}\right] + i\hbar\frac{\partial \widehat{\mathscr{O}}^{(1)}}{\partial z}\right)$$

$$- \frac{1}{3p_0^2}\left(\widehat{\mathscr{O}}^{(1)}\right)^3, \quad (6.57)$$

in which we have to get the expressions for $\widehat{\mathscr{E}}^{(2)}$ and $\widehat{\mathscr{O}}^{(2)}$ by substituting the expressions for $\widehat{\mathscr{E}}^{(1)}$ and $\widehat{\mathscr{O}}^{(1)}$ obtained from the first transformation. We shall stop with the third transformation in this case. With the same prescription,

$$\left|\psi^{(3)}\right\rangle = e^{i\widehat{S}_3}\left|\psi^{(2)}\right\rangle, \qquad \widehat{S}_3 = \frac{i}{2p_0}\beta\widehat{\mathscr{O}}^{(2)}, \quad (6.58)$$

we get

$$i\hbar\frac{\partial}{\partial z}\left|\psi^{(3)}\right\rangle = \widehat{\mathscr{H}}^{(3)}\left|\psi^{(3)}\right\rangle,$$

$$\widehat{\mathscr{H}}^{(3)} = e^{-\frac{1}{2p_0}\beta\widehat{\mathscr{O}}^{(2)}} \widehat{\mathscr{H}}^{(2)} e^{\frac{1}{2p_0}\beta\widehat{\mathscr{O}}^{(2)}} - i\hbar e^{-\frac{1}{2p_0}\beta\widehat{\mathscr{O}}^{(2)}} \frac{\partial}{\partial z}\left(e^{\frac{1}{2p_0}\beta\widehat{\mathscr{O}}^{(2)}}\right)$$

$$\approx -p_0\beta + \widehat{\mathscr{E}}^{(3)},$$

$$\widehat{\mathscr{E}}^{(3)} \approx \widehat{\mathscr{E}}^{(2)} - \frac{1}{2p_0}\beta\left(\widehat{\mathscr{O}}^{(2)}\right)^2$$

$$- \frac{1}{8p_0^2}\left[\widehat{\mathscr{O}}^{(2)}, \left(\left[\widehat{\mathscr{O}}^{(2)}, \widehat{\mathscr{E}}^{(2)}\right] + i\hbar\frac{\partial \widehat{\mathscr{O}}^{(2)}}{\partial z}\right)\right]$$

$$- \frac{1}{8p_0^3}\beta\left(\widehat{\mathscr{O}}^{(2)}\right)^4, \quad (6.59)$$

Spinor Theory for Spin-1/2 Particles

in which we have omitted the odd term, and we have to get the expression for $\widehat{\mathscr{E}}^{(3)}$ by substituting the expressions for $\widehat{\mathscr{E}}^{(2)}$ and $\widehat{\mathscr{O}}^{(2)}$ obtained from the second transformation.

The above transformations make the lower components of the spinor successively smaller and smaller compared to the upper components for a quasiparaxial beam moving in the $+z$-direction. In other words, one has

$$\beta \left| \underline{\psi}^{(3)} \right\rangle \approx \mathbb{I} \left| \underline{\psi}^{(3)} \right\rangle. \tag{6.60}$$

Now, $\widehat{\mathscr{H}}^{(3)}$, without an odd term, is found to be of the form

$$\widehat{\mathscr{H}}^{(3)} = \begin{pmatrix} \widehat{\mathscr{H}}_1^{(3)} + \widehat{\mathscr{H}}_2^{(3)} & 0 \\ 0 & \widehat{\mathscr{H}}_1^{(3)} - \widehat{\mathscr{H}}_2^{(3)} \end{pmatrix}. \tag{6.61}$$

In view of (6.60), we can approximate $\widehat{\mathscr{H}}^{(3)}$ further and write

$$i\hbar \frac{\partial}{\partial z} \left| \underline{\psi}^{(3)} \right\rangle \approx \widehat{\mathscr{H}} \left| \underline{\psi}^{(3)} \right\rangle,$$

$$\widehat{\mathscr{H}} = \begin{pmatrix} \widehat{\mathscr{H}}_1^{(3)} + \widehat{\mathscr{H}}_2^{(3)} & 0 \\ 0 & \widehat{\mathscr{H}}_1^{(3)} + \widehat{\mathscr{H}}_2^{(3)} \end{pmatrix}. \tag{6.62}$$

To enable using directly the original Dirac $|\psi(z)\rangle$, satisfying (6.7), let us retrace the transformations:

$$\left| \underline{\psi}^{(3)} \right\rangle \longrightarrow |\psi(z)\rangle = \mathrm{M}^{-1} e^{-i\widehat{S}_1} e^{-i\widehat{S}_2} e^{-i\widehat{S}_3} \left| \underline{\psi}^{(3)} \right\rangle, \approx \mathrm{M}^{-1} e^{-i\widehat{\mathscr{S}}} \left| \underline{\psi}^{(3)} \right\rangle, \tag{6.63}$$

where

$$\widehat{\mathscr{S}} \approx \widehat{S}_1 + \widehat{S}_2 + \widehat{S}_3 - \frac{i}{2} \left(\left[\widehat{S}_1, \widehat{S}_2 \right] + \left[\widehat{S}_1, \widehat{S}_3 \right] + \left[\widehat{S}_2, \widehat{S}_3 \right] \right) - \frac{1}{4} \left[\left[\widehat{S}_1, \widehat{S}_2 \right], \widehat{S}_3 \right] \cdots, \tag{6.64}$$

obtained by applying the Baker–Campbell–Hausdorff formula

$$e^{\widehat{A}} e^{\widehat{B}} \approx e^{\widehat{A} + \widehat{B} + \frac{1}{2}[\widehat{A},\widehat{B}]}, \tag{6.65}$$

when $\left[\widehat{A}, \widehat{B}\right] \neq 0$ (see Appendix A for details). Implementing the inverse transformation leads to

$$i\hbar \frac{\partial}{\partial z} |\psi(z)\rangle = \widehat{\mathscr{H}_o} |\psi(z)\rangle,$$

$$\widehat{\mathscr{H}_o} = \mathrm{M}^{-1} \left\{ e^{-i\widehat{\mathscr{S}}} \widehat{\mathscr{H}} e^{i\widehat{\mathscr{S}}} - i\hbar e^{-i\widehat{\mathscr{S}}} \frac{\partial}{\partial z} \left(e^{i\widehat{\mathscr{S}}} \right) \right\} \mathrm{M}. \tag{6.66}$$

Calculating $\widehat{\mathscr{H}_o}$, using (3.706), up to fourth order in $\left(\vec{r}_\perp, \vec{p}_\perp \right)$, we have

$$\widehat{\mathscr{H}_o} = \widehat{\mathscr{H}}_{o,p} + \widehat{\mathscr{H}_o'} + \widehat{\mathscr{H}}_o^{(\hbar)} + \widehat{\mathbb{H}}_o^{(\hbar)}, \tag{6.67}$$

where

$$\widehat{\mathcal{H}}_{o,p} = \left(\frac{1}{2p_0}\widehat{p}_\perp^2 + \frac{1}{2}p_0\alpha^2(z)r_\perp^2 - \alpha(z)\widehat{L}_z\right)\mathbb{I}, \tag{6.68}$$

$$\widehat{\mathcal{H}}_o' = \left(\frac{1}{8p_0^3}\widehat{p}_\perp^4 - \frac{\alpha(z)}{2p_0^2}\widehat{p}_\perp^2\widehat{L}_z - \frac{\alpha^2(z)}{8p_0}\left(\vec{r}_\perp\cdot\vec{\widehat{p}}_\perp + \vec{\widehat{p}}_\perp\cdot\vec{r}_\perp\right)^2\right.$$
$$+ \frac{3\alpha^2(z)}{8p_0}\left(r_\perp^2\widehat{p}_\perp^2 + \widehat{p}_\perp^2 r_\perp^2\right) + \frac{1}{8}\left(\alpha''(z) - 4\alpha^3(z)\right)\widehat{L}_z r_\perp^2$$
$$\left. + \frac{p_0}{8}\left(\alpha^4(z) - \alpha(z)\alpha''(z)\right)r_\perp^4\right)\mathbb{I}, \tag{6.69}$$

$$\widehat{\mathcal{H}}_o^{(\hbar)} \approx \left(\frac{\hbar^2}{8p_0}\left(\alpha'(z)^2 - 2\alpha(z)\alpha''(z)\right)r_\perp^2\right.$$
$$\left. + \frac{\hbar^2}{32p_0}\left(\alpha''(z)^2 - \alpha'(z)\alpha'''(z)\right)r_\perp^4\right)\mathbb{I}, \tag{6.70}$$

and, retaining only the Hermitian part,

$$\widehat{\mathbb{H}}_o^{(\hbar)} \approx \frac{3\hbar^2}{64p_0}\alpha'(z)\alpha''(z)yr_\perp^2\left(y\Sigma_x - x\Sigma_y\right)$$
$$+ \left\{\frac{\hbar}{4}\left(\alpha''(z) - 2\alpha(z)^3\right)r_\perp^2 - \frac{\hbar}{2p_0^2}\alpha(z)\widehat{p}_\perp^2\right.$$
$$+ \frac{\hbar}{4p_0}\alpha'(z)\left(\vec{r}_\perp\cdot\vec{\widehat{p}}_\perp + \vec{\widehat{p}}_\perp\cdot\vec{r}_\perp\right) + \frac{\hbar}{p_0}\alpha(z)^2\widehat{L}_z$$
$$+ \frac{\hbar}{16p_0}\alpha''(z)\left(\widehat{p}_\perp^2 r_\perp^2 + r_\perp^2\widehat{p}_\perp^2\right) - \frac{3\hbar}{8p_0}\alpha(z)\alpha''(z)\widehat{L}_z r_\perp^2$$
$$+ \frac{3\hbar}{16}\alpha(z)^2\alpha''(z)r_\perp^4 - \frac{\hbar}{64p_0}\alpha'''(z)\left(\vec{r}_\perp\cdot\vec{\widehat{p}}_\perp + \vec{\widehat{p}}_\perp\cdot\vec{r}_\perp\right)r_\perp^2$$
$$\left. - \frac{\hbar}{64p_0}\alpha'''(z)\left(\{\widehat{p}_x, xr_\perp^2\} + \{\widehat{p}_y, yr_\perp^2\}\right) - \hbar\alpha(z)\right\}\Sigma_z$$
$$- \frac{i\hbar}{48p_0^2}\alpha'(z)\left[\beta\chi, \left\{\vec{\alpha}_\perp\cdot\vec{\widehat{p}}_\perp, \left(\vec{r}_\perp\cdot\vec{\widehat{p}}_\perp + \vec{\widehat{p}}_\perp\cdot\vec{r}_\perp\right)\right\}\right]$$
$$- \frac{i\hbar}{32p_0}\alpha(z)\alpha'(z)\left[\beta\chi, \left\{\vec{\alpha}_\perp\cdot\vec{r}_\perp, \left(\vec{r}_\perp\cdot\vec{\widehat{p}}_\perp + \vec{\widehat{p}}_\perp\cdot\vec{r}_\perp\right)\right\}\right]$$
$$+ \frac{i\hbar}{4}\alpha'(z)\left[\beta\chi, (\alpha_x y - \alpha_y x)\right]$$
$$+ \frac{i\hbar}{32p_0}\left(\hbar\alpha(z) - p_0\right)\alpha'''(z)\left[\beta\chi, (\alpha_x y - \alpha_y x)r_\perp^2\right]$$
$$+ \frac{\hbar^2}{32p_0}\left(2\alpha'(z)^2 r_\perp^2 - \frac{1}{2}\alpha'(z)\alpha'''(z)r_\perp^4\right)\{\beta\chi, \alpha_z\}$$
$$+ \frac{\hbar^2}{128p_0^2}\alpha'''(z)\left[\beta\chi, [\alpha_x\widehat{p}_y - \alpha_y\widehat{p}_x, r_\perp^2]\right], \tag{6.71}$$

with $\alpha(z) = qB(z)/2p_0$.

It should be noted that we are interested in the transfer map for the transverse coordinates and momenta of the beam particle from the field-free input region to the field-free output region. Hence, we can take the transverse position operator to be $\vec{r}_\perp$ and the transverse momentum operator to be $\vec{\hat{p}}_\perp$, as was found earlier in the discussion of free propagation. Calculation of the transfer maps using the above quantum beam optical Dirac Hamiltonian is exactly along the same lines as in the scalar theory, except that now we have to use (6.40) for the definition of averages. In $\widehat{\mathscr{H}_o}$, the quantum beam optical Dirac Hamiltonian for the magnetic round lens, $\widehat{\mathscr{H}_{o,p}}$ is the paraxial term leading to the point-to-point imaging, and $\widehat{\mathscr{H}_o'}$ is the perturbation term responsible for the main third-order aberrations. These two terms and $\widehat{\mathscr{H}_o}^{(\hbar)}$ are essentially scalar terms, $\propto \mathbb{I}$, which act on each component of the Dirac spinor $|\psi(z)\rangle$ individually without mixing any of them. The $\hbar$-dependent $\widehat{\mathbb{H}}_o^{(\hbar)}$ is a matrix term that mixes the components of $|\psi(z)\rangle$ and carries the signature of spin explicitly. Both the $\hbar$-dependent terms, $\widehat{\mathscr{H}_o}^{(\hbar)}$ and $\widehat{\mathbb{H}}_o^{(\hbar)}$, being proportional to the de Broglie wavelength, might be negligible under the conditions in electron microscopy. This justifies ignoring the electron spin and approximating the Dirac theory by the scalar Klein–Gordon theory in electron microscopy (see, *e.g.*, Hawkes and Kasper [72], Ferwerda, Hoenders, and Slump [48, 49], and Lubk [129] for details of such an approximation procedure different from our analysis). However, the following should be pointed out: If the Dirac theory is approximated by a scalar theory, with the inclusion of the $\hbar$-dependent scalar terms, it will differ from the Klein–Gordon theory in the $\hbar$-dependent scalar terms, as is seen by comparing $\widehat{\mathscr{H}_o}^{(\hbar)}$ in (6.70) above and the $\hbar$-dependent scalar Hermitian terms of the Klein–Gordon theory (5.213).

The matrix part in $\widehat{\mathscr{H}_o}$ in the Dirac theory, $\widehat{\mathbb{H}}_o^{(\hbar)}$, adds to the deviation from the Klein–Gordon theory. Without further ado, let us just note that the aberrations of position $((\delta x)(z_{obj}), (\delta y)(z_{obj}))$, given in (5.201) and (5.202), get additional contributions of every type from the matrix part. For example, the additional contributions of spherical aberration type to $(\delta x)(z_{obj})$ and $(\delta y)(z_{obj})$ are, respectively,

$$\frac{1}{p_0^3} \mathbb{C}_s^{(\hbar)} \langle \widehat{p}_x \widehat{p}_\perp^2 \Sigma_z \rangle (z_{obj}) = \frac{\hbar}{8 p_0^4} \int_{z_{obj}}^{z_{img}} dz \left(6\alpha^2 \alpha'' h^4 - \alpha''' h^3 h' + 4\alpha'' h^2 h'^2 \right)$$
$$\times \left(\langle \psi_1(z_{obj}) | \widehat{p}_x \widehat{p}_\perp^2 | \psi_1(z_{obj}) \rangle - \langle \psi_2(z_{obj}) | \widehat{p}_x \widehat{p}_\perp^2 | \psi_2(z_{obj}) \rangle \right.$$
$$\left. + \langle \psi_3(z_{obj}) | \widehat{p}_x \widehat{p}_\perp^2 | \psi_3(z_{obj}) \rangle - \langle \psi_4(z_{obj}) | \widehat{p}_x \widehat{p}_\perp^2 | \psi_4(z_{obj}) \rangle \right), \quad (6.72)$$

and

$$\frac{1}{p_0^3} \mathbb{C}_s^{(\hbar)} \langle \widehat{p}_y \widehat{p}_\perp^2 \Sigma_z \rangle (z_{obj}) = \frac{\hbar}{8 p_0^4} \int_{z_{obj}}^{z_{img}} dz \left(6\alpha^2 \alpha'' h^4 - \alpha''' h^3 h' + 4\alpha'' h^2 h'^2 \right)$$
$$\times \left(\langle \psi_1(z_{obj}) | \widehat{p}_y \widehat{p}_\perp^2 | \psi_1(z_{obj}) \rangle - \langle \psi_2(z_{obj}) | \widehat{p}_y \widehat{p}_\perp^2 | \psi_2(z_{obj}) \rangle \right.$$
$$\left. + \langle \psi_3(z_{obj}) | \widehat{p}_y \widehat{p}_\perp^2 | \psi_3(z_{obj}) \rangle - \langle \psi_4(z_{obj}) | \widehat{p}_y \widehat{p}_\perp^2 | \psi_4(z_{obj}) \rangle \right), \quad (6.73)$$

with $\alpha = \alpha(z)$, $h = h(z,z_{obj})$. Obviously, such contributions, with unequal weights for the four spinor components, would depend on the nature of $\left|\underline{\psi}(z_{obj})\right\rangle$, the spinor wave function at the object plane, with respect to spin.

It is to be noted that in the quantum beam optical Dirac Hamiltonian $\widehat{\mathscr{H}_o}$ in (6.67), the paraxial part $\widehat{\mathscr{H}_{o,p}}$ and the main perturbation part $\widehat{\mathscr{H}_o'}$ are scalar, i.e., $\propto \mathbb{I}$, and are identical to the respective paraxial part and the main perturbation part of the scalar quantum beam optical Hamiltonian (5.181). The additional $\hbar$-dependent scalar and matrix terms of the quantum beam optical Dirac Hamiltonian, though different from the $\hbar$-dependent terms of the scalar quantum beam optical Hamiltonian, are also very small, like the $\hbar$-dependent terms of the scalar quantum beam optical Hamiltonian, compared to the main paraxial and perturbation parts of the Hamiltonian which are responsible for the focusing and main aberration aspects of the magnetic round lens. Thus, we understand the behavior of a quasiparaxial beam of spin-1/2 particles propagating through an axially symmetric magnetic lens on the basis of the Dirac equation.

6.1.1.3 Bending Magnet

We have already discussed the classical mechanics and the scalar Klein–Gordon theory of the bending of a charged particle beam by a dipole magnet. Let us recall. A constant magnetic field in the vertical direction produced by a dipole magnet bends the beam along a circular arc in the horizontal plane. The circular arc of radius of curvature ρ, the design trajectory of the particle, is the optic axis of the bending magnet. It is natural to use the arclength, say S, measured along the optic axis from some reference point as the independent coordinate, instead of z. Let the reference particle moving along the design trajectory carry an orthonormal XY-coordinate frame with it. The X-axis is taken to be perpendicular to the tangent to the design orbit and in the same horizontal plane as the trajectory, and the Y-axis is taken to be in the vertical direction perpendicular to both the X-axis and the trajectory. The curved S-axis is along the design trajectory and perpendicular to both the X- and Y-axes at any point on the design trajectory. The instantaneous position of the reference particle in the design trajectory at an arclength S from the reference point corresponds to $X = 0$ and $Y = 0$. Let any particle of the beam have coordinates (x,y,z) with respect to a fixed right-handed Cartesian coordinate frame with its origin at the reference point on the design trajectory from which the arclength S is measured. Then, the two sets of coordinates of any particle of the beam, (X,Y,S) and (x,y,z), will be related as follows:

$$x = (\rho + X)\cos\left(\frac{S}{\rho}\right) - \rho, \quad z = (\rho + X)\sin\left(\frac{S}{\rho}\right), \quad y = Y \qquad (6.74)$$

To understand the bending of the path of a Dirac particle by a dipole magnet, we have to use the Dirac equation written in the curved (X,Y,S)-coordinate system. To this end, following Jagannathan [80], we borrow the formalism from the theory of general relativity (see, e.g., Brill and Wheeler [16]). Formalism of general relativity has been

Spinor Theory for Spin-1/2 Particles

adopted in particle beam optics by Wei, Li, and Sessler [186] also in the context of studying crystalline beams.

The Dirac equation for a free particle of mass m is

$$i\hbar \frac{\partial}{\partial t} \underline{\psi}(\vec{r},t) = \left[\beta mc^2 + c\vec{\alpha} \cdot \left(-i\hbar \vec{\nabla}\right) \right] \underline{\psi}(\vec{r},t). \tag{6.75}$$

Multiplying both sides from left by $-i\beta/c$, this equation can be written in the relativistically covariant form as

$$\left[\gamma^0 \left(i\hbar \frac{\partial}{\partial x^0} \right) - \sum_{j=1}^{3} \gamma^j \left(-i\hbar \frac{\partial}{\partial x^j} \right) \right] \underline{\psi}(x^0, \{x^j\}) = -imc\underline{\psi}(x^0, \{x^j\}), \tag{6.76}$$

or

$$\left(\sum_{j=0}^{3} \gamma^j \frac{\partial}{\partial x^j} + \frac{mc}{\hbar} \right) \underline{\psi}(\{x^j\}) = 0, \tag{6.77}$$

where $(x^0 = ct, x^1 = x, x^2 = y, x^3 = z)$ and the γ-matrices are defined by

$$\gamma^0 = -i\beta = \begin{pmatrix} -i\mathbb{I} & 0 \\ 0 & i\mathbb{I} \end{pmatrix}, \quad \gamma^1 = -i\beta\alpha_x = \begin{pmatrix} 0 & -i\sigma_x \\ i\sigma_x & 0 \end{pmatrix},$$

$$\gamma^2 = -i\beta\alpha_y = \begin{pmatrix} 0 & -i\sigma_y \\ i\sigma_y & 0 \end{pmatrix}, \quad \gamma^3 = -i\beta\alpha_z = \begin{pmatrix} 0 & -i\sigma_z \\ i\sigma_z & 0 \end{pmatrix}. \tag{6.78}$$

Note that

$$(\gamma^0)^2 = -\mathbb{I}, \quad (\gamma^1)^2 = \mathbb{I}, \quad (\gamma^2)^2 = \mathbb{I}, \quad (\gamma^3)^2 = \mathbb{I},$$
$$\{\gamma^j, \gamma^k\} = 0, \quad \text{for } j \neq k, \tag{6.79}$$

or we can write

$$\{\gamma^j, \gamma^k\} = 2\eta^{jk}\mathbb{I}, \qquad j,k = 0,1,2,3., \tag{6.80}$$

where

$$\left[\eta^{jk}\right] = \begin{pmatrix} -1 & 0 & 0 & 0 \\ 0 & 1 & 0 & 0 \\ 0 & 0 & 1 & 0 \\ 0 & 0 & 0 & 1 \end{pmatrix}, \tag{6.81}$$

is the contravariant metric tensor of the flat space-time. The corresponding covariant metric tensor is

$$\left[\eta_{jk}\right] = \left[\eta^{jk}\right]^{-1} = \left[\eta^{jk}\right] = \begin{pmatrix} -1 & 0 & 0 & 0 \\ 0 & 1 & 0 & 0 \\ 0 & 0 & 1 & 0 \\ 0 & 0 & 0 & 1 \end{pmatrix}, \tag{6.82}$$

such that $[\eta_{jk}][\eta^{jk}] = \mathbb{I}$, or $\sum_{\ell=0}^{3}\eta_{k\ell}\eta^{\ell j} = \delta_k^j$. Defining

$$\gamma_j = \sum_{k=0}^{3}\eta_{jk}\gamma^k, \qquad j=0,1,2,3., \tag{6.83}$$

we find

$$\{\gamma_j,\gamma_k\} = 2\eta_{jk}\mathbb{I}, \qquad j,k=0,1,2,3. \tag{6.84}$$

The coordinates of a space-time point form a contravariant four-vector $(x^j) = (x^0,x^1,x^2,x^3)$. The components of the corresponding covariant four-vector $(x_j) = (x_0,x_1,x_2,x_3)$ are given by

$$x_j = \sum_{k=0}^{4}\eta_{jk}x^k, \qquad j=0,1,2,3,. \tag{6.85}$$

or

$$x_0 = -x^0, \quad x_1 = x^1, \quad x_2 = x^2, \quad x_3 = x^3. \tag{6.86}$$

Using the Einstein summation convention, in which it is understood that repeated indices are to be summed over, we can write $\sum_{k=0}^{4}\eta_{jk}x^k = \eta_{jk}x^k$. Thus, we have $\eta_{k\ell}\eta^{\ell j} = \delta_k^j$. The relativistically invariant distance ds between two points in flat space-time, say, (x^0,x^1,x^2,x^3) and $(x^0+dx^0,x^1+dx^1,x^2+dx^2,x^3+dx^3)$, is given by

$$\begin{aligned}ds^2 &= \eta_{jk}dx^j dx^k = \eta^{jk}dx_j dx_k = dx^j dx_j \\ &= -(dx^0)^2 + (dx^1)^2 + (dx^2)^2 + (dx^3)^2 \\ &= -(dx_0)^2 + (dx_1)^2 + (dx_2)^2 + (dx_3)^2.\end{aligned} \tag{6.87}$$

Note that writing equation (6.77) as

$$\left(\gamma^j\frac{\partial}{\partial x^j}\right)\underline{\psi}(\{x^j\}) = -\left(\frac{mc}{\hbar}\right)\underline{\psi}(\{x^j\}), \tag{6.88}$$

implies immediately, from the algebra of γ-matrices (6.80),

$$\left(\gamma^j\frac{\partial}{\partial x^j}\right)^2\underline{\psi}(\{x^j\}) = \left(-\frac{mc}{\hbar}\right)^2\underline{\psi}(\{x^j\}), \tag{6.89}$$

or,

$$\left(\nabla^2 - \frac{1}{c^2}\frac{\partial^2}{\partial t^2}\right)\underline{\psi}(\vec{r},t) = \left(\frac{mc}{\hbar}\right)^2\underline{\psi}(\vec{r},t), \tag{6.90}$$

the free-particle Klein–Gordon equation for each component of the free-particle Dirac spinor.

Let the curved space-time, in which we are interested, have $(\tilde{x}^\mu) = (\tilde{x}^0, \tilde{x}^1, \tilde{x}^2, \tilde{x}^3)$ and $(\tilde{x}_\mu) = (\tilde{x}_0, \tilde{x}_1, \tilde{x}_2, \tilde{x}_3)$ as the contravariant and covariant coordinates, respectively, and the metric given by

$$ds^2 = \tilde{g}^{\mu\nu} d\tilde{x}_\mu d\tilde{x}_\nu = \tilde{g}_{\mu\nu} d\tilde{x}^\mu d\tilde{x}^\nu, \tag{6.91}$$

with

$$\tilde{g}_{\mu\nu} = \tilde{g}_{\nu\mu}, \quad \tilde{g}^{\mu\nu} = \tilde{g}^{\nu\mu}, \quad \tilde{g}^{\mu\lambda} \tilde{g}_{\lambda\nu} = \delta^\mu_\nu. \tag{6.92}$$

Let

$$dx^j = b^j_\mu d\tilde{x}^\mu, \quad d\tilde{x}^\mu = a^\mu_k dx^k, \quad \text{with } a^\mu_k b^k_\nu = \delta^\mu_\nu. \tag{6.93}$$

Since we should have

$$\eta_{jk} dx^j dx^k = \eta_{jk} b^j_\mu b^k_\nu d\tilde{x}^\mu d\tilde{x}^\nu, \tag{6.94}$$

we get the relation

$$\eta_{jk} b^j_\mu b^k_\nu = \tilde{g}_{\mu\nu}. \tag{6.95}$$

With

$$\gamma_j = \eta_{jk} \gamma^k, \tag{6.96}$$

we see from (6.80) that

$$\{\gamma_j, \gamma_k\} = 2\eta_{jk}. \tag{6.97}$$

Defining

$$\tilde{\gamma}_\mu = b^j_\mu \gamma_j, \tag{6.98}$$

we get

$$\{\tilde{\gamma}_\mu, \tilde{\gamma}_\nu\} = b^j_\mu b^k_\nu \{\gamma_j, \gamma_k\} = 2\eta_{jk} b^j_\mu b^k_\nu = 2\tilde{g}_{\mu\nu}. \tag{6.99}$$

Corresponding to the covariant $\{\tilde{\gamma}_\mu\}$, we have the contravariant $\{\tilde{\gamma}^\mu\}$ given by

$$\tilde{\gamma}^\mu = \tilde{g}^{\mu\nu} \tilde{\gamma}_\nu. \tag{6.100}$$

Now, the free-particle Dirac equation in the curved space-time is taken as

$$\left[\tilde{\gamma}^\mu \left(\frac{\partial}{\partial \tilde{x}^\mu} - \Gamma_\mu \right) + \frac{mc}{\hbar} \right] \psi(\{\tilde{x}^\mu\}) = 0, \tag{6.101}$$

where the spin connection matrices $\{\Gamma_\mu\}$ are given by

$$\Gamma_\mu = \tilde{g}_{\alpha\beta} \left(a^\alpha_j \frac{\partial b^j_\nu}{\partial x^\mu} - \Gamma^\alpha_{\mu\nu} \right) \sigma^{\beta\nu}, \tag{6.102}$$

with

$$\Gamma^\alpha_{\mu\nu} = \frac{1}{2} \tilde{g}^{\alpha\beta} \left(\frac{\partial \tilde{g}_{\nu\beta}}{\partial x^\mu} + \frac{\partial \tilde{g}_{\mu\beta}}{\partial x^\nu} - \frac{\partial \tilde{g}_{\mu\nu}}{\partial x^\beta} \right), \tag{6.103}$$

the Christoffel symbols, and

$$\sigma^{\beta\nu} = \frac{1}{2}\left[\tilde{\gamma}^\beta, \tilde{\gamma}^\nu\right]. \tag{6.104}$$

Note that $\{\Gamma^\alpha_{\mu\nu}\}$ are symmetric in μ and ν, and hence, there are 40 such symbols. In the presence of an electromagnetic field, associated with the four-potential $\{A_\mu\}$, the equation for the Dirac particle of charge q and mass m in the curved space-time becomes

$$\left[\tilde{\gamma}^\mu\left(\frac{\partial}{\partial \tilde{x}^\mu} - \Gamma_\mu - \frac{iq}{\hbar}A_\mu\right) + \frac{mc}{\hbar}\right]\underline{\psi}(\{\tilde{x}^\mu\}) = 0, \tag{6.105}$$

using the principle of minimal coupling.

For the curved space we are interested in

$$\tilde{x}^0 = ct, \quad \tilde{x}^1 = X, \quad \tilde{x}^2 = Y, \quad \tilde{x}^3 = S, \tag{6.106}$$

with the relation between (X, Y, S) and (x, y, z) given by (6.74). Thus, from (6.74) we find that, with $\zeta = 1 + (X/\rho) = 1 + \kappa X$,

$$\begin{pmatrix} dx^0 \\ dx^1 \\ dx^2 \\ dx^3 \end{pmatrix} = \begin{pmatrix} 1 & 0 & 0 & 0 \\ 0 & \cos(\kappa S) & 0 & -\zeta\sin(\kappa S) \\ 0 & 0 & 1 & 0 \\ 0 & \sin(\kappa S) & 0 & \zeta\cos(\kappa S) \end{pmatrix} \begin{pmatrix} d\tilde{x}^0 \\ d\tilde{x}^1 \\ d\tilde{x}^2 \\ d\tilde{x}^3 \end{pmatrix}, \tag{6.107}$$

defining the relation $dx^j = b^j_\mu d\tilde{x}^\mu$. The inverse relation, $d\tilde{x}^\mu = a^\mu_k dx^k$, is given by

$$\begin{pmatrix} d\tilde{x}^0 \\ d\tilde{x}^1 \\ d\tilde{x}^2 \\ d\tilde{x}^3 \end{pmatrix} = \begin{pmatrix} 1 & 0 & 0 & 0 \\ 0 & \cos(\kappa S) & 0 & \sin(\kappa S) \\ 0 & 0 & 1 & 0 \\ 0 & -\frac{1}{\zeta}\sin(\kappa S) & 0 & \frac{1}{\zeta}\cos(\kappa S) \end{pmatrix} \begin{pmatrix} dx^0 \\ dx^1 \\ dx^2 \\ dx^3 \end{pmatrix}. \tag{6.108}$$

From (6.95), we get

$$\tilde{g}_{00} = -1, \quad \tilde{g}_{11} = \tilde{g}_{22} = 1, \quad \tilde{g}_{33} = \zeta^2, \quad \tilde{g}_{\mu\nu} = 0 \text{ for } \mu \neq \nu, \tag{6.109}$$

and from (6.92), it follows that

$$\tilde{g}^{00} = -1, \quad \tilde{g}^{11} = \tilde{g}^{22} = 1, \quad \tilde{g}^{33} = \frac{1}{\zeta^2}, \quad \tilde{g}^{\mu\nu} = 0 \text{ for } \mu \neq \nu. \tag{6.110}$$

Among the 40 Christoffel symbols $\{\Gamma^\alpha_{\mu\nu}\}$, only three are nonvanishing:

$$\Gamma^3_{13} = \frac{\kappa}{\zeta}, \quad \Gamma^1_{33} = -\kappa\zeta, \quad \Gamma^3_{31} = \frac{\kappa}{\zeta}. \tag{6.111}$$

Calculating the spin connection matrices, using (6.102), we find all of them vanishing:
$$\Gamma_0 = \Gamma_1 = \Gamma_2 = \Gamma_3 = 0. \tag{6.112}$$

From (6.83), (6.98), and (6.100), we have
$$\tilde{\gamma}^0 = \gamma^0,$$
$$\tilde{\gamma}^1 = \cos(\kappa S)\gamma^1 + \sin(\kappa S)\gamma^3,$$
$$\tilde{\gamma}^2 = \gamma^2,$$
$$\tilde{\gamma}^3 = \frac{1}{\zeta}\left[-\sin(\kappa S)\gamma^1 + \cos(\kappa S)\gamma^3\right]. \tag{6.113}$$

Thus, the equation for a free Dirac particle in the curved space of interest to us becomes
$$\left[\tilde{\gamma}^0 \frac{\partial}{\partial \tilde{x}^0} + \tilde{\gamma}^1 \frac{\partial}{\partial \tilde{x}^1} + \tilde{\gamma}^2 \frac{\partial}{\partial \tilde{x}^2} + \tilde{\gamma}^3 \frac{\partial}{\partial \tilde{x}^3} + \frac{mc}{\hbar}\right] \underline{\psi}(\{\tilde{x}^\mu\}) = 0. \tag{6.114}$$

Writing explicitly, this equation is
$$\left\{\gamma^0 \frac{1}{c}\frac{\partial}{\partial t} + [\cos(\kappa S)\gamma^1 + \sin(\kappa S)\gamma^3]\frac{\partial}{\partial X} + \gamma^2 \frac{\partial}{\partial Y}\right.$$
$$\left. + [-\sin(\kappa S)\gamma^1 + \cos(\kappa S)\gamma^3]\left(\frac{1}{\zeta}\frac{\partial}{\partial S}\right) + \frac{mc}{\hbar}\right\}\underline{\psi}\left(\vec{R}_\perp, S, t\right) = 0. \tag{6.115}$$

In the case of the dipole magnet used for bending the beam, the required magnetic field is given by
$$B_X = 0, \quad B_Y = B_0, \quad B_S = 0. \tag{6.116}$$

The corresponding vector potential has the components
$$A_X = 0, \quad A_Y = 0, \quad A_S = -B_0\left(X - \frac{\kappa X^2}{2\zeta}\right), \tag{6.117}$$

such that
$$B_X = (\vec{\nabla} \times \vec{A})_X = \frac{1}{\zeta}\left(\frac{\partial(\zeta A_S)}{\partial Y} - \frac{\partial A_Y}{\partial S}\right) = 0,$$
$$B_Y = (\vec{\nabla} \times \vec{A})_Y = \frac{1}{\zeta}\left(\frac{\partial A_X}{\partial S} - \frac{\partial(\zeta A_S)}{\partial X}\right) = B_0,$$
$$B_S = (\vec{\nabla} \times \vec{A})_S = \frac{\partial A_Y}{\partial X} - \frac{\partial A_X}{\partial Y} = 0. \tag{6.118}$$

Now, following (6.105), we get the equation for the Dirac particle moving in the field of the dipole magnet as

$$\left\{ \gamma^0 \frac{1}{c} \frac{\partial}{\partial t} + [\cos(\kappa S)\gamma^1 + \sin(\kappa S)\gamma^3] \frac{\partial}{\partial X} + \gamma^2 \frac{\partial}{\partial Y} \right.$$
$$+ [-\sin(\kappa S)\gamma^1 + \cos(\kappa S)\gamma^3] \left(\frac{1}{\zeta}\right) \left(\frac{\partial}{\partial S} - \frac{iq}{\hbar} \zeta A_S\right)$$
$$\left. + \frac{mc}{\hbar} \right\} \underline{\psi} \left(\vec{R}_\perp, S, t\right) = 0. \qquad (6.119)$$

Multiplying throughout from the left by $-i\hbar c\gamma^0$ and rearranging the terms, we can rewrite this equation as

$$i\hbar \frac{\partial}{\partial t} \underline{\psi}\left(\vec{R}_\perp, S, t\right) = \left\{ mc^2\beta + c\alpha_X \left(-i\hbar \frac{\partial}{\partial X}\right) + c\alpha_Y \left(-i\hbar \frac{\partial}{\partial Y}\right) \right.$$
$$\left. + c\alpha_S \left(-\frac{i\hbar}{\zeta} \frac{\partial}{\partial S} - qA_S\right) \right\} \underline{\psi}\left(\vec{R}_\perp, S, t\right), \qquad (6.120)$$

where

$$\alpha_X = \cos(\kappa S)\alpha_x + \sin(\kappa S)\alpha_z$$
$$= \begin{pmatrix} 0 & \cos(\kappa S)\sigma_x + \sin(\kappa S)\sigma_z \\ \cos(\kappa S)\sigma_x + \sin(\kappa S)\sigma_z & 0 \end{pmatrix},$$
$$\alpha_Y = \alpha_y = \begin{pmatrix} 0 & \sigma_y \\ \sigma_y & 0 \end{pmatrix},$$
$$\alpha_S = -\sin(\kappa S)\alpha_x + \cos(\kappa S)\alpha_z$$
$$= \begin{pmatrix} 0 & -\sin(\kappa S)\sigma_x + \cos(\kappa S)\sigma_z \\ -\sin(\kappa S)\sigma_x + \cos(\kappa S)\sigma_z & 0 \end{pmatrix}. \qquad (6.121)$$

Note that α_X, α_Y, and α_S are Hermitian, odd operators, and anticommute with each other and with β. Further, they obey $\alpha_X^2 = \alpha_Y^2 = \alpha_S^2 = \mathbb{I}$. In other words, α_X, α_Y, and α_S obey exactly the same properties as the standard Dirac matrices α_x, α_y, and α_z.

Let us now consider a monoenergetic paraxial Dirac particle beam passing through the bending magnet. Taking

$$\underline{\Psi}\left(\vec{R}_\perp, S, t\right) = \underline{\psi}\left(\vec{R}_\perp, S\right) e^{-\frac{i}{\hbar}Et}, \qquad (6.122)$$

with $E = \sqrt{m^2c^4 + c^2p_0^2}$ as the total conserved energy of the beam particle with design momentum p_0, we get the time-independent equation

$$\left\{ mc^2\beta + c\alpha_X \left(-i\hbar \frac{\partial}{\partial X}\right) + c\alpha_Y \left(-i\hbar \frac{\partial}{\partial Y}\right) \right.$$
$$\left. + c\alpha_S \left(-\frac{i\hbar}{\zeta} \frac{\partial}{\partial S} - qA_S\right) \right\} \left|\underline{\psi}(S)\right\rangle = E \left|\underline{\psi}(S)\right\rangle. \qquad (6.123)$$

Spinor Theory for Spin-1/2 Particles

Multiplying this equation on both sides from the left by α_S/c and rearranging the terms, we get

$$\frac{i\hbar}{\zeta}\frac{\partial}{\partial S}\left|\underline{\psi}(S)\right\rangle = \widetilde{\mathscr{H}}_\zeta\left|\underline{\psi}(S)\right\rangle,$$

$$\widetilde{\mathscr{H}}_\zeta = -p_0\beta\chi\alpha_S - qA_S\mathbb{I} + \alpha_S\left(\alpha_X\widehat{P}_X + \alpha_Y\widehat{P}_Y\right), \qquad (6.124)$$

where $\widehat{P}_X = -i\hbar(\partial/\partial X)$, $\widehat{P}_Y = -i\hbar(\partial/\partial Y)$, and χ is same as in (6.7):

$$\chi = \begin{pmatrix} \xi\mathbb{I} & 0 \\ 0 & -\xi^{-1}\mathbb{I} \end{pmatrix}, \quad \text{with } \xi = \sqrt{\frac{E+mc^2}{E-mc^2}}. \qquad (6.125)$$

We can now proceed exactly in the same way as we handled (6.7).

Observing that, with

$$\widetilde{M} = \frac{1}{\sqrt{2}}(\mathbb{I} + \chi\alpha_S), \qquad \widetilde{M}^{-1} = \frac{1}{\sqrt{2}}(\mathbb{I} - \chi\alpha_S), \qquad (6.126)$$

one has

$$\widetilde{M}\beta\chi\alpha_S\widetilde{M}^{-1} = \beta, \qquad (6.127)$$

we define

$$\left|\underline{\psi}'(S)\right\rangle = \widetilde{M}\left|\underline{\psi}(S)\right\rangle. \qquad (6.128)$$

Then,

$$\frac{i\hbar}{\zeta}\frac{\partial}{\partial S}\left|\underline{\psi}'(S)\right\rangle = \widetilde{\mathscr{H}}'_\zeta\left|\underline{\psi}'(S)\right\rangle,$$

$$\widetilde{\mathscr{H}}'_\zeta = \widetilde{M}\widetilde{\mathscr{H}}_\zeta\widetilde{M}^{-1} - \frac{i\hbar}{\zeta}\left[\widetilde{M}\left(\frac{\partial\widetilde{M}^{-1}}{\partial S}\right)\right]$$

$$= -p_0\beta + \widetilde{\mathscr{E}} + \widetilde{\mathscr{O}},$$

$$\widetilde{\mathscr{E}} = -qA_S\mathbb{I} + i\left[\cos(\kappa S)\Sigma_x + \sin(\kappa S)\Sigma_z\right]\widehat{P}_Y - \frac{\hbar\kappa}{2\zeta}\beta\Sigma_y,$$

$$\widetilde{\mathscr{O}} = \chi\left[\alpha_X\left(\widehat{P}_X - \frac{i}{2}\hbar\kappa\right) + \alpha_Y\widehat{P}_Y\right], \qquad (6.129)$$

where the even term $\widetilde{\mathscr{E}}$ commutes with β and the odd term $\widetilde{\mathscr{O}}$ anticommutes with β. To obtain the desired quantum beam optical representation of the S-evolution equation, we have to perform the Foldy–Wouthuysen-like transformations. The first transformation is

$$\left|\underline{\psi}^{(1)}\right\rangle = e^{i\widehat{S}_1}\left|\underline{\psi}'\right\rangle,$$

$$\widehat{S}_1 = \frac{i}{2p_0}\beta\widetilde{\mathscr{O}} = \frac{i}{2p_0}\beta\chi\left[\alpha_X\left(\widehat{P}_X - \frac{i}{2}\hbar\kappa\right) + \alpha_Y\widehat{P}_Y\right]. \qquad (6.130)$$

The resulting equation for $\left|\underline{\psi}^{(1)}\right\rangle$ is

$$\frac{i\hbar}{\zeta}\frac{\partial}{\partial S}\left|\underline{\psi}^{(1)}\right\rangle = \widetilde{\mathscr{H}}_\zeta^{(1)}\left|\underline{\psi}^{(1)}\right\rangle,$$

$$\widetilde{\mathscr{H}}_\zeta^{(1)} = e^{-\frac{1}{2p_0}\beta\widehat{\mathscr{O}}}\widetilde{\mathscr{H}}_\zeta' e^{\frac{1}{2p_0}\beta\widehat{\mathscr{O}}} - \frac{i\hbar}{\zeta}\left[e^{-\frac{1}{2p_0}\beta\widehat{\mathscr{O}}}\left(\frac{\partial}{\partial S}e^{\frac{1}{2p_0}\beta\widehat{\mathscr{O}}}\right)\right]$$

$$= -p_0\beta + \widehat{\mathscr{E}}^{(1)} + \widehat{\mathscr{O}}^{(1)},$$

$$\widehat{\mathscr{E}}^{(1)} = \widehat{\mathscr{E}} - \frac{1}{2p_0}\beta\widehat{\mathscr{O}}^2 - \cdots,$$

$$\widehat{\mathscr{O}}^{(1)} = -\frac{1}{2p_0}\beta\left(\left[\widehat{\mathscr{O}},\widehat{\mathscr{E}}\right] + \frac{i\hbar}{\zeta}\frac{\partial\widehat{\mathscr{O}}}{\partial S}\right) - \cdots. \quad (6.131)$$

Let us stop at this first Foldy–Wouthuysen-like transformation and calculate $\widetilde{\mathscr{H}}_\zeta^{(1)}$. The result is

$$\widetilde{\mathscr{H}}_\zeta^{(1)} \approx -p_0\beta - qA_S\mathbb{I} + \frac{1}{2p_0}\left(\widehat{P}_X^2 + \widehat{P}_Y^2 - \frac{1}{4}\hbar^2\kappa^2 - i\hbar\kappa\widehat{P}_X\right)\beta$$
$$+ i\left[\cos(\kappa S)\Sigma_x + \sin(\kappa S)\Sigma_z\right]\widehat{P}_Y - \frac{\hbar\kappa}{2\zeta}\beta\Sigma_y, \quad (6.132)$$

omitting the odd term. Substituting this $\widetilde{\mathscr{H}}_\zeta^{(1)}$ in (6.131), replacing β by $\mathbb{I}$ in the resulting equation since $\left|\underline{\psi}^{(1)}\right\rangle$ will have the lower pair of components very small compared to the upper pair of components, and returning to the original Dirac representation, we get

$$\frac{i\hbar}{\zeta}\frac{\partial}{\partial S}\left|\underline{\psi}(S)\right\rangle \approx \left\{\left[-qA_S - \left(p_0 + \frac{\hbar^2\kappa^2}{8p_0}\right) + \frac{1}{2p_0}\left(\widehat{P}_X^2 + \widehat{P}_Y^2 - i\hbar\kappa\widehat{P}_X\right)\right]\mathbb{I}\right.$$
$$\left. -\frac{\hbar\kappa}{2\zeta}\Sigma_y\right\}\left|\underline{\psi}(S)\right\rangle. \quad (6.133)$$

Multiplying both sides of this equation from the left by ζ, we can write

$$i\hbar\frac{\partial}{\partial S}\left|\underline{\psi}(S)\right\rangle \approx \left\{\zeta\left[-qA_S - \left(p_0 + \frac{\hbar^2\kappa^2}{8p_0}\right) + \frac{1}{2p_0}\left(\widehat{P}_X^2 + \widehat{P}_Y^2 - i\hbar\kappa\widehat{P}_X\right)\right]\mathbb{I}\right.$$
$$\left. -\frac{1}{2}\hbar\kappa\Sigma_y\right\}\left|\underline{\psi}(S)\right\rangle. \quad (6.134)$$

Following (6.117), we take

$$A_S = -B_0\left(X - \frac{\kappa X^2}{2\zeta}\right). \quad (6.135)$$

Spinor Theory for Spin-1/2 Particles

Further, let the curvature of the design trajectory be matched to the dipole magnetic field such that

$$qB_0 = \kappa p_0. \tag{6.136}$$

Then, simplifying and keeping only terms of order up to quadratic in (X,Y) and $\left(\widehat{P}_X, \widehat{P}_Y\right)$, we have

$$i\hbar \frac{\partial}{\partial S} \left|\psi(S)\right\rangle = \left\{ \left[-\left(p_0 + \frac{\hbar^2 \kappa^2}{8p_0}\right) + \frac{1}{2p_0}\left(\widehat{P}_X^2 + \widehat{P}_Y^2\right) + \frac{1}{2}p_0\kappa^2 X^2 \right.\right.$$
$$\left.\left. - \frac{\hbar^2 \kappa^3}{8p_0} X - \frac{i\hbar\kappa}{2p_0}(1+\kappa X)\widehat{P}_X \right]\mathbb{I} - \frac{1}{2}\hbar\kappa\Sigma_y \right\} \left|\psi(S)\right\rangle. \tag{6.137}$$

Now, let us replace the non-Hermitian term $i(1+\kappa X)\widehat{P}_X$ on the right-hand side by its Hermitian part,

$$\frac{1}{2}\left\{ \left(i(1+\kappa X)\widehat{P}_X\right) + \left(i(1+\kappa X)\widehat{P}_X\right)^\dagger \right\}$$
$$= \frac{1}{2}\left\{ i(1+\kappa X)\widehat{P}_X - i\widehat{P}_X(1+\kappa X) \right\} = -\frac{1}{2}\hbar\kappa. \tag{6.138}$$

This leads to

$$i\hbar \frac{\partial}{\partial S}\left|\psi(S)\right\rangle = \widetilde{\mathscr{H}}_{\text{D},d} \left|\psi(S)\right\rangle$$
$$\widetilde{\mathscr{H}}_{\text{D},d} = \left\{ \left[-p_0 + \frac{1}{2p_0}\widehat{P}_\perp^2 + \frac{1}{2}p_0\kappa^2 X^2 \right.\right.$$
$$\left.\left. + \frac{\hbar^2\kappa^2}{8p_0}(1-\kappa X) \right]\mathbb{I} - \frac{1}{2}\hbar\kappa\Sigma_y \right\}. \tag{6.139}$$

Note that this quantum beam optical Dirac Hamiltonian $\widetilde{\mathscr{H}}_{\text{D},d}$, like the scalar quantum beam optical Hamiltonian $\widetilde{\mathscr{H}}_{o,d}$ in (5.279), reproduces the classical beam optical Hamiltonian for the dipole magnet $\mathscr{H}_{o,d}$ in (4.185), when $\vec{P}_\perp$ and $\vec{R}_\perp$ are taken as the corresponding classical variables, and the $\hbar$-dependent terms are dropped. It differs from the scalar quantum beam optical Hamiltonian in the $\hbar$-dependent scalar term $\propto \mathbb{I}$ and has an extra $\hbar$-dependent matrix term that will have different effects on particles in different spin states. However, these $\hbar$-dependent terms are very small compared to the main classical part of the Hamiltonian, which leads to the bending of the particle trajectory by the dipole. Thus, we understand the bending action of the dipole magnet on a paraxial beam of spin-$\frac{1}{2}$ particles on the basis of the Dirac equation.

6.1.2 BEAM OPTICS OF THE DIRAC PARTICLE WITH ANOMALOUS MAGNETIC MOMENT

6.1.2.1 General Formalism

The main framework for studying the spin dynamics and beam polarization in accelerator physics is essentially based on the quasiclassical Thomas–Frenkel–Bargmann–Michel–Telegdi equation, or the Thomas–BMT equation as is commonly called (see Thomas [182, 183], Frenkel [57, 58], Bargmann, Michel, and Telegdi [6]). The Thomas–BMT equation has been understood on the basis of the Dirac equation, independent of accelerator beam optics, in different ways (see Sokolov and Ternov [173], Ternov [180], Corben [29], and references therein). As shown by Derbenev and Kondratenko [32], it is possible to obtain from the Dirac Hamiltonian, using the Foldy–Wouthuysen representation, a quasiclassical effective Hamiltonian accounting for the orbital motion, Stern–Gerlach effect, and the Thomas–BMT spin evolution (see Barber, Heinemann, and Ripken [4, 5]). The same quasiclassical effective Hamiltonian has also been justified by reducing the Dirac theory to the Pauli theory (see Jackson [77]). Based on such a quasiclassical Hamiltonian, a completely classical approach to accelerator beam dynamics has also been developed (see Barber, Heinemann, and Ripken [4, 5]) in which an extended classical canonical formalism is used by adding to the classical phase space two new real canonical variables describing all the three components of spin. Following Conte, Jagannathan, Khan, and Pusterla [26], we present here the spinor quantum charged particle beam optical formalism which, at the level of single particle theory, gives a unified account of orbital motion, Stern–Gerlach effect, and the Thomas–BMT spin evolution for a paraxial beam of Dirac particles with anomalous magnetic moment.

Let us consider the propagation of a paraxial beam of charged spin-$\frac{1}{2}$ particles with an anomalous magnetic moment through an optical element with straight optic axis. We shall take the optic axis of the system to be along the z-direction. As we have already seen, the z-evolution equation for the time-independent spinor wave function of the beam particle in the $\left|\psi'\right\rangle$-representation, (6.11), is

$$i\hbar \frac{\partial}{\partial z}\left|\psi'(z)\right\rangle = \widehat{\mathscr{H}}\left|\psi'(z)\right\rangle,$$

$$\widehat{\mathscr{H}} = -p_0\beta + \widehat{\mathscr{E}} + \widehat{\mathscr{O}},$$

$$\widehat{\mathscr{E}} = -qA_z\mathbb{I} + \frac{q\phi}{2c}\left(\xi + \xi^{-1}\right)\beta$$
$$\quad - \frac{\mu_a}{2c}\left[\left(\xi + \xi^{-1}\right)\vec{B}_\perp \cdot \vec{\Sigma}_\perp + \left(\xi - \xi^{-1}\right)\beta B_z \Sigma_z\right],$$

$$\widehat{\mathscr{O}} = \chi\left\{\vec{\alpha}_\perp \cdot \vec{\pi}_\perp + \frac{\mu_a}{2c}\left[\left(\xi - \xi^{-1}\right)(\sigma_y \otimes (B_x\sigma_y - B_y\sigma_x))\right.\right.$$
$$\left.\left.\quad + \left(\xi + \xi^{-1}\right)B_z(\sigma_x \otimes \mathbb{I})\right]\right\},$$

$$\text{with } \chi = \begin{pmatrix} \xi\mathbb{I} & 0 \\ 0 & -\xi^{-1}\mathbb{I} \end{pmatrix}, \quad \xi = \sqrt{\frac{E+mc^2}{E-mc^2}}, \quad (6.140)$$

Spinor Theory for Spin-1/2 Particles

and $E = \sqrt{m^2c^4 + c^2p_0^2}$ as the conserved total energy of a particle of the beam with the design momentum p_0. Let us recall that $\left|\underline{\psi}'\right\rangle$ is related to the Dirac $\left|\underline{\psi}\right\rangle$ by

$$\left|\underline{\psi}'\right\rangle = \mathsf{M}\left|\underline{\psi}(z)\right\rangle, \qquad \mathsf{M} = \frac{1}{\sqrt{2}}\left(\mathbb{I} + \chi\alpha_z\right). \tag{6.141}$$

To obtain the desired quantum beam optical representation of the z-evolution equation, we have to perform the Foldy–Wouthuysen-like transformations outlined earlier. The first transformation is

$$\left|\underline{\psi}^{(1)}\right\rangle = e^{i\widehat{S}_1}\left|\underline{\psi}'\right\rangle, \qquad \widehat{S}_1 = \frac{i}{2p_0}\beta\widehat{\mathcal{O}}. \tag{6.142}$$

The resulting equation for $\left|\underline{\psi}^{(1)}\right\rangle$ is

$$i\hbar\frac{\partial}{\partial z}\left|\underline{\psi}^{(1)}\right\rangle = \widehat{\mathscr{H}}^{(1)}\left|\underline{\psi}^{(1)}\right\rangle,$$

$$\widehat{\mathscr{H}}^{(1)} = e^{-\frac{1}{2p_0}\beta\widehat{\mathcal{O}}}\widehat{\mathscr{H}}e^{\frac{1}{2p_0}\beta\widehat{\mathcal{O}}} - i\hbar\left[e^{-\frac{1}{2p_0}\beta\widehat{\mathcal{O}}}\frac{\partial}{\partial z}\left(e^{\frac{1}{2p_0}\beta\widehat{\mathcal{O}}}\right)\right]$$

$$= -p_0\beta + \widehat{\mathcal{E}}^{(1)} + \widehat{\mathcal{O}}^{(1)},$$

$$\widehat{\mathcal{E}}^{(1)} \approx \widehat{\mathcal{E}} - \frac{1}{2p_0}\beta\widehat{\mathcal{O}}^2 - \cdots,$$

$$\widehat{\mathcal{O}}^{(1)} \approx -\frac{1}{2p_0}\beta\left(\left[\widehat{\mathcal{O}},\widehat{\mathcal{E}}\right] + i\hbar\frac{\partial\widehat{\mathcal{O}}}{\partial z}\right) - \cdots. \tag{6.143}$$

Substituting the expressions for $\widehat{\mathcal{E}}$, $\widehat{\mathcal{O}}$, and ξ from (6.140), we have,

$$\widehat{\mathscr{H}}^{(1)} = \left[-p_0 - qA_z + \frac{1}{2p_0}\widehat{\pi}_\perp^2 + \frac{m^2\mu_a^2}{2p_0^3}\left(B_\perp^2 + \frac{E^2}{m^2c^4}B_z^2\right)\right]\mathbb{I}$$

$$+ \frac{qE\phi}{c^2p_0}\beta + \frac{m\hbar\mu_a}{2p_0^2}\left(\nabla\times\vec{B}\right)_z\beta$$

$$- \frac{1}{2p_0}\left((\hbar q + 2m\mu_a)B_z\Sigma_z + \frac{E\mu_a}{c^2}\vec{B}_\perp\cdot\vec{\Sigma}_\perp\right)$$

$$+ \frac{m\mu_a}{2p_0^2}\left[\frac{E}{mc^2}\left(B_z\vec{\Sigma}_\perp\cdot\vec{\pi}_\perp + \vec{\Sigma}_\perp\cdot\vec{\pi}_\perp B_z\right) - \left(\vec{B}_\perp\cdot\vec{\pi}_\perp + \vec{\pi}_\perp\cdot\vec{B}_\perp\right)\beta\Sigma_z\right]. \tag{6.144}$$

Let us now approximate $\widehat{\mathscr{H}}^{(1)}$ keeping only terms of order $1/p_0$, consistent with the paraxial approximation. Then, with $2m\mu_a = aq\hbar$, $E = \gamma mc^2$, and $\vec{\mathbb{S}} = \frac{\hbar}{2}\vec{\Sigma}$, we have

$$i\hbar \frac{\partial}{\partial z} \left| \psi^{(1)} \right\rangle = \widehat{\mathscr{H}}^{(1)} \left| \psi^{(1)} \right\rangle,$$

$$\widehat{\mathscr{H}}^{(1)} \approx \left(-p_0 - qA_z + \frac{1}{2p_0} \widehat{\pi}_\perp^2 \right) \mathbb{I} + \frac{q\gamma m\phi}{p_0} \beta$$

$$- \frac{q}{p_0} \left((1+a) B_z \mathbb{S}_z + \gamma a \vec{B}_\perp \cdot \vec{\mathbb{S}}_\perp \right). \quad (6.145)$$

As we have seen already in (6.17), for a paraxial plane wave, $\left| \psi^{(1)} \right\rangle$ has almost vanishing lower pair of components compared to the upper pair of components.

Up to now all the observables, the field components, time, etc., are defined with reference to the laboratory frame. But, in the covariant description, the spin of the Dirac particle has simple operator representation in terms of the Pauli matrices only in a frame in which the particle is at rest. So, it is usual in accelerator physics to define spin with reference to the instantaneous rest frame of the particle while keeping the other observables, field components, etc., defined with reference to the laboratory frame. In the Dirac representation, the operator for spin as defined in the instantaneous rest frame of the particle is given by (see Sokolov and Ternov [173] for details)

$$\vec{\mathbb{S}}_R = \frac{1}{2} \hbar \left[\beta \left(\vec{\Sigma} - \frac{c^2 \left(\vec{\Sigma} \cdot \vec{\widehat{\pi}} + \vec{\widehat{\pi}} \cdot \vec{\Sigma} \right)}{2E(E+mc^2)} \right) + \frac{c\vec{\widehat{\pi}}}{E} \alpha_x \right]. \quad (6.146)$$

The relation between the Dirac $\left| \psi \right\rangle$-representation and the $\left| \psi^{(1)} \right\rangle$-representation being

$$\left| \psi^{(1)} \right\rangle = e^{-\frac{1}{2p_0} \beta \widehat{\mathcal{O}}} \mathrm{M} \left| \psi \right\rangle, \quad (6.147)$$

the operator for spin in the rest frame of the particle in the $\left| \psi^{(1)} \right\rangle$-representation is

$$\vec{\mathbb{S}}_R^{(1)} = e^{-\frac{1}{2p_0} \beta \widehat{\mathcal{O}}} \mathrm{M} \vec{\mathbb{S}}_R \mathrm{M}^{-1} e^{\frac{1}{2p_0} \beta \widehat{\mathcal{O}}}, \quad (6.148)$$

as follows from (3.134) since in this case the transformation is not unitary. Let us now make another transformation,

$$\left| \psi_A \right\rangle = \widehat{U}^{(A)} \left| \psi^{(1)} \right\rangle, \qquad \widehat{U}^{(A)} = e^{-\frac{i}{2p_0} \left(\widehat{\pi}_x \Sigma_y - \widehat{\pi}_y \Sigma_x \right)}. \quad (6.149)$$

Note that this transformation preserves the smallness of the lower pair of components compared to the upper pair of components. In the $\left| \psi_A \right\rangle$-representation, the operator for spin in the instantaneous rest frame of the particle becomes

$$\vec{\mathbb{S}}_R^{(A)} = \widehat{U}^{(A)} \vec{\mathbb{S}}_R^{(1)} \left(\widehat{U}^{(A)} \right)^\dagger \approx \frac{\hbar}{2} \vec{\Sigma} = \vec{\mathbb{S}}, \quad (6.150)$$

as we desire, considering the paraxial approximation and the effect of β on $\left| \psi_A \right\rangle$ being the same as $\mathbb{I}$. Using the transformation (6.149) in (6.145), and calculating up to the paraxial approximation, we get the z-evolution equation for $\left| \psi_A \right\rangle$ as

Spinor Theory for Spin-1/2 Particles

$$i\hbar \frac{\partial}{\partial z} \left| \underline{\psi}_A \right\rangle = \widehat{\mathscr{H}_A} \left| \underline{\psi}_A \right\rangle,$$

$$\widehat{\mathscr{H}_A} = \widehat{U}^{(A)} \widehat{\mathscr{H}}^{(1)} \left(\widehat{U}^{(A)} \right)^\dagger - i\hbar \widehat{U}^{(A)} \left[\frac{\partial}{\partial z} \left(\widehat{U}^{(A)} \right)^\dagger \right]$$

$$\approx \left(-p_0 - qA_z + \frac{1}{2p_0} \widehat{\pi}_\perp^2 \right) \mathbb{I} + \frac{q\gamma m\phi}{p_0} \beta$$

$$- \frac{q}{p_0} \left\{ \vec{B} \cdot \vec{S} + a \left(B_z S_z + \gamma \vec{B}_\perp \cdot \vec{S}_\perp \right) \right\}. \quad (6.151)$$

In the previous treatments of optical elements we used to return to the original Dirac representation by retracing the transformations made on it and the Dirac Hamiltonian. This led to the expansion of the Dirac Hamiltonian as a paraxial part plus a series of nonparaxial aberration parts so that we can work with the Dirac Hamiltonian approximated up to any desired level of accuracy. In the present case, we shall follow an equivalent procedure. We shall not return to the original Dirac representation. We shall continue to work with the $\left| \underline{\psi}_A \right\rangle$-representation. Thus, we have to model the wave function of any system in the $\left| \underline{\psi}_A \right\rangle$-representation. Since $\left| \underline{\psi}_A \right\rangle$ has the lower pair of components almost vanishing compared to the upper pair of components, we can as well take $\left| \underline{\psi}_A \right\rangle$ to be a two-component spinor. Since $\widehat{\mathscr{H}_A}$ is even, i.e., block-diagonal with the odd part zero, its 11-block element becomes the Hamiltonian in the two-component $\left| \underline{\psi}_A \right\rangle$-representation. We shall denote the two-component $\left| \underline{\psi}_A \right\rangle$ as $\left| \underline{\psi}^{(A)} \right\rangle$, which satisfies the z-evolution equation

$$i\hbar \frac{\partial}{\partial z} \left| \underline{\psi}^{(A)} \right\rangle = \widehat{\mathscr{H}}^{(A)} \left| \underline{\psi}^{(A)} \right\rangle,$$

$$\widehat{\mathscr{H}}^{(A)} \approx \left(-p_0 - qA_z + \frac{1}{2p_0} \widehat{\pi}_\perp^2 + \frac{q\gamma m\phi}{p_0} \right) \mathbb{I}$$

$$- \frac{q}{p_0} \left\{ \vec{B} \cdot \vec{S} + a \left(B_z S_z + \gamma \vec{B}_\perp \cdot \vec{S}_\perp \right) \right\}, \quad (6.152)$$

where $\vec{S} = (\hbar/2)\vec{\sigma}$. We shall now rewrite this equation as

$$i\hbar \frac{\partial}{\partial z} \left| \underline{\psi}^{(A)} \right\rangle = \widehat{\mathscr{H}}^{(A)} \left| \underline{\psi}^{(A)} \right\rangle,$$

$$\widehat{\mathscr{H}}^{(A)} \approx \left(-p_0 - qA_z + \frac{1}{2p_0} \widehat{\pi}_\perp^2 + \frac{q\gamma m\phi}{p_0} \right) \mathbb{I} + \frac{\gamma m}{p_0} \vec{\Omega} \cdot \vec{S},$$

$$\text{with } \vec{\Omega} = -\frac{q}{\gamma m} \left((1+a)\vec{B}_\parallel + (1+\gamma a)\vec{B}_\perp \right), \quad (6.153)$$

where $\vec{B}_\parallel$ and $\vec{B}_\perp$ are the components of $\vec{B}$ along the optic axis and perpendicular to it. We shall refer to the $\left| \underline{\psi}^{(A)} \right\rangle$-representation as the quantum accelerator optical representation.

Note that $\widehat{\mathcal{H}}^{(A)}$, the quantum accelerator optical Hamiltonian, is Hermitian. In other words, $\left| \underline{\psi}^{(A)} \right\rangle$ has unitary z-evolution, and hence it can be normalized as

$$\left\langle \underline{\psi}^{(A)}(z) \middle| \underline{\psi}^{(A)}(z) \right\rangle = \sum_{j=1}^{2} \int d^2 r_\perp \left| \psi_j^{(A)}(\vec{r}_\perp, z) \right|^2 = 1, \qquad (6.154)$$

at any z. When the beam is described by a 2×2 density matrix

$$\rho^{(A)}(z) = \begin{pmatrix} \rho_{11}^{(A)}(z) & \rho_{12}^{(A)}(z) \\ \rho_{21}^{(A)}(z) & \rho_{22}^{(A)}(z) \end{pmatrix}, \qquad (6.155)$$

at any z, the quantum accelerator optical z-evolution equation is

$$i\hbar \frac{\partial \rho^{(A)}(z)}{\partial z} = \left[\widehat{\mathcal{H}}^{(A)}, \rho^{(A)}(z) \right]. \qquad (6.156)$$

If the beam can be described as a pure state, we would have $\rho^{(A)}(z) = \left| \underline{\psi}^{(A)}(z) \right\rangle \left\langle \underline{\psi}^{(A)}(z) \right|$.

6.1.2.2 Lorentz and Stern–Gerlach Forces, and the Thomas–Frenkel–BMT Equation for Spin Dynamics

The average of any observable, represented by the operator $\widehat{O}$ in the quantum accelerator optical representation, is given by

$$\begin{aligned} \left\langle \widehat{O} \right\rangle(z) &= \mathrm{Tr}\left(\rho^{(A)}(z) \widehat{O} \right) \\ &= \sum_{j,k=1}^{2} \int \int d^2 r_\perp d^2 r'_\perp \left\langle \vec{r}_\perp \middle| \rho_{jk}^{(A)}(z) \middle| \vec{r}'_\perp \right\rangle \left\langle \vec{r}'_\perp \middle| \widehat{O}_{kj} \middle| \vec{r}_\perp \right\rangle, \end{aligned} \qquad (6.157)$$

in the transverse plane at z. It follows from (6.156) that the equation for z-evolution of $\left\langle \widehat{O} \right\rangle(z)$ is given by

$$\frac{d}{dz} \left\langle \widehat{O} \right\rangle(z) = \frac{i}{\hbar} \left\langle \left[\widehat{\mathcal{H}}^{(A)}, \widehat{O} \right] \right\rangle(z) + \left\langle \frac{\partial \widehat{O}}{\partial z} \right\rangle(z). \qquad (6.158)$$

Thus,

$$\frac{d \langle \vec{r}_\perp \rangle}{dz} = \frac{i}{\hbar} \left\langle \left[\widehat{\mathcal{H}}^{(A)}, \vec{r}_\perp \right] \right\rangle = \frac{1}{p_0} \langle \widehat{\pi}_\perp \rangle. \qquad (6.159)$$

Quasiclassically, for any observable $\mathcal{O}$, we can write

$$\frac{d \langle \widehat{O} \rangle}{dt} \approx v_z \frac{d \langle \widehat{O} \rangle}{dz} \approx \frac{p_0}{\gamma m} \frac{d \langle \widehat{O} \rangle}{dz}. \qquad (6.160)$$

Spinor Theory for Spin-1/2 Particles

For example, it follows from the above two equations that

$$\frac{d\langle\vec{r}_\perp\rangle}{dt} \approx \frac{p_0}{\gamma m}\frac{d\langle\vec{r}_\perp\rangle}{dz} = \frac{1}{\gamma m}\langle\vec{\hat{\pi}}_\perp\rangle, \tag{6.161}$$

as should be. For $\vec{\hat{\pi}}_\perp$ we have, with $\hat{\pi}_z \approx p_0$,

$$\begin{aligned}
\frac{d\langle\vec{\hat{\pi}}_\perp\rangle}{dz} &= \frac{i}{\hbar}\left\langle\left[\widehat{\mathscr{H}}^{(A)},\vec{\hat{\pi}}_\perp\right]\right\rangle + \left\langle\frac{\partial\vec{\hat{\pi}}_\perp}{\partial z}\right\rangle \\
&= \frac{i}{\hbar}\left\langle\left[\widehat{\mathscr{H}}^{(A)},\vec{\hat{\pi}}_\perp\right]\right\rangle - q\left\langle\frac{\partial\vec{A}_\perp}{\partial z}\right\rangle \\
&\approx \frac{q\gamma m}{p_0}\left\langle-\vec{\nabla}_\perp\phi\right\rangle + \frac{q}{2p_0}\left\langle\left(\vec{\hat{\pi}}\times\vec{B}-\vec{B}\times\vec{\hat{\pi}}\right)_\perp\right\rangle - \frac{\gamma m}{p_0}\left\langle\vec{\nabla}_\perp\left(\vec{\Omega}\cdot\vec{S}\right)\right\rangle \\
&= \frac{q\gamma m}{p_0}\left\langle\vec{E}_\perp\right\rangle + \frac{q}{2p_0}\left\langle\left(\vec{\hat{\pi}}\times\vec{B}-\vec{B}\times\vec{\hat{\pi}}\right)_\perp\right\rangle - \frac{\gamma m}{p_0}\left\langle\vec{\nabla}_\perp\left(\vec{\Omega}\cdot\vec{S}\right)\right\rangle. \tag{6.162}
\end{aligned}$$

Hence,

$$\frac{d\langle\vec{\hat{\pi}}_\perp\rangle}{dt} \approx q\langle\vec{E}_\perp\rangle + \frac{q}{2\gamma m}\left\langle\left(\vec{\hat{\pi}}\times\vec{B}-\vec{B}\times\vec{\hat{\pi}}\right)_\perp\right\rangle - \left\langle\vec{\nabla}_\perp\left(\vec{\Omega}\cdot\vec{S}\right)\right\rangle. \tag{6.163}$$

The quantum accelerator optical Hamiltonian $\widehat{\mathscr{H}}^{(A)}$ corresponds to $i\hbar(\partial/\partial z) = -\hat{p}_z$. Hence, we have to take the operator corresponding to π_z, the kinetic momentum in the longitudinal direction, as

$$\hat{\pi}_z = \hat{p}_z - qA_z = -\left(\widehat{\mathscr{H}}^{(A)} + qA_z\right). \tag{6.164}$$

Then, we have

$$\begin{aligned}
\frac{d\langle\hat{\pi}_z\rangle}{dz} &= \frac{i}{\hbar}\left\langle\left[\widehat{\mathscr{H}}^{(A)},-\left(\widehat{\mathscr{H}}^{(A)}+qA_z\right)\right]\right\rangle - \left\langle\frac{\partial}{\partial z}\left(\widehat{\mathscr{H}}^{(A)}+qA_z\right)\right\rangle \\
&= \frac{i}{\hbar}\left\langle\left[\widehat{\mathscr{H}}^{(A)},-qA_z\right]\right\rangle - \frac{q\gamma m}{p_0}\left\langle\frac{\partial\phi}{\partial z}\right\rangle \\
&\quad - \frac{1}{2p_0}\left\langle\frac{\partial}{\partial z}\hat{\pi}_\perp^2\right\rangle - \frac{\gamma m}{p_0}\left\langle\frac{\partial}{\partial z}\left(\vec{\Omega}\cdot\vec{S}\right)\right\rangle \\
&= \frac{q\gamma m}{p_0}\langle E_z\rangle + \frac{q}{2p_0}\left\langle\left(\vec{\hat{\pi}}\times\vec{B}-\vec{B}\times\vec{\hat{\pi}}\right)_z\right\rangle - \frac{\gamma m}{p_0}\left\langle\frac{\partial}{\partial z}\left(\vec{\Omega}\cdot\vec{S}\right)\right\rangle, \tag{6.165}
\end{aligned}$$

and

$$\frac{d\langle\hat{\pi}_z\rangle}{dt} \approx q\langle E_z\rangle + \frac{q}{2\gamma m}\left\langle\left(\vec{\hat{\pi}}\times\vec{B}-\vec{B}\times\vec{\hat{\pi}}\right)_z\right\rangle - \frac{\gamma m}{p_0}\left\langle\frac{\partial}{\partial z}\left(\vec{\Omega}\cdot\vec{S}\right)\right\rangle. \tag{6.166}$$

Combining the two equations (6.163) and (6.166), we get

$$\frac{d\langle\vec{\pi}\rangle}{dt} = \left\langle q\left[\vec{E} + \frac{1}{2}\left(\frac{\vec{\pi}}{\gamma m}\times\vec{B} - \vec{B}\times\frac{\vec{\pi}}{\gamma m}\right)\right]\right\rangle - \left\langle\vec{\nabla}\left(\vec{\Omega}\cdot\vec{S}\right)\right\rangle. \tag{6.167}$$

The first term of this equation represents the Lorentz force, $q\left[\vec{E}+\left(\vec{v}\times\vec{B}\right)\right]$. The second term represents the Stern–Gerlach force. In the instantaneous rest frame of the particle where $\gamma = 1$, the second term is seen to reduce to the familiar Stern–Gerlach force

$$\vec{F}_{SG} = -\vec{\nabla}V, \qquad V = -\mu\vec{\sigma}\cdot\vec{B}, \tag{6.168}$$

in which μ is the total spin magnetic moment of the particle, and apart from the spin, the field components, the coordinates, etc. are also defined in the rest frame of the particle. Equations (6.162) and (6.165), respectively, correspond to the transverse and longitudinal Stern–Gerlach kicks. For a discussion of relativistic Stern–Gerlach force using the Derbenev–Kondratenko semiclassical Hamiltonian, see Heinemann [74].

In the case of spin, we have

$$\begin{aligned}\frac{d\langle\vec{S}\rangle}{dz} &= \frac{i}{\hbar}\left\langle\left[\widehat{\mathscr{H}}^{(A)},\vec{S}\right]\right\rangle = \frac{i}{\hbar}\frac{\gamma m}{p_0}\left\langle\left[\vec{\Omega}\cdot\vec{S},\vec{S}\right]\right\rangle \\ &= \frac{\gamma m}{p_0}\left\langle\vec{\Omega}\times\vec{S}\right\rangle.\end{aligned} \tag{6.169}$$

As a result, we can write

$$\frac{d\langle\vec{S}\rangle}{dt} \approx \left\langle\vec{\Omega}\times\vec{S}\right\rangle, \tag{6.170}$$

which corresponds to the classical equation for spin dynamics,

$$\frac{d\vec{S}}{dt} = \vec{\Omega}\times\vec{S}, \tag{6.171}$$

the Thomas–Frenkel–Bargmann–Michel–Telegdi equation, or the Thomas–BMT equation. For more details on the Thomas–BMT equation, see, *e.g.*, Conte and MacKay [27], Lee [127], and Montague [136]. Thus, Ω in the quantum accelerator optical Hamiltonian $\widehat{\mathscr{H}}^{(A)}$ is the Thomas–BMT spin precession frequency. The polarization of the beam is characterized by the vector $\vec{P}$ defined by $(\hbar/2)\vec{P} = (\hbar/2)\langle\vec{\sigma}\rangle = \langle\vec{S}\rangle$. Thus, the quantum accelerator optical z-evolution equation (6.153) accounts for the orbital and spin motions of the particle, up to the paraxial approximation. To get higher order corrections, one has to go beyond the paraxial approximation by continuing the process of Foldy–Wouthuysen-like transformations up to the desired order.

Spinor Theory for Spin-1/2 Particles

6.1.2.3 Phase Space and Spin Transfer Maps for a Normal Magnetic Quadrupole

Let us now consider a paraxial beam of Dirac particles of charge q, mass m, and anomalous magnetic moment μ_a, propagating through a normal magnetic quadrupole with the optic axis along the z-direction. The practical boundaries of the lens can be taken to be at z_ℓ and z_r with $z_\ell < z_r$. Since there is no electric field, $\phi = 0$. The field of the normal magnetic quadrupole is

$$\vec{B}(\vec{r}) = (-Q_n y, -Q_n x, 0), \tag{6.172}$$

where

$$Q_n = \begin{cases} \text{constant in the lens region} & (z_\ell \le z \le z_r) \\ 0 \text{ outside the lens region} & (z < z_\ell, z > z_r) \end{cases} \tag{6.173}$$

The corresponding vector potential can be taken as

$$\vec{A}(\vec{r}) = \left(0, 0, \frac{1}{2} Q_n \left(x^2 - y^2\right)\right). \tag{6.174}$$

Now, from (6.153), the quantum accelerator optical Hamiltonian for the normal magnetic quadrupole is seen to be, with $w = z_r - z_\ell$, $K_n = qQ_n/p_0$, and $\eta_n = q(1+\gamma a)Q_n w/p_0^2$,

$$\widehat{\mathscr{H}}(z) = \begin{cases} -p_0 + \frac{\hat{p}_\perp^2}{2p_0}, & \text{for } z < z_\ell \text{ and } z > z_r, \\ -p_0 + \frac{\hat{p}_\perp^2}{2p_0} - \frac{1}{2} p_0 K_n \left(x^2 - y^2\right) \\ \quad + \frac{\eta_n p_0}{w} (y S_x + x S_y), & \text{for } z_\ell \le z \le z_r. \end{cases} \tag{6.175}$$

Here, we have denoted $\widehat{\mathscr{H}}^{(A)}$ simply as $\widehat{\mathscr{H}}$, and we shall do so in the rest of this section also. Similarly, we will write $\hat{\rho}(z)$ instead of $\hat{\rho}^{(A)}(z)$.

Let us consider the propagation of the beam from the transverse xy-plane at $z = z_i < z_\ell$ to the transverse plane at $z > z_r$. The beam propagates in free space from z_i to z_ℓ, passes through the lens from z_ℓ to z_r, and propagates through free space again from z_r to z. Let us now write the Hamiltonian $\widehat{\mathscr{H}}(z)$ as a core part $\widehat{\mathscr{H}}_0(z)$ plus a perturbation part $\widehat{\mathscr{H}}'(z)$:

$$\widehat{\mathscr{H}}(z) = \widehat{\mathscr{H}}_0(z) + \widehat{\mathscr{H}}'(z)$$

$$\widehat{\mathscr{H}}_0(z) = \begin{cases} -p_0 + \frac{\hat{p}_\perp^2}{2p_0}, & \text{for } z < z_\ell \text{ and } z > z_r, \\ -p_0 + \frac{\hat{p}_\perp^2}{2p_0} - \frac{1}{2} p_0 K_n \left(x^2 - y^2\right), & \text{for } z_\ell \le z \le z_r, \end{cases}$$

$$\widehat{\mathscr{H}}'(z) = \begin{cases} 0, & \text{for } z < z_\ell \text{ and } z > z_r, \\ \frac{\eta_n p_0}{w} (x S_y + y S_x), & \text{for } z_\ell \le z \le z_r. \end{cases} \tag{6.176}$$

Note that $\widehat{\mathscr{H}}(z)$ is Hermitian. A formal integration of (6.156), the z-evolution equation for $\hat{\rho}(z)$, gives

$$\hat{\rho}(z) = \hat{U}(z, z_i) \hat{\rho}(z_i) \hat{U}^\dagger(z, z_i), \tag{6.177}$$

with
$$\widehat{U}(z,z_i) = \mathsf{P}\left(e^{-\frac{i}{\hbar}\int_{z_i}^{z}dz.\widehat{\mathscr{H}}(z)}\right), \qquad \widehat{U}(z_i,z_i) = I, \tag{6.178}$$

and
$$i\hbar\frac{\partial}{\partial z}\widehat{U}(z,z_i) = \widehat{\mathscr{H}}(z)\widehat{U}(z,z_i),$$
$$i\hbar\frac{\partial}{\partial z}\widehat{U}^\dagger(z,z_i) = -\widehat{U}^\dagger(z,z_i)\widehat{\mathscr{H}}(z). \tag{6.179}$$

As we know, a convenient expression for $\widehat{U}(z,z_i)$ is given by the Magnus formula:

$$\widehat{U}(z,z_i) = e^{-\frac{i}{\hbar}\widehat{T}(z,z_i)},$$
$$\widehat{T}(z,z_i) = \int_{z_i}^{z}dz_1\,\widehat{\mathscr{H}}(z_1) + \frac{1}{2}\left(-\frac{i}{\hbar}\right)\int_{z_i}^{z}dz_2\int_{z_i}^{z_2}dz_1\left[\widehat{\mathscr{H}}(z_2),\widehat{\mathscr{H}}(z_1)\right]$$
$$+ \frac{1}{6}\left(-\frac{i}{\hbar}\right)^2\int_{z_i}^{z}dz_3\int_{z_i}^{z_3}dz_2\int_{z_i}^{z_2}dz_1\left\{\left[\left[\widehat{\mathscr{H}}(z_3),\widehat{\mathscr{H}}(z_2)\right],\widehat{\mathscr{H}}(z_1)\right]\right.$$
$$\left.+ \left[\left[\widehat{\mathscr{H}}(z_1),\widehat{\mathscr{H}}(z_2)\right],\widehat{\mathscr{H}}(z_3)\right]\right\} + \cdots. \tag{6.180}$$

Going to the interaction picture, let us define
$$\widehat{\rho}_\mathrm{i}(z) = \widehat{U}_0^\dagger(z,z_i)\widehat{\rho}(z)\widehat{U}_0(z,z_i), \tag{6.181}$$

with
$$\widehat{U}_0(z,z_i) = \mathsf{P}\left(e^{-\frac{i}{\hbar}\int_{z_i}^{z}dz.\widehat{\mathscr{H}_0}(z)}\right). \tag{6.182}$$

It follows that
$$i\hbar\frac{\partial\widehat{\rho}_\mathrm{i}(z)}{\partial z} = \left(i\hbar\frac{\partial}{\partial z}\widehat{U}_0^\dagger(z,z_i)\right)\widehat{\rho}(z)\widehat{U}_0(z,z_i) + \widehat{U}_0^\dagger(z,z_i)\left(i\hbar\frac{\partial\widehat{\rho}(z)}{\partial z}\right)\widehat{U}_0(z,z_i)$$
$$+ \widehat{U}_0^\dagger(z,z_i)\widehat{\rho}(z)\left(i\hbar\frac{\partial}{\partial z}\widehat{U}_0(z,z_i)\right)$$
$$= -\widehat{U}_0^\dagger(z,z_i)\widehat{\mathscr{H}_0}(z)\widehat{\rho}(z)\widehat{U}_0(z,z_i) + \widehat{U}_0^\dagger(z,z_i)\left[\widehat{\mathscr{H}}(z),\widehat{\rho}(z)\right]\widehat{U}_0(z,z_i)$$
$$+ \widehat{U}_0^\dagger(z,z_i)\widehat{\rho}(z)\widehat{\mathscr{H}_0}(z)\widehat{U}_0(z,z_i)$$
$$= \widehat{U}_0^\dagger(z,z_i)\left\{\left(\widehat{\mathscr{H}}(z) - \widehat{\mathscr{H}_0}(z)\right)\widehat{\rho}(z) - \widehat{\rho}(z)\left(\widehat{\mathscr{H}}(z) - \widehat{\mathscr{H}_0}(z)\right)\right\}$$
$$= \widehat{U}_0^\dagger(z,z_i)\left[\widehat{\mathscr{H}'}(z),\widehat{\rho}(z)\right]\widehat{U}_0(z,z_i)$$
$$= \left[\widehat{\mathscr{H}'_\mathrm{i}}(z),\widehat{\rho}_\mathrm{i}(z)\right] \tag{6.183}$$

where
$$\widehat{\mathscr{H}'_\mathrm{i}}(z) = \widehat{U}_0^\dagger(z,z_i)\widehat{\mathscr{H}'}(z)\widehat{U}_0(z,z_i). \tag{6.184}$$

Spinor Theory for Spin-1/2 Particles 325

Formally integrating (6.183), the z-evolution equation for $\widehat{\rho}_i(z)$, we get

$$\widehat{\rho}_i(z) = \widehat{U}'_i(z, z_i)\, \widehat{\rho}_i(z_i)\, \widehat{U}'^\dagger_i(z, z_i), \qquad (6.185)$$

where

$$\widehat{U}'_i(z, z_i) = \mathrm{P}\left(e^{-\frac{i}{\hbar} \int_{z_i}^{z} dz\, \widehat{\mathscr{H}}'_i(z)} \right)$$

$$= \mathrm{P}\left(e^{-\frac{i}{\hbar} \int_{z_i}^{z} dz\, \left(\widehat{U}_0^\dagger(z,z_i) \widehat{\mathscr{H}}'(z) \widehat{U}_0(z,z_i) \right)} \right). \qquad (6.186)$$

From (6.181), we find that

$$\widehat{\rho}_i(z_i) = \widehat{\rho}(z_i), \qquad (6.187)$$

and hence the equation (6.185) becomes

$$\widehat{\rho}_i(z) = \widehat{U}'_i(z, z_i)\, \widehat{\rho}(z_i)\, \widehat{U}'^\dagger_i(z, z_i). \qquad (6.188)$$

From (6.181), we have

$$\widehat{\rho}(z) = \widehat{U}_0(z, z_i)\, \widehat{\rho}_i(z)\, \widehat{U}_0^\dagger(z, z_i). \qquad (6.189)$$

Thus,

$$\widehat{\rho}(z) = \widehat{U}_0(z, z_i)\, \widehat{U}'_i(z, z_i)\, \widehat{\rho}(z_i)\, \widehat{U}'^\dagger_i(z, z_i)\, \widehat{U}_0^\dagger(z, z_i). \qquad (6.190)$$

The transfer maps for the observables of the system are to be calculated using the formula for the average

$$\left\langle \widehat{O} \right\rangle(z) = \mathrm{Tr}\left(\widehat{\rho}(z) \widehat{O} \right)$$

$$= \mathrm{Tr}\left(\widehat{U}_0(z, z_i)\, \widehat{U}'_i(z, z_i)\, \widehat{\rho}(z_i)\, \widehat{U}'^\dagger_i(z, z_i)\, \widehat{U}_0^\dagger(z, z_i)\, \widehat{O} \right)$$

$$= \mathrm{Tr}\left[\widehat{\rho}(z_i)\left(\widehat{U}'^\dagger_i(z, z_i)\, \widehat{U}_0^\dagger(z, z_i)\, \widehat{O}\, \widehat{U}_0(z, z_i)\, \widehat{U}'_i(z, z_i) \right) \right]. \qquad (6.191)$$

Note that in the field-free regions ($z < z_\ell$, $z > z_r$) and within the lens region ($z_\ell \leq z \leq z_r$), the Hamiltonian $\widehat{\mathscr{H}}(z)$ is z-independent. In view of this and the semi-group property of the z-evolution operator $\widehat{U}$, we have

$$\widehat{U}_0(z, z_i) = \widehat{U}_0(z, z_r)\, \widehat{U}_0(z_r, z_\ell)\, \widehat{U}_0(z_\ell, z_i)$$

$$= \exp\left\{ -\frac{i}{\hbar}(z - z_r)\left(-p_0 + \frac{\widehat{p}_\perp^2}{2p_0} \right) \right\}$$

$$\times \exp\left\{ -\frac{i}{\hbar}(z_r - z_\ell)\left(-p_0 + \frac{\widehat{p}_\perp^2}{2p_0} - \frac{1}{2}p_0 K_n\left(x^2 - y^2\right) \right) \right\}$$

$$\times \exp\left\{ -\frac{i}{\hbar}(z_\ell - z_i)\left(-p_0 + \frac{\widehat{p}_\perp^2}{2p_0} \right) \right\},$$

$$\widehat{U}'_i(z, z_i) = \widehat{U}'_i(z, z_r)\, \widehat{U}'_i(z_r, z_\ell)\, \widehat{U}'_i(z_\ell, z_i)$$

$$= \exp\left\{-\left(\frac{i\eta_n p_0}{\hbar w}\right)\widehat{U}_0^\dagger(z,z_i)(xS_y+yS_x)\widehat{U}_0(z,z_i)\right\}$$

$$= \exp\left\{-\frac{i\eta_n}{\hbar}\left[\left(\frac{1}{w\sqrt{K_n}}\sinh\left(w\sqrt{K_n}\right)p_0 x\right.\right.\right.$$
$$\left.+\frac{1}{wK_n}\left(\cosh\left(w\sqrt{K_n}\right)-1\right)\widehat{p}_x\right)S_y$$
$$+\left(\frac{1}{w\sqrt{K_n}}\sin\left(w\sqrt{K_n}\right)p_0 y\right.$$
$$\left.\left.\left.+\frac{1}{wK_n}\left(\cos\left(w\sqrt{K_n}\right)-1\right)\widehat{p}_y\right)S_x\right]\right\}. \tag{6.192}$$

It is straightforward to calculate the required transfer maps using (6.191) and (6.192). For the transverse position coordinates and momentum components, we get

$$\begin{pmatrix}\langle x\rangle(z)\\ \frac{1}{p_0}\langle\widehat{p}_x\rangle(z)\\ \langle y\rangle(z)\\ \frac{1}{p_0}\langle\widehat{p}_y\rangle(z)\end{pmatrix}\approx\begin{pmatrix}T_{11}^{(X)}&T_{12}^{(X)}&0&0\\ T_{21}^{(X)}&T_{22}^{(X)}&0&0\\ 0&0&T_{11}^{(Y)}&T_{12}^{(Y)}\\ 0&0&T_{21}^{(Y)}&T_{22}^{(Y)}\end{pmatrix}\begin{pmatrix}\langle x\rangle(z_i)\\ \frac{1}{p_0}\langle\widehat{p}_x\rangle(z_i)\\ \langle y\rangle(z_i)\\ \frac{1}{p_0}\langle\widehat{p}_y\rangle(z_i)\end{pmatrix}$$
$$+\eta_n\begin{pmatrix}\frac{1}{wK_n}\left(\cosh\left(w\sqrt{K_n}\right)-1\right)\langle S_y\rangle(z_i)\\ -\frac{1}{w\sqrt{K_n}}\sinh\left(w\sqrt{K_n}\right)\langle S_y\rangle(z_i)\\ -\frac{1}{wK_n}\left(\cos\left(w\sqrt{K_n}\right)-1\right)\langle S_x\rangle(z_i)\\ -\frac{1}{w\sqrt{K_n}}\sin\left(w\sqrt{K_n}\right)\langle S_x\rangle(z_i)\end{pmatrix}, \tag{6.193}$$

where

$$\begin{pmatrix}T_{11}^{(X)}&T_{12}^{(X)}\\ T_{21}^{(X)}&T_{22}^{(X)}\end{pmatrix}=\begin{pmatrix}1&z-z_r\\ 0&1\end{pmatrix}$$
$$\times\begin{pmatrix}\cosh\left(w\sqrt{K_n}\right)&\frac{1}{\sqrt{K_n}}\sinh\left(w\sqrt{K_n}\right)\\ \sqrt{K_n}\sinh\left(w\sqrt{K_n}\right)&\cosh\left(w\sqrt{K_n}\right)\end{pmatrix}$$
$$\times\begin{pmatrix}1&z_\ell-z_i\\ 0&1\end{pmatrix},$$

$$\begin{pmatrix}T_{11}^{(Y)}&T_{12}^{(Y)}\\ T_{21}^{(Y)}&T_{22}^{(Y)}\end{pmatrix}=\begin{pmatrix}1&z-z_r\\ 0&1\end{pmatrix}$$
$$\times\begin{pmatrix}\cos\left(w\sqrt{K_n}\right)&\frac{1}{\sqrt{K_n}}\sin\left(w\sqrt{K_n}\right)\\ -\sqrt{K_n}\sin\left(w\sqrt{K_n}\right)&\cos\left(w\sqrt{K_n}\right)\end{pmatrix}$$
$$\times\begin{pmatrix}1&z_\ell-z_i\\ 0&1\end{pmatrix}. \tag{6.194}$$

Spinor Theory for Spin-1/2 Particles

For the components of spin, we have

$$\langle S_x \rangle (z) \approx \langle S_x \rangle (z_i) + p_0 \eta_n \left(\frac{1}{w\sqrt{K_n}} \sinh\left(w\sqrt{K_n}\right) \langle x S_z \rangle (z_i) \right.$$

$$\left. + \frac{1}{p_0 w K_n} \left(\cosh\left(w\sqrt{K_n}\right) - 1\right) \langle \widehat{p}_x S_z \rangle (z_i) \right),$$

$$\langle S_y \rangle (z) \approx \langle S_y \rangle (z_i) - p_0 \eta_n \left(\frac{1}{w\sqrt{K_n}} \sin\left(w\sqrt{K_n}\right) \langle y S_z \rangle (z_i) \right.$$

$$\left. - \frac{1}{p_0 w K_n} \left(\cos\left(w\sqrt{K_n}\right) - 1\right) \langle \widehat{p}_y S_z \rangle (z_i) \right),$$

$$\langle S_z \rangle (z) \approx \langle S_z \rangle (z_i) - p_0 \eta_n \left(\frac{1}{w\sqrt{K_n}} \sinh\left(w\sqrt{K_n}\right) \langle x S_x \rangle (z_i) \right.$$

$$- \frac{1}{w\sqrt{K_n}} \sin\left(w\sqrt{K_n}\right) \langle y S_y \rangle (z_i)$$

$$+ \frac{1}{p_0 w K_n} \left(\cosh\left(w\sqrt{K_n}\right) - 1\right) \langle \widehat{p}_x S_x \rangle (z_i)$$

$$\left. + \frac{1}{p_0 w K_n} \left(\cos\left(w\sqrt{K_n}\right) - 1\right) \langle \widehat{p}_y S_y \rangle (z_i) \right). \quad (6.195)$$

Thus we have derived the transfer maps for the quantum averages of transverse position and momentum components, and the spin components, for a monoenergetic paraxial beam of Dirac particles propagating through a normal magnetic quadrupole lens with a straight optic axis. It is evident that the transfer maps for position and momentum components include the Stern–Gerlach effect. The lens is focusing (defocusing) in the yz-plane and defocusing (focusing) in the xz-plane when $K_n > 0$ ($K_n < 0$). The transverse Stern–Gerlach kicks to the trajectory slope $\left(\delta \langle \vec{\widehat{p}}_\perp \rangle / p_0 \sim \eta_n \hbar \right)$ are seen to disappear at relativistic energies, varying like $\sim 1/\gamma$. At nonrelativistic energies, with $\gamma \sim 1$, the kicks are $\sim w m K_n \mu / p_0^2$, where μ is the total magnetic moment. If the spin of the particle is ignored, the transfer maps for transverse position and momentum components become the classical transfer maps. The transfer maps for the averages of the spin components, or the components of the polarization vector, obtained above correspond to the quantum beam optical version of the Thomas–BMT equation for spin dynamics in the paraxial approximation and are linear in the polarization components only when there is no spin-space correlation, i.e., for the classical results to hold, one should have $\langle x S_z \rangle = \langle x \rangle \langle S_z \rangle$, $\langle y S_z \rangle = \langle y \rangle \langle S_z \rangle$, $\langle \widehat{p}_x S_z \rangle = \langle \widehat{p}_x \rangle \langle S_z \rangle$, etc.

6.1.2.4 Phase Space and Spin Transfer Maps for a Skew Magnetic Quadrupole

Propagation of a paraxial beam of Dirac particles with anomalous magnetic moment through a skew magnetic quadrupole and the corresponding phase space and spin transfer maps have been discussed by Khan [92]. A skew magnetic quadrupole is associated with the magnetic field

$$\vec{B}(\vec{r}) = (-Q_s x, Q_s y, 0), \tag{6.196}$$

corresponding to the vector potential

$$\vec{A}(\vec{r}) = (0, 0, -Q_s xy), \tag{6.197}$$

in which Q_s is nonzero in the quadrupole region ($z_\ell \leq z \leq z_r$) and zero in the field-free outside region. For the propagation of a paraxial beam of Dirac particles with anomalous magnetic moment, the quantum accelerator optical Hamiltonian of the skew magnetic quadrupole will be, as seen from (6.153), with $w = z_r - z_\ell$, $K_s = qQ_s/p_0$, and $\eta_s = q(1+\gamma a)Q_s w/p_0^2$,

$$\widehat{\mathcal{H}_s}(z) = \begin{cases} -p_0 + \frac{\hat{p}_\perp^2}{2p_0}, & \text{for } z < z_\ell \text{ and } z > z_r, \\ -p_0 + \frac{\hat{p}_\perp^2}{2p_0} + p_0 K_s xy \\ + \frac{\eta_s p_0}{w}(xS_x - yS_y), & \text{for } z_\ell \leq z \leq z_r. \end{cases} \tag{6.198}$$

Now, let us make the following transformations:

$$\begin{pmatrix} x \\ \hat{p}_x \\ y \\ \hat{p}_y \end{pmatrix} = \frac{1}{\sqrt{2}} \begin{pmatrix} 1 & 0 & 1 & 0 \\ 0 & 1 & 0 & 1 \\ -1 & 0 & 1 & 0 \\ 0 & -1 & 0 & 1 \end{pmatrix} \begin{pmatrix} x' \\ \hat{p}'_x \\ y' \\ \hat{p}'_y \end{pmatrix},$$

$$\begin{pmatrix} S_x \\ S_y \end{pmatrix} = \frac{1}{\sqrt{2}} \begin{pmatrix} 1 & 1 \\ -1 & 1 \end{pmatrix} \begin{pmatrix} S'_x \\ S'_y \end{pmatrix}. \tag{6.199}$$

Then, the Hamiltonian $\widehat{\mathcal{H}_s}(z)$ is seen to become

$$\widehat{\mathcal{H}_s}(z) = \begin{cases} -p_0 + \frac{\hat{p}_\perp'^2}{2p_0}, & \text{for } z < z_\ell \text{ and } z > z_r, \\ -p_0 + \frac{\hat{p}_\perp'^2}{2p_0} - \frac{1}{2}p_0 K_s \left(x'^2 - y'^2\right) \\ + \frac{\eta_s p_0}{w}\left(y'S'_x + x'S'_y\right), & \text{for } z_\ell \leq z \leq z_r. \end{cases} \tag{6.200}$$

same as for a normal magnetic quadrupole (6.175) with Q_s, K_s, and η_s instead of Q_n, K_n, and η_n, respectively, and (x,y), $(\hat{p}_x, \hat{p}_y)$, and (S_x, S_y) replaced by (x',y'), $(\hat{p}'_x, \hat{p}'_y)$, and (S'_x, S'_y), respectively. Note that S_z is unchanged and $\left(\sigma'_x = 2S'_x/\hbar, \sigma'_y = 2S'_y/\hbar, \sigma_z = 2S_z/\hbar\right)$ obey the same algebra as $(\sigma_x, \sigma_y, \sigma_z)$, i.e.,

$$\sigma'^2_x = I, \quad \sigma'^2_y = I, \quad \sigma'_x \sigma'_y = -\sigma'_y \sigma'_x,$$
$$\sigma_z \sigma'_x = -\sigma'_x \sigma_z, \quad \sigma_z \sigma'_y = -\sigma'_y \sigma_z. \tag{6.201}$$

Spinor Theory for Spin-1/2 Particles

This shows that the transfer maps for $(\langle x' \rangle, \langle \widehat{p}'_x \rangle/p_0, \langle y' \rangle, \langle \widehat{p}'_y \rangle/p_0)$, and $(\langle S'_x \rangle, \langle S'_y \rangle, \langle S_z \rangle)$ will be the same as in (6.193 and 6.194) and (6.195), respectively, except for the replacement: $(\langle x \rangle, \langle \widehat{p}_x \rangle/p_0, \langle y \rangle, \langle \widehat{p}_y \rangle/p_0) \longrightarrow (\langle x' \rangle, \langle \widehat{p}'_x \rangle/p_0, \langle y' \rangle, \langle \widehat{p}'_y \rangle/p_0)$, $(S_x, S_y) \longrightarrow (S'_x, S'_y)$, $Q_n \longrightarrow Q_s$, $K_n \longrightarrow K_s$, $\eta_n \longrightarrow \eta_s$. Once the transfer maps are obtained in terms of $(\langle x' \rangle, \langle \widehat{p}'_x \rangle/p_0, \langle y' \rangle, \langle \widehat{p}'_y \rangle/p_0)$, and (S'_x, S'_y, S_z), the required transfer maps for $(\langle x \rangle, \langle \widehat{p}_x \rangle/p_0, \langle y \rangle, \langle \widehat{p}_y \rangle/p_0)$, and (S_x, S_y, S_z) can be obtained using the inverse of the transformations (6.199). The resulting transfer maps are as follows:

$$\begin{pmatrix} \langle x \rangle(z) \\ \frac{1}{p_0}\langle \widehat{p}_x \rangle(z) \\ \langle y \rangle(z) \\ \frac{1}{p_0}\langle \widehat{p}_y \rangle(z) \end{pmatrix} \approx \mathcal{T}_{sq}(z, z_i) \left(\begin{pmatrix} \langle x \rangle(z_i) \\ \frac{1}{p_0}\langle \widehat{p}_x \rangle(z_i) \\ \langle y \rangle(z_i) \\ \frac{1}{p_0}\langle \widehat{p}_y \rangle(z_i) \end{pmatrix} \right.$$

$$\left. + \frac{\eta_s}{2} \begin{pmatrix} \frac{1}{wK_s}(-\mathrm{C}^-\langle S_x \rangle(z_i) + (\mathrm{C}^+ - 2)\langle S_y \rangle(z_i)) \\ \frac{1}{w\sqrt{K_s}}(-\mathrm{S}^+\langle S_x \rangle(z_i) + \mathrm{S}^-\langle S_y \rangle(z_i)) \\ \frac{1}{wK_s}(-(\mathrm{C}^+ - 2)\langle S_x \rangle(z_i) + \mathrm{C}^-\langle S_y \rangle(z_i)) \\ \frac{1}{w\sqrt{K_s}}(-\mathrm{S}^-\langle S_x \rangle(z_i) + \mathrm{S}^+\langle S_y \rangle(z_i)) \end{pmatrix} \right),$$

(6.202)

and

$$\langle S_x \rangle(z) \approx \langle S_x \rangle(z_i) + \frac{1}{2}p_0\eta_s \left[-\frac{1}{w\sqrt{K_s}}(\mathrm{S}^-\langle xS_z \rangle(z_i) + \mathrm{S}^+\langle yS_z \rangle(z_i)) \right.$$

$$\left. + \frac{1}{p_0 w K_s}((\mathrm{C}^+ - 2)\langle \widehat{p}_x S_z \rangle + \mathrm{C}^-\langle \widehat{p}_y S_z \rangle) \right]$$

$$\langle S_y \rangle(z) \approx \langle S_y \rangle(z_i) + \frac{1}{2}p_0\eta_s \left[-\frac{1}{w\sqrt{K_s}}(\mathrm{S}^+\langle xS_z \rangle(z_i) + \mathrm{S}^-\langle yS_z \rangle(z_i)) \right.$$

$$\left. + \frac{1}{p_0 w K_s}((\mathrm{C}^-)\langle \widehat{p}_x S_z \rangle + (\mathrm{C}^+ - 2)\langle \widehat{p}_y S_z \rangle) \right]$$

$$\langle S_z \rangle(z) \approx \langle S_z \rangle(z_i) + \frac{1}{2}p_0\eta_s \left[\frac{1}{w\sqrt{K_s}}(\mathrm{S}^-\langle xS_x \rangle(z_i) + \mathrm{S}^+\langle xS_y \rangle(z_i) \right.$$

$$+ \mathrm{S}^+\langle yS_x \rangle(z_i) + \mathrm{S}^-\langle yS_y \rangle(z_i))$$

$$- \frac{1}{p_0 w K_s}((\mathrm{C}^+ - 2)\langle \widehat{p}_x S_x \rangle(z_i) + \mathrm{C}^-\langle \widehat{p}_x S_y \rangle(z_i)$$

$$\left. + \mathrm{C}^-\langle \widehat{p}_y S_x \rangle(z_i) + (\mathrm{C}^+ - 2)\langle \widehat{p}_y S_y \rangle)(z_i) \right] \qquad (6.203)$$

where

$$\mathcal{T}_{sq}(z,z_i) = \begin{pmatrix} 1 & z-z_r & 0 & 0 \\ 0 & 1 & 0 & 0 \\ 0 & 0 & 1 & z-z_r \\ 0 & 0 & 0 & 1 \end{pmatrix}$$

$$\times \frac{1}{2} \begin{pmatrix} C^+ & \frac{1}{\sqrt{K_s}}S^+ & C^- & \frac{1}{\sqrt{K_s}}S^- \\ -\sqrt{K_s}S^- & C^+ & -\sqrt{K_s}S^+ & C^- \\ C^- & \frac{1}{\sqrt{K_s}}S^- & C^+ & \frac{1}{\sqrt{K_s}}S^+ \\ -\sqrt{K_s}S^+ & C^- & -\sqrt{K_s}S^- & C^+ \end{pmatrix}$$

$$\times \begin{pmatrix} 1 & z_\ell-z_i & 0 & 0 \\ 0 & 1 & 0 & 0 \\ 0 & 0 & 1 & z_\ell-z_i \\ 0 & 0 & 0 & 1 \end{pmatrix}, \qquad (6.204)$$

and

$$C^\pm = \cos(w\sqrt{K_s}) \pm \cosh(w\sqrt{K_s}),$$
$$S^\pm = \sin(w\sqrt{K_s}) \pm \sinh(w\sqrt{K_s}). \qquad (6.205)$$

6.2 NONRELATIVISTIC QUANTUM CHARGED PARTICLE BEAM OPTICS: SPIN-$\frac{1}{2}$ PARTICLES

The nonrelativistic Schrödinger–Pauli equation for a spin-$\frac{1}{2}$ particle of charge q and mass m moving in the time-independent electromagnetic field of an optical system is

$$i\hbar \frac{\partial \underline{\Psi}(\vec{r},t)}{\partial t} = \left\{ \left[\frac{1}{2m} \left(\vec{\hat{p}} - q\vec{A}(\vec{r}) \right)^2 + q\phi(\vec{r}) \right] \mathbb{I} - \mu \vec{\sigma} \cdot \vec{B} \right\} \underline{\Psi}(\vec{r},t), \qquad (6.206)$$

where $\vec{A}(\vec{r})$ and $\phi(\vec{r})$ are the magnetic vector potential and the electric scalar potential of the field, $\underline{\Psi}(\vec{r},t)$ is the two-component wave function of the particle, and μ is the total magnetic moment of the particle. Let us consider a particle of a nonrelativistic monoenergetic quasiparaxial beam with total energy $E = p_0^2/2m$, where p_0 is the design momentum with which it enters the optical system from the free space outside and $|\vec{p}_{0\perp}| \ll p_0$. The total energy E is conserved when the beam propagates through any optical system with time-independent electromagnetic field we are considering. Choosing the wave function of the particle as

$$\underline{\Psi}(\vec{r},t) = e^{-iEt/\hbar} \underline{\psi}(\vec{r}), \qquad (6.207)$$

we get the time-independent Schrödinger–Pauli equation obeyed by $\underline{\psi}(\vec{r}_\perp, z)$:

$$\left\{ \left[\frac{\hat{\vec{\pi}}^2}{2m} + q\phi(\vec{r}) \right] \mathbb{I} - \mu \vec{\sigma} \cdot \vec{B} \right\} \underline{\psi}(\vec{r}) = E \underline{\psi}(\vec{r}), \qquad (6.208)$$

Spinor Theory for Spin-1/2 Particles   331

where $\vec{\widehat{\pi}} = \vec{\tilde{p}} - q\vec{A}$. Multiplying the equation on both sides by $2m$, taking

$$2m\mathrm{E} = \mathrm{p}_0^2, \qquad 2mq\phi = \tilde{\mathrm{p}}^2, \qquad (6.209)$$

and rearranging the terms, we have

$$\widehat{\pi}_z^2 \underline{\psi} = \left(\mathrm{p}_0^2 - \tilde{p}^2 - \widehat{\pi}_\perp^2 + 2m\mu\vec{\sigma} \cdot \vec{B} \right) \underline{\psi}, \qquad (6.210)$$

in which we have dropped the identity matrix I as to be understood. Note that this equation is identical, except for the additional term $2m\mu\vec{\sigma} \cdot \vec{B}$ and ψ being a two-component wave function, to the time-independent nonrelativistic Schrödinger equation (5.298), which is the starting point for our derivation of the nonrelativistic quantum beam optical Hamiltonian (5.299). Now, starting with (6.210), we can follow the same procedure, going through the Feshbach–Villars-like formalism and the Foldy–Wouthuysen-like transformations. So, let us introduce a four-component wave function

$$\begin{pmatrix} \underline{\psi}_1 \\ \underline{\psi}_2 \end{pmatrix} = \begin{pmatrix} \underline{\psi} \\ \frac{\widehat{\pi}_z}{\mathrm{p}_0}\underline{\psi} \end{pmatrix}. \qquad (6.211)$$

Then, we can write (6.210) equivalently as

$$\frac{\widehat{\pi}_z}{\mathrm{p}_0} \begin{pmatrix} \underline{\psi}_1 \\ \underline{\psi}_2 \end{pmatrix} = \begin{pmatrix} 0 & 1 \\ \frac{1}{\mathrm{p}_0^2}\left(\mathrm{p}_0^2 - \tilde{p}^2 - \widehat{\pi}_\perp^2 + 2m\mu\vec{\sigma} \cdot \vec{B}\right) & 0 \end{pmatrix} \begin{pmatrix} \underline{\psi}_1 \\ \underline{\psi}_2 \end{pmatrix}. \qquad (6.212)$$

Now, let

$$\begin{pmatrix} \underline{\psi}_+ \\ \underline{\psi}_- \end{pmatrix} = M \begin{pmatrix} \underline{\psi}_1 \\ \underline{\psi}_2 \end{pmatrix} = \frac{1}{2} \begin{pmatrix} 1 & 1 \\ 1 & -1 \end{pmatrix} \begin{pmatrix} \underline{\psi}_1 \\ \underline{\psi}_2 \end{pmatrix}$$

$$= \frac{1}{2} \begin{pmatrix} \underline{\psi} + \frac{\widehat{\pi}_z}{\mathrm{p}_0}\underline{\psi} \\ \underline{\psi} - \frac{\widehat{\pi}_z}{\mathrm{p}_0}\underline{\psi} \end{pmatrix}. \qquad (6.213)$$

Then,

$$\frac{\widehat{\pi}_z}{\mathrm{p}_0} \begin{pmatrix} \underline{\psi}_+ \\ \underline{\psi}_- \end{pmatrix} = \frac{1}{\mathrm{p}_0}\left(-i\hbar\frac{\partial}{\partial z} - qA_z\right)\begin{pmatrix} \underline{\psi}_+ \\ \underline{\psi}_- \end{pmatrix}$$

$$= M \begin{pmatrix} 0 & 1 \\ \frac{1}{\mathrm{p}_0^2}\left(\mathrm{p}_0^2 - \tilde{p}^2 - \widehat{\pi}_\perp^2 + 2m\mu\vec{\sigma} \cdot \vec{B}\right) & 0 \end{pmatrix} M^{-1} \begin{pmatrix} \underline{\psi}_+ \\ \underline{\psi}_- \end{pmatrix}$$

$$= \begin{pmatrix} 1 - \frac{1}{2\mathrm{p}_0^2}\left(\widehat{\pi}_\perp^2 + \tilde{p}^2 - 2m\mu\vec{\sigma} \cdot \vec{B}\right) & -\frac{1}{2\mathrm{p}_0^2}\left(\widehat{\pi}_\perp^2 + \tilde{p}^2 - 2m\mu\vec{\sigma} \cdot \vec{B}\right) \\ \frac{1}{2\mathrm{p}_0^2}\left(\widehat{\pi}_\perp^2 + \tilde{p}^2 - 2m\mu\vec{\sigma} \cdot \vec{B}\right) & -1 + \frac{1}{2\mathrm{p}_0^2}\left(\widehat{\pi}_\perp^2 + \tilde{p}^2 - 2m\mu\vec{\sigma} \cdot \vec{B}\right) \end{pmatrix} \begin{pmatrix} \underline{\psi}_+ \\ \underline{\psi}_- \end{pmatrix}.$$

$$(6.214)$$

Rearranging this equation, we get

$$i\hbar \frac{\partial}{\partial z}\begin{pmatrix}\psi_+ \\ \psi_-\end{pmatrix} = \widehat{H}\begin{pmatrix}\psi_+ \\ \psi_-\end{pmatrix},$$

$$\widehat{H} = -\mathrm{p}_0\beta + \widehat{\mathscr{E}} + \widehat{\mathscr{O}},$$

$$\widehat{\mathscr{E}} = -qA_z\mathbb{I} + \frac{1}{2\mathrm{p}_0}\left[\sigma_z \otimes \left(\widehat{\pi}_\perp^2 + \tilde{p}^2 - 2m\mu\vec{\sigma}\cdot\vec{B}\right)\right],$$

$$\widehat{\mathscr{O}} = \frac{i}{2\mathrm{p}_0}\left[\sigma_y \otimes \left(\widehat{\pi}_\perp^2 + \tilde{p}^2 - 2m\mu\vec{\sigma}\cdot\vec{B}\right)\right]. \quad (6.215)$$

Note that $\widehat{H}$ has the same structure as the Dirac equation, with the leading term $-\mathrm{p}_0\beta$, an even term $\widehat{\mathscr{E}}$ which does not couple ψ_+ and ψ_- and an odd term $\widehat{\mathscr{O}}$ which couples ψ_+ and ψ_-. Further, the odd term anticommutes with β. It is also clear that the lower pair of components ψ_- will be small compared to the upper pair of components ψ_+. Foldy–Wouthuysen-like transformations will further make the lower pair of components smaller compared to the upper pair of components. If we stop with one Foldy–Wouthuysen-like transformation, we can get the quantum beam optical Hamiltonian up to paraxial approximation. If we do this following the same procedure as done earlier, and approximate the resulting four-component wave function by a two-component wave function consisting of only the upper pair of components, we get the nonrelativistic quantum beam optical Hamiltonian for a spin-$\frac{1}{2}$ particle, in the paraxial approximation, as

$$\widehat{\mathscr{H}}_{o,\mathrm{NR}} = \left[-\mathrm{p}_0 - qA_z + \frac{1}{2\mathrm{p}_0}\left(\widehat{\pi}_\perp^2 + \tilde{p}^2\right)\right]\mathbb{I} - \frac{m\mu}{\mathrm{p}_0}\vec{\sigma}\cdot\vec{B}$$

$$= \left(-\mathrm{p}_0 - qA_z + \frac{\widehat{\pi}_\perp^2}{2\mathrm{p}_0} + \frac{mq\phi}{\mathrm{p}_0}\right)\mathbb{I} - \frac{m\mu}{\mathrm{p}_0}\vec{\sigma}\cdot\vec{B}. \quad (6.216)$$

Thus, the nonrelativistic quantum beam optical z-evolution equation for the two-component wave function of a spin-$\frac{1}{2}$ particle is, in the paraxial approximation,

$$i\hbar\frac{\partial \psi(\vec{r}_\perp,z)}{\partial z} = \left[\left(-\mathrm{p}_0 - qA_z + \frac{\widehat{\pi}_\perp^2}{2\mathrm{p}_0} + \frac{mq\phi}{\mathrm{p}_0}\right)\mathbb{I} - \frac{m\mu}{\mathrm{p}_0}\vec{\sigma}\cdot\vec{B}\right]\psi(\vec{r}_\perp,z), \quad (6.217)$$

to which the relativistic paraxial quantum beam optical equation (6.153) reduces while taking the nonrelativistic limit

$$\gamma \longrightarrow 1, \qquad p_0 \longrightarrow \mathrm{p}_0, \quad (6.218)$$

and noting that

$$\frac{q(1+a)}{p_0}\vec{B}\cdot\vec{S} = \frac{mq(1+a)\hbar}{2mp_0}\vec{\sigma}\cdot\vec{B} = \frac{m}{\mathrm{p}_0}\left(\frac{qg\hbar}{4m}\right)\vec{\sigma}\cdot\vec{B} = \frac{m\mu}{\mathrm{p}_0}\vec{\sigma}\cdot\vec{B}. \quad (6.219)$$

This shows that one can get the nonrelativistic expressions and formulae by taking the nonrelativistic limit of the results derived from the quantum beam optical Dirac equation exactly like in the scalar theory based on the Klein–Gordon equation. However, it is not possible to get the relativistic quantum beam optical equation (6.153) by replacing m by γm and p_0 by p_0 in the nonrelativistic quantum beam optical equation (6.217). Thus, it should be noted that converting the expressions and formulae derived in the nonrelativistic theory to relativistic expressions and formulae by replacing the rest mass m by γm, the so-called relativistic mass, as has been the common practice in electron optics, is misleading in the spinor theory of quantum charged particle beam optics.

7 Concluding Remarks and Outlook on Further Development of Quantum Charged Particle Beam Optics

In conclusion, we have to make several remarks. Analogy between optics and mechanics and its role in progress of physics is well known (for historical details see, *e.g.*, Hawkes and Kasper [70], Lakshminarayanan, Ghatak, and Thyagarajan [125], Khan [108, 113, 115]). In the early days of electron optics, analogy with light optics provided much guidance. In the development of the formalism of spinor theory of charged particle beam optics presented here, the inspiration came from the work of Sudarshan, Simon, and Mukunda in light optics (see Sudarshan, Simon, and Mukunda [175], Mukunda, Simon, and Sudarshan [137, 138], Simon, Sudarshan, and Mukunda [170, 171], Khan [120]). The formalism of quantum charged particle beam optics in turn has led to a similar approach, called quantum methodology, to the Helmholtz scalar and the Maxwell vector optics. In analogy with the derivation of the scalar theory of quantum charged particle beam optics starting with the Klein–Gordon equation, scalar theory of light beam optics is derived starting with the Helmholtz equation and using the Feshbach–Villars-like form and the Foldy–Wouthuysen-like transformations. In analogy with the derivation of the spinor theory of quantum charged particle beam optics starting with the Dirac equation, the vector theory of light beam optics, including polarization, is derived starting with the Maxwell equations expressed in a matrix form and using the Foldy–Wouthuysen-like transformations. This methodology leads to scalar and vector theories of light beam optics expressed as paraxial approximation followed by nonparaxial aberrations (for details, see Khan, Jagannathan, and Simon [100], and Khan [101, 102, 103, 104, 105, 106, 107, 109, 110, 111, 112, 114, 115, 116, 117, 118, 119]).

It is certain that at some point in future, quantum mechanics would become important in accelerator beam optics. For example, Hill has discussed in [76] how quantum mechanics places limits on the achievable transverse beam spot sizes in accelerators, Kabel has considered in [89] how the Pauli exclusion principle will limit the minimum achievable emittance for a beam of fermions in a circular accelerator, Venturini and Ruth have suggested in [185] a framework to compute the wave function of a single charged particle confined to a magnetic lattice transport line or storage ring, Chao and Nash have examined in [19] the possible measurable macroscopic effect of quantum mechanics on beams through its echo effect using the Wigner phase space distribution function, and Heifets and Yan have pointed out in [73] that quantum uncertainties affect substantially the classical results of tracking for trajectories close to the separatrix and hence the quantum maps can be useful in quick findings of the nonlinear resonances. It is clear that several quantum mechanical problems related to the beam optics, particularly the nonlinear beam optics, will have to be handled in the future particle accelerators built to explore particle physics beyond the standard model (For an overview of the currently envisaged landscape of future particle accelerators, see, *e.g.*, Syphers and Chattopadhyay [176]).

In the field of small-scale electron beam devices, electron beam lithography is the primary tool in micro- and nanofabrications, particularly in the fabrication of semiconductor devices and mesoscopic devices. In electron beam lithography for nanoscale fabrication, it is required to generate high-resolution electron beams to facilitate high-speed drawing of fine patterns on material substrates (see, *e.g.*, Manfrinato et al. [132], Wu, Makiuchi, and Chen [193]). Though classical electron beam optics may be successful in designing the beam optical systems of the present-day electron beam lithography, future advances in the technology would certainly require the quantum theory of electron beam optics. Advancements in high-resolution electron microscopy take place using various aspects of quantum mechanics. Scanning tunneling microscopy was invented using the quantum tunneling effect to image directly the atoms of a sample (see, *e.g.*, Chen [21]). Recently, Kruit et al. [123] had a proposal to implement the suggestion of the so-called quantum electron microscope (Putnam and Yanik [153]) that would use the principle of interaction-free measurement based on the quantum Zeno effect (Misra and Sudarshan [133]). Another suggestion is electron microscopy based on quantum entanglement (Okamoto and Nagatani [140]). Highest ever resolution in transmission electron microscopy has been reported by Jiang et al. [87] using ptychographic analysis of the diffraction patterns of low-energy electrons by a two-dimensional specimen. It should be emphasized that the formalism of quantum charged particle beam optics presented here might not be related to the quantum aspects of the operational principles of such novel electron microscopes, but will benefit the designing of electron beam optical elements like the electron lenses used in the system.

It may be noted that the quantum charged particle beam optics presented here is only at a preliminary stage of development with the quantum theory of several concepts and topics of classical charged particle beam dynamics, particularly the accelerator beam dynamics, waiting to be developed. The single most important task would be to develop the quantum theory of particle acceleration and the longitudinal beam dynamics and integrate it with the formalism of quantum beam optics to help the treatment of energy spread, energy loss, etc. To this end, one has to find the quantum mechanical analog of the classical six-dimensional phase space in which the relative energy deviation and the longitudinal distance relative to the reference particle are taken as the longitudinal phase space coordinates in addition to the four transverse phase space coordinates. Next, one has to extend the theory beyond the single particle dynamics with the generalization of the concepts like emittances and the Twiss (or Courant-Snyder) parameters. In this respect, phenomenological models of quantum theory of charged particle beam dynamics based on Schrödinger-like equations, often with the Planck constant replaced by an emittance, could provide some guidance. Fedele et al. have developed extensively such quantum-like theories of charged particle beam physics, called thermal wave model and quantum wave model, to deal with different situations, including collective effects, space charge, etc. (see, *e.g.*, Fedele and Miele [40], Fedele and Shukla [41], Fedele et al. [42], Fedele, Man'ko, and Man'ko [43], Fedele et al. [45, 46, 47] and references therein). A model similar to the thermal wave model of Fedele et al. has been considered also by Dattoli et al. [30]. Quantum-like approach has been used to study the phenomenon of halo formation in accelerator beams by Pusterla [152], and Khan and Pusterla [94, 96, 97, 98]. Similarly, a semiclassical stochastic model for the collective dynamics of charged particle beams has been developed by Petroni et al. [143]. May be it will be profitable to formulate the quantum charged particle beam optics using the Wigner function (see, *e.g.*, Chao and Nash [19], Dragt and Habib [37], Jagannathan and Khan [81], Fedele, Man'ko, and Man'ko [44]). Heinemann and Barber have studied the dynamics of spin-$\frac{1}{2}$ polarized beams using the Wigner function in the context of accelerator physics (see [75]; see also Mitra and Ramanathan [134] for a discussion of the Wigner function for the Klein–Gordon and Dirac particles).

In fine, we have presented the elements of *Quantum Charged Particle Beam Optics* showing why and how classical charged particle beam optics works very successfully in the design and operation of charged particle beam devices from low-energy electron microscopes to high-energy particle accelerators. In future, developments in technology would certainly require quantum theory for the purpose of design and operation of various kinds of charged particle beam devices. We hope that the formalism of quantum charged particle beam optics presented here would be the basis for the development of the quantum theory of future charged particle beam devices.

Bibliography

1. Acharya, R. and Sudarshan, E. C. G., Front description in relativistic quantum mechanics, *J. Math. Phys.*, **1** (1960) 532–536.
2. Agarwal, G. S. and Simon, R., A simple realization of fractional Fourier transform and relation to harmonic oscillator Green's function, *Optics Comm.*, **110** (1994) 23–26.
3. Arfken, G. B., Weber, H. J., and Harris, F. E., *Mathematical Methods for Physicists: A Comprehensive Guide*, (7th Edition, Academic Press, Waltham, MA, 2012).
4. Barber, D. P., Heinemann, K., and Ripken, G., A canonical 8-dimensional formalism for classical spin-orbit motion in storage rings, *Z. Phys. C*, **64**(1) (1994) 117–142.
5. Barber, D. P., Heinemann, K., and Ripken, G., A canonical 8-dimensional formalism for classical spin-orbit motion in storage rings, *Z. Phys. C*, **64**(1) (1994) 143–167.
6. Bargmann, V., Michel, L., and Telegdi, V. L., Precession of the polarization of particles moving in a homogeneous electromagnetic field, *Phys. Rev. Lett.*, **2** (1959) 435–436.
7. Bauer, E., LEEM basics, *Surface Rev. and Lett.*, **5**(6) (1998) 1275–1286.
8. Bell, J. S., in: *Progress in Particle Physics*: *Proceedings of the XIII. Internationale Universitätswochen für Kernphysik 1974 der Karl-Franzens-Universität Graz at Schladming*, Austria, 1974, Ed. P. Urban, *Acta Physica Austriaca*, **13** (1974) 395.
9. Bell, J. S. and Leinass, J. M., The Unruh effect and quantum fluctuations of electrons in storage rings, *Nucl. Phys. B*, **284** (1987) 488–508.
10. Bellman, R. and Vasudevan, R., *Wave Propagation: An Invariant Imbedding Approach*, (D. Reidel, Dordrecht, 1986).
11. Berz, M., Modern map methods in particle beam physics, in *Advances in Imaging and Electron Physics*, Vol. 108, Ed. P. W. Hawkes, (Academic Press, San Diego, 1999) pp. 1–318.
12. Berz, M., Makino, K., and Wan, W., *An Introduction to Beam Physics*, (Taylor & Francis, Boca Raton, FL, 2016).
13. Bjorken, J. D. and Drell, S. D., *Relativistic Quantum Mechanics*, (McGraw-Hill, New York, 1994).
14. Blanes, S., Casas, F., Oteo, J. A., and Ros, J., The Magnus expansion and some of its applications, *Phys. Rep.*, **470** (2009) 151–238.
15. Boxem, R. V., Partoens, B., and Verbeeck, J., *Dirac Kirchhoff diffraction theory*, arXiv:1303.0954[quant-ph].
16. Brill, D. R. and Wheeler, J. A., Interaction of neutrinos and gravitational fields, *Rev. Mod. Phys.*, **29**(3) (1957) 465–479.
17. Busch, H., On the action of the concentration coil in a Braun tube, *Arch. Elektrotech.* **18** (1927) 583–594.
18. Byron, F. W. and Fuller, R. W., *Mathematics of Classical and Quantum Physics*, (Dover, New York, 1992).
19. Chao, A. and Nash, B., Possible quantum mechanical effect on beam echo, in *Proceedings of the 18th Advanced ICFA Beam Dynamics Workshop on Quantum Aspects of Beam Physics—2000, Italy*, Ed. P. Chen, (World Scientific, 2002), pp. 90–94.
20. Chao, A. W., Mess, K.H., Tigner, M., and Zimmerman, F., (Eds.), *Handbook of Accelerator Physics and Engineering*, (2nd Edition, World Scientific, Singapore, 2013).

21. Chen, C. J., *Introduction to Scanning Tunneling Microscopy*, (2nd Edition, Oxford University Press, New York, 2008).
22. Chen, P., (Ed.), *Proceedings of the 15th Advanced ICFA Beam Dynamics Workshop on Quantum Aspects of Beam Physics—1998, USA*, (World Scientific, 1999).
23. Chen, P., (Ed.), *Proceedings of the 18th Advanced ICFA Beam Dynamics Workshop on Quantum Aspects of Beam Physics—2000, Italy*, (World Scientific, 2002).
24. Chen, P., and Reil, P., (Eds.), *Proceedings of the 28th ICFA Advanced Beam Dynamics and Advanced & Novel Accelerators Workshop—2003, Japan*, (World Scientific, 2004).
25. Cohen-Tannoudji, C., Diu, B., and Loloë, F., *Quantum Mechanics*—Vols. 1&2, (Wiley, New York, 1992).
26. Conte, M., Jagannathan, R., Khan, S. A., and Pusterla, M., Beam optics of the Dirac particle with anomalous magnetic moment, *Part. Accel.*, **56** (1996) 99–126.
27. Conte, M. and MacKay, W. W., *An Introduction to the Physics of Particle Accelerators*, (2nd Edition, World Scientific, Singapore, 2008).
28. Corben, H. C. and Stehle, P., *Classical Mechanics*, (2nd Edition, Wiley, New York, 1964).
29. Corben, H. C., *Classical and Quantum Theories of Spinning Particle*, (Holden-Day, San Francisco, CA, 1968).
30. Dattoli, G., Giannessi, L., Mari, C., Rihetta, M., and Torre, A., Formal quantum theory of electronic rays, *Opt. Commun.* **87** (1992) 175–180.
31. Dattoli, G., Renieri, A., and Torre, *Lectures on the Free Electron Laser Theory and Related Topics*, (World Scientific, Singapore, 1993).
32. Derbenev, Ya. S. and Kondratenko, A. M., Polarization kinetics of particles in storage rings, *Sov. Phys. JETP*, **37**(6) (1973) 968–973.
33. Dragt, A. J., A Lie Algebraic theory of geometrical optics and optical aberrations, *J. Opt. Soc. America*, **72**(3) (1982) 372–379.
34. Dragt, A. J., *Lie Methods for Nonlinear Dynamics with Applications to Accelerator Physics*, (University of Maryland, 2018); https://physics.umd.edu/dsat/dsatliemeth
35. Dragt, A. J. and Forest, E., Lie algebraic theory of charged particle optics and electron microscopes, in *Advances in Electronics and Electron Physics*, Vol. 67, Ed. P. W. Hawkes, (Academic Press, Cambridge, MA, 1986) pp. 65–120.
36. Dragt, A. J., Neri, F., Rangarajan, G., Healy, L. M., and Ryne, R. D., Lie algebraic treatment of linear and nonlinear beam dynamics, *Ann. Rev. Nucl. Part. Sci.*, **38** (1988) 455–496.
37. Dragt, A. J. and Habib, S., How Wigner functions transform under symplectic maps, in *Proceedings of the 15th Advanced ICFA Beam Dynamics Workshop on Quantum Aspects of Beam Physics—1998, USA*, Ed. P. Chen (World Scientific, 1999), pp. 651–669.
38. Durand, E., Le principe de Huygens et la diffraction de l'électron en théorie de Dirac, *Comptes rendus de l'Académie des Sciences, Paris*, **236** (1953) 1337–1339.
39. Esposito, G., Marmo, G., Miele, G., and Sudarshan, E. C. G., *Advanced Concepts in Quantum Mechanics*, (Cambridge University Press, Cambridge, 2014).
40. Fedele, R. and Miele, G., A thermal-wave model for relativistic-charged-particle beam propagation, *Il Nuovo Cimento D*, **13**(12) (1991) 1527–1544.
41. Fedele, R. and Shukla, P. K., Self-consistent interaction between the plasma wake field and the driving relativistic electron beam, *Phys. Rev. A*, **45**(6) (1992) 4045–4049.
42. Fedele, R., Miele, G., Palumbo, L., and Vaccaro, V. G., Thermal wave model for nonlinear longitudinal dynamics in particle accelerators, *Phys. Lett. A*, **179**(6) (1993) 407–413.
43. Fedele, R., Man'ko M. A., and Man'ko, V.I., Wave-optics applications in charged-particle-beam transport, *J.Russ. Laser Res.*, **21**(1) (2000) 1–33.

44. Fedele, R., Man'ko M. A., and Man'ko, V.I., Charged-particle-beam propagator in wave-electron optics: Phase-space and tomographic pictures, *J. Opt. Soc. America A*, **17** (2000) 2506–2512.
45. Fedele, R., Tanjia, F., De Nicola, S., Jovanović, D., and Shukla, P. K., Quantum ring solitons and nonlocal effects in plasma wake field excitations, *Phys. Plasmas*, **19**(10) (2012) 102106.
46. Fedele, R., Tanjia, F., Jovanovic, D., De Nicola, S., and Ronsivalle, C., Wave theories of non-laminar charged particle beams: From quantum to thermal regime, *J. Plasma Phys.*, **80** (2014) 133–145.
47. Fedele, R., Jovanovi, D., De Nicola, S., Mannan, A., and F. Tanjia, F., Self-modulation of a relativistic charged-particle beam as thermal matter wave envelope, *J. Phy. Conf. Ser.*, **482** (2014) 012014.
48. Ferwerda, H. A., Hoenders, B. J., and Slump, C. H., Fully relativistic treatment of electron-optical image formation based on the Dirac equation, *Opt. Acta*, **33** (1986) 145–157.
49. Ferwerda, H. A., Hoenders, B. J., and Slump, C. H., The fully relativistic foundation of linear transfer theory in electron optics based on the Dirac equation, *Opt. Acta*, **33** (1986) 159–183.
50. Feshbach, H. and Villars, F., Elementary relativistic wave mechanics of spin 0 and spin $\frac{1}{2}$ particles, *Rev. Mod. Phys.*, **30** (1958) 24–45.
51. Fishman, L., de Hoop, M. V., and van Stralen, M. J. N., Exact constructions of square-root Helmholtz operator symbols: The focusing quadratic profile, *J. Math. Phys.* **41** (2000) 4881–4938.
52. Fishman, L., One-way wave equation modeling in two-way wave propagation problems, in *Mathematical Modeling of Wave Phenomena 2002, Mathematical Modeling in Physics, Engineering, and Cognitive Sciences*, Vol. 7, Eds. B. Nilsson and L. Fishman, (Växjö University Press, Växjö, Sweden, 2004) pp. 91–111.
53. Foldy, L. L. and Wouthuysen, S. A., On the Dirac theory of spin $\frac{1}{2}$ particles and its non-relativistic limit, *Phys. Rev.*, **78** (1950) 29–36.
54. Forest, É., Berz, M., and Irwin, J., Normal form methods for computational periodic systems, *Part. Accel.*, **24** (1989) 91–97.
55. Forest, É. and Hirata, K., *A contemporary guide to beam dynamics*, KEK Report 92-12 (KEK, Tsukuba, 1992).
56. Forest, É., *Beam Dynamics: A New Attitude and Framework*, (Taylor & Francis, New York, 1998).
57. Frenkel, J., Die Elektrodynamik des rotierenden Elektrons, *Z. Phys.*, **37**(4-5) (1926) 243–262.
58. Frenkel, J., Spinning Electrons, *Nature*, **117** (1926) 653–654.
59. Glaser, W., *Grundlagen der Elecktronenoptik* (Springer, New York, 1952).
60. Glaser, W. and Schiske, P., Elektronenoptische Abbildung auf Grund der Wellenmechanik. I, *Annalen der Physik*, **12** (1953) 240–266.
61. Glaser, W. and Schiske, P., Elektronenoptische Abbildungen auf Grund der Wellenmechanik. II, *Annalen der Physik*, **12** (1953) 267–280.
62. Glaser, W., Elektronen und Ionenoptik in *Handbuch der Physik*, Vol. 33, Ed. S. Flügge, (Springer, Berlin, 1956) 123–395.
63. Goldstein, H., Poole, C. P., and Safko, J. L., *Classical Mechanics*, (3rd Edition, Addison Wesley, Boston, MA, 2002).
64. Greiner, W., *Quantum Mechanics*, (4th Edition, Springer, New York, 2001).

65. Greiner, W., *Relativistic Quantum Mechanics: Wave Equations*, (3rd Edition, Springer, New York, 2000).
66. Griffiths, D. J., *Introduction to Electrodynamics*, (4th Edition, Cambridge University Press, Cambridge, 2017).
67. Griffiths, D. J. and Schroeter, D. F., *Introduction to Quantum Mechanics*, (3rd Edition, Cambridge University Press, Cambridge, 2018).
68. Groves, T. R., *Charged Particle Optics Theory: An Introduction*, (Taylor & Francis, Boca Raton, FL, 2015).
69. Hand, I. N. and Skuja, A., in *Proceedings of the Conference on Spin and Polarization Dynamics in Nuclear and Particle Physics*, Trieste, 1988, Eds. A. O. Barut, Y. Onel, and A. Penzo, (World Scientific, 1990) p.185.
70. Hawkes, P. W. and Kasper, E., *Principles of Electron Optics*—Vol. 1: *Basic Geometrical Optics*, (2nd Edition, Elsevier, San Diego, CA, 2017).
71. Hawkes, P. W. and Kasper, E., *Principles of Electron Optics*—Vol. 2: *Applied Geometrical Optics*, (2nd Edition, Elsevier, San Diego, CA, 2017).
72. Hawkes, P. W. and Kasper, E., *Principles of Electron Optics*—Vol. 3: *Wave Optics*, (3rd Edition, Elsevier, San Diego, CA, 1994).
73. Heifets, S. and Yan, Y. T., Quantum effects in tracking, in *Proceedings of the 15th Advanced ICFA Beam Dynamics Workshop on Quantum Aspects of Beam Physics—1998, USA*, Ed. P. Chen (World Scientific, 1999), pp. 121–130.
74. Heinemann, K., *On Stern–Gerlach force allowed by special relativity and the special case of the classical spinning particle of Derbenev and Kondratenko*, arXiv:physics/9611001.
75. Heinemann, K. and Barber, D. P., The semiclassical FoldyWouthuysen transformation and the derivation of the Bloch equation for spin-$\frac{1}{2}$ polarised beams using Wigner functions, in *Proceedings of the 15th Advanced ICFA Beam Dynamics Workshop on Quantum Aspects of Beam Physics—1998, USA*, Ed. P. Chen (World Scientific, 1999), pp. 695–701.
76. Hill, C. T., The diffractive quantum limits of particle colliders, in *Proceedings of the 18th Advanced ICFA Beam Dynamics Workshop on Quantum Aspects of Beam Physics—2000, Italy*, Ed. P. Chen, (World Scientific, 2002), pp. 3–18.
77. Jackson, J. D., On understanding spin-flip synchrotron radiation and the transverse polarization of electrons in storage rings, *Rev. Mod. Phys.*, **48**(3) (1976) 417–434.
78. Jackson, J. D., *Classical Electrodynamics*, (3rd Edition, Wiley, New York, 1998).
79. Jagannathan, R., Simon, R., Sudarshan, E. C. G., and Mukunda, N. Quantum theory of magnetic electron lenses based on the Dirac equation, *Phys. Lett. A*, **134** (1989) 457–464.
80. Jagannathan, R., Quantum theory of electron lenses based on the Dirac equation, *Phys. Rev. A*, **42** (1990) 6674–6689.
81. Jagannathan, R. and Khan, S. A., Wigner functions in charged particle optics, in *Selected Topics in Mathematical Physics—Professor R. Vasudevan Memorial Volume*, Eds. R. Sridhar, K. Srinivasa Rao, and V. Lakshminarayanan, (Allied Publishers, New Delhi, 1995), pp. 308–321.
82. Jagannathan, R. and Khan, S. A., Quantum theory of the optics of charged particles, in *Advances in Imaging and Electron Physics*, Vol. 97, Eds. P. W. Hawkes, B. Kazan, and T. Mulvey, (Academic Press, San Diego, CA, 1996), pp. 257–358.
83. Jagannathan, R., The Dirac equation approach to spin-$\frac{1}{2}$ particle beam optics, in *Proceedings of the 15th Advanced ICFA Beam Dynamics Workshop on Quantum Aspects of Beam Physics—1998, USA*, Ed. P. Chen (World Scientific, 1999), pp. 670–681.

84. Jagannathan, R., Quantum mechanics of Dirac particle beam optics: Single-particle theory, in *Proceedings of the 18th Advanced ICFA Beam Dynamics Workshop on Quantum Aspects of Beam Physics—2000, Italy*, Ed. P. Chen, (World Scientific, 2002), pp. 568–577.
85. Jagannathan, R., Quantum mechanics of Dirac particle beam transport through optical elements with straight and curved axes, in *Proceedings of the 28th ICFA Advanced Beam Dynamics and Advanced & Novel Accelerators Workshop—2003, Japan*, Eds. P. Chen and K. Reil, (World Scientific, 2003), pp. 13–21.
86. Jagannathan, R., On generalized clifford algebras and their physical applications, in *The Legacy of Alladi Ramakrishnan in the Mathematical Sciences*, Eds. K. Alladi, J. Klauder, and C. R. Rao (Springer, New York, 2010) pp. 465–489.
87. Jiang, Y., Chen, Z., Han, Y., Deb, P., Gao, H., Xie, S., Purohit, P., Tate, M. W., Park, J., Gruner, S. M., Elser, V., and Muller, D. A., Electron ptychography of 2D materials to deep sub-Ångström resolution, *Nature*, **559** (2018) 343–349.
88. Johnson, M. H. and Lippmann, B. A., Motion in a constant magnetic field, *Phys. Rev.* **76** (1949) 828–832.
89. Kabel, A. C., Quantum ground state and minimum emittance of a fermionic particle beam in a circular accelerator, in *Proceedings of the 18th Advanced ICFA Beam Dynamics Workshop on Quantum Aspects of Beam Physics—2000, Italy*, Ed. P. Chen, (World Scientific, 2002), pp. 67–75.
90. Khan, S. A. and Jagannathan, R., Quantum mechanics of charged-particle beam transport through magnetic lenses, *Phys. Rev. E* **51**(3) (1995) 2510–2515.
91. Khan, S. A., *Quantum Theory of Charged-Particle Beam Optics*, Ph.D Thesis, (University of Madras, Chennai, India, 1997). https://imsc.res.in/xmlui/handle/123456789/75
92. Khan, S. A., Quantum theory of magnetic quadrupole lenses for spin-$\frac{1}{2}$ particles, in *Proceedings of the 15th Advanced ICFA Beam Dynamics Workshop on Quantum Aspects of Beam Physics - 1998, USA*, Ed. P. Chen (World Scientific, 1999), pp. 682–694.
93. Khan, S. A., Quantum aspects of accelerator optics, in *Proceedings of the 1999 Particle Accelerator Conference (PAC99)*, March 1999, New York, Eds. A. Luccio and W. MacKay, (1999), pp. 2817–2819.
94. Khan, S. A., and Pusterla, M., Quantum mechanical aspects of the halo puzzle, in *Proceedings of the 1999 Particle Accelerator Conference (PAC99)*, March-April, 1999, New York, Eds. A. Luccio and W. MacKay, pp. 3280–3281.
95. Khan, S. A., and Pusterla, M., On the form of Lorentz-Stern-Gerlach force, arXiv:physics/9910034 [physics.class-ph] (1999).
96. Khan, S. A. and Pusterla, M., Quantum-like approach to the transversal and longitudinal beam dynamics: The halo problem, *Eur. Phys. J. A*, **7** (2000) 583–587.
97. Khan, S. A., and Pusterla, M., Quantum-like approaches to the beam halo problem, in *Proceedings of the 6th International Conference on Squeezed States and Uncertainty Relations (ICSSUR'99)*, May 1999, Napoli, Italy, Eds. D. Han, Y.S Kim, and S. Solimeno, NASA Conference Publication Series, 2000-209899 (2000) pp. 438–441.
98. Khan, S. A., and Pusterla, M., Quantum approach to the halo formation in high current beams, *Nucl. Instrum. Methods Phys. Res. Sect. A*, **464** (2001) 461–464.
99. Khan, S. A., Quantum formalism of beam optics, in *Proceedings of the 18th Advanced ICFA Beam Dynamics Workshop on Quantum Aspects of Beam Physics—2000, Italy*, Ed. P. Chen, (World Scientific, 2002), pp. 517–526.
100. Khan, S. A., Jagannathan, R., and Simon, R., Foldy–Wouthuysen transformation and a quasiparaxial approximation scheme for the scalar wave theory of light beams, arXiv:physics/0209082 [physics.optics], (2002).

101. Khan, S. A., Wavelength-dependent modifications in Helmholtz optics, *Int. J. Theor. Phys.*, **44** (2005) 95–125.
102. Khan, S. A., An exact matrix representation of Maxwell's equations, *Phys. Scr.*, **71**(5) (2005) 440–442.
103. Khan, S. A., The Foldy–Wouthuysen transformation technique in optics, *Optik*, **117**(10) (2006) 481–488.
104. Khan, S. A., Wavelength-dependent effects in light optics, in *New Topics in Quantum Physics Research*, Eds. V. Krasnoholovets and F. Columbus, (Nova Science Publishers, New York, 2006), pp. 163–204.
105. Khan, S. A., The Foldy–Wouthuysen transformation technique in optics, in *Advances in Imaging and Electron Physics*, Vol. 152, Ed. P. W. Hawkes, (Academic Press, San Diego, CA, 2008) pp. 49–78.
106. Khan, S. A., Maxwell optics of quasiparaxial beams, *Optik*, **121** (2010) 408–416.
107. Khan, S. A., Aberrations in Maxwell optics, *Optik*, **125** (2014) 968–978,
108. Khan, S. A., *International Year of Light and Light-based Technologies*, (LAP LAMBERT Academic Publishing, Germany, 2015).
109. Khan, S. A., Passage from scalar to vector optics and the Mukunda–Simon–Sudarshan theory for paraxial systems, *J. Mod. Opt.*, **63** (2016) 1652–1660.
110. Khan, S. A., Quantum aspects of charged-particle beam optics, in: *Proceedings of the Fifth Saudi International Meeting on Frontiers of Physics 2016 (SIMFP 2016)*, February 2016, Gizan, Saudi Arabia, Eds. A. Al-Kamli, N. Can, G. O. Souadi, M. Fadhali, A. Mahdy and M. Mahgoub, AIP Conference Proceedings, **1742** (2016) pp. 030008-1–030008-4.
111. Khan, S. A., Quantum methodologies in Helmholtz optics, *Optik*, **127** (2016) 9798–9809.
112. Khan, S. A, Quantum methods in light beam optics, *Opt. Photonics News*, **27** (2016) 47.
113. Khan, S. A., International year of light and history of optics, in *Advances in Photonics Engineering, Nanophotonics and Biophotonics*, Ed. T. Scott, (Nova Science Publishers, New York, 2016), pp. 1–56.
114. Khan, S. A., Quantum methodologies in Maxwell optics, in *Advances in Imaging and Electron Physics*, Vol. 201, Ed. P. W. Hawkes, (Academic Press, New York, 2017) pp. 57–135.
115. Khan, S. A., Hamilton's optical-mechanical analogy in the wavelength-dependent regime, *Optik*, **130** (2017) 714–722.
116. Khan, S. A., Linearization of wave equations, *Optik*, **131C** (2017) 350–363.
117. Khan, S. A., Polarization in Maxwell optics, *Optik*, **131** (2017) 733–748.
118. Khan, S. A., Aberrations in Helmholtz optics, *Optik*, **153C** (2018) 164–181.
119. Khan, S. A., Quantum mechanical techniques in light optics, in *Proceedings of the Sixth Saudi International Meeting on Frontiers of Physics 2018 (SIMFP 2018)*, February–March 2018, Gizan, Saudi Arabia, Eds. Al-Kamli, A., Souadi, G. O., Hakami, J., Can, N., Mahdy, A., Mahgoub, M., and Mujahid, Z., AIP Conference Proceedings, **1976** (American Institute of Physics, 2018) 020016.
120. Khan, S. A., E. C. G. Sudarshan and the quantum mechanics of charged-particle beam optics, *Curr. Sci.*, **115**(9) (2018) 1813–1814.
121. Khandekar, D. C. and Lawande, S. V., *Path Integral Methods and Their Applications*, (World Scientific, Singapore, 1993).
122. Kleinert, H., *Path Integrals in Quantum Mechanics, Statistics, Polymer Physics, and Financial Markets*, (5th Edition, World Scientific, Singapore, 2009).

123. Kruit, P., Hobbs, R. G., Kim, C.-S., Yang, Y., Manfrinato, V. R., Hammer, J., Thomas, S., Weber, P., Klopfer, B., Kohstall, C., Juffmann, T., Kasevich, M. A., Hommelhoff, P., and Berggren, K. K., Designs for a quantum electron microscope, *Ultramicroscopy*, **164** (2016) 31–45.
124. Lakshminarayanan, V., Sridhar, R., and Jagannathan, R., Lie algebraic treatment of dioptric power and optical aberrations, *J. Opt. Soc. Am. A*, **15**(9) (1998) 2497–2503.
125. Lakshminarayanan, L., Ghatak, A., and Thyagarajan, K., *Lagrangian Optics*, (Springer, New York, 2002).
126. Lakshminarayanan, V. and Varadharajan, L. S., *Special Functions for Optical Science and Engineering*, (SPIE: The International Society for Optics and Photonics, 2015).
127. Lee, S. Y., *Accelerator Physics*, (2nd Edition, World Scientific, Singapore, 2004)
128. Liñares, J., Optical propagators in vector and spinor theories by path integral formalism, in *Lectures on Path Integration:Trieste*, Ed. A. Cerdeira, et al., (World Scientific, Singapore, 1993) 378–397.
129. Lubk, A., Paraxial quantum mechanics, in *Advances in Imaging and Electron Physics*, Vol. 206, Ed. P. W. Hawkes, (Academic Press, London, 2018).
130. Magnus, W., On the exponential solution of differential equations for a linear operator, *Comm. Pure Appl. Math.*, **7** (1954) 649–673.
131. Mais, H., *Nonlinear Problems in Accelerator Physics*, (DESY Report, 93-058).
132. Manfrinato, V. R., Stein, A., Zhang, L., Nam, C.-Y., Yager, K. G., Stach, E. A., and Black, C. T., Aberration-corrected electron beam lithography at the one nanometer length scale, *Nano Lett.* **17** (2017) 4562–4567.
133. Misra, B. and Sudarshan, E. C. G., The Zeno's paradox in quantum theory, *J. Math. Phys.*, **18** (1977) 756–763.
134. Mitra, A. N. and Ramanathan, R., On simulating Liouvillian flow from quantum mechanics via Wigner functions, *J. Math. Phys.*, **39**(9) (1998) 4492–4498.
135. Mondragón, J. S. and Wolf, K. B., *Lie Methods in Optics: Proceedings of the Workshop*—1985, Mexico, Eds. J. S. Mondragón and K. B. Wolf (Springer, New York, 1986).
136. Montague, B. W., Polarized beams in high energy storage rings, *Phys. Rep.*, **113**(1) (1984) 1–96.
137. Mukunda, N., Simon, R., and Sudarshan, E. C. G., Paraxial-wave optics and relativistic front description II: The vector theory, *Phys. Rev. A*, **28** (1983) 2933–2942.
138. Mukunda, N., Simon, R., and Sudarshan, E. C. G., Fourier optics for the Maxwell field: Formalism and applications, *J. Opt. Soc. Am. A*, **2** (1985) 416–426.
139. Newton, T. D. and Wigner, E. P., Localized states for elementary systems, *Rev. Mod. Phys.*, **21**(3) (1949) 400–405.
140. Okamoto, H. and Nagatani, Y., Entanglement-assisted electron microscopy based on a flux qubit, *Appl. Phys. Lett.*, **104** (2014) 062604.
141. Orloff, J., (Ed.), *Handbook of Charged Particle Optics*, (2nd Edition, Taylor & Francis, NewYork, 2009)
142. Parthasarathy, R., *Relativistic Quantum Mechanics*, (Narosa (India), 2011).
143. Petroni, N. C., De Martino, S., De Siena, S., and Illuminati, F., Stochastic collective dynamics of charged-particle beams in the stability regime,*Phys. Rev. E*, **63** (2000) 016501.
144. Phan-Van-Loc, Sur le principe de Huygens en Théorie de l'électron de Dirac, *Comptes rendus de l'Académie des Sciences, Paris*, **237** (1953) 649–651.
145. Phan-Van-Loc, Principe de Huygenes Théorie de l'électron de Dirac et intégrales de contour, *Comptes rendus de l'Académie des Sciences, Paris*, **238** (1954) 2494–2496.

146. Phan-Van-Loc, Diffractions des ondes Ψ_n de l'électron de Dirac, *Ann. Fac. Sci. Univ. Toulouse*, **18** (1955) 178–192.
147. Phan-Van-Loc, Interprétation physique de l'expression mathématique du principe de Huygens en théorie de l'électron de Dirac, *Cahiers Physique*, No. 97 **12** (1958) 327–340.
148. Une nouvelle manière d'etablir l'expression mathématique du principe de Huygens en théorie de l'electron de Dirac, *C. R. Acad. Sci. Paris*, **246** (1958) 388–390.
149. Phan-Van-Loc, *Principes de Huygens en theorie de l'electron de Dirac*, (Thesis, Toulouse, 1960).
150. Pozzi, G., Particles and waves in electron optics and microscopy, in *Advances in Imaging and Electron Physics*, Vol. 194, Ed. P. W. Hawkes, (Academic Press, New York, 2016) pp. 1–234.
151. Pryce, M. H. L., The mass-centre in the restricted theory of relativity and its connexion with the quantum theory of elementary particles, *Proc. Roy. Soc. Ser. A*, **195** (1948) 62–81.
152. Pusterla, M., Quantum-like approach to beam dynamics: Application to the LHC and HIDIF projects, in *Proceedings of the 18th Advanced ICFA Beam Dynamics Workshop on Quantum Aspects of Beam Physics—2000, Italy*, Ed. P. Chen, (World Scientific, Singapore, 2002), pp. 561–567.
153. Putnam, W. P. and Yanik, M. F., Noninvasive electron microscopy with interaction-free quantum measurements, *Phys. Rev. A* **80** (2009) 040902.
154. Radlička, T., Lie algebraic methods in charged particle optics, in *Advances in Imaging and Electron Physics*, Vol. 151, Ed. P.W. Hawkes, (Academic Press, New York, 2008) pp. 241–362.
155. Ramakrishnan, A., *L-Matrix Theory or the Grammar of Dirac Matrices*, (Tata-McGraw Hill, New Delhi, 1972).
156. Rangarajan, G., Dragt, A. J., and Neri, F., Solvable map representation of a nonlinear symplectic map, *Part. Accel.*, **28** (1990) 119–124.
157. Rangarajan, G. and Sachidanand, M., Spherical aberrations and its correction using Lie algebraic methods, *Pramana*, **49**(6) (1997) 635–643.
158. Reiser, M., *Theory and Design of Charged Particle Beams*, (2nd Edition, Wiley, New York, 2008).
159. Rosenzweig, J. B., *Fundamentals of Beam Physics*, (Oxford University Press, Oxford, 2003).
160. Rougemaille, N. and Schmid, A., Magnetic imaging with spin-polarized low-energy electron microscopy, *Europ. Phys. J.: Appl. Phys.*, **50**(2) (2010) 20101.
161. Rubinowicz, A., Über das Kirchhoffsche Beugungsproblem für Elektronenwellen, *Acta. Phys. Polon.*, **3** (1934) 143–163.
162. Rubinowicz, A., Die Beugungswelle in der Kirchhoffschen Theorie der Beugung, *Acta. Phys. Polon.*, **23** (1963) 727–744.
163. Rubinowicz, A., The Miyamoto–Wolf diffraction wave, *Progress in Optics*, Vol. 4, Ed. E. Wolf, (Elsevier, 1965) pp. 199–240.
164. Rubinowicz, A., *Die Beugungswelle in der Kirchhoffschen Theorie der Beugung*, (2nd Edition, Panstwowe Wydawnictwo Naukowe, Warsaw; Springer, Berlin, 1966).
165. Ryne, R. D. and Dragt, A. J., Magnetic optics calculations for cylindrically symmetric beams, *Part. Accel.*, **35** (1991) 129–165.
166. Sakurai, J. J. and Napolitano, J., *Modern Quantum Mechanics*, (2nd Edition, Cambridge University Press, Cambridge, 2017).

167. Scherzer, O., Über einige Fehler von Elektronenlinsen, *Z. Physik*, **101**(9-10) (1936) 593–603.
168. Seryi, A., *Unifying Physics of Accelerators, Lasers, and Plasma*, (Taylor & Francis, Boca Raton, FL, 2016).
169. Shankar, R., *Principles of Quantum Mechanics*, (2nd Edition, Springer, New York, 1994).
170. Simon, R., Sudarshan, E. C. G., and Mukunda, N., Gaussian-Maxwell beams, *J. Opt. Soc. Am. A* **3** (1986) 536–540.
171. Simon, R., Sudarshan, E. C. G., and Mukunda, N., Cross polarization in laser beams, *Appl. Opt.*, **26** (1987) 1589–1593.
172. Singh, A. and Carroll, S. M., *Modeling position and momentum in finite-dimensional Hilbert spaces via generalized Clifford algebra*, arXiv:1806.10134 [quant-ph]
173. Sokolov, A. A. and Ternov, I. M., *Radiation from Relativistic Electrons*, Translated by S. Chomet, Ed. C. W. Kilmister, (American Institute of Physics, New York, 1986).
174. Sudarshan, E. C. G. and Mukunda, N., *Classical Dynamics: A Modern Perspective*, (Wiley, New York, 1974).
175. Sudarshan, E. C. G., Simon, R., and Mukunda, N., Paraxial-wave optics and relativistic front description I: The scalar theory, *Phys. Rev. A*, **28** (1983) 2921–2932.
176. Syphers, M. J. and Chattopadhyay, S., Landscape of future accelerators at the energy and intensity frontiers, in *Proceedings of the 38th International Conference on High Energy Physics*, USA, 2016, *Proceedings of Science*: PoS(ICHEP2016) 043.
177. Tani, S., Connection between particle models and field theories. I. The case spin $\frac{1}{2}$, *Prog. Theor. Phys.* **6** (1951) 267–285.
178. Tekumalla, A. R., A class of transformations on the Dirac Hamiltonian, *Physica* **78** (1974) 191–197.
179. Tekumalla, A. R. and Santhanam, T. S., Unitary Foldy–Wouthuysen transformations for particles of arbitrary spin, *Lett. Nuovo Cimento* **10** (1974) 737–740.
180. Ternov, I. M., Evolution equation for the spin of a relativistic electron in the Heisenberg representation, *Sov. Phys. JETP*, **71**(4) (1990) 654–656.
181. Ternov, I. M., Synchrotron radiation, *Phys. Uspekhi*, **38**(4) (1995) 409–434.
182. Thomas, L. H., The motion of the spinning electron, *Nature*, **117** (1926) 514–514.
183. Thomas, L. H., The kinematics of an electron with an axis, *Phil. Mag.*, **3** (1927) 1–22.
184. Todesco, E., *Overview of single particle nonlinear dynamics*, (CERN-LHC-99-001). *AIP Conference Proceedings* **468** (1999), pp. 157–172.
185. Venturini, M. and Ruth, R. D., Single-particle quantum dynamics in a magnetic lattice, in *Proceedings of the 18th Advanced ICFA Beam Dynamics Workshop on Quantum Aspects of Beam Physics—2000, Italy*, Ed. P. Chen, (World Scientific, Singapore, 2002), pp. 551–560.
186. Wei, J., Li, X. P., and Sessler, A. M., Low energy states of circulating stored ion beams: Crystalline beams, *Phys. Rev. Lett.*, **73** (1994) 3089–3092.
187. Weidemann, H., *Particle Accelerator Physics*, (4th Edition, Springer, Berlin, 2015).
188. Wilcox, R. M., Exponential operators and parameter differentiation in quantum physics, *J. Math. Phys.*, **8** (1967) 962–982.
189. Wolf, K. B., *Integral Transforms in Science and Engineering*, (Plenum, New York, 1979).
190. Wolf, K. B., On time-dependent quadratic quantum hamiltonians, *SIAM J. Appl. Math.*, **40**(3) (1981) 419–431.

191. Wolf, K. B., *Lie Methods in Optics—II: Proceedings of the Second Workshop—1988*, Mexico, Ed. K. B. Wolf (Springer, 1989).
192. Wolski, A., *Beam Dynamics in High Energy Particle Accelerators*, (Imperial College Press, London, 2014).
193. Wu, C. S., Makiuchi, Y., and Chen, C., High energy electron beam lithography for nanoscale fabrication, in *Lithography*, Ed. M. Wang, (InTech Publ., 2010) pp. 241–266.
194. Ximen, J., Canonical theory in electron optics, in *Advances in Electronics and Electron Physics*, Vol. 81, Ed. P. W. Hawkes, (Academic Press, London, 1991) pp. 231–277.

Index

A

Aberrations
    coefficients, 7, 258, 262, 277
    effect of quantum uncertainties on, 277–281
    paraxial approximation, 250–259
Anomalous magnetic moment
    beam optics of the Dirac particle
        general formalism, 316–320
        Lorentz and Stern–Gerlach forces, 320–322
        phase space and spin transfer maps, 323–330
        Thomas–Frenkel–BMT equation, 320–322
Anti-Hermitian term, 46, 222, 260, 261, 263, 267
Axially symmetric electrostatic lens, 204–205, 269–270
Axially symmetric magnetic lens, 178–197, 300–306
    aberrations, 250–259
    point-to-point imaging, 232–250
    quantum corrections, 260–262

B

Baker–Campbell–Hausdorff (BCH) formula, 112, 303
Beam optical Hamiltonian
    classical, 5, 7, 174, 179, 197, 202, 204, 205, 207–211, 276, 293, 315
    quantum, 5–8, 172, 223, 230, 233, 250, 262, 263, 266, 268–271, 276, 279, 281, 292, 306, 315, 331, 332
Beam optics
    anomalous magnetic moment
        general formalism, 316–320
        Lorentz and Stern–Gerlach forces, 320–322
        phase space and spin transfer maps, 323–330
        Thomas–Frenkel–BMT equation, 320–322
Bending magnet, 206–210, 271–277, 306–315
Bohr magneton, 117
Bound states, 94, 110
Busch formula, 187, 248

C

Canonical transformations
    electromagnetic optical element, 29–30
    symplecticity
        Poisson bracket, 35–36
        time-dependent, 32–35
        time-independent, 31–32
    theory, 25–29
Cauchy–Schwarz inequality, 83
Christoffel symbols, 310
Classical charged particle beam optics
    curved optic axis, optical elements, 206–210
    free propagation, 174–178
    nonrelativistic, 210–212
    relativistic, 173–174
    straight optic axis, optical elements
        axially symmetric electrostatic lens, 204–205
        axially symmetric magnetic lens, 178–197
        electrostatic quadrupole, 205–206
        normal magnetic quadrupole, 197–201
        skew magnetic quadrupole, 202–204

Classical Hamiltonian, 51, 78, 90, 139, 213
Constant magnetic field
  charged particle in, 18–20, 22–25
  Dirac equation, 154–156
  Klein-Gordon particle, charged, 139–141
  quantum mechanics, 103–106
Curved optic axis, optical elements, 206–210
Cyclic property, 127
Cyclotron frequency, 20, 104–106

D

de Broglie wavelength, 84, 87, 231, 277, 297, 299, 305
Density operator, 126–132
Derbenev–Kondratenko semiclassical Hamiltonian, 322
Diffraction
  Dirac-Pauli equation, 293–300
  Klein–Gordon equation, 230–232
Dipole magnet, 6, 173, 206, 208–210, 271, 274, 276, 277, 306, 311, 315
Dirac delta function, 43, 48
Dirac equation
  constant magnetic field, 154–156
  Dirac-Pauli equation, 153–154
  Foldy–Wouthuysen transformation, 156–167
  free-particle equation, 141–148
  spin and helicity, 150–153
  spin magnetic moment, 153–154
  spin-1/2 particle, 217
  Zitterbewegung, 149–150
Dirac Hamiltonian, 4, 8, 143, 150, 152, 154, 158, 160, 162, 165, 168, 217, 218, 293, 295, 297, 298, 305, 306, 315, 316, 319

Dirac–Pauli equation, 4, 7, 8, 153–154
  quantum charged particle beam optics, 287–292
  axially symmetric magnetic lens, 300–306
  bending magnet, 306–315
  diffraction, 293–300
  general formalism, 316–320
  Lorentz and Stern–Gerlach forces, 320–322
  normal magnetic quadrupole, 330–333
  phase space and spin transfer maps, 323–330
  Thomas–Frenkel–BMT equation, 320–322
Dirac probability current density, 145
Dirac probability density, 145
Dirac's classical–quantum correspondence rule, 68, 227
Dirac spinor, 144, 145, 154, 156, 162, 287, 288, 291, 292, 295, 305, 308
Dynamics
  constant magnetic field, 18–20, 22–25
  system of particles, 36–39
Dyson time-ordered exponential, 114

E

Ehrenfest's theorem, 4, 68–70, 87, 88, 224, 232
Electromagnetic field, 173–174
  Hamiltonian formalism, 14–15
  Lagrangian formalism, 10–12
Electromagnetic optical element, canonical transformations, 29–30
Electron
  in constant magnetic field, 154–156
  Schrödinger–Pauli equation for, 117–119

Index

spin magnetic moment of, 153–154
Electron microscopy
 axially symmetric magnetic lens, 178–197
 effect of quantum uncertainties, 277–281
Electron optical imaging
 aberrations, 250–259
 point-to-point imaging, 232–250
 quantum corrections, 260–262
Electron optical technology, 201
Electrostatic quadrupole lens, 205–206, 270–271
Euler–Lagrange equations, 10, 36
Extended canonical transformation, 32, 35

F

Fermi–Dirac statistics, 77
Feshbach–Villars formalism, 4, 5, 215, 223, 281, 335
Feshbach–Villars representation, 137–139, 168–170
Feynman path integral method, 237
F(ocusing)O(nonfocusing)D(efocusing)O(nonfocusing) (FODO-channel), 201, 265
Foldy–Wouthuysen representation, 4, 165–167, 169, 170, 296, 316
Foldy–Wouthuysen theory, 218
Foldy–Wouthuysen transformation technique, 4–7, 217, 218, 223, 275, 281, 293–295, 301, 313–314, 317, 322, 335
 Dirac equation, 156–167
 Klein–Gordon equation, Feshbach–Villars form of, 168–170
Free-particle equation, 78–90, 141–148
Free propagation, 174–178
 Dirac–Pauli equation, 293–300

quantum charged particle beam optics, 230–232
Frequency of gyration, 20
Fresnel diffraction formula, 231, 299

G

Gaussian function, 84, 94
Green's function, 67

H

Hamiltonian formalism
 in electromagnetic field, 14–15
 optical, 29–30
 Poisson brackets
  in constant magnetic field, 18–20
  theory, 16–18
 theory, 12–13
 time-dependent, 32–35, 114–115, 283–285
Hamiltonian operator, 51, 78, 92, 213, 223, 263, 270
Hamilton's stationary action, 9
$\hbar$–dependent anti-Hermitian terms, 260, 261, 263, 267
$\hbar$–dependent Hermitian terms, 251, 260, 261, 266, 267
Heisenberg
 canonical commutation relation, 45, 47
 commutation relation, 47
 equation of motion, 4, 67, 68, 90, 150, 229
 operator, 67
 picture, 4, 67, 68, 81–82, 115, 116, 127, 130, 149, 165, 166, 170, 172, 224, 226, 296
 uncertainty principle, 52, 84, 86–89, 98, 279
Helicity operator, 152
Helicity, spin and, 150–153
Helmholtz scalar optics, 335
Hermite polynomials, 93–94

Hermitian conjugate, 42, 144,
149, 160
Hermitian Hamiltonian, 139, 222
Hermitian operator, 4, 42, 43, 46, 47,
49, 52, 57–59, 67, 70, 73,
76, 82, 83, 95, 115,
126, 220
Hermiticity, 44, 45, 55, 144, 159, 165

I

Identical particles system, 119–125
Independent variable
in constant magnetic field,
22–25
theory, 21–22
Interaction picture, 115–116

J

Jacobian matrix, 31–35, 38
Jacobi identity, 17, 46
Jeffreys–Wentzel–Kramers–Brillouin
(JWKB) approximation
technique, 111

K

Kinetic energy, 211
Klein–Gordon equation, 6, 7, 84,
155, 213, 333
aberrations, 250–259
constant magnetic field,
139–141
diffraction, 230–232
Feshbach–Villars
representation, 137–139,
168–170
formalism, 214–230
free particle, 271–273
free-particle equation and
difficulties, 132–137
point-to-point imaging, 232–250
quantum corrections, 260–262
time-independent, 273–274
Klein–Gordon–Fock equation, 133
Klein–Gordon theory, 305, 306

L

Ladder operators, 95
Lagrangian formalism
in electromagnetic field, 10–12
theory, 9–10
Laguerre polynomial, 102
Landau levels, 106
Landé g-factor, 117
Larmor radius, 20
Lie operator, 17
Lie transfer operator method, 3, 18,
189, 197, 230, 251
Linear differential equation, 170–172
Linear harmonic oscillator, 90–100
Lorentz force, 320–322

M

Magnetic quadrupole, 5, 8, 174
normal, 197–201, 262–267
phase space and spin transfer
maps, 323–327
skew, 202–204, 267–269
phase space and spin transfer
maps, 327–330
Magnetic rigidity, 209
Magnus formula, 114, 191, 283
linear differential equation,
170–172
Maxwell vector optics, 335
Minimal coupling principle, 273,
288, 310
Mixed state density operator,
126–132

N

Negative-energy states, 153
Newton–Wigner position
operator, 296
Nodes, 94
Non-Hermitian terms, 138, 222, 297,
298, 315
Nonrelativistic classical charged
particle beam optics,
210–212

Index

Nonrelativistic probability current density, 134, 135
Nonrelativistic probability density, 134, 135
Nonrelativistic quantum charged particle beam optics
    spin-0 and spinless particles, 281–282
    spin-1/2 particles, 330–333
Nonrelativistic Schrödinger equation, 103
Nonrelativistic single particle quantum mechanics
    constant magnetic field, 103–106
    free particle, 78–90
    interaction picture, 115–116
    JWKB approximation technique, 111
    linear harmonic oscillator, 90–100
    scattering states, 107–111
    Schrödinger–Pauli equation, 117–119
    time-dependent systems, 111–116
    two-dimensional isotropic harmonic oscillator, 100–103
Normal magnetic quadrupole, 197–201, 262–267, 323–327
Number operator, 96

O

Optical Hamiltonian, 3, 5–8, 29–30, 34, 172, 174, 179, 189, 197, 200, 202, 204–213, 220, 223, 230, 233, 234, 238, 250, 251, 260, 262, 263, 266–271, 276, 279, 281, 292, 293, 299, 306, 315, 320–323, 328, 331, 332
Orbital angular momentum quantum number, 75–76

Oscillator
    linear harmonic, 90–100
    two-dimensional isotropic harmonic, 100–103
Oscillatory motion, 91–92, 150

P

Paraxial approximation, 232–259
Paraxial beam, 1, 174, 177, 178, 180, 184, 188, 189, 197, 202, 204, 230, 231, 233, 234, 245, 250, 268, 299, 316, 323, 327, 328
Pauli–Dirac equation, 154
Pauli equation, 117–119
Pauli exclusion principle, 77, 335
Phase space and spin transfer maps
    normal magnetic quadrupole, 323–327
    skew magnetic quadrupole, 327–330
Point-to-point imaging, 232–250
Poisson brackets
    canonical transformations, 35–36
    Hamiltonian formalism in constant magnetic field, 18–20
    theory, 16–18
Position probability density, 42, 84, 92
Positive operator, 95
Potential energy, 90, 91, 104, 110, 154, 211
Propagator, 66, 89, 90, 99, 231, 237, 283–285, 299
Pryce–Tani–Foldy–Wouthuysen transformation, 157
Ptychographic analysis, 336
Pure state density operator, 126–132

Q

Quantum
    corrections, 7, 224, 260–262
    dynamics, 4, 42, 51–68, 214

Quantum (*cont.*)
  electrodynamics, 117
  kinematics, 42–50
  uncertainties, 277–281, 336
Quantum charged particle beam
    optics
  axially symmetric electrostatic
    lens, 269–270
  bending magnet, 271–277
  electron microscopy and
    nonlinearities, 277–281
  electrostatic quadrupole lens,
    270–271
  formalism, 213–214
  Klein–Gordon Equation
    aberrations, 250–259
    diffraction, 230–232
    formalism, 214–230
    point-to-point imaging,
      232–250
    quantum corrections,
      260–262
    normal magnetic quadrupole,
      262–267
    skew magnetic quadrupole,
      267–269
    spin-0 and spinless particles,
      281–282
    time-dependent quadratic
      Hamiltonian, 283–285
Quantum mechanics
  density operator, 126–132
  nonrelativistic single particle
    constant magnetic field,
      103–106
    free particle, 78–90
    interaction picture, 115–116
    JWKB approximation
      technique, 111
    linear harmonic oscillator,
      90–100
    scattering states, 107–111
    Schrödinger–Pauli equation,
      117–119
    time-dependent systems,
      111–116
    two-dimensional isotropic
      harmonic oscillator,
      100–103
  relativistic (*see* relativistic
    quantum mechanics)
  single particle
    Ehrenfest's theorem, 68–70
    quantum dynamics, 51–68
    quantum kinematics, 42–50
    spin, 70–77
    system of identical particles,
      119–125
Quasiclassical Hamiltonian, 316

R

Ray coordinates, 178, 277
Reference particle, 1, 2, 30, 206, 271,
    306, 336
Relativistic classical charged particle
    beam optics, 173–174
Relativistic mass, 2, 5, 212, 282, 333
Relativistic quantum charged particle
    beam optics
  Dirac–Pauli equation, 287–292
    axially symmetric magnetic
      lens, 300–306
    bending magnet, 306–315
    diffraction, 293–300
    general formalism, 316–320
    Lorentz and Stern–Gerlach
      forces, 320–322
    normal magnetic quadrupole,
      330–333
    phase space and spin transfer
      maps, 323–330
    Thomas–Frenkel–BMT
      equation, 320–322
Relativistic quantum mechanics, 137
  Dirac equation
    constant magnetic field,
      154–156
    Dirac–Pauli equation,
      153–154

free-particle equation, 141–148
spin and helicity, 150–153
spin magnetic moment, 153–154
Zitterbewegung, 149–150
Foldy–Wouthuysen transformation
Dirac equation, 156–167
Feshbach–Villars form of the Klein–Gordon equation, 168–170
Klein–Gordon equation
constant magnetic field, 139–141
Feshbach–Villars representation, 137–139
free-particle equation and difficulties, 132–137
Rest energy, 12, 14, 78, 133, 136, 141, 154, 161, 165, 211
Right handed, 10, 153
Round electrostatic lens, 204–205
Round magnetic lenses, 186–187, 247–248

S

Scanning tunneling microscopy, 110, 336
Scattering states, 107–111, 110
Schrödinger equation, 4, 7, 51, 231
energy eigenvalues of the system, 107, 110
nonrelativistic, 2–4, 6, 7, 84, 103, 132, 136, 213–214, 231, 281, 287, 296, 331
time-dependent, 4, 81, 93, 94, 183, 223, 244
Schrödinger–Pauli equation, 117–119, 130, 154, 156, 330
Schwarz inequality, 83
Semigroup property, 65, 191, 198, 224, 263, 325
S-evolution equation, 276

Sextupoles, 279
Single particle quantum mechanics
Ehrenfest's theorem, 68–70
quantum dynamics, 51–68
quantum kinematics, 42–50
spin, 70–77
Skew magnetic quadrupole, 202–204, 267–269, 327–330
Slater determinant, 125
Spin, 70–77
and helicity, 150–153
Spin-0 and spinless particles, 281–282
Spin magnetic moment, 153–154
Spinor theory for spin-1/2 particles
Dirac–Pauli equation, quantum charged particle beam optics, 287–292
axially symmetric magnetic lens, 300–306
bending magnet, 306–315
diffraction, 293–300
general formalism, 316–320
Lorentz and Stern–Gerlach forces, 320–322
normal magnetic quadrupole, 330–333
phase space and spin transfer maps, 323–330
Thomas–Frenkel–BMT equation, 320–322
nonrelativistic quantum charged particle beam optics, 330–333
Stern–Gerlach effect, 316, 327
Stern–Gerlach force, 320–322
Straight optic axis, optical elements
axially symmetric electrostatic lens, 204–205
axially symmetric magnetic lens, 178–197
electrostatic quadrupole, 205–206

Straight optic axis, optical elements *(cont.)*
  normal magnetic quadrupole, 197–201
  skew magnetic quadrupole, 202–204
Symplectic matrix, 32, 34
System of identical particles, 119–125
System of particles, dynamics, 36–39

T

Taylor series expansion, 61
Thomas–BMT equation, 316, 322, 327
Thomas–Frenkel–BMT equation (Thomas–Frenkel–Bargmann–Michel–Telegdi equation), 316, 320–322
Time-dependent
  canonical transformations, 32–35
  Dirac equation, 127, 144, 153
  perturbation theory, 4, 114, 116, 250–252
  quadratic Hamiltonia, 283–285
  systems, 111–116
Time-independent
  canonical transformations, 31–32
  Dirac equation, 146, 154–155
  free particle, 149

Hamiltonian, 65, 81, 93, 111, 123
perturbation theory, 111, 124
Schrödinger equation, 92, 93, 101–104, 107, 124, 281, 330–331
Two-component spinor theory, 4, 8, 117, 165, 293, 319
Two-dimensional isotropic harmonic oscillator, 100–105, 237

U

Uncertainty principle, 4, 52, 82, 84, 86–89, 98, 279
Unitary basis, 53, 55, 58
Unitary matrix, 54, 56, 59

V

Variance, 82
von Neumann equation, 126, 129

W

Wave function, 6, 41, 43–44, 47–51, 57, 66, 79, 81, 83–85, 92, 94, 98, 106, 109, 110, 125, 130, 131, 215, 220, 223, 225, 249, 276, 289, 319, 330, 335
Wigner function, 336, 337

Z

Zitterbewegung phenomenon, 149–150, 165–167, 296